International Scientific-Technical Conference on Innovative Engineering Technologies, Equipment and Materials (ISTC-IETEM 2017)

AA006456

IOP Conference Series: Materials Science and Engineering
Volume 412

Kazan, Russia
6 – 8 December 2017

ISBN: 978-1-5108-7368-1
ISSN: 1757-8981

Printed from e-media with permission by:

Curran Associates, Inc.
57 Morehouse Lane
Red Hook, NY 12571

CURRAN ASSOCIATES INC. proceedings.com

Some format issues inherent in the e-media version may also appear in this print version.

Copyright© (2017) by the Institute of Physics
All rights reserved. The material featured in this book is subject to IOP copyright protection, unless otherwise indicated.

Printed by Curran Associates, Inc. (2018)

For permission requests, please contact the Institute of Physics at the address below.

Institute of Physics
Dirac House, Temple Back
Bristol BS1 6BE UK

Phone: 44 1 17 929 7481
Fax: 44 1 17 920 0979

techtracking@iop.org

Additional copies of this publication are available from:

Curran Associates, Inc.
57 Morehouse Lane
Red Hook, NY 12571 USA
Phone: 845-758-0400
Fax: 845-758-2633
Email: curran@proceedings.com
Web: www.proceedings.com

Table of contents

Volume 412

International Scientific-technical Conference on Innovative Engineering Technologies, Equipment and Materials 2017 (ISTC-IETEM-2017)

6–8 December 2017, Kazan, Russian Federation

Accepted papers received: 27 July 2018
Published online: 23 October 2018

Preface

Papers

IOP Conf. Series: Materials Science and Engineering **412** (2018) 011001 doi:10.1088/1757-899X/412/1/011001

Abstract

International Scientific Conference "Innovative engineering technologies, equipment and materials - 2017" has been hosted by the exhibition center "Kazan Fair" within the framework of the 17th international specialized exhibition "Mechanical engineering. Metalworking. Kazan »and 11th specialized exhibition" TechnoWelding" on the 6th and 8th of December, 2017.

The conference program was including 6 sections of work:

Section 1 - Innovative materials, technologies and equipment in mechanical engineering

Section 2 - Innovative developments and economics in engineering

Section 3 - Chemical Engineering

Section 4 - Mathematical and physical modeling of information, technical, technological and management systems and processes

Section 5 - Innovative Welding Technologies in Industry

Section 6 - Innovative developments of small and medium enterprises

Participants and guests of the conference discussed the scientific and technical achievements in the field of development, design and manufacture of engineering products. See the results of studies of the country's leading universities in the field of robotics, new materials (including on the basis of production waste), waste management and recycling, mechanical engineering, energy, fuel cells, manufacturing economy.

Presidium of the session "Innovative development of mechanical engineering and economics" headed by Director of the Institute of Engineering Prof. Nail Faikovich Kashapov. The Presidium also included: Deputy. Director of the Institute for educational activities Ildar Hafizov Ilsurovich and Deputy. Director of the Institute for Science Gilmanshin Iskander Rafailevich.

Research and Innovation have been presented by the staff of the Engineering Institute of KFU, Institute of Naberezhnye Chelny, Institute of Chemistry named after A.M. Butlerov, Institute of Management, Economics and Finance, Kazan State Power Engineering University and Kazan National Research Technical University named after A. N. Tupolev – KAI

Active participation in the conference were graduate students and students of the Engineering Institute. The experience of live communication with scientists, practitioners, representatives of the real economy will allow buduyuschim experts to determine its scientific direction for a successful solution of important problems of import substitution and enhancing competitiveness assigned to the engineering industry of our country.

Content from this work may be used under the terms of the Creative Commons Attribution 3.0 licence. Any further distribution of this work must maintain attribution to the author(s) and the title of the work, journal citation and DOI.
Published under licence by IOP Publishing Ltd

Peer review statement

All papers published in this volume of *IOP Conference Series: Materials Science and Engineering* have been peer reviewed through processes administered by the proceedings Editors. Reviews were conducted by expert referees to the professional and scientific standards expected of a proceedings journal published by IOP Publishing.

Content from this work may be used under the terms of the Creative Commons Attribution 3.0 licence. Any further distribution of this work must maintain attribution to the author(s) and the title of the work, journal citation and DOI.
Published under licence by IOP Publishing Ltd

Educational laboratorial facility for researching the technology of plasma gasification of production and consumption wastes including medical wastes

T M Abdullin, I R Gilmanshin[1,3], N F Kashapov[1], S I Gilmanshina[1], A I Galeeva[1,2], D R Krainova[1]

[1] Kazan Federal University, 18 Kremlyovskaya street, Kazan, 420008, Russian Federation
[2]Kazan State Power Engineering University, Kazan, 420066, Russian Federation
[3] Kazan national research technical university named after A N Tupolev, K. Marx St. 10, Kazan, 420111 Tatarstan, Russia

is-er@yandex.ru

Abstract. The article deals with the problems of waste management methods, which do not cope with the task of ecological utilization of solid waste. Based on the analysis of literature data, the following is presented: the morphological composition of solid waste, the block diagram of a laboratory waste gasification plant. And also a set of equipment characteristics for the modernization of a laboratory installation of plasma gasification.

The usual scheme for handling solid domestic waste for Russia is the collection of solid household waste without separation and burial (> 90%) leading to land degradation (employing over 50 thousand hectares of those present).
That allows increasing the migration in the siege space (the amount of solid waste exceeds the capacity of natural ecosystems). Burning was quite able to cope with the mission. But due to increased incomes and their variety, the disposal of old methods inevitably carries a serious threat to the environment.
In (Table 1), comparison of the methods of thermal destruction of solid waste.

Table 1. Comparison of methods of thermal destruction

	Plazma gazification	Usual gazification	Burning
Destruction	99 %	90 %	70%
Temperature	≥ 1200C	≤ 1200C	≤ 1000C
The presence of resins and furans at the outlet	No	Yes	A lot
Ash	1%	10%	30%, toxic
Type of waste	any	Besides of individual inorganic materials	Besides of individual inorganic materials

Content from this work may be used under the terms of the Creative Commons Attribution 3.0 licence. Any further distribution of this work must maintain attribution to the author(s) and the title of the work, journal citation and DOI.
Published under licence by IOP Publishing Ltd

IOP Conf. Series: Materials Science and Engineering **412** (2018) 012001 doi:10.1088/1757-899X/412/1/012001

Sorting	Not required	required	required
Emissions of flue gases	low	average	high

Plasma gasification has limitations:
1) Power consumption.
Because there are many costly and energy-consuming steps in the purification of exhaust gases, in addition to supplying plasmatrons, in order to comply with environmental standards.
2) In most cases, accurate calculation of plasmachemical reactions in the reactor is not possible.
Because SDW has a different morphological composition and a high temperature in the reactor, which makes it difficult to control the reactions to the reactor and simulate the ongoing reactions.
Therefore, we consider (in Table 2) the elemental morphological composition of various fractions of solid domestic waste, including medical ones.
Table 2. The content of individual elements in the composition of solid waste

Component	Element content, %					Specific heat of combustion (lowest), MJ / kg
	C	H	O	N	S	
Textile	56,1	6,8	32,2	4,8	0,1	22,6
Wood	51	6,1	42,6	0,2	0,1	20,4
Paper	46,2	6,2	47,1	0,3	0,23	16,9
Leather	77,9	6,0	15,1	0,3	0,7	31,2
Rubber	77,9	6,0	15,1	0,3	0,7	31,2
Plastic	67,7	9,3	21,5	1,1	0,4	30,3
On average, for all waste materials	50,5	6	31	2,5	0,2	20,2

On Fig. 1 the block diagram of a laboratory installation of plasma gasification is presented.

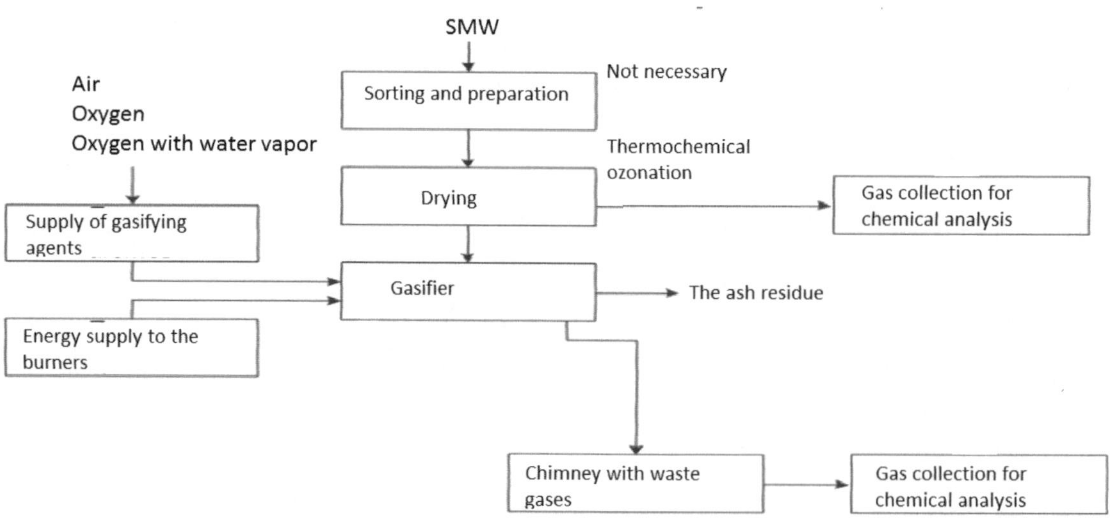

Figure 1. Block diagram of a laboratory waste gasification plant

Modernization of the laboratory installation (Figure 1) is due to a change in the working gas in the gasifier chamber. Air is replaced by a more active oxidant (oxygen) or a gas mixture (oxygen with steam). A mixture of oxygen with water vapor is used because hydroxyl radicals (OH •) are formed (which is the strongest known oxidant). In the gasifier, the mixture decomposes to form hydroxyl radicals. The lifetime of the hydroxyl radical in the biological medium is approximately 10-9 sec, and the diffusion radius 10 -8 m. The main types of damage to biomolecules by hydroxyl radicals are: the detachment of the hydrogen atom (rupture any CH bond)

In chemistry, the main source of hydroxyl radical formation is the Fenton reaction (1) and the Haber-Weiss reaction (2) [1, 2].

$$Fe^{2+} + H_2O_2 \rightarrow OH^{\cdot} + OH^{-} + Fe^{3+} \quad (1)$$

$$O_2^{\cdot -} + H_2O_2 \rightarrow OH^{-} + OH^{\cdot} + O_2 \quad (2)$$

And instead of thermal drying, we use thermochemical ozonization.

Ozonization is used to reduce the organic mass of solid waste. For example, as a result of the oxidation of oil products by ozone in water, 48 elements of 51 organics are removed (mainly alkanes, olefins, esters, ethers, etc.) [3,4,5,6,7,8].

Ozone is toxic and decomposes at high temperatures instantaneously, which excludes the possibility of getting into the atmosphere. For example, the ozone half-life in air (Table 3) depends mainly on the air temperature.

Table 3. Ozone half-life in air

Air temperature, C	-50	-35	-25	20	120	250
Half life	3 months	18 days	8 days	3 days	1.5 hours	1.5 seconds

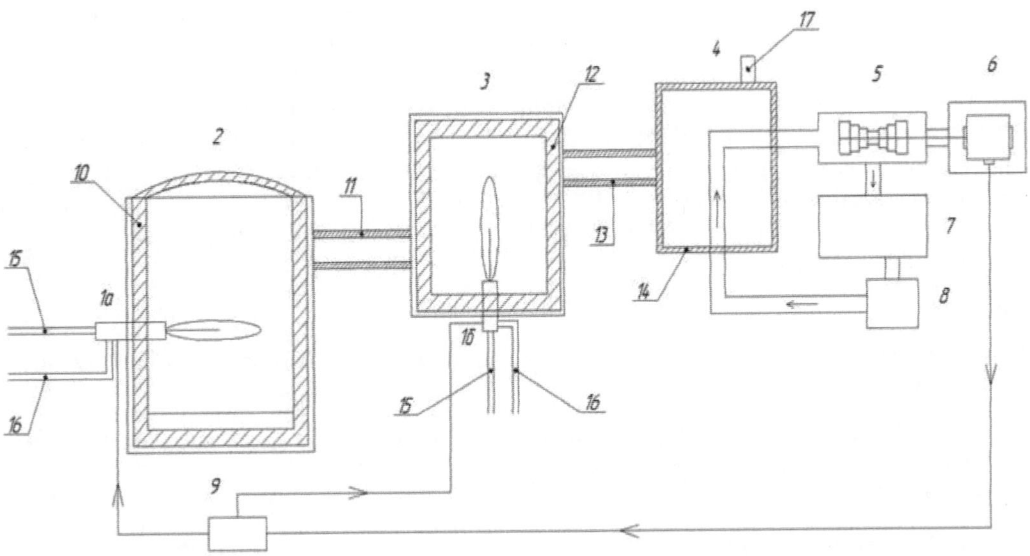

Figure 2. Diagram of a laboratory installation of plasma gasification of medical waste of classes A, B, and B.

Upgrade the chamber of the gasifier by adding a layer of carbonized graphite. Since the temperature in the chamber, when the working gas changes, reaches up to 3500 ° C.

Upgrade the drying chamber with an ozonator built in, with a range of characteristics (Table 4).

Table 4. Range of technical characteristics of the ozonizer

Ozone productivity, g / h	20 - 100
Fan capacity, m^3 / h	360 - 750
Supply voltage	220 B
Power, W	250 – 1040
Price, rubles	39 000 – 92 000

Conclusions.

Lack of plasma gasification, energy consumption, can be solved by selecting the working gas in the gasifier and reducing the mass of solid waste at the preparation / drying stage. This was revealed during the literary review of the chemical component of processes in the gasifier and in the stages of plasma gasification. The main difficulties in studying the chemical component of the processes - the lack of reliable data on the mechanism of the chemical process, the parameters of its individual stages and kinetic transport coefficients, etc. Ozonization was chosen as the preparation of solid waste (the stage of grinding, drying, etc.). Ozone destruction of various materials is a new direction and only developing, there are studies and installations for the destruction of tires for example. But because of the uncertain composition of solid waste, ozone depletion requires research. Since ozone is toxic and explosive with certain components, which requires experiments and the development of technology / mode of ozonation of solid waste. And also when analyzing the chemical component of the processes in the gasifier, a decision was taken to select and study the working gas in the gasifier. As there are foreign studies of increasing the efficiency of processes in the gasifier due to changes in the working gas (a mixture of oxygen and water vapor). A study will be made of the change in efficiency from changes in the composition of the working gas and the development of a technology / operating mode of the gasifier with various gas mixtures. The development of a model for the installation of a laboratory for the study of thermochemical destruction of production and consumption wastes, including medical ones. The choice of the design of the ozonizer is made and the range of required characteristics is determined.

References

[1] Mitteler R 2002 *Oxidative stress, antioxidants and stress tolerance* Trends in Plant Science. V 7 (9) pp 405-10

[2] Dat J, Vandenabeele S, Vranova E, Van Montagu M, Inze D, Van Breusegem F 2000 *Dual action of the active oxygen species during plant stress responses* Cellular and Molecular Life Sciences V 57 pp 779-95

[3] Chen G 2002 *Treatment of Oil-Polluted Water* Beijing: Chemical Industry Press pp 5-27

[4] Wu J, Doan H, Upreti S 2007 *Decolorization of aqueous textile reactive dye by ozone* ChemicalEngineering Journal V 11(5) pp 1-5
[5] Jin Anotai, Rosawan Wuttiponga Chettiyappan Visvanathan 2007 *Oxidation and detoxification of pentachlorophenol in aqueous phase by ozonation* Journal of Environmental Management V 85 pp 345-49
[6] Lin C K, Liu J C, Chen M C and Tsai T Y 2001 *Enhanced biodegradation of petrochemical wastewater using ozonation and BAC advanced treatment system* Water Research V 35(3) pp 699-704
[7] Nemes A, Fabian I and Gordon G 2000 *Experimental aspects of mechanistic studies on aqueous ozone decomposition in alkaline solution* Ozone Science and Engineering V 22(3) pp 287-304
[8] Rischbieter E, Stein H and Schumpe A 2000 *Ozone solubilities in water and aqueous salt solutions* Journal of Chemical Engineering Data V 45 pp 338-40
[9] Gonopolsky A M, Fedorov O L 1999 *Disposal of waste of medical institutions in a pressurized plasma furnace* Clean city No 1 (5) pp 28-31
[10] Gilmanshina S I, Gilmanshin I R, Sagitova R N, Galeeva A I 2016 *The feature of scientific explanation in the teaching of chemistry in the environment of new information of school students' developmental education* International Journal of Environmental and Science Education 11(4) pp 349-58
[11] Gilmanshin I, Kashapov N, Gilmanshina S, Galeeva A 2016 *IOP Conference Series: Materials Science and Engineering* 134(1) **012007**
[12] Azimov Y, Gilmanshin I, Gilmanshina S 2016 *IOP Conference Series: Materials Science and Engineering* 134(1) **012003**

IOP Conf. Series: Materials Science and Engineering **412** (2018) 012002 doi:10.1088/1757-899X/412/1/012002

THE TECHNIQUE FOR PRODUCING ANOTECHNOLOGIES WITH CARBON AND FERROMAGNETIC METAL OR ALLOY

G M Ahmadiev

Kazan Federal University, Naberezhnye Chelny Institute, 420008, Russia, Kazan, Kremlyovskaya St., 18

GMAhmadiev@kpfu.ru

Abstract. Nanoparticles according to the method and technology including metal-carbon particles, particularly particles of a ferromagnetic metal or alloy encapsulated in graphitic carbon are suitable for use as a detectable moiety for imaging by the method of a magnetic resonance and fluorescence, drug delivery, cell marking and for local thermal therapeutic treatment such as hyperthermia.

Introduction. The given technology relates to nanoparticles with metal ferromagnetic nanocrystals combined with amorphous or graphite carbon which has chemical groups capable of dissociation in aqueous solution. The scope of the given research includes detectable moiety for imaging by the method of a magnetic resonance and fluorescence, drug delivery devices, cell labeling, and local thermal therapeutic treatment such as hyperthermia.

Currently one of the most important applications of ferromagnetic nanoparticles belongs to detectable moietyfor imaging by the method of a magnetic resonance. The presence of an inhomogeneity in a magnetic field leads to shorter triggering time of T1 and T2 in a magnetic resonance. Therefore, the presence of ferromagnetic particles leads to black spots in displaying the magnetic resonance of protons. A good resolution requires small ferromagnetic particles with rather high magnetization.

The interest fordetectable moiety for imaging by the method of a magnetic resonance shows the number of recently filed applications for patents. WO-A-2004/107368 describes magnetic particles of iron-based oxide with sizes less than 20 nm, which surface is modified by amino groups. The isoelectric point is greater than 10 or equal to 10. WO-A-2009/109588 depictsparticles of iron-based oxide with two different ligands, where the first ligand contains an electrostatically charged group and the second ligand is a hydrophilic group. WO-A-2009/135937 describes a linker bound at the first end with a polyethyleneimine polymer and at the other end with a nanoparticle core or a polyethylene glycol polymer grafted to a polyimine polymer. Such ferromagnetic oxides can have a wide distribution of particles by their sizes, agglomeration of individual particles, instability due to reaction or recrystallization into non-ferromagnetic iron-based oxide and toxic properties. The fact that ferromagnetic particles are strongly clustered is especially problematic [1-15].

The given researchinvolves scientific justificationof the method for obtaining nanoparticles with carbon and ferromagnetic metal or alloy.

Research Methodology

Particles of a ferromagnetic oxide are used to produce a magnetic resonance imaging. The oxide particles are relatively stable in the air. The most known ferromagnetic iron-based oxides are magnetite, Fe_3O_4 or Fe (II) Fe (III) $2O_4$ and maghemite, γ-Fe_2O_3. The combination with other atoms of divalent metal such as cobalt or nickel also yields ferromagnetic oxides, for example $CoFe_2O_3$ and $NiFe_2O_3$.

Content from this work may be used under the terms of the Creative Commons Attribution 3.0 licence. Any further distribution of this work must maintain attribution to the author(s) and the title of the work, journal citation and DOI.
Published under licence by IOP Publishing Ltd

Study Results and their Discussion

Small particles of magnetite are usually obtained by mixing solutions containing Fe (II) and Fe (III) compounds. The method can lead to small cluster particles of magnetite depending on mixing. The formation of magnetic domains leads to the fact that magnetic particles do not have a magnetic moment in the absence of an external magnetic field. Sufficiently small ferromagnetic particles do not form a set of domains. It should be noted that small ferromagnetic particles are single-domain particles, which indicate that moments of magnetic atoms in individual particles are not ordered into different domains, but they are oriented in the same direction. Therefore, single-domain particles have a ferromagnetic moment even in the absence of an external magnetic field. When particles are suspended in a liquid without forming clusters, they can rotate freely. Then the orientation of magnetic moments of individual particles reaches thermodynamic equilibrium which depends on the magnetic moment of particles, the strength of the external magnetic field and thermal energy (temperature). Since in contrast to paramagnetic materials magnetic moments of ferromagnetic particles include thousands or millions of atomic magnetic moments, this paramagnetic response is referred to superparamagnetism. Magnetic particles that can move more or less freely in the liquid do not have retained magnetization; in the absence of an external magnetic field the magnetic moment of single-domain particles suspended in the liquid is negligibly small. When the energy of magnetic anisotropy of ferromagnetic particles has the order of KG, i.e. the order of thermal energy, the orientation of magnetic moments of individual particles can reach thermodynamic equilibrium also in cases when particles cannot rotate.

The magnetic coupling between single-domain particles suspended in a liquid leads to the fact that particles usually form clusters in which magnetic moments of individual particles are oriented in a way the external magnetic field not to exert influence on clusters. The cluster is for biomedical applications.

As a result, the current state in the field of ferromagnetic metal particles involves the application of inert layers to metal particles. The United States Patent No.4855091 describes the obtaining of small particles of nickel, iron, or cobalt by reducing corresponding precursors applied to a highly porous ceramic carrier and then undesirable subsequent exposure to small particles of a gas stream with carbon of magnetic particles [2].

Results

Nanoparticles by primary technologies including metal-carbon particles, particularly particles of a ferromagnetic metal or alloy encapsulated in graphitic carbon are suitable for use as a detectable moiety for imaging by the method of a magnetic resonance and fluorescence, drug delivery, cell marking and for local thermal therapeutic treatment such as hyperthermia.

The research can be used in medicine, biology and veterinary medicine in manufacturinga detectable moiety for imaging by the method of a magnetic resonance or fluorescence, drug delivery, cell marking. Carbon-contained particles from microcrystalline cellulose, colloidal carbon and mixtures are infiltrated with an aqueous solution of at least one ferromagnetic metal precursor, such as a citric, acetic or formic acid salt, hydroxy acids or ammonium citrate. The infiltrated particles are dried and heated in an inert and substantially oxygen-free atmosphere at a temperature of 450-600 ° C or more than 700 ° C. The resulting nanoparticles contain graphite carbon and 3-100 particles of at least one ferromagnetic metal, at least partially encapsulated in a graphite carbon. Ferromagnetic particles comprise a metal selected from the group consisting of nickel, cobalt, nonreactive metals and their combinations. The particles of ferromagnetic metal are distributed uniformly and do not form clusters.

References

[1] Patent for an invention No. 002567620.*Nanochasticy, soderzhashchieuglerodiferromagnitnyjmetallilisplav*[Nanotechnologies with carbon and ferromagnetic metal or alloy] //EnneskensLeonardusVejnand (nl),Gyos John Vilhelm (nl), Resink Bernard Hendrik (nl),BerbenPiterHildegArdus (nl),HukstraYAkobus (nl).
[2] The United States Patent No. 4855091
[3] Karnegi-Mellon. The United States Patent for an invention No. 5456986
[4] Patent for an invention WO-A-2004/107368
[5] Patent for an invention WO-A-2009/109588
[6] Patent for an invention WO-A-2009/135937
[7] Patent for an invention WO-A-99/46782
[8] Patent for an invention USA-2008/0057001
[9] D. Tasis, etc. Chem. Rev. (2006) 106, 1105-1136

[10] Harris p.j.f. Et al, *A simple technique for the synthesis of filled carbon nanoparticles*, Chem. Phys. Lett., 1998, v. 293, pp.53-58;
[11] US 2006/0137487 A1, 29.06.2006;
[12] US 2006/0116343 A1, 01.06.2006;
[13] WO 2007/146426 A2, 21.12. 2007;
[14] US 2008/0213189 A1, 04.09.2008;
[15] Shitova N.B. etc. 2009 *Rutenij-uglerodnyjnanokompozit* [Ruthenium-carbon nanocomposite material].Zhurnalstrukturnojhimii t 50 No 2 pp 283-87

IOP Conf. Series: Materials Science and Engineering **412** (2018) 012003 doi:10.1088/1757-899X/412/1/012003

The practice of a mobile pulsation unit (MPU) to increase the efficiency of oil wells

A V Akhmerov[1], D P Iovlev[1], L S Sabitov[1,2], A A Sinyavin[1], A R Hayrullin[1]

[1]Kazan State Power Engineering University, 51 Krasnoselskaya street, Kazan, 420066, Russian Federation
[2]Kazan Federal University, 18 Kremlyovskaya street, Kazan, 420008, Russian Federation

diow@inbox.ru

Annotation. The pulsatile stimulation technique (near-wellbore area) is an effective method of hydroimpulsive well treatment in oil production. The results of testing a pilot sample of a mobile pulsation unit at injection wells are presented, with an increase in injection capacity 3-4 times and other positive effects (well pressure drop, chemical composition change). The obtained results confirm the effectiveness of pulsation effects and create the prerequisites for the development of complex energy-saving ways of oil well treatment.

1. Introduction

The increased volume of extraction of hard oil (tight oil) and the current trend of a general decrease in oil recovery (oil recovery) determine the urgency of restoring well performance. Due to colmatation processes of various nature, the main well repair activities are aimed at near-wellbore area (NWA) treatment, to increase the permeability of the reservoir, to increase the depth of work, to maintain production (or to increase injectivity, etc.) [1].

Numerous studies and tests have shown that the greatest efficiency in cleaning PZP and activation of in-situ phenomena is achieved by creating a non-stationary treatment in the treatment zone. Today, there are many ways of creating non-stationary effects due to periodic waves (impulses) of pressure at the wellhead (top of well) [2,3]. In most ways, a one-time pressure pulse of a critical force (from 100atm to 250atm.) Is used, so they are related to destructive methods of stimulation.

Various hydrodynamical and hydroimpulsive effects on the reservoir are used as methods for the development of low permeability formation, as a technology for stabilizing and increasing the flow rate of wells [4-8].

The idea of the hydroimpulse effect on the PPP has found its embodiment in the creation of wellhead equipment for the formation of alternating sign pressure pulses in the near-wellbore zone due to the formation of a standing wave in the well [9]. In the tests of this method, the technological result was achieved by 70% injection and 95% production wells. In this case, after treatment, the injectivity of injection wells increased by 2-5 times, and the production well rate increased by 2-4 times, with water cutting decreasing by 15-30%.

The industrial application of the technology of complex wave and chemical effects on the PZP [10] made it possible to obtain additional production after processing at 40 wells of the order of 3.5-5.0

Content from this work may be used under the terms of the Creative Commons Attribution 3.0 licence. Any further distribution of this work must maintain attribution to the author(s) and the title of the work, journal citation and DOI.
Published under licence by IOP Publishing Ltd

IOP Conf. Series: Materials Science and Engineering **412** (2018) 012003 doi:10.1088/1757-899X/412/1/012003

thousand tons per year.

Considering technologies using the infrasound range of impulse action (less than 1 Hz), one should also mention the variable pressure method [11], which is well known and is used mainly to intensify the inflow of oil fluid (fluid), including during drilling.

The success of research and development in this area and the results of industrial tests determine the feasibility of using currently unsteady hydrodynamic (bottom-hole) treatment methods (bottom-hole area treatment), and confirm their prospects in order to restore and improve the operating performance of oil wells.

In this connection, the pulsating method of action on the bottomhole zone and the mobile pulsating unit (MPA) for its realization is an effective method of hydroimpulse treatment of oil wells. The pulsation unit provides a time and power pulse response to the PZP and performs operations to lift the contamination (inclusion) at the wellhead [12].

2. Field tests

The pilot-industrial sample of the MPU installation (Fig.1, 2) is a multifunctional test bench. For the given oil field conditions the stand allows to conduct debugging of technological and regime parameters of pulsating methods of processing and repair of wells, and also to intensify the operating technologies.

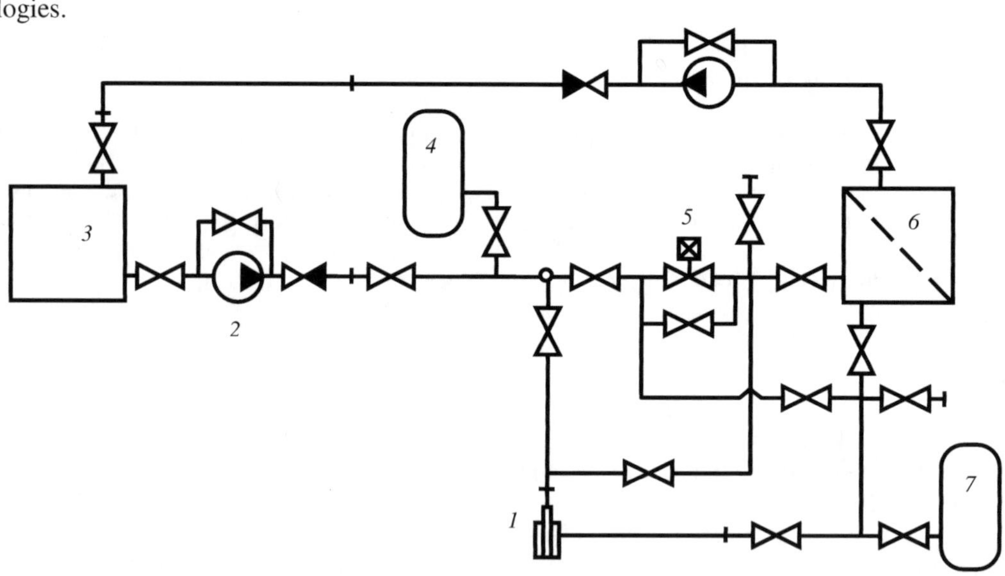

Fig. 1. Scheme MPU: 1 - well; 2 - the pump; 3,6 - tanks; 4.7 - hydraulic accumulators; 5 - control valve

Fig. 2. Field tests of MPU

The pulsating (hydropulse) regime is realized due to the creation of low-frequency (0.02-0.001 Hz) pressure pulses (up to 80 atm) at the wellhead (Fig.3, 4). At the same time, the following factors are determining the intensive cleaning of the CAP (including the removal of the colmatant (deposition of pollution) from the porous space of the reservoir): repeatedly repeated depressions to the reservoir; high velocity of fluid motion in the perforation interval; formation of a time-dependent non-stationary hydrodynamic regime with constant updating of contact zones, mixing of working agent and fluid. Cyclically repeated, undamped, turbulent pulsation stimulation (pulsatile stimulation) regulates the filtration properties and stabilizes the hydraulic conductivity of the bottomhole zone. Comparison of the calculated effectiveness of continuous and pulsation drainage regimes (drainage) shows that the energy costs in the pulsation regime are lower by 3-3.5 times [13].

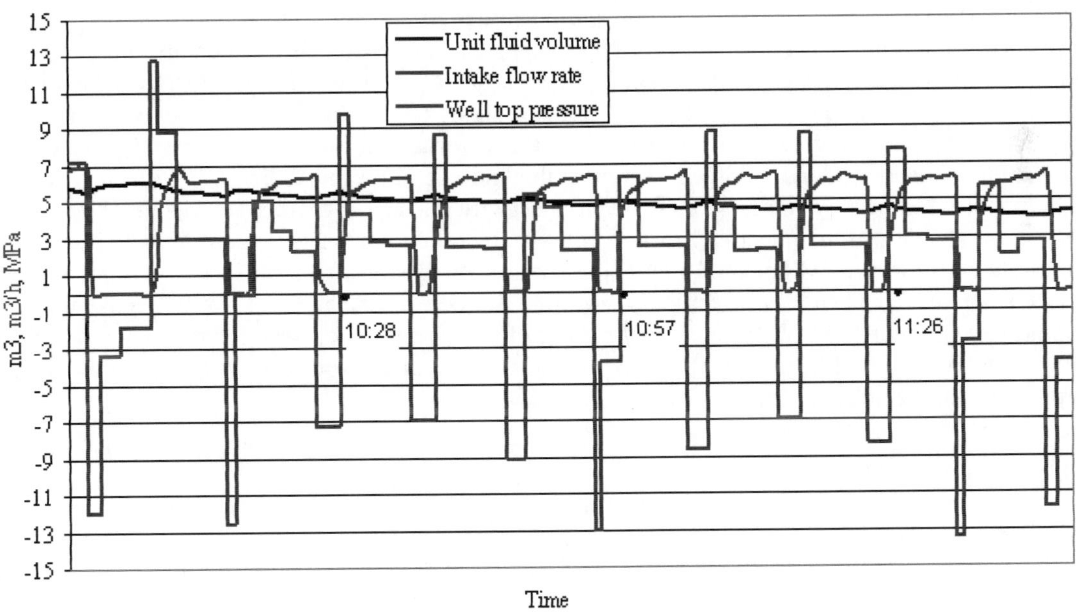

Fig. 3. Fragment of operational data at the beginning of well treatment

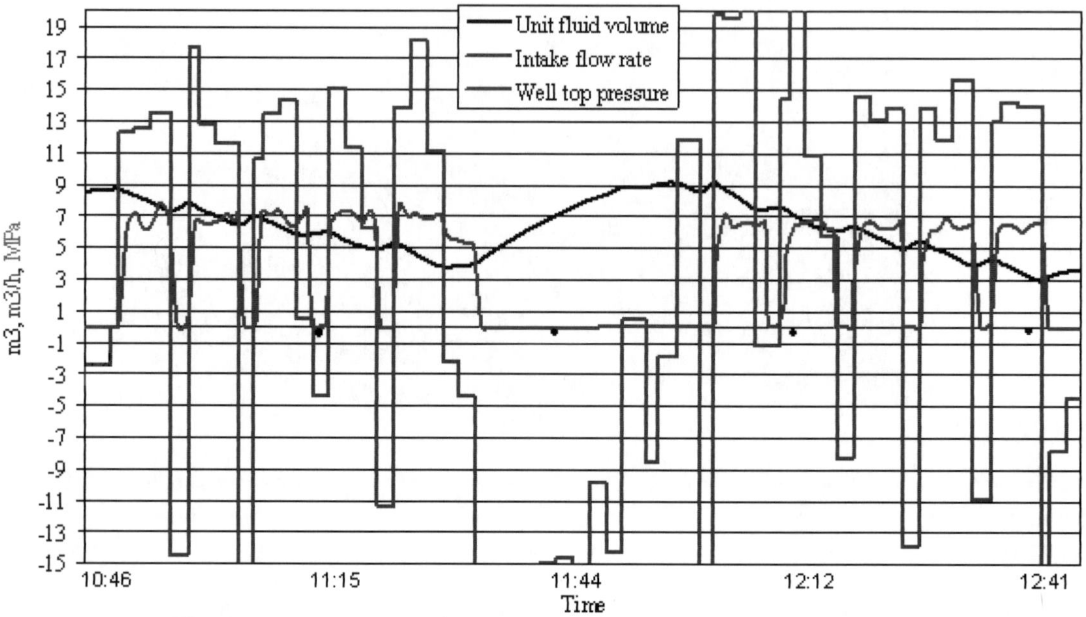

Fig. 4. Fragment of operational data at the end of the well treatment

Experimental-industrial tests of the MPA plant confirmed its operability in various technological regimes. The pulsating treatment of 9 wells (deposits in the territory of the Republic of Tatarstan, Russian Federation) showed the pattern of the appearance of characteristic positive effects (pressure decrease, flow increase, etc.) [14]. So, for example, when testing the well of the Urustamak field (depth 1550 m, reservoir thickness 3.2 m, permeability 70-100 mD, porosity 17%), the injectivity increased 4-fold (Fig. 5, well 1). A significant change in the chemical composition of the working fluid occurred (a decrease in rigidity of 2-3 times, a decrease in the chloride ion by a factor of 2), which indicates the influence of the elastic forces of pulsations on the bottomhole formation zone. In Fig. 3.4 shows the fragments of operational data recorded during the processing wells of the Urmyshlinsky deposit (depth 1280 m, reservoir thickness 7 m, permeability 137-1340 mD, porosity 15-24%), of which the dynamics of increase in injectivity during processing are visible. As a result of the test, an increase in the injectivity of the well was recorded 4.2 times (Figure 5, well 2), while the rate of acceleration over 24 hours of treatment increased from 0.66 to 3.57 m3 / (day • atm). When the well was surveyed, it was also established to open the perforation intervals and increase the actual thickness of the beds (up to 50%), reduce the mark of the current face, clean the reservoir and take out the colmatizing particles in the form of metal oxides, and significantly change the chemical composition of the working fluid.

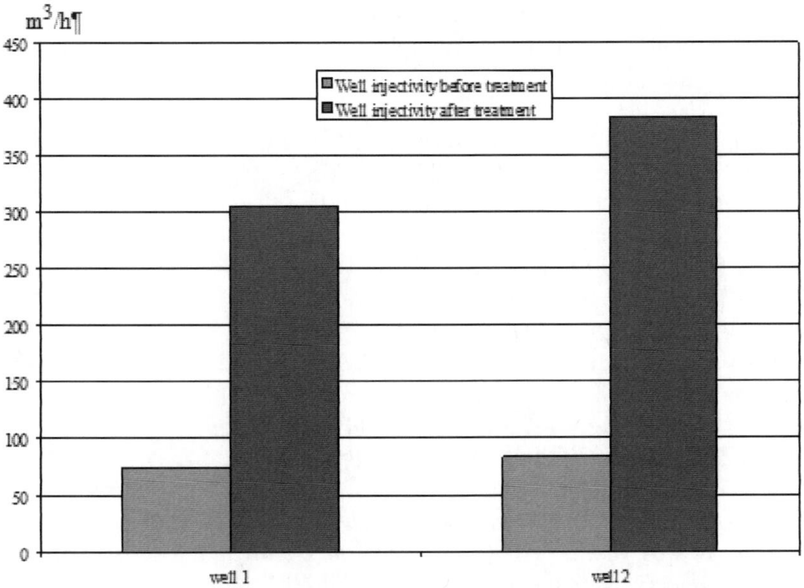

Fig. 5. Change in injection capacity of injection wells during pulsating treatment

3. Discussion

The obtained results unequivocally confirm that the influence of the elastic forces of pulsations influence the downhole and bottomhole zones. In this case, the pulsating equipment, namely the installation of MPA, is located at the wellhead. In contrast to existing technologies, pulsation processing does not use any downhole devices, machinery or equipment. To implement technological operations in pulsating mode, additional land special equipment or special equipment is not used either. Particularly it should be noted that all effects in pulsation processing are obtained due to hydrodynamic regimes of exposure to commercial water, without the use of active reagents. The obtained results of field tests are of great interest for the development of efficient technologies for pulsating processing of oil wells. The practice of servicing oil wells and the success of technological operations has already unequivocally proven that complex processing methods are the most effective in achieving repair objectives and the most economically viable [15]. Functional and technological capabilities of the MPA unit allow successfully combining the advantages of pulsating effects with the capabilities of complex or combined well treatment technologies, including chemical, physical, thermal, and others.

In the pulsation treatment of producing wells, the conditions for the formation of characteristic regularities and effects will appear similarly, as in the treatment of injectors. The combination of the resulting hydrodynamic effect from the pulsating effect and the activity of the reagents will enhance the efficiency of the reaction and heat and mass exchange processes taking place in the PPP. At the same time, the intensification of these processes by keeping them in the non-stationary mode will ensure a reduction in the required amount of active substances or a replacement for cheaper ones, increase the depth of its penetration into the formation, reduce processing time, or increase success, all other things being equal.

Conclusion

The results (changes in the chemical composition, removal of the colmatite, increase in injectivity, reduction in pressure, increase in the thickness of the reservoir, etc.) obtained during the pilot industrial tests confirm the effectiveness of the pulsational (hydroimpulse) effect in the treatment of the bottomhole zone of the wells, and the technological advantages of the method and the functionality of the apparatus allow us to consider pulsation processing as a kind of effective energy resource saving method of repairing oil wells, especially when implementing reagent, complex or combined methods.

References

[1] Hu X *et al* 2017 *Physics of petroleum reservoirs* (Springer Mineralogy) p 506
[2] Veliev F G *et al* 1988 *USSR Patent* No. 4483064/03
[3] Hamzin A A *et al* 1998 *RF Patent* No. 96112379/03
[4] Balashkand *et al* 2000 *Karotazhnik* 79 77–85
[5] Bazhaluk YA M *et al* 2000 *Karotazhnik* 64 91–4
[6] Yanturin A SH *et al* 1986 *Oil Economy* 2 63–6
[7] Uryupin V A *et al* 1995 *Oil Economy* 2 78–9
[8] Darishcheva V I *et al* 2016 *Mathematical Models and Computer Simulations* 8 725–733
[9] Nurislamov N B *et al* 1999 *RF Patent* No. 2159326
[10] Kozhin N S *et al* 2016 *Engineering practice* 7 110–15
[11] Stefankevich Z B *et al* 1992 *Oil And Gas Geology* 8 18–21
[12] Guryanov A I *et al* 2014 *RF Patent* No. 2555718
[13] PROSHCHEKALNIKOV D V *et al* 2015 *Bulletin of the Technological University* 18 131–3
[14] Iovlev D P *et al* 2016 *Mater. of the Int. Sci. and Practical Con.: Dostizheniya, problemy i perspektivy razvitiya neftegazovoy otrasli* 1 383–7
[15] Mullakaev M S *et al* 2017 *J. of Petroleum Sci. and Eng.* 159 1–7

IOP Conf. Series: Materials Science and Engineering **412** (2018) 012004 doi:10.1088/1757-899X/412/1/012004

The application of plasma-electrolysis boronizing of parts

V I Astashchenko[1], D V Emelyanov[2], I N Gilazov[1]

[1] Kazan Federal University, Naberezhnye Chelny Institute, 423812, Russia, Naberezhnye Chelny, Prospekt Syuyumbike 10A

[2]Kazan National Research Technical University named after A.N. Tupolev – KAI

astvi-52@mail.ru

Abstract. The article deals with methods of increasing the working capacity of parts due to the improvement of structural condition of the surface. There were analyzed the characteristic features of the borating method and plasma-electrolysis processing, the necessity for joint use of these methods. On the basis of the review the author suggests the application of plasma-electrolysis boronizing of parts. By increasing the working capacity of parts due to the improvement of structural condition of the surface by means of borating method and plasma-electrolysis processing, it is possible to use more rational application of alloyed materials for manufacturing various parts that have high physical and mechanical, operational characteristics, which is certainly actual for modern machinery production.

The growth of working capacity of machines and mechanisms directly depends on the performance of its components. Therefore, increasing the wear resistance, life cycle of parts such as bearing assemblies, gears, camshafts of internal combustion engines and other parts that often experience large shock loads, strain and work in friction, it is possible to increase the overall mechanism.

It is necessary to have a number of properties such as high strength and hardness of the surface layer combining ductility of the core for life cycle of parts. Such properties can be achieved by applying various methods of surface hardening and surface impregnation. One of less studied methods is the plasma-electrolysis processing of parts and the borating method. The possibility of joint use of the given methods makes it possible to use more rational application of alloyed materials for manufacturing various parts that have high physical and mechanical, operational characteristics, which is certainly actual for modern machinery production.

In modern engineering there used methods for creating various wear-resistant coatings of functional materials (powder coatings, ceramics and heat-resistant polymers, single-component and multicomponent metal and composite galvanized carbide coatings, etc.) for hardening the surface of structural and alloyed steels and components of machines [1].

The surface impregnation is one of the most effective and simple methods of hardening steel parts surface used in final stages of processing and easily combined with component heat treatment.

The saturation of steel surface with boron – boronizing is the most applicable methods of surface impregnation. Boron as well as nitrogen and carbon forms:
– solid solutions for limited solubility;
– intermediate phases: carbides, nitrides, carbonitrides, borides;
– plays the role of reinforcing phases.

Boronizing is the process of surface impregnation consisting in saturation of surface layers of parts with boron, when they are heated in a particular boron-containing medium to increase hardness and resistance.

Content from this work may be used under the terms of the Creative Commons Attribution 3.0 licence. Any further distribution of this work must maintain attribution to the author(s) and the title of the work, journal citation and DOI.
Published under licence by IOP Publishing Ltd

Solid-phase mixtures of various structures, molten salt, gas environment and pastes or coatings are used as a saturating medium when borating instruments [2].

The required condition for formation of a diffusion layer is the presence of active atomic boron at a saturable surface. In addition, the temperature and duration of soaking in a saturating medium must ensure the diffusion of atomic boron in steel with the formation of chemical compounds - iron borides. The ecologically clean method of plasma-electrolysis processing of metal parts, tools, etc. has been intensively introduced for last 15 years. The analysis showed the presence of boride phases FeB, Fe_2B in electrolyte compositions containing ammonium borofluoride - NH_4BF_3, sodium tetraborate sodium - $Na_2B_4O_7$ and boron carbide - B_4C after boronizing.

The generator is the source of direct current on boronizing, for example, ND 1500/700. By applying it, the current is set to 50-80A. The current increases by approximately 50A for every 100/150 mm, when added to the crucible the level of molten borax also increases. Filling the crucible by borax, the system is resisted for 3-5 hours at a temperature of 950 ° C and current in the electrolysis circuit, depending on the size of the parts 400-600 A [3].

The external characteristics of the beginning of electrolysis are the vigorous release of gas bubbles at the anode and flashes of combusted metallic sodium over the cathode.

On the basis of thermodynamic calculations, it is shown that in the entire range of boron temperature during the electrolysis of borax, the potential of sodium evolution is greater than that of boron, i.e. boron is the primary product.

By operating borax burns out and the part of it is removed with slag and parts. Therefore, portions of fresh borax are added to the bath periodically and refining is carried out after 5-6 hours. After processing a certain number of products, it is necessary to check the chemical composition of the borax, since the iron content in the bath exceeds 0.5%, which greatly impairs boronage.

The voltage on the electrodes is set to 12-20V, the current density is 0.15-0.3 A / cm2. By increasing current density, the depth and surface hardness rise, i.e. the separation of atomic boron on the cathode accelerates. Thus, increasing the current density from 0.1 to 0.3 A / cm2 (bath temperature 950 ° C, processing time 3 h) is accompanied by an increase in layer hardness from 0.13 to 0.16 mm [3].

When surface treatment of steels is processed by an electrolytic-plasma method, the entire surface in the solution is heated. The part is placed on the bracket and immersed in the electrolyte solution. Then the voltage is applied and the solution temperature is maintained within specified limits. In the electrolyte when a current flows through the electrolyte at the cathode (quenched part), a gas jacket of hydrogen is formed (the current strongly increases and the part heats up). After switching off the current, the part can be immediately hardened in the electrolyte. In this case, the heating occurs rapidly and depending on the heating time it is possible to control the depth of the layer processing and hardening [4].

Heating of the parts occurs throughout the product area. There is a 5-10% increase in temperature growth only in places of large protrusions due to the technology peculiarity.

Parameters of equipment for surface hardening by electroplasma method [4]:

– equipment capacity: calculated on the basis of the area of the workpiece. For quenching 1 square approximately 8 kW is needed.

– current density 30 A / sq.dm.

– operating voltage 380 V.

– the quenching time is directly proportional to the depth of the hardened layer: from 3 to 10 seconds.

As a result, the combination of two methods of conducting plasma-electrolysis processing with boron-containing solutions, it is possible to obtain maximum wear resistance and surface hardness of the parts.

References

[1] Mishustin N M 2012 *Iznosostojkie boridnye pokrytiya, poluchaemye na konstrukcionnyh i legirovannyh stalyah s ispol'zovaniem TVCH-nagreva* [Wear-resistant boride coatings obtained on structural and alloy steels using HD heating] (Barnaul: Thesis for a scientific degree) 198 p
[2] Okolovich G A 2010 *Shtampovye stali dlya holodnogo deformirovaniya metallov* [Stamping steels for cold deformation of metals] (Barnaul: Altai State Technical University named after I.I. Polzunov) 202 p
[3] Voinov B A 2010 *Iznosostojkie splavy i pokrytiya* [Wear-resistant alloys and coatings] (Moscow: Mechanical Engineering) pp 37-50
[4] Emelyanov D V, Gilazov I N, Astashchenko V I 2017 *O vozmozhnosti ispol'zovaniya borsoderzhashchih diffuzionnyh pokrytij na rezhushchih instrumentah iz bystrorezhushchih stalej* [The application of boron-containing diffusion coatings on cutting tools of high-speed steels] (Rostov-on-Don: North-Caucasian Scientific Center of Southern Federal University) 116 p

IOP Conf. Series: Materials Science and Engineering **412** (2018) 012005 doi:10.1088/1757-899X/412/1/012005

Optimal geometrical parameters of trihedral steel support's cross section

I R Badertdinov[1,4], I L Kuznetsov[1], N F Kashapov [2], I R Gilmanshin[2], L S Sabitov [2,3]

[1] Kazan State University of Architecture and Civil Engineering, Kazan, 420043, Russia
[2] Kazan Federal University, 18 Kremlyovskaya street, Kazan, 420008, Russian Federation
[3] Kazan State Power Engineering University, 51 Krasnoselskaya street, Kazan, 420066, Russian Federation
[4] SUE кЕ "Engineering Center" Industrial Safety ", Kazan, Russia

l.sabitov@bk.ru

Abstract. The article deals with steel triangular lattice towers with a triangular or diagonal bars for the construction of supports of transmission lines, poles wind generator installations, supports for the placement of lighting equipment. For these constructions have become consumption depends on the geometric parameters of cross-section. Therefore, to reduce the consumption of steel must be assigned the optimal value of parameters such as the height of the cross-section, angle braces. For these supports, an analytical expression of the masses. From the condition of a minimum weight of a formula for determining the optimal height of the cross-section triangular lattice towers. the equations and the graphs allow to determine these parameters to determine the optimal tilt angle braces and supports a triangular lattice truss. The value of the optimal angle for the supports with a triangular lattice is $\alpha_{opt} = 68^{\circ}$, and for bearings with diagonal bars $\alpha_{opt} = 63^{\circ}$.

In practice, the construction of steel poles of power transmission lines, supports of wind generators, supports for placing light-signal equipment are widely used lattice supports [1, 4]. In this case, the most rational type of section of lattice supports are supports with a trihedral cross-section [5,6].

It was shown in [7, 8] that the steel consumption depends on the geometric parameters of the cross section. Therefore, in order to reduce the consumption of steel, it is necessary to assign the optimum value of such parameters as the height of the cross section, the angle of inclination of the braces.

Consider the lattice support of a triangular cross-section (Fig. 1).

Content from this work may be used under the terms of the Creative Commons Attribution 3.0 licence. Any further distribution of this work must maintain attribution to the author(s) and the title of the work, journal citation and DOI.
Published under licence by IOP Publishing Ltd

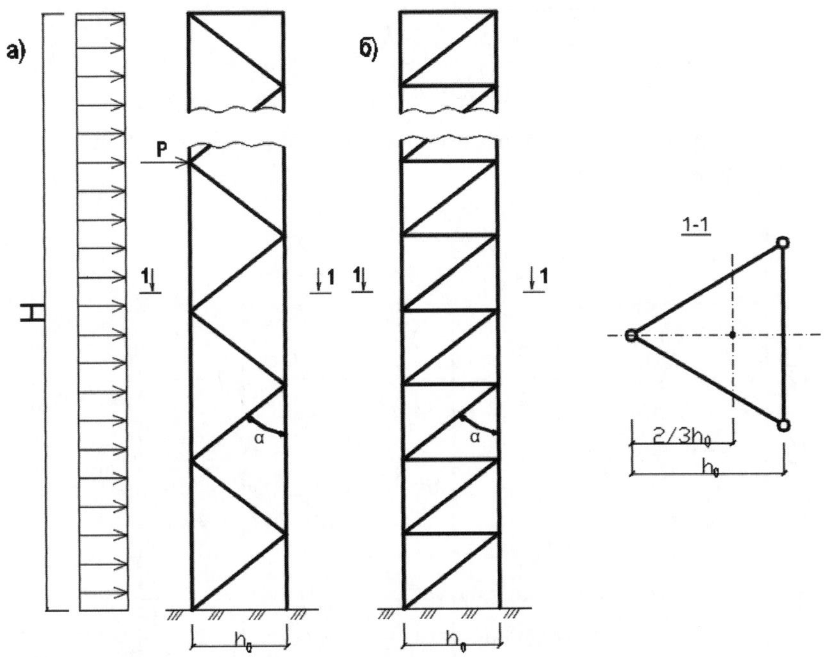

Fig. 1. The design scheme of a support of a triangular cross-section
a) with a triangular lattice; b) with a diagonal grating.

For this support, in [8] the author obtained in general form an analytic expression for its mass, which can be written as follows:

$$G = \psi_{TP} \cdot \gamma \cdot H \cdot \left\{ \frac{\psi_{\Pi}}{R_y \cdot \gamma_c} \cdot \left[N + \frac{3M}{h_0} + \frac{3 \cdot n \cdot k_{\Pi} \cdot R_y \cdot \gamma_c \cdot h_0^2}{m^2} \right] + 3\psi_P \cdot C \cdot h_0^2 \cdot \xi \right\}, (1)$$

Here $\psi_{тр}$ is the constructive mass factor of the triangular support; $\psi_п$ is the constructive mass ratio of the support belts; γ is the density of steel; H - height of the support; R_y is the design resistance of steel; γ_c - the coefficient of working conditions of the bearing belts; N - longitudinal force in the support; M is the value of the bending moment; h_0 is the height of the cross section of the support; n = 1 for a support with a diagonal grating; n = 2 - for supports with triangular lattice; $k_п$ is a coefficient that takes into account longitudinal bending; m = tgα, where α is the angle of inclination of the brace; ψ_p is the constructive mass factor of the lattice;

$$C = \frac{\mu^2}{n_c \cdot \lambda_{pred}},$$

where μ – coefficient of reduction of the calculated length; λ_{pred} – limiting value of the flexibility of the lattice rod; $\xi = \frac{(1+m^2) \cdot \sqrt{1+m^2}}{m^2}$ – for supports with a triangular lattice;

$$\xi = \frac{\left(m + \sqrt{1+m^2}\right) \cdot \left(1+m^2\right)}{m^2}$$ – for supports with a diagonal grille.

The height of the cross section (h_0) and the slope angle (α), given by m = tgα, are present in the expression of the support mass. Therefore, in order to find the optimum height of the cross-section of the support (h_{opt}), it is necessary to take the derivative, i.e. dG / dh = 0, from where

$$h_{opt} = \sqrt[3]{\frac{\psi_{\Pi} \cdot m^2 \cdot M}{2 \cdot \gamma_c \cdot R_y \cdot \left(n \cdot \psi_{\Pi} \cdot k_{\Pi} + \psi_P \cdot m^2 \cdot \xi \cdot C\right)}}, \tag{2}$$

In Fig. 2 is a graph of the dependence of the mass on the height of the cross section of the support. The graph is compiled for example in Table 1.

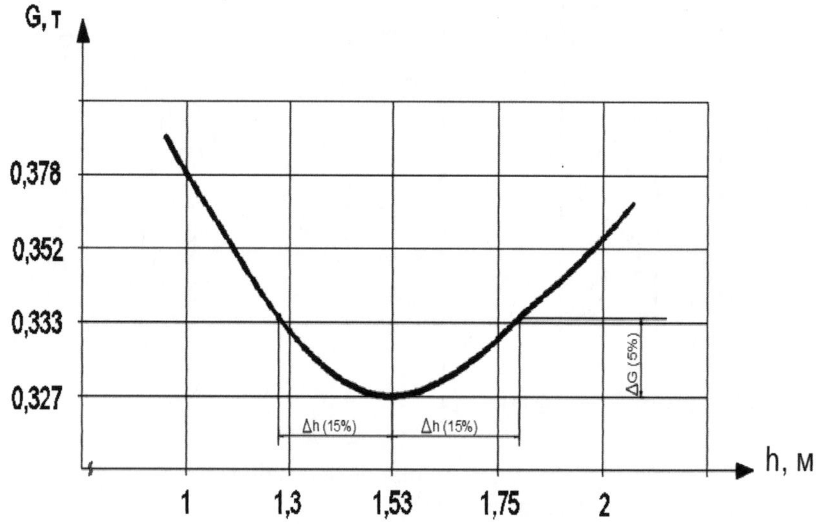

Fig. 2. Graph of the dependence of the mass of the support on the height of the cross section.

Parameter table for the example in question

Table 1

Height of support H	20 m
Equivalent uniformly distributed wind load	50 kg/m
Concentrated horizontal load at height 15 m	300 kg
The support is made of steel with a triangular lattice	C245
Cross-section of support elements	steel pipe
$\psi_{тр}$=1,1; $\psi_п$=1,05; $\psi_р$=1,05; n=2; R_y=2450kg/cm^2; γ_c=1; ξ=2.828; $k_п$=3,2*10^{-5}; m=1; n_c=1; c=0.66*10^{-5}; M_{max}= 14.5 t*m; h_{opt}=153 sm (at m=1)	

From the graph (Figure 2) it follows that the deviation of the height of the section of the support from the optimum value by ± 15% changes the mass of the support by 5%. This makes it possible, when assigning the final height of the section of the support, to apply it taking into account other technological factors.

Another geometric parameter that affects the mass of the support is the angle of inclination of the braces. We find the optimal value of this parameter, also taking the derivative with respect to the mass of the support (1) by substituting ξ into its value. From the solution of the equations obtained from the condition dG / dm = 0, we obtain the equations for finding the optimal value of the angle of inclination of the braces, namely:

- for supports with a triangular lattice

$$(m^2 - 2)\left(\sqrt{m^2 + 1}\right) - A_1 = 0, \tag{3}$$

IOP Conf. Series: Materials Science and Engineering **412** (2018) 012005 doi:10.1088/1757-899X/412/1/012005

- for supports with a diagonal grille

$$(m^2-2)\left(\sqrt{m^2+1}\right)+m(m^2-1)-A_1=0,(4)$$

Here

$$A_1=\frac{\psi_{\Pi}nk_{\Pi}}{\psi_p C},(5)$$

For practical calculations, it is advisable to carry out the optimal angle of inclination of the braces according to the graphs given in Fig. 3 and Fig. 4.

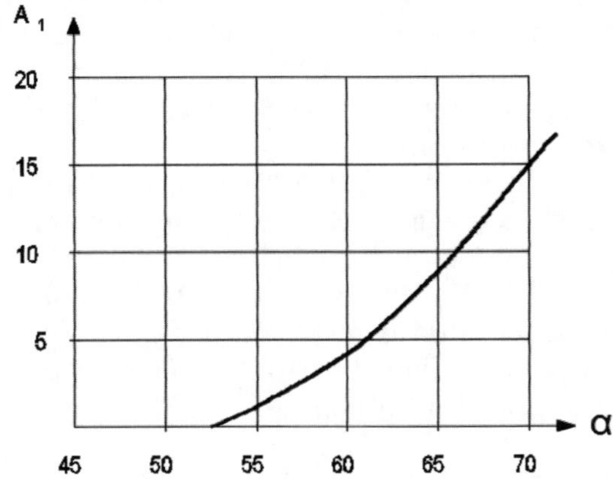

Fig. 3. Graph for determining the optimal angle of inclination for supports with a triangular lattice

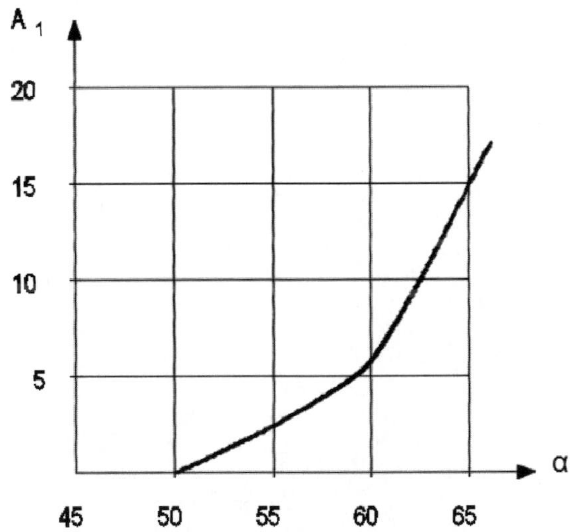

Fig. 4. Graph for determining the optimal angle of inclination for supports with a triangular lattice.

For the support considered in Table 1, the value of A_1 = 9.7, so the value of the optimal angle of inclination according to the given graphs for supports with a triangular lattice is equal to $\alpha_{opt} = 68^o$,

IOP Conf. Series: Materials Science and Engineering **412** (2018) 012005 doi:10.1088/1757-899X/412/1/012005

and for supports with a diagonal grating, $\alpha_{opt} = 63^{o}$ o. Optimal section height at the optimal angle of inclination of the brace is 68 ° to 45 °, as in the example from 153 cm to 149 cm.

Conclusions

Industrial Safety A formula is obtained for determining the optimal height of supports of a triangular cross-section. Analytical conditions and graphs are obtained for determining the optimal slope angle.

References

[1] Kuznetsova V V 1999 *Metal Constructions. The 3 volumes* (M.: Publishing House DIA) p 528
[2] Sabitov L S, Khamidullin I N 2015 *Optimal steel multifaceted supports of overhead lines elektroperdachi* Collection of materials of the conference of young scientists «Youth and Innovation Tatarstan» pp 35-38
[3] Sabitov L S, Kuznetsov I L, Khamidullin I N *Development and research of supports for wind generator systems* Scientific Bulletin VGASU. Construction and architecture No 2 (38) pp 34-40
[4] Sabitov L S, Kuznetsov I L, Badertdinov I R 2014 *Development and research of grid triangular supports* Proceedings of the 14th International Scientific and Technical Conference «Effective constructions: theory and practice» pp 15-17
[5] Kuznetsov I L, Sabitov L S, Badertdinov I R 2008 *Trihedral latticework* RF Patent № 2378469. Kuznetsov I L, Sabitov L S, Badertdinov I R 2015 *Triangular lattice towers* RF patent under the application № 2015105647/03
[6] Sabitov L S, Kuznetsov I L, Penkovtsev S A 2014 *Choosing the type of cross-section management supports* Volga scientific journal 4 (32) pp 90-94
[7] Sabitov L S, Kuznetsov I L, Badertdinov I R 2016 *On the question of the rational type of cross-section of support* Power of Tatarstan No 1 pp 47-51

Evaluation of the applicability of pre-evaporative cooling of air in front of the cooling tower in the conditions of the third climatic zone of the Russian Federation

D S Balzamov[1], L S Sabitov[1,2], B F Timershin[3], E Yu Balzamova[1]

[1]Kazan State Power Engineering University, Krasnoselskaya street, 51, Kazan, 420066, Russian Federation
[2]Kazan Federal University, Kremlyovskaya street, 18, Kazan, 420008, Russian Federation
[3]LTD "Engineering Center" Energoprogress ", Volgogradskaya street, 34, Kazan, 420044, Russian Federation

dbalzamov@mail.ru

Abstract. This article provides an evaluation of the applicability of the evaporative cooling system of air supplied to the cooling tower in order to reduce the temperature of the cooled water at the outlet from the cooling tower in the conditions of the third climatic zone of the Russian Federation.

The problem of effective functioning of circulating water supply is especially acute in the warm period of the year. The high temperature of the circulating water before the condensers leads to a decrease in the generation of electrical energy in the turbine generators. Increasing the efficiency of water cooling devices is an urgent task for energy companies. Increasing the efficiency of the cooling towers can be carried out in two directions: a change in the design of the sprinklers and the cooling tower itself and a change in the external conditions (change in the parameters of the outside air and return water fed to the cooling tower) [2].

The system of evaporative air cooling, developed by the resident of the Skolkovo Foundation, is being considered, which is mounted in the air intake windows of the tower cooler BG-2600. The cooling tower provides cooling of the condenser of turbine PT60-130 / 13 with the density of irrigation q_{op} = 8 $m^3/m^2 \cdot$hour), the width of the cooling zone Δt=8°C and the outside temperature t_{air} = 20°C. The technical solution is that the external air entering the cooling tower is cooled, passing through the evaporator panels installed in the cooling tower air intake windows (Fig. 1).

Cooling of air occurs due to the instant evaporation of finely dispersed drops of water (fog) created by injectors with an outlet section of channels of 2-5 mm at a pressure of 70 atm. The process is accompanied by the absorption of the so-called latent heat of evaporation from the outside air.

Content from this work may be used under the terms of the Creative Commons Attribution 3.0 licence. Any further distribution of this work must maintain attribution to the author(s) and the title of the work, journal citation and DOI.
Published under licence by IOP Publishing Ltd

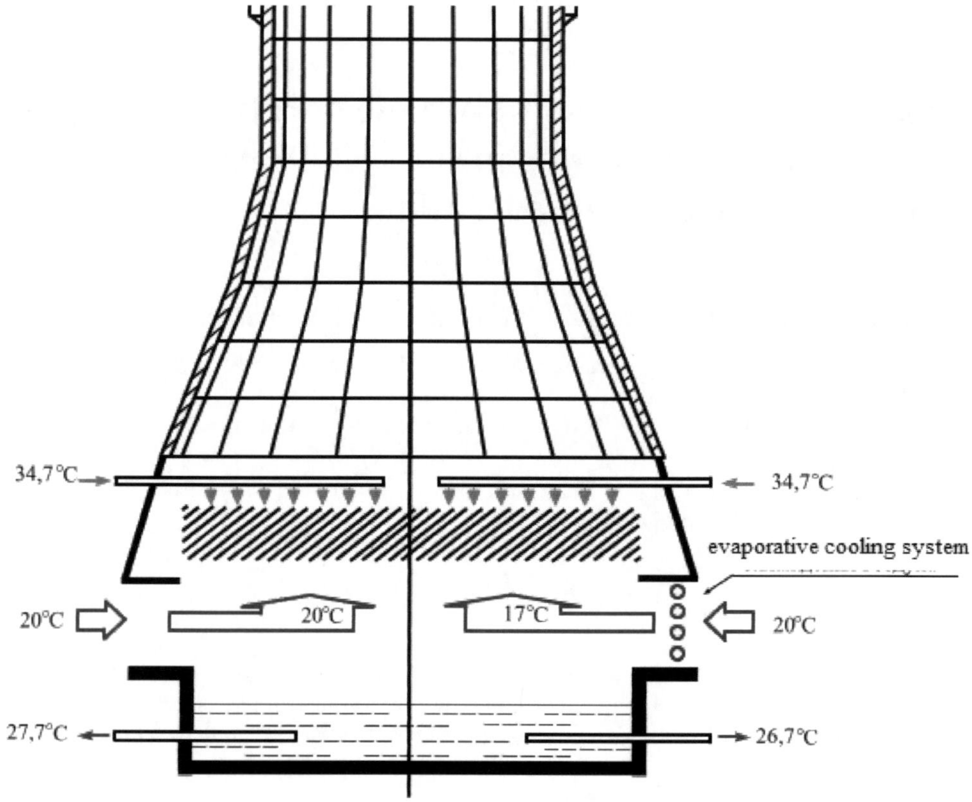

Fig.1 Flow pattern of air in the cooling tower

The evaporative cooling system allows cooling of the external air supplied to the cooling tower by 2°C when spraying 1 g/m^3 of air. A spray version of 1.5 g/m^3 is considered, with cooling of the intake air at 3 °C. It is worth noting that in order not to violate the process of evaporative cooling in the cooling tower, the humidity of the air supplied to the cooling tower should not exceed 80%. Thus, there is a limit on the amount of water sprayed, so for Tatarstan's climatic conditions this is 1.5 g/m^3. In this regard, the actual outdoor temperatures during the period from May to September 2014 for Naberezhnye Chelny city were analyzed. As an example, Fig. 2-3 shows the graphs of the distribution of outdoor air temperature.

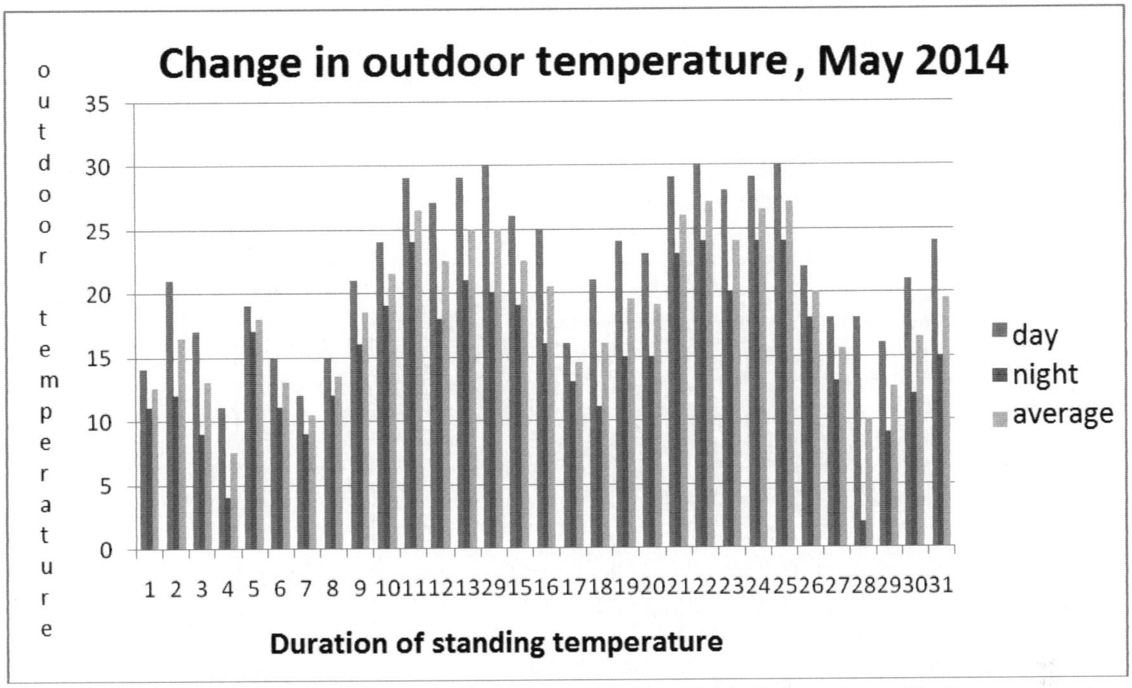

Fig. 2 Change in outdoor temperature in May 2014, Naberezhnye Chelny

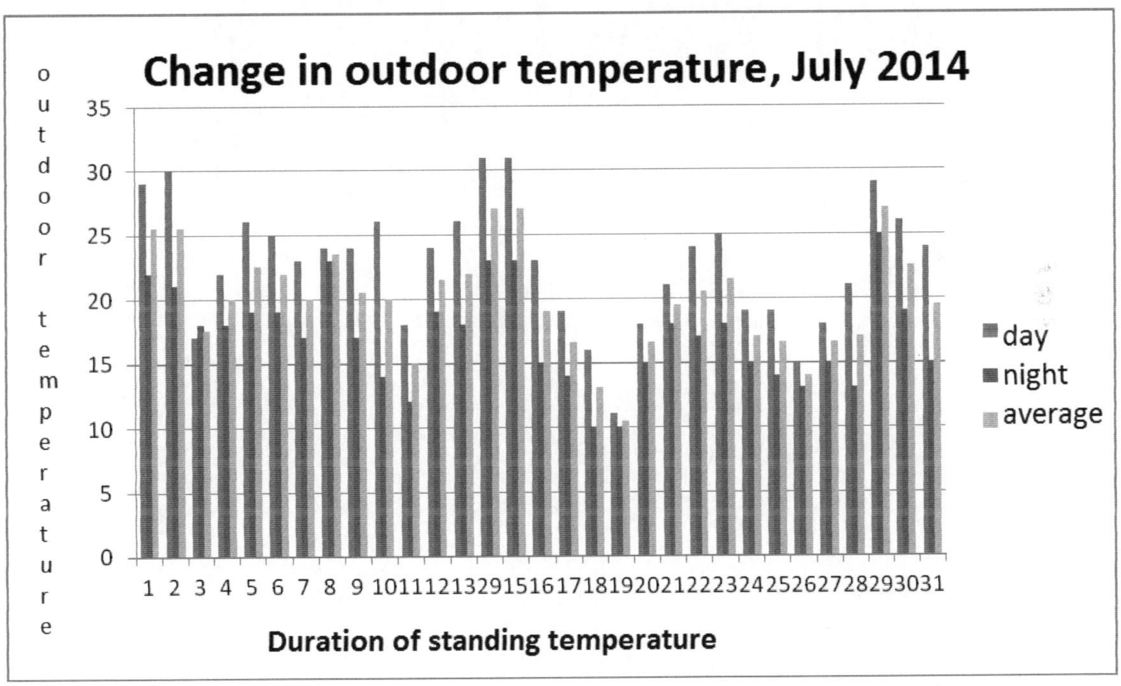

Fig. 3 Change in outdoor temperature in July 2014. for one of the cities of the Republic of Tatarstan.

An analysis of the change in ambient air temperatures during the period from June to August showed that the average daily temperature was 19.45 °C. In May, the outside air temperature was 18.71 °C, and in September it was 13.57 °C. The mean daily and mean temperatures for the hottest period (July) were 22.71 °C and 17.06 °C, respectively. The maximum air temperature during the summer period rose to 31 °C and lasted only two days. Thus, during the summer period of 2014, two modes of operation of the evaporative cooling system were possible: spraying 1 g/m^3 and 1.5 g/m^3.

IOP Conf. Series: Materials Science and Engineering **412** (2018) 012006 doi:10.1088/1757-899X/412/1/012006

In view of the above, an important condition for the efficient operation of the evaporative cooling system is to provide precise control of the parameters of the cooled air sucked into the cooling tower and to maintain them at a given level, which leads to more complicated operation of the system.

As follows from the nomogram (Fig. 2), the temperature of the cooled circulating water after the evaporative cooling system of air will be 26.7 °C at the temperature of water supplied for cooling tgor. = 34.7 °C. Thus, the temperature of chilled water decreased by 1 °C.

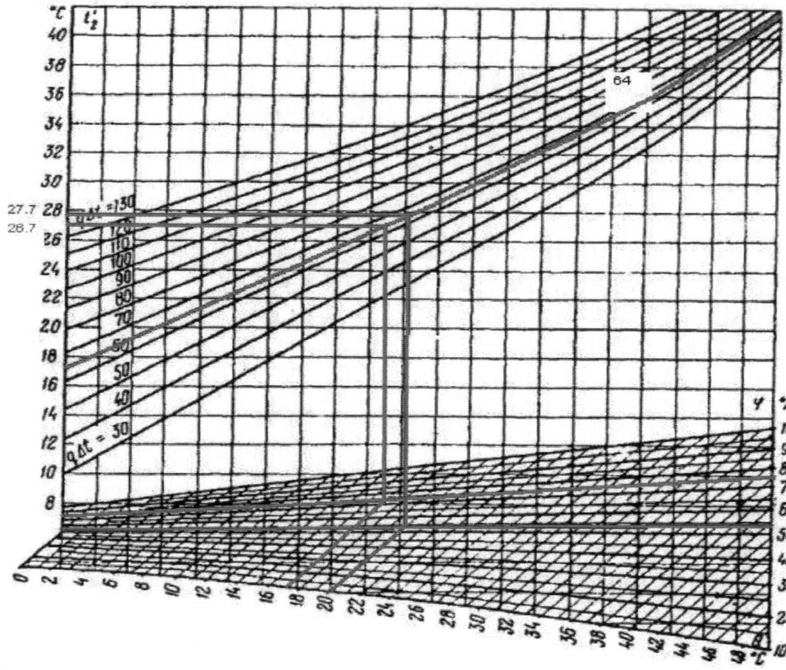

Fig. 4 The process of cooling water at an outdoor temperature of + 20 °C.

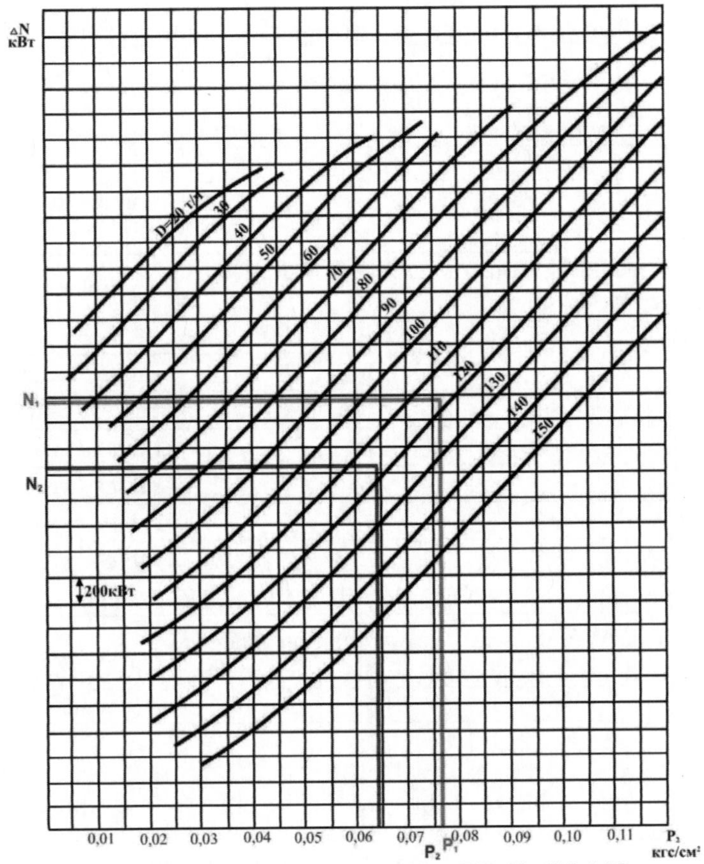

Fig. 5. Characteristics of the turbine PT-60-130 / 13

For the PT60-130/13 turbine, the number of hours of operation during the summer period is τ= 3600 hours. During the summer period, the temperature was set at + 17 °C and above. A decrease in the circulating water temperature at the inlet to the condenser by 1 °C leads to an increase in the output of electricity by 250 kWh [1]. The economic effect for the whole period of work will be:

$$N_E = \tau\ 250 = 3600 \cdot 250 = 900\ 000 \text{ kWh}.$$

The specific heat consumption for turbine PT60-130/13 according to the PTO station data for 2014 is 1577 kcal / kWh, while the savings in specific heat consumption are as follows:

$$q_T^H = 1577 \cdot 900\ 000 = 1\ 419\ 300\ 000 \text{ kcal/year} = 1419{,}3 \text{ Gcal/year}$$

1 t.c.f. = 7 Gcal, have:

$$q_{ud} = 1419{,}3/7 = 202{,}758 \text{ t.c.f.}$$

At a cost of 1 t.c.f. = 3527 rubles (according to the station for the period of 2014), the economic effect will be:

$$Э = 202{,}758 \cdot 3527 = 715\ 124 \text{ rubles}$$

The flow rate of the return water for the turbine condenser of the brand PT60-130/13 is 8000 m^3 at the cooling tower's flow rate of 20,800 m^3/h. Thus, the BG 2600 cooling tower can operate in a cycle with 2 turbine condensers PT60-130/13. In this regard, the economic effect of reducing the temperature of the circulating water will be twice as much and amount to 1430248 rubles. Table 1 presents the results of a simple payback period for the evaporative cooling system for the BG-2600 cooling tower. The calculations are carried out for the actual summer conditions of 2014, as a period with the lowest summer temperatures.

IOP Conf. Series: Materials Science and Engineering **412** (2018) 012006 doi:10.1088/1757-899X/412/1/012006

Table 1. The results of calculating the simple payback period of the evaporative cooling water system

Name	Unit of measurement	Value
Cost of evaporative water cooling system	rubles	11 110 000
Tariff for electricity	rubles kWh	1,256
Tariff for water	rubles /m^3	6,3
Calculated values		
Cost of construction (including construction and installation works, design and survey work)	rubles	17 039 000
Annual operating costs:		
energy costs	rubles / year	52 243
average annual overhaul costs	rubles/ year	277 750
average annual maintenance costs (parts + service)	rubles / year	286 000
Depreciation deductions	rubles / year	1 703 900
Expenditure on evaporated water	rubles / year	148 599,36
Total annual operating costs:	rubles / year	2 468 492,36
Annual savings during operation AIR EVOPARATIVE COOLING SYSTEM:		
economic effect with a decrease in the circulating water temperature by 1 °C	rubles / year	1 430 248
Payback period	year	7,8

Conclusion

In view of the limited amount of water sprayed on the process of evaporation in the evaporative cooling system of air in the climatic zone of the Republic of Tatarstan, the proposed technical solution is inefficient.

It should also be noted that the pre-evaporative cooling system is demanding for the quality of the water being sprayed due to the small output cross-section of the 0.2 mm nozzles. It is necessary to strictly control the degree of saturation of air supplied to the cooling tower.

The most preferred conditions for the application of the evaporative cooling system are the southern regions with a long standing of high temperatures and low humidity of air [3].

References

[1] Ruzankov V N, Laditskaya N V, Kruglov G A, Danilin V M 1974 *Typical normative characteristic of the turbine unit PT-60-130/13* LMZ: approved Ch. Ing. Glavtekhupravleniya on the operation of power systems V. Zubenko

[2] Laptev A G, Vigayeva I A 2004 *Design and calculation of industrial cooling towers*: Monograph (Kazan: KSPEU) p 180

[3] *Determination of the economic effect of the cooling tower Number 2 of Astrakhanskaya CHPP-2 from the introduction of the evaporative cooling of outdoor air* 2007 S. Petersburg: JSC "All-Russian Scientific Research Institute of Hydraulic Engineering" B.E. Vedeneeva " p 25

IOP Conf. Series: Materials Science and Engineering **412** (2018) 012007 doi:10.1088/1757-899X/412/1/012007

Increase of efficiency of heat sources work due to application of condensation economizers on an example of a boiler PTVM-180

D S Balzamov[1], L S Sabitov[1,2], B F Timershin[3], E Yu Balzamova[1]

[1]Kazan State Power Engineering University, Krasnoselskaya street, 51, Kazan, 420066, Russian Federation
[2]Kazan Federal University, Kremlyovskaya street, 18, Kazan, 420008, Russian Federation
[3]LTD "Engineering Center" Energoprogress ", Volgogradskaya street, 34, Kazan, 420044, Russian Federation

dbalzamov@mail.ru

Abstract. In this article, a method is considered to increase the efficiency of boiler room operation with PTVM-180 boilers by the use of a flue gas flue in a condensing economizer. The efficiency evaluation of the utilization of the heat exchanger in accordance with the operating mode of the boiler house was evaluated, and the technical and economic parameters of the project were assessed, depending on the unplanned changes in such key factors as capital costs, operating time, operating mode.

The economy of fuel burned in boiler units is an urgent task when solving the problem of reducing the cost of produced heat energy.
At present, one of the ways to increase the efficiency of boiler houses is the use of deep utilization of the heat of flue gases using condensing economizers.
As an object for the introduction of condensation technology, the boiler house of Kazan is being considered. On the boiler-house are installed water-heating boilers of the type PTVM-180 in the amount of 2 pcs. In the summer period the boiler-house does not work.
The main indicators of the boiler house are presented in Table. 1.

Table 1 Main indicators of boiler plant operation.

Indicator	Output, Gcal		Operating time, h		Average load, Gcal / h		Excess air, α		Number of combustion air for α> 1, nm³/h		Gas consumption, nm³/h		
	BK-1	BK-2	BK-1	BK-2	BK-1	BK-2	BK-1	BK-2	BK-1	BK-2	BK-1	BK-2	General
January	94 085	79 542	728	740	129,24	107,49	1,24	1,14	154 007	119 701	12 420	10 500	22 920
February	90 772	71 188	672	672	135,08	105,93	1,23	1,14	144 753	105 216	11 769	9 230	20 998
March	80 920	60 969	744	744	108,76	81,95	1,27	1,18	130 756	91 536	10 296	7 757	18 053
April	65 133	60 490	704	720	92,52	84,01	1,28	1,17	103 902	88 203	8 117	7 538	15 656

Content from this work may be used under the terms of the Creative Commons Attribution 3.0 licence. Any further distribution of this work must maintain attribution to the author(s) and the title of the work, journal citation and DOI.
Published under licence by IOP Publishing Ltd

Indicator	Output, Gcal		Operating time, h		Average load, Gcal / h		Excess air, α		Number of combustion air for α> 1, nm³/h		Gas consumption, nm³/h		
	ВК-1	ВК-2	ВК-1	ВК-2	ВК-1	ВК-2	ВК-1	ВК-2	ВК-1	ВК-2	ВК-1	ВК-2	General
May	-	14 671	-	106	-	138,4 1	-	1,13	-	18 995	-	1 681	1 681
Jun	-	-	-	-	-	-	-	-	-	-	-	-	-
Jul	-	-	-	-	-	-	-	-	-	-	-	-	-
Aug	-	-	-	-	-	-	-	-	-	-	-	-	-
Sept	-	21 133	-	263	-	80,35	-	1,19	-	180 92,8	-	15 204	15 204
Octr	43 446	76 591	443	745	98,07	102,8 1	1,18	1,16	79 408	137 616	6 730	11 863	18 593
Nov	70 093	76 294	720	720	97,35	105,9 6	1,18	1,16	105 052	112 408	8 903	9 690	18 593
Dec	79 631	87 320	744	744	107,0 3	117,3 7	1,18	1,15	119 567	127 779	10 133	11 111	21 244

The installation of condensing economizers on boilers allows to solve such tasks as deep cooling of exhaust gases, thereby reducing thermal pollution of the environment, and also to increase fuel efficiency. The utilized heat energy can be used to heat raw water, preheat the return network water before entering the hot water boiler.

In the condensation economizer, the combustion products are deeply cooled to a temperature at which the maximum possible part of the water vapor contained in the gases can be condensed and the latent heat of condensation can be used. In addition, deep cooling of gases allows fuller use of their physical heat [1].

The heat exchanger is installed behind the smoke exhausters of the PTVM-180 boilers through new flues, some gases (about 25%) are taken from the common flue, pass through the heat exchanger, cool and return to the common flue.

Smoke gases after the heat exchanger have a temperature of 50°C and, when mixed with the main stream, are heated, thereby preventing the condensation of water vapor in the common flue. The heat exchanger and part of the flue after it are made of stainless steel to prevent low-temperature corrosion.

The newly installed equipment will create additional aerodynamic resistance. The increase in resistance in the heat exchanger is partially compensated by the decrease in the volume of flue gases after the heat exchanger, which occurs due to a decrease in their temperature, condensation and removal of a part of the water vapor, which will eliminate the need to replace the operating smoke exhausters.

The condensate formed in the body of the utilizer is desalted water. The presence of weakly basic carbonic acid in the condensate (CO_2 binding with water) gives pH- 3.5. Condensate is removed from the body of the utilizer by gravity into a special tank through the body of the heat exchanger, equipped with a hydraulic seal. Further from the tank through the calciner the condensate pump is fed to the chemical water treatment.

The value of "dewpoint" of combustion products depends on the amount of excess air, which can be determined from the graph shown in Fig. 1.

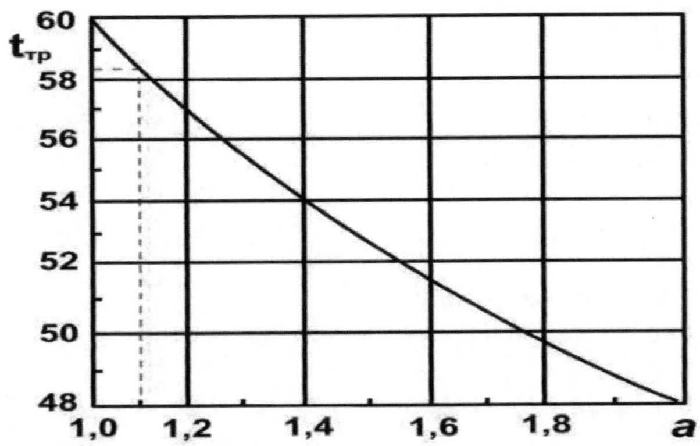

Fig.1 Dependence of the "dew point" of natural gas combustion products on the excess air factor

The amount of moisture condensing during the cooling of flue gases also depends on the value of the excess air factor and can be determined from the graph shown in Fig. 2.

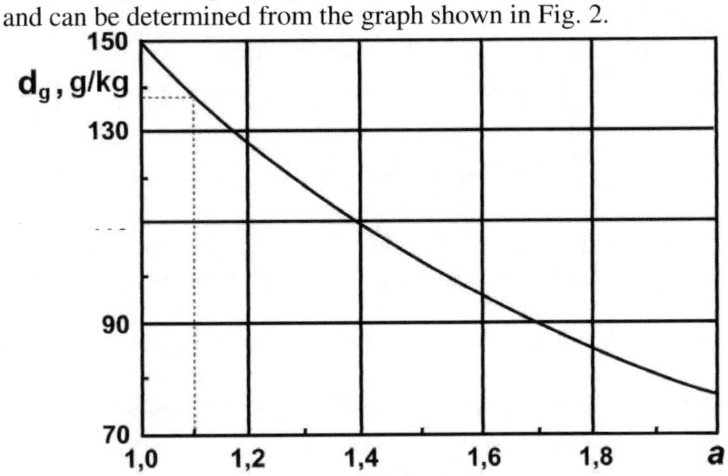

Fig. 2. Dependence of the moisture content of products of complete combustion of natural gas on the excess air factor (weighted average for gas fields in Russia and CIS countries)

The amount of heat that is utilized in the condensation economizer is the sum of the physical heat of the flue gases and the latent heat of condensation of the water vapor.
The amount of heat can be determined from equation (1).

$$0{,}9 \cdot \Sigma Q_{ke} = (G_{dg} \cdot c_p \cdot (t_{dg1} - t_{dg2}) + (G_{vp} \cdot r_{vp})/4{,}186 \text{, Gcal/h} \quad (1)$$

Where G_{dg} – flue gas consumption, nm^3/h; c_p – averaged heat capacity of flue gases, kJ/kg•°C; t_{dg1} – flue gas temperature at the condenser economizer inlet, °C; t_{dg2} – flue gas temperature at the outlet of the condensing economizer, °C; G_{vp} – the amount of condensing water vapor in the economizer, kg / h; r_{vp} – latent heat of condensation of water vapor, kJ/kg; 0,9 – efficiency of condensing economizer; 4,186 – the conversion factor from kJ to kcal.
It is proposed to use a condensation economizer to preheat make-up water [3].
The make-up temperature is assumed at 40 ° C in connection with the technological requirements of HVO. The heating temperature of the feed in the heat exchanger is regulated by the flow rate of the flue gases.
The amount of condensate formed during the operation of the utilizer for the period October-April is determined by the expression (2):

$$G_k = G_{vp} \cdot \sum \tau \quad (2)$$

The amount of thermal energy contained in the resulting condensate formed when the part of the boiler gas is cooled is determined by the formula (3):

$$Q_k^{p/m} = ((t_g - t_{v1}) \cdot G_{vp} \cdot /1000) \cdot \tau \quad (3)$$

The results of the calculation are given in Table. 3.
The consumption of flue gases is determined by formula (4):

$$G_{dg} = B \cdot K_1 + K_2 \cdot Q_i + (-1) \cdot (k_3 + K_4 \cdot Q_i) \cdot (273 + t_p)/273 \quad (4)$$

where B – second consumption of natural fuel, kg/s (nm^3/s); K_1,K_2,K_3,K_4 – numerical coefficients selected for each type of fuel by the least squares method; K_1 – -0,739 for natural gas; K_2 – 0,278 for natural gas; K_3 – 0,0864 for natural gas; K_4 – 0,267 for natural gas; Q_i – lower heat of combustion of fuel MJ / nm^3; α - coefficient of excess air; t_p – flue gas temperature.
The amount of moisture formed during condensation is determined by formula (5) and nomogram (Fig. 2).

$$G_{vp} = \frac{d_{vp}}{1000} \cdot \rho_{dg} \cdot G_{dg}\text{, kg/h} \quad (5)$$

where d_{vp} – moisture content of flue gases (determined from the nomogram of Fig. 2), g / kg;
ρ_{dg} – averaged density of flue gases (in the temperature range 50-145 ° C), kg / m^3;

It is intended to use a condensing economizer to preheat the return network water before feeding it to the hot water boiler.
The temperature of the return network water at the outlet of the condensing economizer is determined by formula (6).

$$t_{2в} = t_{1в} + \frac{\Sigma Q_e}{(G_в \cdot c_в \cdot 1000)/1000000} \quad (6)$$

where $t_{1в}$ – return water temperature at the condenser economizer inlet, °C; $G_в$ – return water flow rate, t / h; $c_в$ – heat capacity of water, kcal / kg ·°C; 1000 – density of water, kg / m^3;1000 000 – transfer to Gcal.
The amount of heat received in the economizer is determined by the equation (7):

$$Q_т = \frac{G_в \cdot c_в \cdot (t_{2в} - t_{1в}) \cdot \tau}{1000}\text{, Gcal} \quad (7)$$

where τ - operating time of the boiler within a month, hour.
Further, an increase in the production of thermal energy is determined, the values of which are given in Table. 2.

Table 2. The results of calculating the amount of thermal energy that is utilized in the condensation economizer [2].

Period	Water temperature at the heat recovery, °C	The water temperature at the output of the heat recovery boiler, ° C	Amount of heat, Gcal	Increase in output,%
October	50	40	295,65	0,29

Period	Water temperature at the heat recovery, ^{0}C	The water temperature at the output of the heat recovery boiler, ° C	Amount of heat, Gcal	Increase in output,%
November	48,41	40	1 066,69	0,74
December	50,81	40	1 169,47	0,65
January	48,57	40	1 175,58	1,33
February	44,34	40	1 102,59	0,84
March	39,8	40	1 181,69	0,9
April	40,63	40	892,89	1,14
Total			6884,55	0,84 (average.)

Table 3. The results of calculating the thermal energy of the condensate formed in the utilizer.

Period	Amount of heat, Gcal
October	27,30
November	90,68
December	97,73
January	98,95
February	92,73
March	95,45
April	72,94
Total	575,78

To assess the natural draft of the chimney in the conditions of reducing the temperature of the flue gases, an aerodynamic calculation of the chimney was performed, taking into account the work of the condensing heat exchanger. Calculations showed that with the introduction of heat recovery, the natural draft of the existing chimney will have a reserve of 7.29 mm.vod.st.

Conclusion

As a result of the conducted studies, the introduction of a condensing heat recovery unit for flue gas at the boiler house with PTVM-180 boilers will reduce the cost of heating raw water, as well as reduce the amount of feed-through due to the beneficial use of condensate formed during deep cooling of a part of the flue gases. A simple payback period for the introduction of a condensing economizer will be 5.76 years with a capital expenditure of 17,849,724 rubles. The economic effect will amount to 3,096,651 rubles.

References

[1] Zhikhar G I, Zakrevsky V A 2010 *Use of heat of exhaust gases of boiler units in a contact heat exchanger* Heat power engineering pp 41-49

[2] Balzamov D S, Timershin B F 2017 *Overview of advanced technologies for generating companies* Modern Science V 2 pp 26-29

[3] Balzamov D S, Ziganshina A R 2016 *Increasing the efficiency of heat sources In the collection* Energy, electromechanics and energy-efficient technologies through the eyes of youth, the materials of the IVth Russian youth scientific school-conference: in 2 volumes pp 36-38

IOP Conf. Series: Materials Science and Engineering **412** (2018) 012008 doi:10.1088/1757-899X/412/1/012008

Investigation of a nickel coating deposition processes from solid nickel electrolyte

R S Kashapova[1], N M Lyadov[2], R N Kashapov[1,2], L N Kashapov[1], N F Kashapov[1] and I A Faizrakhmanov[2]

[1] Engineering Institute, Kazan Federal University, Kazan, 420008, Russia

[2] Laboratory of Radiation Physics, Kazan Physical-Technical Institute of the Kazan Scientific Center of the Russian Academy of Sciences, Kazan, 420029, Russia

kashramiln@gmail.com

Abstract. Nickel is the basis of most of the heat-resistant materials used in the aerospace industry for power plant parts. Recently, the method of electrolytic coating with nickel is used to create protective coatings on aluminum, magnesium, zinc and cast irons. It is known to use the method of nickel plating of aluminum and magnesium alloys, in particular to protect the duralumin blades of screw aircraft. The lifetime of nickel-plated cast-iron drums for drying in paper production is significantly higher than in conventional cast iron, and paper quality is also improving. The aim of the work is to influence the concentration of hypophosphite on the kinetics of the cathode process and on the properties of the resulting nickel coating.

To increase the wear resistance of rubbing surfaces of parts and to restore their dimensions, hard nickel plating is often used. Nickel coatings have less hardness than chromium plating and have the following advantages: they are relatively easy to process, have a high viscosity with a layer thickness of up to 2 mm; the coefficient of nickel linear expansion is close to the coefficient of steel linear expansion, while in chromium it is several times higher. For solid nickel plating, the power of direct current sources is 3 to 4 times lower than for chromium plating, and the energy consumption is about 20 times less. In the paper, the task is to find a technique for obtaining a nickel coating on properties not inferior to chromium plating.

The following composition was chosen as the electrolyte of solid nickel plating (g/l): $NiSO_4*7H_2O$ - 250; $NiCl_{2*}6H_2O$ -50; H_3BO_3 -40; $NaH_2PO_{2*}2H_2O$ – 25. This composition is usually used to apply the top layer of a bi-nickel coating. It is interesting to determine the influence of hypophosphite concentration on the kinetics of the cathodic process and on the properties of the coating.

Content from this work may be used under the terms of the Creative Commons Attribution 3.0 licence. Any further distribution of this work must maintain attribution to the author(s) and the title of the work, journal citation and DOI.
Published under licence by IOP Publishing Ltd

IOP Conf. Series: Materials Science and Engineering **412** (2018) 012008 doi:10.1088/1757-899X/412/1/012008

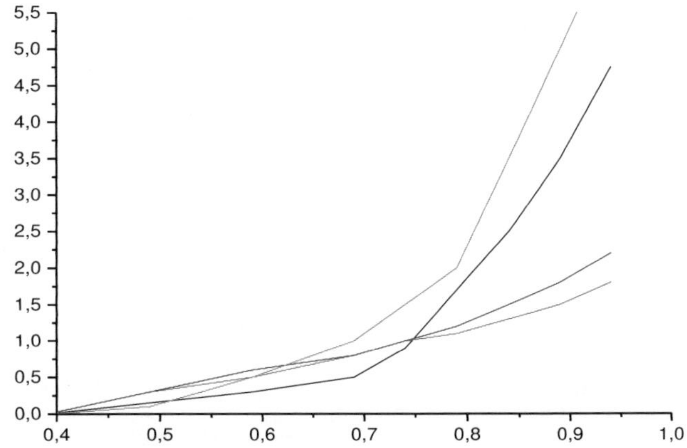

Figure 1. Cathodic polarized curves (CPC) in the electrolyte of solid nickel plating (g/l): $NiSO_4*7H_2O$ -250; $NiCl_2*6H_2O$ 50; H_3BO_3 -40; at concentration NaH_2PO_2 (g/l): 1-7; II -15; III-25 at a temperature 25 ^{O}C and pH=1,8.

Figure 1 shows the total cathodic polarized curves in the potentiodynamic regime at various concentrations of sodium hypophosphite. As can be seen from the CPC, the introduction of sodium hypophosphite into the electrolyte leads to a decrease in the polarization of the electrode. And the speed of the cathode process in the potential range 0.4-0.85 V varies sympathetically the concentration of hypophosphite. Depolarization of the cathodic process in the presence of sodium hypophosphite in the electrolyte can be explained, according to the literature data, by the flow on the cathode of two independent electrochemical reactions: discharge of nickel ions $Ni^{2+}+2\bar{e} \rightarrow Ni$ and restoration of phosphorus donors $H_3PO_2^{-}+H^{+}+\bar{e} \rightarrow P+2H_2O$ or an anion of hypophosphite $H_3PO_2^{-}+2H^{+}+\bar{e} \rightarrow P+2H_2O$. This mechanism of the process is confirmed by the data presented in Figure 2.

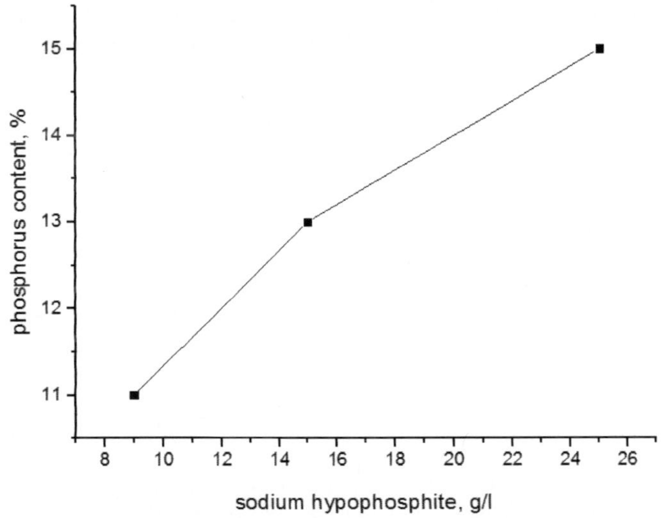

Figure 2. Dependence of the phosphorus content in the cathode deposit on the concentration of NaH_2PO_2 in the electrolyte of solid nickel plating.

It follows from Fig. 2. that an increase in the concentration of sodium hypophosphite leads to an increase in the phosphorus content in the nickel-phosphorus alloy. With a current density of more than 2.5 A / dm2 on the CPC (Figure 1), a limit current area appears, associated with a decrease in the concentration of hypophosphite in the near cathode space (NCS). The temperature of the

IOP Conf. Series: Materials Science and Engineering **412** (2018) 012008 doi:10.1088/1757-899X/412/1/012008

electrolyte is one of the most important factors affecting the electrodeposition of nickel from the nickel-hypophosphite electrolyte.

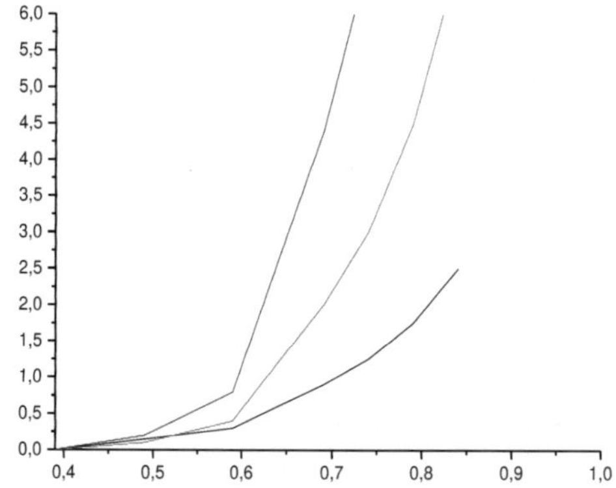

Figure 3. Cathode polarized curves in electrolyte of solid nickel at temperatures (^{O}C): I-25; 2-45; 3-65.

From the CPC shown in Figure 3, we see a decrease in the polarization of the cathodic process as the temperature of the electrolyte increases. This may depend on the reduction of the concentration limitations of electrode processes due to the enhancement of the discharging ions diffusion. Установлено, что качественные никель-фосфатные осадки получаются при температуре электролита 50° и более. It is established that qualitative nickel-phosphate precipitates are obtained at an electrolyte temperature of 50 ° and more. The current yield of nickel under these conditions is 85-88%, however, it is impossible to exclude the imposition of non-current processes on the main electrochemical reactions according to the equations: $H_2PO_2^{-}$+ H_2O= $H_3PO_2^{-}$+H_2 and $2H_2PO_2^{-}$= H_2PO_3+P+OH^{-}+H. The influence of electrolyte mixing on the coating quality was also studied. When moving, ion diffusion accelerates to the electrode, the solution and acid composition in the NCS are equalized, and galvanic precipitation is uniform in thickness and appearance.

The working capacity or amount of passed electricity (Q) is a criterion for the life of the electrolyte. The electrolyte of solid nickel plating is determined by the concentration of ions Ni^{2+} $H_2PO_2^{-}$. As the Q increases, the acidity of the solution decreases, diffusion difficulties occur, and impoverishment by the phosphor of the galvanic coating may occur.

The change in NaH_2PO_2 concentration as a function of transmitted electricity amount is shown in Figure 4.

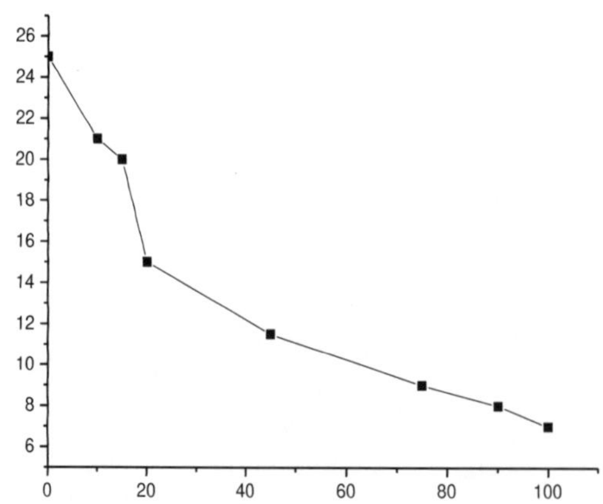

Figure 4. Change in the sodium hypophosphite concentration from the amount of transmitted electricity

When Q 100 A * h/l is passed, the NaH_2 PO_2 content is reduced to 7 g/l. This has no significant effect on the phosphorus content in the plating and the quality of the coating. It is shown that the current yield of a metal depends little on Q and the content of sodium hyposite (Fig. 5) and when the concentration is adjusted, the electrolyte's performance is not limited by other and components of the solution.

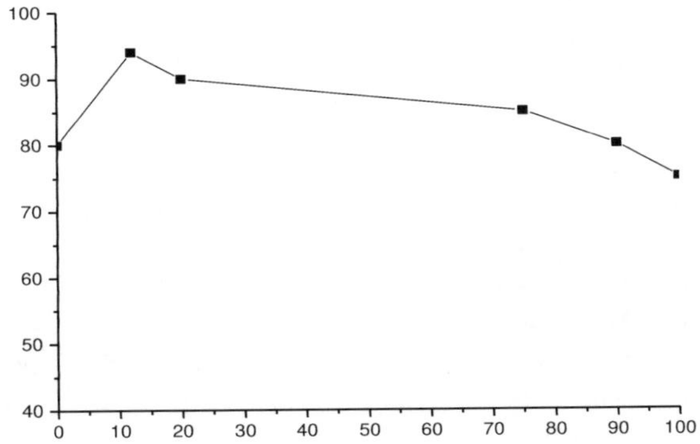

Figure 5. Dependence of the yield of metal on the amount of transmitted electricity

Table 1 Dependence of the yield of metal on sodium hypophosphite content

Content NaH_2PO_2, g/l	25	20,3	15	8,2	7,9	7,4
the yield of metal, %	80	95	91	86	82	72

The study showed that a multiple adjustment of the sodium hyposphite concentration to the initial value (25 g / l) makes it possible to obtain non-porous and wear-resistant nickel-phosphoric coatings. When controlling the components of the electrolyte and the correct electrolysis regime, the electrolyte of solid nickel plating is not different from the usual sulfate nickel electrolytes.

The combined electrocrystallization of nickel and phosphorus causes the galvanic coating to be more hard than ordinary nickel precipitates From the literature data it is known that freshly deposited coatings are a supersaturated solid solution of phosphorus in $\mathcal{L}$-nickel. Microhardness of the coating depends on the content of phosphorus in the sediment.

Figure 6 shows that when the content of phosphorus in the coating varies from 11 to 14% of the weight, the microhardness of the nickel-phosphorus precipitate fluctuates between 550-669 kg / mm^2, and in some cases reaches 700 kg / mm^2.

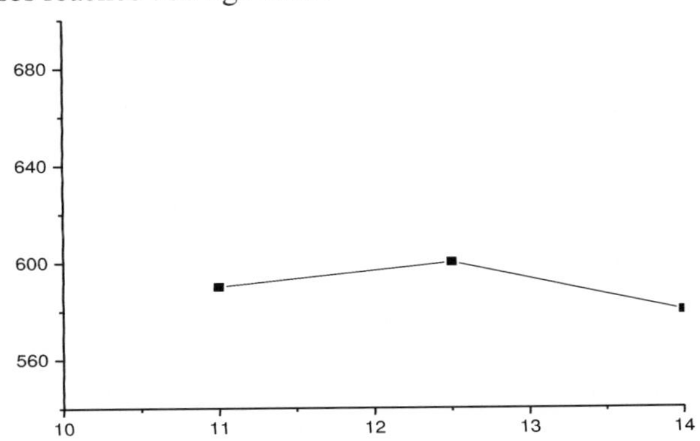

Figure 6. Dependence of the nickel-phosphorus coating microhardness on the content of phosphorus

The inclusion of phosphorus in the galvanic sediment changes not only the microhardness of the coating, but also imparts antifriction properties and high corrosion resistance to the coating.

This work was funded by the subsidy allocated to Kazan Federal University for the state assignment in the sphere of scientific activities (3.9399.2017/8.9).

IOP Conf. Series: Materials Science and Engineering **412** (2018) 012008 doi:10.1088/1757-899X/412/1/012008

References

[1] Kashapov R N, Kashapov L N, Kashapov N F 2017 *Analysis and development of methods for obtaining metallic powders for selective laser melting* IOP Conference Series: Materials Science and Engineering Vol 240 Is 1 **012071**
[2] Kashapov R N, Kashapov L N, Kashapov N F 2015 *The research of anodic microdischarges in plasma-electrolyte processing* IOP Conference Series: Materials Science and Engineering Vol 86 Is 1 **012019**
[3] Kashapov L N, Kashapov N F, Kashapov R N 2013 *Research of the impact acidity of electrolytic cathode on the course of the plasma-electrolytic process* Journal of Physics: Conference Series. Vol 479 Is 1 **012011**
[4] Kashapov L N, Kashapov N F, Kashapov R N 2014 *Influence of plasma-electrolyte discharge to the glass surface* Journal of Physics: Conference Series. Vol 567 Is 1 **012024**
[5] Kashapov L N, Kashapov N F, Kashapov R N, Denisov D G 2016 *Plasma electrolytic treatment of products after selective laser melting* Journal of Physics: Conference Series Vol 669 Is 1 **012029**

IOP Conf. Series: Materials Science and Engineering **412** (2018) 012009 doi:10.1088/1757-899X/412/1/012009

The influence of iron microstructure on tool capacity during cutting process

G K Davletshina, A G Kondrashov, B F Zairov and L.F. Zairov

Kazan Federal University, Naberezhnye Chelny Institute, 423812, Russia, Naberezhnye Chelny, Prospekt Syuyumbike 10A

kpfu.ktomp@yandex.ru

Abstract. An overview of machinability by cutting cast irons is given. The effect of the microstructure of cast iron on the tool's working capacity during cutting has been studied. The reasons for the tool failure during drilling are revealed.

Despite a significant amount of research on the study of machinability of materials by cutting, currently conventional units of measurement of this characteristic have not been established. It is generally accepted [1-6] that the material has good workability if during cutting, the cutting force and depreciation of the tool are negligible, and the tool life and the quality of the treated surface are sufficiently high.

The machinability parameters depend on the structure and properties of the cast iron, which are determined by its grade, the composition of the raw materials, the production technology, heat treatment regimes and other technological factors. It is impossible to evaluate the machinability for only one of the parameters of the cutting process, since processing of different materials with even one hardness can be characterized by different values of cutting forces or temperatures [7-9].

The machinability of cast irons is determined by the specific features of the form of castings, the conditions of casting and, but above all, their microstructure.

No less important is the microstructure of the metal base (matrix) of cast iron. The features of the structure of the matrix of high-strength cast iron with nodular graphite are: a) the location of ferrite mainly in the form of rims around inclusions of globular graphite; b) thinner than gray cast iron, the structure of lamellar perlite, often resembling sorbitolike perlite. The microstructure of the matrix affects the stability of the cutting tool, the level of optimum cutting conditions and the processing efficiency [10, 11].

The main drawback of high-strength cast iron during machining is the varying resistance of the cutting tool. The processing of high-strength cast iron products requires a three-fold increase in the number of tools compared to the treatment of gray cast iron. High-strength cast iron contains more silicon and alloying elements in the form of hard-to-digest carbides. This results in more intensive abrasive depreciation of the working surfaces of the cutting tool and the release of a significant amount of heat, which further reduces the resistance of the tool due to a decrease in the ability to resist

Content from this work may be used under the terms of the Creative Commons Attribution 3.0 licence. Any further distribution of this work must maintain attribution to the author(s) and the title of the work, journal citation and DOI.
Published under licence by IOP Publishing Ltd

depreciation. With a low rigidity of the technological system, the tool also wears out as quickly because of the unevenness of the cut layer, high shock loads and oscillations in cutting forces. [12,13].

Studies have been conducted on the machinability of parts made of cast iron which is widely used in automotive industry. The experiment included the processing of 10 samples. Cutting tools – drill diameter 10...16 mm. Chemical composition of cast iron was determined by spectroscopy by using microphotometry MFS-51 and spectrograph, AFS-51. Selection and preparation of samples for investigations were conducted in accordance with State Standard 3443-87. A 4% solution of nitric acid was used as a reagent for etching pig iron.

As a result of the research, the influence of the microstructure of the cast iron on the operability of the cutting tool was revealed. Thus, when drilling with high-strength cast iron, with a different microstructure, a metal base and the presence of graphite of a spherical regular and irregular shape, which is uniformly and unevenly distributed in an amount of 8-12% (State Standart3443-87), the cutting tool breaks down (Table).

Table 1

Parameters of cast iron properties in castings

Partname	Hardness, HB	Metalbase	Extrafactors
Body 6520-3104055	187-197	Perliteplate 30 ... 60% and ferrite	Inclusions of vermicular graphite, unevenly distributed
Plate 4326-3105040	477	Needle martensite and residual austenite and ferrite (F6-10%) perlite in some places 80-90%	Martensite in a metal base. Hardness is toohigh
Bearingcover 5320-2402079	285	Perlite plate is approximately 60% and ferrite.	Hardness is toohigh
Body 6520-3104055	156	Perlite lamellar in an amount of 10 to 30% and of 30 to 60%	From the surface to a depth of 1.0 mm, clusters of rectilinear graphite are observed, joined together by a film of particles of the undecomposed modifier; in the defect zone, there is a duster and pearlitic rim - "oxide deposit" - a defect in the casting.
Cover 53205-2502209	143-149	Ferrite	From the surface to a depth of 0.18 mm there is a ferrite rim. Also there are accumulations of rectilinear graphite joined together by a film with particles of undecomposed modifier propagating to a depth of up to 4.5 mm - defect of casting production - "oxide deposit"
Body 5320-2402112	217	Perlite granular ≈40%; ferrite ≈ 40%;	Troostite with microhardness HV 349-371

		troostite≈20%.	
Crankcase 5320-2506111	156	Perlite plate + ferrite from 40% to 70%	The pores on the walls are clean, decarburized damaging metal to a depth up to 7.5 mm - the "gas porosity" of the casting

The results of the study are as follows:the structural factors of the non-diffusive decomposition of austenite (troostite, martensite, etc.) with high microhardness are the main factors of the cutting tool failure during drilling operations. In cast iron with a ferritic base, defects in casting are observed, which also have an effect on the performance of the tool during machining.
Thus, in order to reduce breakage of tools when drilling cast iron, it is necessary to provide stable characteristics in structure, hardness and purity of the alloy.

References

[1] Lyubimov V E and others 1991 *Guide to metal cutting processing* (K.:Technics) 239 p

[2] Baranchikov V I and others 1990 *Progressive cutting tools and conditions* (M.: Engineering) 285 p

[3] Kas'yanov S V, Kondrashov A G, and Safarov D T 2017 *Rapid Assessment of Wear-Resistant Tool Coatings* Russian Engineering Research Vol 37 No 11 pp 969–73

[4] Safarov D T, Kondrashov A G, Safarova L R and Glinina G F 2017 *Energy planning in production shops with numerically controlled machine tools Russian Engineering Research* Vol 37 No 9 pp 827–34

[5] Golovko A.N., I.V. *Kinematic calculation of the error in gear shaving* Russian Engineering Research Vol 31 Is10 pp 1034–35

[6] Golovko A. *Determination of the profile of the worm-type tool* MATEC Web of Conferences 129 **01043**

[7] Batygin Yu V, Lalazarova N A, Plotnikov I V 2011 *Research ofhigh-strength cast iron machinability with spherical graphite* Vestnik KhNADU Issue 54

[8] Trenev D V 2007 *Processing of cast iron by cutting* Tool, technology, equipment No 3 pp 56-57

[9] Grechishnikov V A, Petukhov Y E, Pivkin P M, Romanov V B, Ryabov E A, Yurasov O I , Yurasova S Y 2017 *Trochoidal slot milling* Russian Engineering Research Vol 37 No 9 pp 821-23

[10] *Machining of Cast Irons* 1989Metals Handbook Ninth Edition V 16 pp 648-65

[11] Petrov S M, Davletshina G K, Zairov B F, Zairov L F 2017 *Cutter profile of a mill for machining screw channels* Russian Engineering Research Vol 37 No 8 pp 728-29

[12] Davletshina G K, Zairov B F, Petrov S M 2014 *The use of carbide inserts in the processing of high-strength cast iron* Collection of articles based on the XVIII-XIX International scientific-practical conference No 11-12 (17) 120 p

[13] Davletshina G K, Astashchenko V I, Zamarayeva T A 2010 *Properties and prospects of products' application from high-strength cast iron in mechanical engineering* Education and science-production: International scientific-technical and educational conference Part 1 book 3 273 p

IOP Conf. Series: Materials Science and Engineering **412** (2018) 012010 doi:10.1088/1757-899X/412/1/012010

Assessment of the state of the technical system in the presence of parametric perturbations and the absence of information about the input actions

D N Demyanov

Kazan Federal University, Naberezhnye Chelny Institute, 423812, Russia, Naberezhnye Chelny, Prospekt Syuyumbike 10A

demyanovdn@mail.ru

Abstract. The paper deals with the problem of assessment of the state of a technical system described by first-order differential equations whose parameters change over time according to a certain law. The conditions of solvability of the synthesis problem in terms of matrix canonization technology are formed. An algorithm is proposed for calculating the coefficients of the state observer of the luenberger type, providing an asymptotic tendency to zero of the assessment error.

1. Introduction

It is known that one of the most important conditions for the effective functioning of complex technical systems is a regular monitoring of their state[1]. Most often, it is possible to assess the state with a sufficient degree of accuracy by processing the readings of sensors or statistical analysis of performance characteristics. However, in a number of cases, these methods are not effective enough for a number of technical and economic reasons, therefore so-called state observers are used to evaluate the state of technical systems [2]. At present, the theory of state observers for systems described by first-order linear differential equations is well developed and is set out in the relevant literature. In this case, the problems of designing state observers for nonlinear or non-stationary dynamical systems are much worse and contain a large number of unresolved problems [3, 4], the solution of which is a very urgent problem.

2. Statement of the problem

Let the mathematical model of the considered technical system be given by equations in the state space:

$$\begin{cases} \dot{x}_1 = A_{11}x_1 + A_{12}x_2 + vF_{11}x_1 + vF_{12}x_2 + B_1u; \\ \dot{x}_2 = A_{21}x_1 + A_{22}x_2 + vF_{21}x_1 + vF_{22}x_2 + B_2u; \\ y = x_1. \end{cases} \quad (1)$$

Here, x_1, x_2 are the elements of the state vector of the system under consideration; u is the input effect; y is output signal; v is a time function that determines the law of changing the parameters of the system; A_{11}, A_{12}, A_{21}, A_{22}, F_{11}, F_{12}, F_{21}, F_{22} are numerical matrices of the coefficients.

It is assumed that the output signal is accessible to measurement with a high degree of accuracy,

Content from this work may be used under the terms of the Creative Commons Attribution 3.0 licence. Any further distribution of this work must maintain attribution to the author(s) and the title of the work, journal citation and DOI.
Published under licence by IOP Publishing Ltd

the function v is known, the vectors x_2 and u can not be measured.

From the substantial point of view, the formulated conditions represent the description of a linear dynamic system with parametric disturbances (matrix coefficients of the equation of dynamics change over time by a certain law). The most obvious example is the wear and tear of elements of the technical system during operation.

It is necessary to synthesize the observer of the state of the dynamic system (1), which allows for known y and v to form a vector of evaluation z such that:

$$\lim_{t\to\infty}\varepsilon(t)=\lim_{t\to\infty}\left[z(t)-x_2(t)\right]=0. \tag{2}$$

3. Synthesis of the equation of the observer

Let's compose the equation of the observer on the basis of the second equation of the system (1), having supplemented it with the component proportional to a difference of an assessment and true value of a vector x_2

$$\dot{z}=A_{21}y+A_{22}z+vF_{21}y+vF_{22}z+B_2u+L\left(A_{12}z+vF_{12}z-A_{12}x_2-vF_{12}x_2\right). \tag{3}$$

Substituting in equation (3) the true values of the terms $A_{12}x_2$ and $vF_{12}x_2$ from the first equation of the system (1), we obtain:

$$\dot{z}=\left(A_{22}+LA_{12}\right)z+v\left(F_{22}+LF_{12}\right)z+\left(A_{21}+LA_{11}\right)y+v\left(F_{21}+LF_{11}\right)y+\left(B_2+LB_1\right)u-L\dot{y}. \tag{4}$$

The equation of dynamics of the error of estimation of the vector x_2 can be obtained by subtracting from the equation (3) the second equation of the system (1):

$$\dot{\varepsilon}=\left(A_{22}+LA_{12}\right)\varepsilon+v\left(F_{22}+LF_{12}\right)\varepsilon. \tag{5}$$

In general, the solution of equation (5) depends on function v, and the law of change of the estimation error can be quite complicated. However, the problem can be significantly simplified by choosing a matrix of observations from the condition:

$$F_{22}+LF_{12}=0. \tag{6}$$

In this case, the equation of the dynamics of the estimation error will look like the following:

$$\dot{\varepsilon}=\left(A_{22}+LA_{12}\right)\varepsilon. \tag{7}$$

If the pair (A_{22}, A_{12}) is completely observable by Kalman, then the corresponding choice of the matrix L can ensure the asymptotic tending to zero of the estimation error, that is, the condition (2) is satisfied.

According to the accepted restrictions, the input signal u is inaccessible to measurement, therefore it should also be excluded from the dynamics equation of the observer. For this it is sufficient to ensure the following condition:

$$B_2+LB_1=0. \tag{8}$$

When constraints (6) and (8) are satisfied, equation (4) is essentially simplified and takes the following form:

$$\dot{z}=\left(A_{22}+LA_{12}\right)z+\left(A_{21}+LA_{11}\right)y+v\left(F_{21}+LF_{11}\right)y-L\dot{y}. \tag{9}$$

It should be noted that equation (9) includes a component proportional to the derivative of the output signal. At the same time, it is known that in practice, the operation of differentiation, as a rule, leads to the appearance of additional errors and it is recommended to avoid it. To do this, we modify the calculation scheme by introducing an auxiliary variable $w(t)$:

$$w(t)=z(t)+Ly(t). \tag{10}$$

We rewrite equation (9) taking into account relation (10):

$$\dot{w} = \left(A_{22} + LA_{12}\right)w + \left(A_{21} + LA_{11} - A_{22}L - LA_{12}L\right)y + v\left(F_{21} + LF_{11}\right)y. \tag{11}$$

The equation to estimate the vector x_2 will look like the following:

$$z(t) = w(t) - Ly(t). \tag{12}$$

Thus, in the case of the observability of the pair (A_{22}, A_{12}) under the conditions (6) and (8), it is possible to make the observer equations ensuring the asymptotic tending to zero of the estimation error of the unmeasured component of the state vector of the technical system (1).

4. Analysis of solvability conditions

We define the conditions under which the observer equations have the form (9) or (11) – (12). To do this, we combine the conditions (6) and (8), presenting them as one linear matrix equation:

$$K_1 + LK_2 = 0. \tag{13}$$

Here is denoted: $K_1 = \left[F_{22} \quad B_2\right]; K_2 = \left[F_{12} \quad B_1\right]$.

The solution of equation (13) exists if condition [5] is satisfied:

$$K_1\bar{K}_2^R = 0. \tag{14}$$

Here and in what follows, for an arbitrary matrix M, the symbols $\bar{M}^R$ and $\bar{M}^L$ will denote non-zero matrices of maximal rank such that $M\bar{M}^R = 0$ and $\bar{M}^L M = 0$ respectively. In addition, we define a matrix construction $\tilde{M} = \tilde{M}^R\tilde{M}^L$ in which $\tilde{M}^R$ and $\tilde{M}^L$ are matrices of maximal rank such that $\tilde{M}^L M\tilde{M}^R = I$. Detailed information on the properties of these matrices and a description of the algorithm for their calculation are given in the paper [5].

In the general case, the solution set of equation (13) is given by the formula:

$$\{L\}_\eta = -K_1\tilde{K}_2 + \eta\bar{K}_2^L. \tag{15}$$

Here η is an arbitrary numerical matrix of the corresponding dimension.

We rewrite the matrix of coefficients from the right-hand side of equation (7), taking into account condition (15):

$$A_{22} + LA_{12} = A_{22} - K_1\tilde{K}_2A_{12} + \eta\bar{K}_2^L A_{12}.$$

Let's introduce new symbols:

$$A^* = \left(A_{22} - K_1\tilde{K}_2A_{12}\right)^T; \; B^* = \left(\bar{K}_2^L A_{12}\right)^T. \tag{16}$$

Then the condition (2) is satisfied by the required arrangement of eigenvalues of the matrix $A^* + B^*\eta^T$, the setting of which by the matrix selection η is a classical problem of modal control.

Thus, the solution of the problem is possible if the condition (14) is satisfied, and the pair (A^*, B^*) defined by formulas (16) is fully controlled by Kalman.

5. Synthesis algorithm

Summarizing the results obtained, we formulate an algorithm for calculating the observer's coefficients for the state of the dynamic system in the presence of parametric perturbations and the lack of information on input effects.

1. To set the numerical values of the matrices of coefficients of model (1), the desired distribution

of the poles of the observer, information about the permissibility or non-permissibility of differentiation of the output signal.

2. To verify the fulfillment of condition (14). If this is done, you can proceed to the next step. Otherwise, the proposed method is not applicable.

3. To calculate the matrix of coefficients by formulas (16) and estimate the controllability of the pair (A^*, B^*). If it is fully controled, then you can proceed to the next step. Otherwise, it is recommended to change the requirements for the dynamics of the observer or use a different calculation algorithm.

4. Solve the problem of modal control by providing the choice of a matrix η with the required arrangement of the eigenvalues of the matrix $A^* + B^*\eta^T$.

5. To calculate a matrix of observations according to the formula (15).

6. To make the equations of the observer. If differentiation of the output signal is allowed, then equation (9) is used. Otherwise, the equation of the dynamics of the observer is given by (11), and the expression for estimating the vector x_2 is given by formula (12).

The end of algorithm.

6. Conclusions

A new method of assessing the state of the technical system in the presence of parametric perturbations and the absence of information about the input actions is proposed. The solvability conditions of the problem and a step-by-step algorithm for calculating the coefficients of the observer providing an asymptotic tendency to zero evaluation errors are formulated. The obtained results can be used in practice to build diagnostic systems of technical objects whose characteristics change over time according to the known law.

7. Acknowledgment

The work is executed at financial support of RFBR (grant No. 17-08-00516).

References

[1] Mironovskij L A 1998 *Functional diagnostics of dynamic systems* (SPb.: Nauchnoe izdanie) p 256

[2] Kuzovkov N T 1976 *Modal control and monitoring devices* (M.: Mashinostroenie) p 184

[3] Zheng G, Boutat D, Wang H 2017 *A nonlinear Luenberger-like observer for nonlinear singular systems* Automatica Vol 86 pp 11–17

[4] Asanov A Z, Dem'yanov D N 2017 *Analytical synthesis of the functional observer of the state of bilinear dynamical system* Avtometriya No 4 pp 26–34

[5] Bukov V N 2006 *Systems investment. Analytical approach to the analysis and synthesis of matrix systems* (Kaluga: Publ. N.F. Bochkareva) p 720

Application of the theory of compound cores for the assessment of stress pattern in the cross section of a strengthened beam column

R F Fardiev[1], L S Sabitov[2,3], N F Kashapov[2] and I R Gilmanshin[2,4]

[1] OOO "PKF KARKAS", Kazan, 420043, Russian Federation
[2] Kazan Federal University, 18 Kremlyovskaya street, Kazan, 420008, Russian Federation
[3] Kazan State Power Engineering University, 51 Krasnoselskaya street, Kazan, 420066, Russian Federation
[4] Kazan national research technical university named after A N Tupolev, K. Marx St. 10, Kazan, 420111 Tatarstan, Russia

l.sabitov@bk.ru

Abstract. The article considers the results of theoretical studies of the stress-strain state of an eccentrically compressed element reinforced with a reinforced concrete cage. The theory of composite rods is used as a theoretical basis. As a result of the solution of the theoretical problem, analytical expressions are obtained for determining the normal stresses in the cross-section of the reinforced element and tangential stresses along the contact seam. The results of the calculations for the proposed approach are compared with the previously performed research results.

To assess the strength of the reinforced element, the important point is to determine the magnitude and nature of the distribution of forces applied to the cage and to the main element. To date, there is a deformation method for determining stresses and forces [1, 2], but it is based on the hypothesis of flat sections, which can be broken for the section reinforced with a reinforced concrete cross-section.

To solve this problem, in the case of eccentric compression for long elements, it is proposed to use the theory of compound rods [3], introducing some changes related to the complex stress-strain state of the reinforced element.

Content from this work may be used under the terms of the Creative Commons Attribution 3.0 licence. Any further distribution of this work must maintain attribution to the author(s) and the title of the work, journal citation and DOI.
Published under licence by IOP Publishing Ltd

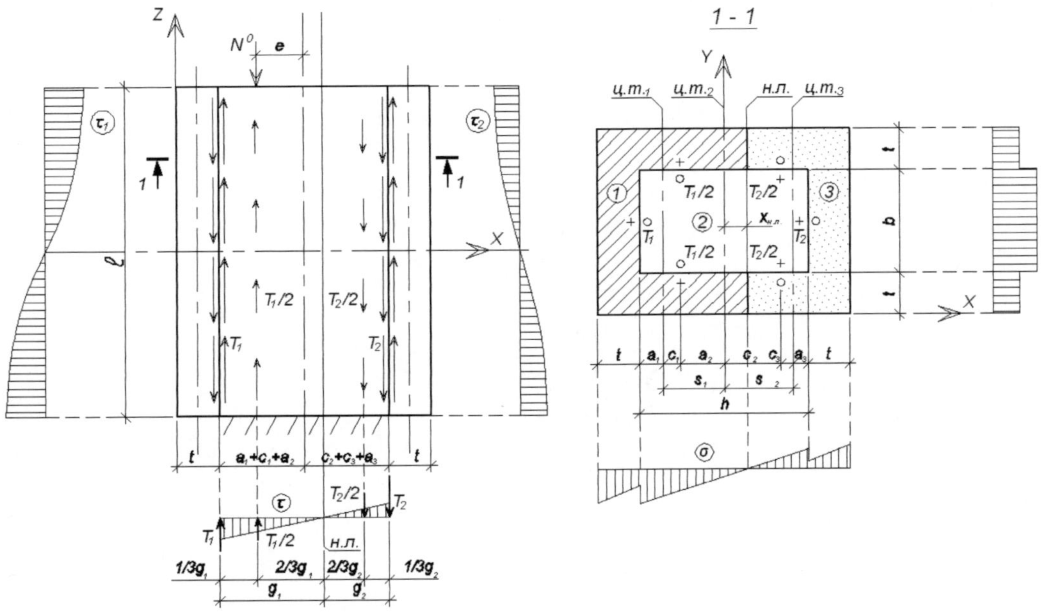
Fig. 1. The calculation scheme

An eccentrically compressed element in the cross section is represented in the form of a packet of three cores (Figure 1): a rod from the main reinforced element, a core from the part of the cage located to the left of the zero section line, a rod from the part of the cage located to the right of the zero section line. The coordinate of the zero line is preliminarily determined from the known equation of the course of the resistance of materials:

$$x_{н.л.} = \frac{(h+2t)^2}{12e}, \tag{1}$$

where e – co-ordinate of force application N, h – section height.

The considered system is calculated by the method of forces. As the main system, a core devoid of shear bonds whose action is replaced by unknowns T_1 and T_2.

The total rigidity of a rod devoid of shear bonds will be equal to:

$$\Sigma EI = E_1 I_y^1 + E_2 I_y^2 + E_3 I_y^3. \tag{2}$$

he tangential forces cause bending moments in the constituent cores:

$$\begin{aligned} M_1 &= T_1 a_1 - 2\frac{T_1}{2}c_1 = T_1 a_1 - T_1 c_1, \\ M_2 &= -T_1(a_2 + c_1 + a_1)b - 2\frac{T_1}{2}a_2 - T_2(c_2 + c_3 + a_3) - 2\frac{T_2}{2}c_2 = \\ &= -T_1 a_1 - T_1 c_1 - 2T_1 a_2 - T_2 c_3 - T_2 a_3 - 2T_2 c_2, \\ M_3 &= T_2 a_3 - 2\frac{T_2}{2}c_3 = T_2 a_3 - T_2 c_3. \end{aligned} \tag{3}$$

The total bending moment according to the method of forces will be equal to the sum of the bending moment from the external load and the moments in each of the cores:

$$M = M^0 + N^0 e + M_1 + M_2 + M_3 = Ne - 2Ts_1 - 2T_2 s_2, \tag{4}$$

where s_1 and s_2 – the distance between the centers of gravity of the cores.
Longitudinal forces in the constituent cores will be equal:

$$\begin{aligned} N_1 &= -2T_1, \\ N_2 &= -N^0 + 2T_1 - 2T_2, \\ N_3 &= 2T_2. \end{aligned} \tag{5}$$

The differential equation of the curved axis of the pack of rods will be written in the same way as for the bent element:

$$x'' = -\frac{M}{\Sigma EI}. \tag{6}$$

The total shift of the outer fibers of the two cores will be equal to the difference in shear of the two adjacent outermost fibers located on either side of the contact seam. The first derivative of the absolute elongation along the length of the seam is equal to the relative elongation. Taking this expression into account (8) takes the form:

$$\Delta' = (\Delta_i)' - (\Delta_{i+1})' = \varepsilon_i - \varepsilon_{i+1}. \tag{7}$$

The relative elongation of the outermost fibers of the cores is determined from the expressions:

$$\begin{aligned} &\varepsilon_1^{раст} = x''\frac{t}{2} = -\frac{Mt/2}{\Sigma EI} \text{ - растянутых волокон 1-го стержня;} \\ &\varepsilon_1^{сжат} = \varepsilon_2^0 - x''(a_1 + c_1 + a_2) = \frac{N_2}{E_2 A_2} + \frac{M(a_1 + c_1 + a_2)}{\Sigma EI} \text{ - сжатых волокон 2-го стержня;} \\ &\varepsilon_2^{раст} = \varepsilon_2^0 + x''(c_2 + c_3 + a_3) = \frac{N_2}{E_2 A_2} - \frac{M(c_2 + c_3 + a_3)}{\Sigma EI} \text{ - растянутых волокон 2-го стержня;} \\ &\varepsilon_2^{сжат} = -x''\frac{t}{2} = \frac{Mt/2}{\Sigma EI} \text{ - сжатых волокон 3-го стержня.} \end{aligned} \tag{8}$$

The total shift of the outermost fibers of the first and second core according to expression (7) taking into account (8) will be equal to:

$$\begin{aligned} &\Delta_1' = \varepsilon_1^{раст} - \varepsilon_1^{сжат} = -\frac{N_2}{E_2 A_2} - \frac{M(1)}{\Sigma EI} \text{ - первого и второго стержня;} \\ &\Delta_2' = \varepsilon_2^{раст} - \varepsilon_2^{сжат} = \frac{N_2}{E_2 A_2} - \frac{M(2)}{\Sigma EI} \text{ - второго и третьего стержня.} \end{aligned} \tag{9}$$

In the expressions (9) the following notation is introduced:

$$\begin{aligned}(1)&=(a_1+c_1+a_2+t/2),\\(2)&=(c_2+c_3+a_3+t/2).\end{aligned} \tag{10}$$

In order for the basic system to become equivalent to the specified one, it is necessary to equate the shear in the seams with the shear shear forces divided by the stiffness coefficient of the seam:

$$\begin{aligned}\Delta_i&=\frac{\tau_i}{\xi}=\frac{T_i'}{\xi},\\\Delta_i'&=\frac{T_i''}{\xi}.\end{aligned} \tag{11}$$

The coefficient of stiffness of the seam ξ depends on the compliance of the joint of the prefabricated and monolithic concrete, the presence of a keyed joint, the transverse reinforcement that crosses the contact seam, the presence of treatment of the contact surface.

To obtain equations with the known solution for three cores, we find the sum and difference of the shifts Δ_1 and Δ_2:

$$\begin{cases}\dfrac{T_1''}{\xi}+\dfrac{T_2''}{\xi}=-\dfrac{M\left((1)+(2)\right)}{\Sigma EI}\\\dfrac{T_1''}{\xi}-\dfrac{T_2''}{\xi}=\dfrac{2N^0}{E_2A_2}-\dfrac{M\left((2)-(1)\right)}{\Sigma EI}.\end{cases} \tag{12}$$

From equations (10) it follows that:

$$\begin{aligned}(1)+(2)&=h+t,\\(2)-(1)&=0.\end{aligned} \tag{13}$$

We introduce the notation: T_a is the antisymmetric part of the set of forces, T_c is the antisymmetric part of the set of efforts:

$$\begin{aligned}T_a&=T_1+T_2,\\T_c&=T_1-T_2.\end{aligned} \tag{14}$$

Taking into account (14), the system of equations (12) takes the form:

$$\begin{cases}\dfrac{T_a''}{\xi}=\dfrac{2(h+t)(T_1s_1+T_2s_2)}{\Sigma EI}-\dfrac{N^0e}{\Sigma EI},\\\dfrac{T_c''}{\xi}=-\dfrac{4T_c}{E_2A_2}+\dfrac{2N^0}{E_2A_2},\end{cases} \tag{15}$$

where the expression in parentheses of the i_{th} equation can be represented in the form:

$$T_1s_1+T_2s_2=\frac{T_a(s_1+s_2)}{2}+\frac{T_c(s_1-s_2)}{2}\approx\frac{T_a(s_1+s_2)}{2}. \tag{16}$$

Taking into account (16), equations (15) take the form:

IOP Conf. Series: Materials Science and Engineering **412** (2018) 012011 doi:10.1088/1757-899X/412/1/012011

$$\begin{cases} \dfrac{T_a''}{\xi} = \dfrac{(h+t)(s_1+s_2)}{\Sigma EI} T_a - \dfrac{N^0 e}{\Sigma EI} = \gamma_a T_a + \delta_a \\ \dfrac{T_c''}{\xi} = -\dfrac{4}{E_2 A_2} T_c + \dfrac{2N^0}{E_2 A_2} = \gamma_c T_c + \delta. \end{cases} \tag{17}$$

where γ_a, δ_a, γ_a, δ_a,– coefficients defined from expressions:

$$\gamma_a = \frac{(h+t)(s_1+s_2)}{\Sigma EI}, \ \delta_a = -\frac{N^0 e}{\Sigma EI}$$
$$\gamma_c = -\frac{4}{E_2 A_2}, \ \delta_c = \frac{2N^0}{E_2 A_2}. \tag{18}$$

The system of equations (17) has a known solution:

$$\begin{cases} T_a = \dfrac{\delta_a}{\gamma_a}\left(\dfrac{ch(\lambda_a x)}{ch(\lambda_a l)} - 1\right), \\ T_c = \dfrac{\delta_c}{\gamma_c}\left(\dfrac{ch(\lambda_c x)}{ch(\lambda_c l)} - 1\right). \end{cases} \tag{19}$$

The tangential stresses are determined from expressions:

$$\begin{cases} \tau_a = T_a' = \dfrac{\delta_a \lambda_a}{\gamma_a}\left(\dfrac{sh(\lambda_a x)}{ch(\lambda_a l)} - 1\right), \\ \tau_a = T_c' = \dfrac{\delta_c \lambda_c}{\gamma_c}\left(\dfrac{sh(\lambda_c x)}{ch(\lambda_c l)} - 1\right), \end{cases} \tag{20}$$

where $sh\lambda x$ – hyperbolic sine, $ch\lambda x$ – hyperbolic cosine, defined from expressions:

$$ch(\lambda x) = \frac{e^{\lambda x} + e^{-\lambda x}}{2},$$
$$sh(\lambda x) = \frac{e^{\lambda x} - e^{-\lambda x}}{2}; \tag{21}$$

λ_a, λ_c – coefficients defined by expressions:

$$\lambda_a = \sqrt{\gamma_a \xi_a}$$
$$\lambda_c = \sqrt{\gamma_c \xi_c} \tag{22}$$

With a sufficiently rigid contact seam at $\xi > 10$, expressions (22) take the form:

IOP Conf. Series: Materials Science and Engineering **412** (2018) 012011 doi:10.1088/1757-899X/412/1/012011

$$ch\lambda x = sh\lambda x \approx \frac{e^{\lambda x}}{2}. \tag{23}$$

In this case, the system of equations (20) takes the form:

$$\begin{cases} \tau_a = \dfrac{\delta_a}{\gamma_a}\left(e^{\lambda_a(x-l)} - 1\right) \\ \tau_c = \dfrac{\delta_c}{\gamma_c}\left(e^{\lambda_c(x-l)} - 1\right) \end{cases} \tag{24}$$

For known τa and τ_c, the required stresses τ_1 and τ_2 can be expressed from expressions (14):

$$\begin{cases} \tau_1 = \dfrac{\tau_c + \tau_a}{2}, \\ \tau_2 = \dfrac{\tau_a - \tau_c}{2}. \end{cases} \tag{25}$$

Unknown stresses in each of the cores can be determined from expression:

$$\sigma_i = \frac{N_i}{A_i} + \frac{M_i x_i}{I_i}, \tag{26}$$

where N_i – the forces determined in accordance with expression (5), M_i-bending moments in the cores, determined from expression:

$$M_i = -x'' E_i I_i = \frac{M}{\Sigma EI} E_i I_i. \tag{27}$$

Taking equations (5, 6) into account, the expressions for determining the unknown stresses for the component cores will take the form:

$$\begin{aligned} \sigma_1 &= -\frac{2T_1}{A_1} + \frac{(T_1a_1 - T_1c_1)E_1x}{\Sigma EI}, \\ \sigma_2 &= \frac{-N^0 + 2T_1 - 2T_2}{A_2} + \frac{(-T_1a_1 - T_1c_1 - 2T_1a_2 - T_2c_3 - T_2a_3 - 2T_2c_2)E_2x}{\Sigma EI}, \\ \sigma_3 &= -\frac{2T_2}{A_3} + \frac{(T_2a_3 - T_2c_3)E_3x}{\Sigma EI}. \end{aligned} \tag{28}$$

With known tangential forces, the refined position of the zero line is determined from expression (35):

$$x_{н.л.} = \frac{\left(N^0 - 2T_1 + 2T_2\right)\Sigma EI}{A_2E_2\left(-T_1a_1 - T_1c_1 - 2T_1a_2 - T_2c_3 - T_2a_3 - 2T_2c_2\right)}. \tag{29}$$

The calculation is made by successive approximations with the refinement of the position of the zero line until the required accuracy of the determination of effort is achieved.

IOP Conf. Series: Materials Science and Engineering **412** (2018) 012011 doi:10.1088/1757-899X/412/1/012011

To verify the analytical expressions obtained, the normal stresses in the cross section of the reinforced element were compared (see Figure 2) from the analytical expressions obtained and according to the approach based on the piecewise stress distribution law [4, 5].

For the calculation, an element 1 m long with a cross-section of 150 × 200 cm, reinforced with a reinforced concrete cage of 3 cm thickness.

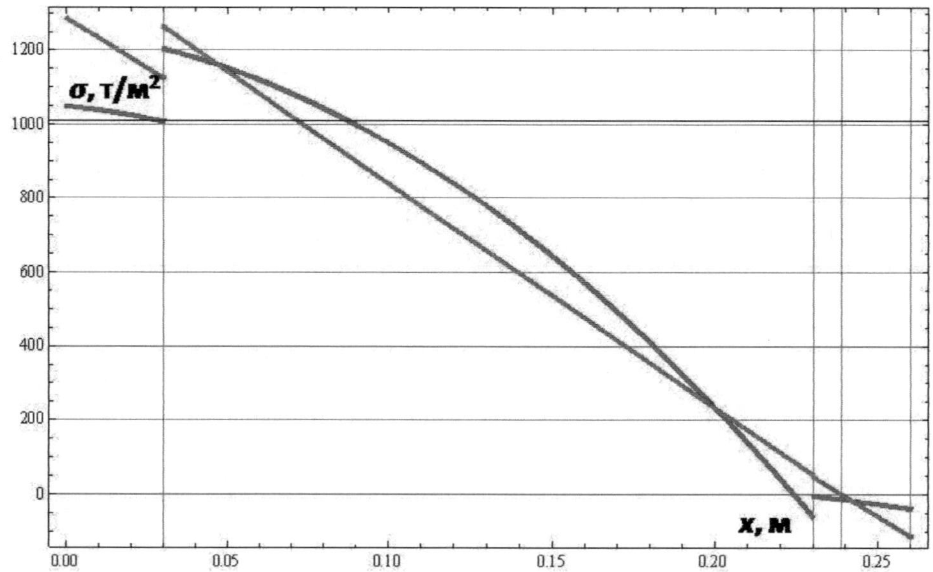

Fig. 2. Graph of the dependence of normal stresses in the cross section of the element: at z = 0, according to the proposed method (linear dependence) and according to the previously proposed approach (nonlinear dependence)

The analysis of the obtained results showed that the proposed calculation approach is rather well correlated with the results of previous studies.

According to the accepted calculation scheme, the tangential stresses between the cores must be maximal at z = 0, and minimal at z = 1, which is confirmed by the obtained graphs (Figure 3).

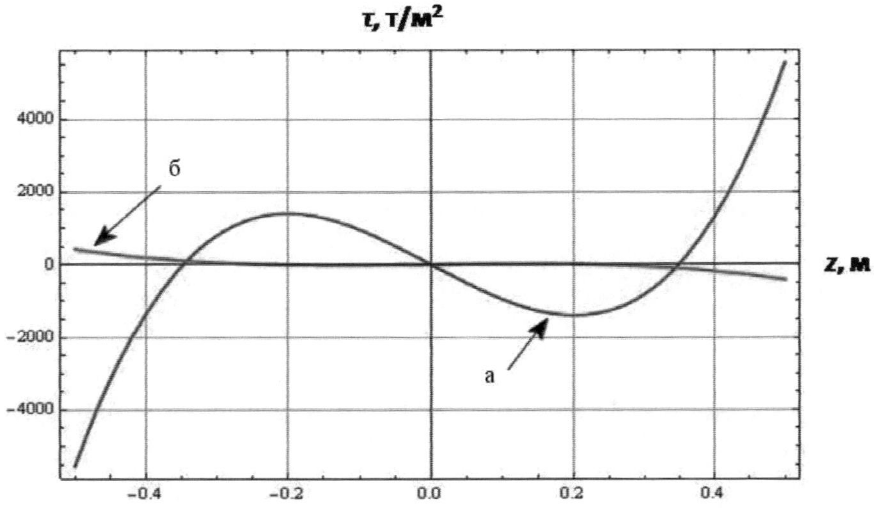

Fig. 3. Graph of the dependence of tangential stress along the contact seam:
a - between the 1st and 2nd elements, b - between the 2nd and 3rd elements

Data on the nature of the distribution of stresses on the contact seam allow one to evaluate its strength by known methods [6, 7, 8] to perform a test of the strength of the cross section.

Thus, on the basis of the theory of compound cores, a solution was obtained to determine the nature of the distribution of forces between the cage and the column in the case of reinforcement of an eccentrically loaded element with a reinforced concrete cage.

References

[1] Babich V I, Kochkarev D V 2004 *Calculation of reinforced concrete structures by deformation* Concrete and reinforced concrete No 2 pp 12-16

[2] Zalesov A S, Chistyakov E A, Laricheva I Y 1997 *New methods of normal section calculation on the basis of a deformation design model* Concrete and reinforced concrete No 5 pp 31-34

[3] Rzhanitsyn A R 1986 *Compound bars and plates (*M.: Stroiizdat) p 316

[4] Fardiev R F, Kayumov R A, Mustafin I I 2011 *Calculation of a beam column strengthened by a reinforced concrete collar with allowance for previous load history and nonlinear properties of concrete* Izvesiya KGASU pp 109-114

[5] Fardiev R F, Mustafin A I 2012 *The study on the reinforced concrete beam columns strengthened by a collar with allowance for stress state before strenthening* Integration, partnership and innovation in construction science and education pp 152-57

[6] Valeev G S, Fatkhullin V S 1994 *Application of prefabricated reinforced concrete in reconstruction of buildings and structures* News of Higher Educational Institutions. Construction pp 114-17

[7] Grozdov V T, Sergeev S L 1996 *The question of allowance for strength of the contact zone under the calculation of reinforced concrete bending structures strengthened by the extension of cross sections* News of Higher Educational Institutions. Construction No 4 pp 34-38

[8] Sungatullin Y G 1975 *Experimental and theoretical basis of the calculation of shear resistance of reinforced and non-reinforced contacts of cast-in-place and precast structures* Cast-in-place and precast structures: collection of research papers pp 90-146

IOP Conf. Series: Materials Science and Engineering **412** (2018) 012012 doi:10.1088/1757-899X/412/1/012012

Results of experimental investigations of extra-compressible compression elements enhanced by reinforced concrete

R F Fardiev[1], L S Sabitov[2,3], N F Kashapov[2] and I R Gilmanshin[2,4], I K Kiyamov[2]

[1] OOO “PKF KARKAS”, Kazan, 420043, Russian Federation
[2] Kazan Federal University, 18 Kremlyovskaya street, Kazan, 420008, Russian Federation
[3] Kazan State Power Engineering University, 51 Krasnoselskaya street, Kazan, 420066, Russian Federation
[4] Kazan national research technical university named after A N Tupolev, K. Marx St. 10, Kazan, 420111 Tatarstan, Russia

l.sabitov@bk.ru

Abstact. The article considers a range of issues related to ensuring the compatibility of prefabricated and monolithic concrete in the case of reinforcement of eccentrically compressed elements with reinforced concrete cage. A calculation expression is presented to determine the required thickness of the gain from the condition of ensuring the strength of the shear joint. The main design expressions for testing the strength of the contact seam in the case of using keys and reinforcement by transverse reinforcement are considered. Based on the proposed theoretical premises, a program of experimental studies has been developed.

This method of amplification is used when it is necessary to increase the bearing capacity of eccentrically loaded reinforced concrete columns during the reconstruction of civil and industrial buildings, as well as in the case of "new" construction in the event of errors or miscalculations made at the design and construction stage.
The purpose of the experimental studies was the experimental evaluation of the actual operation of the reinforced elements. Experimental studies were carried out according to the developed program of experimental studies (Figure 1).

Content from this work may be used under the terms of the Creative Commons Attribution 3.0 licence. Any further distribution of this work must maintain attribution to the author(s) and the title of the work, journal citation and DOI.
Published under licence by IOP Publishing Ltd

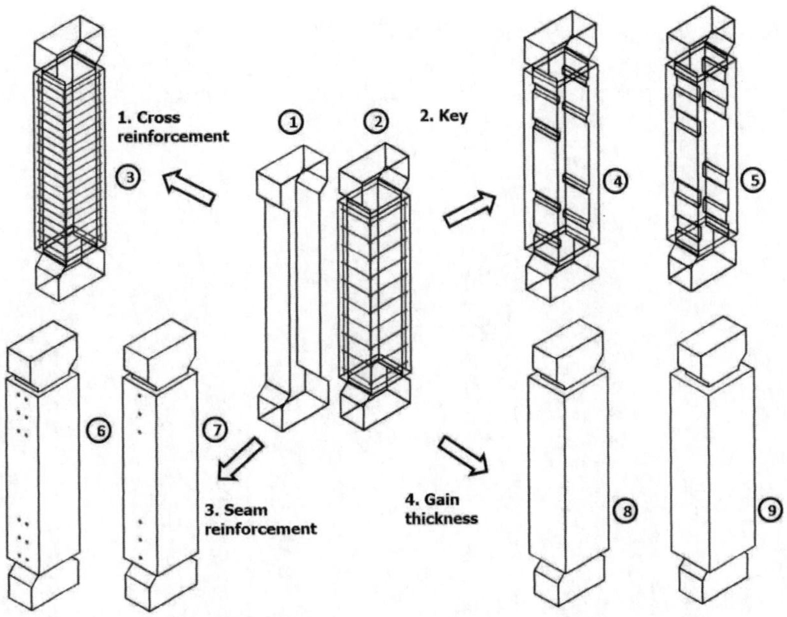

Fig. 1. Program of experimental research.

In addition to testing samples without amplification and with simple reinforcement of the clip (1, 2 series), the program included the study of factors such as: the effect of transverse reinforcement (2, 3 series), the presence of keys on the contact seam (4, 5 series), the presence of reinforcement seam (6, 7 series), the influence of the gain thickness (8, 9 series). Each of the series includes 3-5 samples of twins, a total of 35 samples. In this article, we only consider the analysis of the operation of the elements in the 1st and 2nd series.

As the material of all elements, concrete of class B15 is adopted. The main element is reinforced with 4 stems of longitudinal reinforcement Ø 10 A-III and transverse - Ø 5 Vp-I with a pitch of 200 mm. To prevent premature failure, the console of the main element is reinforced with additional reinforcement and reinforced by the outer corners. The reinforced concrete clips are reinforced with 4 stems of longitudinal reinforcement Ø 10 A-III, transverse - Ø 5 Vp-I with a step of 100 mm. All geometric parameters of the samples were taken with a similarity coefficient to real constructions equal to 0.5: the length of the elements is assumed to be 1.5 m, the cross-sectional dimensions of the main element are 15 × 20 cm, the thickness of the cage is 3 cm.

Fig. 2. Testing of one of the reinforced samples.

The tests were carried out on an IPS-200 press with a maximum load of 200 tons. Strain gauges were used to fix longitudinal and transverse deformations. Measurements of deformations were duplicated by indications of dial gauge indicators. The deflection of the elements from the eccentric compression was measured using a specially designed frame (Fig. 2).

Virtually all samples (33 pieces) were tested with a load eccentricity of 50 mm, two samples were tested with an eccentricity of 70 mm. The load was applied in steps with fixation of the monitored parameters according to GOST [1]. Since real reinforcement of columns of residential and public buildings is impossible to completely transfer the load from all overlaps to the reinforcement element, the application of the load in the studies also occurred directly only on the reinforced element.

The destruction of the elements without amplification occurred in the second case of the third stage of the stress-strain state of reinforced concrete elements, the destruction of the compressed zone over the concrete, when the stresses in the stretched reinforcement do not reach the yield point. At the same time, the value of the theoretical destructive load obtained from the calculation of the allowance [2], taking into account the safety factor, was 23% lower compared to the experimental value (Table 1).

Analysis of the cracking pattern, destruction and processing of the instrument readings showed that the operation of an eccentrically compressed element reinforced with a reinforced concrete cage is more complex. This is due to the fact that the load on the reinforcement element is transmitted along the contact seam through tangential stresses and within the cross-section the reinforced concrete has different strength and deformation characteristics.

At loads of 80-85% of the destructive components, a sharp decrease in the build-up of longitudinal deformations of compression in the cage was observed (Figure 3), which indicates the manifestation of nonlinear properties of reinforced concrete and the appearance of slippage of the cage relative to the main element in the compressed zone.

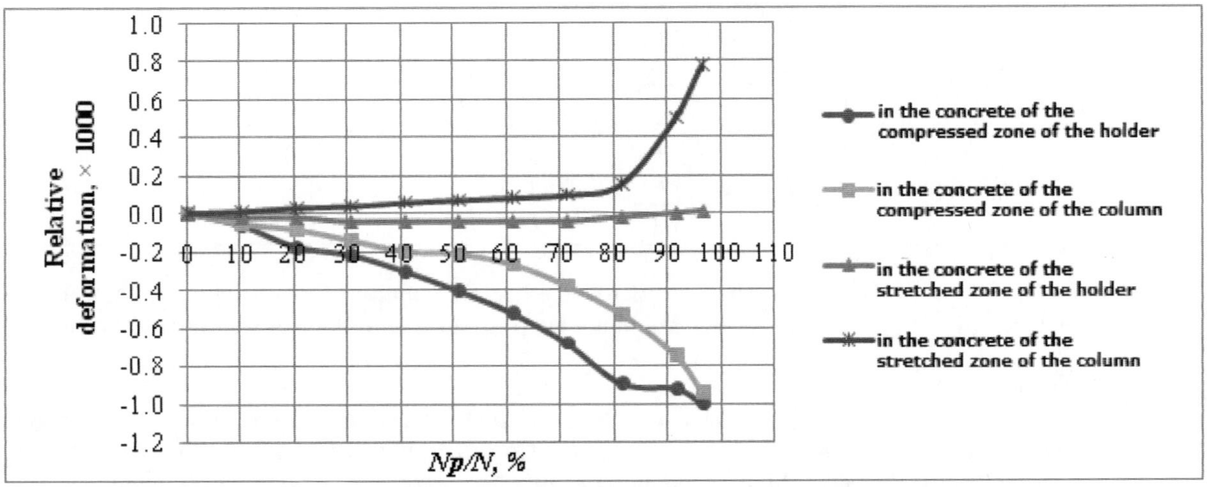

Fig. 3. Dependence of longitudinal relative strain on load.

In the presence of slippage, the transverse deformations increase sharply (Figure 4) and the bending of the main element of the Clip aims to contain the bending of the main element, therefore considerable tensile stresses arise in it, causing longitudinal cracks in the lower and upper parts of the casing (Figure 2).

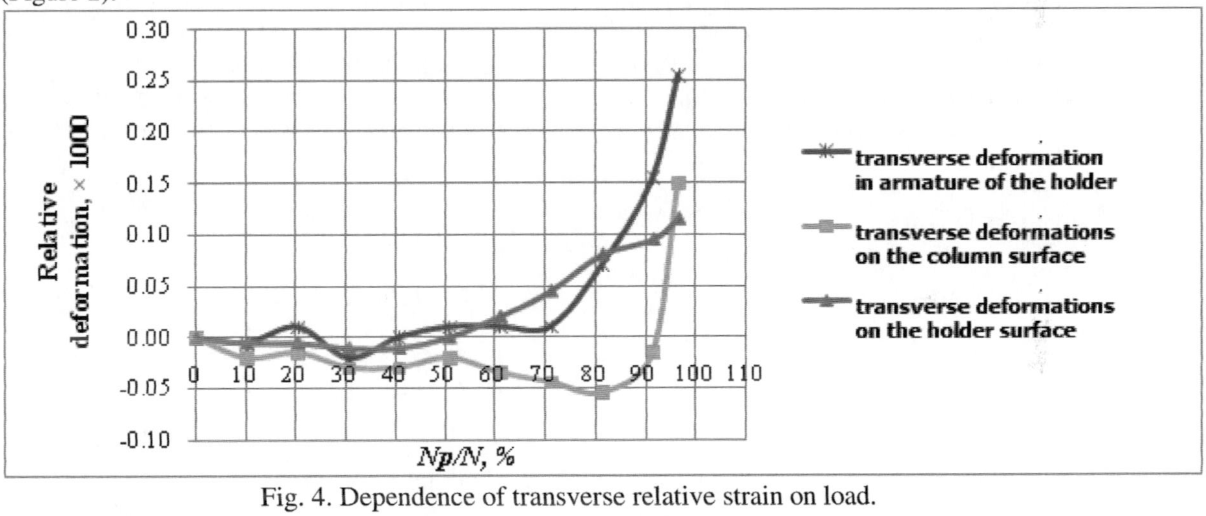

Fig. 4. Dependence of transverse relative strain on load.

A sharp increase in bending also causes an increase in longitudinal deformations in the stretched zone of the cage (Fig. 3) and the formation of transverse cracks.

A further increase in the load causes an increase in the width of the opening of the longitudinal cracks in the cage and the development of them from the top and bottom of the cage to its central part along the length.

The destruction of the reinforced element occurred over the concrete of the compressed zone of the main element inside the cage while the casing was destroyed by the tensile forces (Figure 2).

Data on the carrying capacity of samples. Table. 1.

Factor under investigation	№	N	N_{cp}	N_m	$\frac{N_m-N_{cp}}{N_{cp}}$, %	$\frac{N_{cp2}-N_{cp1}}{N_{cp1}}$, %

IOP Conf. Series: Materials Science and Engineering **412** (2018) 012012 doi:10.1088/1757-899X/412/1/012012

Samples without amplification	1	66.7	55.9	43	23	–
	2	49.2				
	3	51.7				
Reinforced samples	1	94.1	94.6	58	39	69
	2	93.8				
	3	96.0				

Due to the effect of the "cage", the considered method of amplification by increasing the cross-sectional area by 22% resulted in an increase in the bearing capacity of the element by 69% (Table 1).

References

[1] Grozdov V T, Teryanik V V 1991 *On the strength and deformation of eccentrically compressed columns reinforced with cages* Izvestiya VUZov..Construction and architecture No 2 pp 128 -130
[2] Lazovsky D N, Petzold T M *Strengthening of reinforced concrete structures. Allowance П 1-98 to SNIP 2.03.01-84* (Minsk, State Enterprise "Stroytechnorm") - p 198
[3] *Manual on the design of concrete and reinforced concrete structures from heavy and lightweight concrete without prestressing the reinforcement* (to SNiP 2.03.01-84) 1984 (TSNIpromzdany)
[4] *Manual on the design of stone and reinforced structures* (to SNiP II-22-81) 1989 (Moscow: Central Institute for Standard Design)
[5] *Design of reinforced concrete prefabricated-monolithic structures* 1991 Reference guide to SNiP 2.03.01-84 *. (Moscow. Stroyizdat)
[6] *Manual for the design of residential buildings* (to SNIP 2.08.01-89 *)M. 2000
[7] *Recommendations on designing reinforcement of reinforced concrete structures of buildings and structures of the reconstructed enterprises. Overground structures and structures* 1992 (M. Stroyizdat) p 191

Mathematical modeling of binding material coalescence at the production of syntactic carbon foams

V L Fedyaev[1,2], E R Galimov[1], N Ya Galimova[1], V M Samoylov[3] and M A Orlov[4]

[1]IME - Subdivision of FIC KazanSC of RAS, Lobachevsky str. 2/3, Kazan, 420111 Tatarstan, Russia
[2]Kazan national research technical university named after A N Tupolev, K. Marx str. 10, Kazan, 420111 Tatarstan, Russia
[3]JSC NIIgrafit, Elektrodnaya str. 2, Moscow, 111524, Russia
[3]Bauman Moscow state technical university, Baumanskaya 2-ya str., building 1, house 5, Moscow, 105005 Russia
engelgalimov@yandex.ru

Abstract. The processes of agglutination and coalescence of a softened (liquefied) coating material of microspheres are considered both in the absence of an external influence and in its presence. The periodicity cell is singled out, the dynamics of the change in the form of the interface between the liquid and gas phases is described. The thermal and mechanical effects are marked out as external factors leading to shrinkage of the microspheres layer with coatings. Relationships for estimating the shrinkage of the periodicity cell and determining the influence of the named factors, viscosity, and surface tension on the porosity of the layer material are presented.

1. Introduction

Foams are one of the varieties of porous materials widely represented in nature and engineering. As in any porous structure, in the foam, you can mark out two main components: a solid framework (matrix, skeleton) and a liquid (usually gaseous) medium, located in the pores (cavities) formed by the framework material. In the extreme case, the situation in which there is no medium in the pores (vacuum) or its content (concentration) is so small that it can be neglected is possible. At the same time, cases in which both the liquid and gas (vapor) are in the pores are quite common. Foams according to one of the definitions are highly concentrated coarsely dispersed systems in which the dispersion medium is a solid material or liquid, and the dispersed phase – gas. The main distinguishing feature of foam is that it has a high porosity (its macroporosity is $m_s > 0.5$).

Depending on the conditions of production, foams can be divided into two types: natural and technical. In the surrounding nature, it is a marine foam, less often river foam often observed near the coastal edge; magma the streams of which erupt from the volcanoes.

2. Body text

In the solidified state the magma material is a pumice stone. Here, it is necessary to note one more difference of foam from porous materials itself: the foam framework material can be liquid (fluidal) and solid deformable. Perhaps, only by a long stretch of the imagination, the biological tissues of the lungs, the middle part of the human and animal bones, plant stems and etc. are foams. Undoubtedly, the conditions for their formation, operation and their properties are specific.

Content from this work may be used under the terms of the Creative Commons Attribution 3.0 licence. Any further distribution of this work must maintain attribution to the author(s) and the title of the work, journal citation and DOI.
Published under licence by IOP Publishing Ltd

There are much more examples of foam materials used in everyday life at the implementation of technological processes, production and operation of various technical objects: fire extinguishing systems; flotation technologies for purification of liquid media, extracting of valuable components from them; sponges, heat-insulating and building materials (foam plastic, foam concrete, expanded clay, etc.); foam materials used for softening, damping the loads, etc.

Syntactic carbon foams have a special place among the foam materials. Firstly, this is due to the fact that carbon itself has unique performance properties secondly, the framework material of this foam consists of carbon serving as a binder for hollow microspheres of about 10-50 μm in size located in it. In turn, microsphere shells can be made of metal ceramics, carbon, and other materials. Thus, these foams are characterized by the above-mentioned macroporosity m_p (pore sizes of about 1 ÷ 5 mm) and microporosity m_g the micropore sizes are, of course, slightly smaller than the microsphere sizes.

By now, there are a number of production technologies of syntactic carbon foams [1].This paper focuses on one of the technology stages in which the microspheres with a layer of the polymer coating are raw material. The main technological task of the implementation of this stage is the production of a monolithic polymer material containing to the extent possible uniformly distributed microspheres Let us mark out two variants of the solution to this problem: due to particles agglutination when the coating material is softened and by coalescence of the liquefied layers of the coating.

Agglutination of particles containing microspheres

For a mathematical description of the agglutination process of the named particles in a bulk layer, let us move from a real macroporous medium to a fictitious one, whose voids are the gas spherical cavities of diameter d_s, distributed uniformly over the volume [2]. If the real layer of particles with the diameter d_p has the rhombohedral laying, then it is easy to make sure that the macropores diameter of the fictitious material is:

$$d_s = d_p\left(\left(6\sin^2\theta\sqrt{1+\cos\theta}\big/\left(\pi\left(1+\cos\theta\right)\right)\right)-1\right)^{1/3}, \tag{1}$$

where θ – the angle characterizing the laying peculiarities of spherical particles (60° ≤ θ ≤90°). Approximately

$$d_s \approx \left(1-0.22\cos\theta\right)d_p. \tag{2}$$

Hence the initial macroporosity of the medium

$$m_{s0} \approx 0.5-0.33\cos\theta. \tag{3}$$

It should be noted that the given spherical macropores are only equivalent to cavities of a very complicated form of the original real bulk medium.

Following [3] to describe the dynamics of the change of the equivalent radius of the spherical pore $a = a(\tau)$ over time τ we use the equation:

$$da/d\tau = -3\kappa\sigma/\left(4\eta\right), \tag{4}$$

where η – the viscosity of the polymer coating material, σ – the surface tension coefficient at the interphase boundary, κ – correction coefficient taking into account the influence of the external factors on the particles agglutination process.

Integrating the equation (4) under the condition that at the initial time the radius $a = a_0$ for a time τ the values of η, σ are close to their average values $\bar{\eta}$, $\bar{\sigma}$, we obtain:

$$a = a(\tau) = a_0 - K_0\tau, \tag{5}$$

where $K_0 = 3\kappa\bar{\sigma}/4\bar{\eta}$.

It can be seen from (5) that the time for complete agglutination of particles containing the microspheres ($a = 0$) will be:

$$\tau_{sc} = a_0/K_0.$$

In the case when at the initial time $\tau = 0$ the laying of spherical particles of the bulk layer with radius r_p is such that $\theta = 90°$, $a_0 = 0.5d_s$ (see (1)) we have

$$\tau_{sc} \approx 1.33(1-0.22\cos\theta)\bar{\eta}r_p/(\kappa\bar{\sigma}).$$

As follows from this formula the smaller is the particle size r_p, the average viscosity of the polymer material $\bar{\eta}$ and the greater is the average surface tension $\bar{\sigma}$, the smaller will be the time of particles agglutination. This is completely consistent with the existing physical representations. As for the parameter κ, it can be assumed that it depends on the pressure ratio $\bar{p} = p_c/p_s$, where p_s, p_c – the pressure in the threshold space and softened polymer material of the microsphere coating layer, respectively.

In particular

$$\kappa = k \cdot \bar{p}.$$

Here k – the empirical coefficient the pressures p_s, p_c characterize the external impact on the medium.

It must also be kept in mind that along the process under consideration the temperature T of the porous medium can change accordingly $\bar{\eta} = \bar{\eta}(T)$, $\bar{\sigma} = \bar{\sigma}(T)$, $\bar{p}_n = \bar{p}_n(T)$, $\kappa = \kappa(T)$.

If the change of the given quantities is relatively small during the allotted time, then the above relationships are applicable with a certain accuracy degree.

Coalescence of the particles with the liquefied layer of the coating

At the moment of contact of the spherical surfaces of the liquefied surface layers of particles, we select the periodicity cell enclosed between six planes passing through the centers of neighboring particles. This cell includes eight spherical segments and a gas cavity bounded by eight spherical surfaces six sections of the secant planes.

Immediately after the neighboring spherical segments come in contact, depending on magnitude of the pressure p_{s0} in the pore space at that moment, there are two possible variants of the coalescence process of the particles. If the pressure p_{s0} is small the gas in the macrocavities is discharged the interphase surface remains convex the pore space will have a complicated formula. In this situation for a mathematical description of the particles coalescence, it is convenient to use the model of an ideal (tubular) porous medium [2]. At the initial time the channels radius of the named medium according to the Kozeni formula will be:

$$r_{s0} = 2m_{so}r_p/(3(1-m_{so})). \tag{6}$$

Another estimate of the radius r_{so} is possible [2, 4]:

$$r_{so} \approx 2.83\sqrt{k_s/m_{so}}, \tag{7}$$

where k_s – permeability coefficient of the bulk layer.

Further assuming that the dependence analogous to (4) operates in the circular cross-section of the pore channel where the sphere radius a is substituted by the current radius r_s of the channel cross-section, we find:

$$r_s = r_s(\tau) = r_{s0} - K_0\tau. \tag{8}$$

Hence the time of complete coalescence of the polymer material particles

$$\tau_{sc} = r_{so}/K_0 = 8m_{s0}\bar{\eta}r_p/(9\kappa\bar{\sigma}(1-m_{s0})). \tag{9}$$

Substituting the dependence (3) in the relationship (9), we approximately get:

$$\tau_{sc} \approx 0.89(1-0.66\cos\theta)^2\bar{\eta}r_p/(\kappa\bar{\sigma}). \tag{10}$$

Comparing the formula for the agglutination time and (10), we make sure that the time for complete coalescence of particles is generally less than time for their agglutination.

In the case when the pressure p_{s0} in the pore space is increased after the contact of the neighboring spherical segments near the points of contact the polymer layer material is closed down under the action of capillary forces, the sections with curvature inverse to the curvature of the original spherical surfaces are formed. Obviously, the interphase boundary with the curvature of unlike sings will be unstable and in a relatively short period of time will take a form close to spherical one. It is easy to verify that its radius

$$r_{so} = r_p \sqrt[3]{(6/\pi)-1} . \tag{11}$$

According to the Laplace formula on a given spherical surface

$$p_{s0} - p_{c0} = 2\sigma / r_{s0} . \tag{12}$$

Here p_{c0} is the pressure in the liquid material of the coating layer by the time when the external factors start acting.

Since the radius r_{s0} of the spherical cavity located in the center of the selected periodicity cell is smaller than its transverse dimensions these cavities are closed (bubbles) the gas pressure in them during the processing step can, in general, be changed only by heating or cooling the porous medium. In the absence of phase transitions at the interface boundary according to the Clapeyron-Mendeleev law the pressure in the cavity is

$$p_s = RM_s T / V_s , \tag{13}$$

where R – gas constant, M_s – the gas mass in the cavity, V_s – its volume.

In (12) replacing the values of p_{s0}, p_{c0}, r_{s0} by the current values of p_s, p_c, r_s, substituting the pressure p_s (13) we get the equation to find the cavity radius r_s:

$$4\pi p_c r_s^3 + 2\sigma r_s^2 = 3RM_s T . \tag{14}$$

Obviously, the change of radius r_s is small, so we write $r_s = (1+\varepsilon_s) r_{s0}$ where ε is a small quantity. After a series of transformations and discarding of small quantities for estimating ε_s we find the relation:

$$\varepsilon_s \approx 0.25\left(\left(3RM_s T / r_{s0}^3\right) - 2\left(2\pi p_c + \sigma / r_{s0}\right)\right) / \left(3\pi p_c + \sigma / r_{s0}\right) . \tag{15}$$

It can be seen from (15) that with increasing the temperature T and other parameters being constant ε_s and hence the radius r_s increase. Conversely, when surface tension coefficient σ and pressure p_c increase, the parameter ε_s decreases.

3. Conclusions

Thus, the relationships to determine the time of complete agglutination and coalescence of the microsphere coating polymer material and to estimate the influence of initial particle sizes, their laying, material viscosity, and surface tension at the interphase boundary on these processes are obtained. The dependence characterizing the change of the gas cavity dimensions in the periodicity cell at the particles coalescence in the course of the heating (cooling) of the porous medium and the influence of mechanical loads on it is established. These relationships and dependencies can be useful when choosing the rational technological modes of production of the articles made of syntactic carbon foams and in other cases [4-10].

References

[1] Fedyaev V L, Galimov E R, Samoylov V M, Danilov E A and Orlov M A 2016 Production of thermally conductive carbon foams and their application in automobile transport *Proceedings of the International Scientific and technical conference "Innovative Engineering Technologies, Equipment and Materials – 2016" part 2* pp. 211 – 214

[2] Polubarinova – Kochina P Ya 1977 *Theory of ground water motion* (Moscow: Publishing house Nauka)

[3] Yakovlev A D 1987 *Powder paints* (Leningrad, Publishing house Khimiya)

[4] Belov S V 1981 *Porous metals in mechanical engineering* (Moscow: Publishing house Mashinostroenie)

[5] Kashapov N F, Luchkin G S and Luchkin A G 2010 *Vacuum technologies for the application of functional coatings* "Vestnik" of Kazan State Technological University No 2 pp 340-34

[6] Kashapov N F, Luchkin G S and Tagirov R F 2010 *Vacuum test bench for the application of functional coatings* "Vestnik" of Kazan State Technological University No 2 pp 346-52

[7] Mazo A B 1996 Mathematical modeling of processes of hot metal processing (Kazan Math. Soc.) p 209

[8] Adamson A 1979 Physical chemistry of surfaces (Moscow: Mir)
[9] Betchelor Dzh 1973 *Introduction to fluid dynamics* (Moscow: Mir)
[10] Galimov E R, Fedyaev V L, Galimova N Ya, Takhaviev M S and Siraev A R 2017 *Mathematical modeling of processes occurring during deposition of sprayed particles of polymeric powder* Journal of Physics : Conf. Series vol 789 **012007**

IOP Conf. Series: Materials Science and Engineering **412** (2018) 012014 doi:10.1088/1757-899X/412/1/012014

Heat transfer in syntactyc carbon porous materials

V L Fedyaev[1,2], E R Galimov[2], N Ya Galimova[2], A V Belyaev[2] and V M Samoylov[3]

[1]IME - Subdivision of FIC KazanSC of RAS, Lobachevsky str. 2/3, Kazan, 420111 Tatarstan, Russia
[2]Kazan national research technical university named after A N Tupolev, K. Marx str. 10, Kazan, 420111 Tatarstan, Russia
[3]JSC NIIgrafit, Elektrodnaya str. 2, Moscow, 111524, Russia

engelgalimov@yandex.ru

Abstract. Heat transfer in the syntactic carbon porous materials is considered. A number of assumptions regarding the microstructure of these materials and the corresponding mechanisms of heat transfer are made. Relations to estimate the effective density, heat capacity and thermal conductivity of the materials taking into account the radiant heat exchange between the matrix elements are given. The behavior of these parameters depending on the structure, properties, and components of syntactic porous material is analyzed.

1. Introduction
It is known that carbon materials have unique properties in consequence of which they are widely used in engineering. At present, highly porous carbon foams are of great interest [1-6, 11]. When matrix material of the foam consists of carbon and the hollow microspheres distributed in it, it is syntactic carbon foam (SCF).

In fact, SCFs are heterogeneous systems consisting of carbon, a microsphere shell material, a gaseous medium in the cavity of microspheres, and gaseous or vaporous medium or a fluid impregnating the SCF. A distinctive feature of these materials is that SCF properties can be varied in a wide range by changing the porosity due to the matrix itself, proportions between binding carbon material and the microspheres, their dimensions, the shell material (carbon, ceramics, metal and etc.) the gaseous medium composition. According to what has been said the establishment of behavior regularities of SCF thermophysical properties is of great interest.

2. Body text
At first, we turn to the SCF matrix. If the properties of the microsphere shell material differ from the properties of the carbon, then the matrix can be considered as a porous composite. Let us suppose that in its control volume V_k n_g microspheres with an average volume of cavities V_g are located sufficiently uniformly. It follows that the microporosity of the composite material will be

$$m_g = n_g \varepsilon_g ,$$

where $\varepsilon_g = V_g / V_k$ – the fraction of the microsphere cavity volume in the control volume.

Content from this work may be used under the terms of the Creative Commons Attribution 3.0 licence. Any further distribution of this work must maintain attribution to the author(s) and the title of the work, journal citation and DOI.
Published under licence by IOP Publishing Ltd

Further denoting the fraction of microsphere shell material as $\varepsilon_m = V_m/V_k$, we get $\varepsilon_c = V_c/V_k = 1 - n_g(\varepsilon_g + \varepsilon_m)$ (V_c – the carbon volume in V_k). Hence the effective density of the porous composite material will be:

$$\rho_k = m_g \rho_g + \rho_{mc}, \tag{1}$$

where $\rho_{mc} = n_g \varepsilon_m \rho_m + \varepsilon_c \rho_c$ – the effective density of the solid phase; ρ_g, ρ_m, ρ_c – density of the gas, microsphere shell material, carbon.

Assuming that in the singled out controlled volume the initial material is three-layered perpendicularly to the direction of heat propagation in order to estimate the effective thermal conductivity λ_{mc} of the layer consisting of microsphere shell material and carbon, we get the dependence:

$$\lambda_{mc} = (1 - \mathrm{m}_g) \Big/ \left(\frac{n_g \varepsilon_m}{\lambda_m} + \frac{\varepsilon_c}{\lambda_c} \right). \tag{2}$$

Here λ_m, λ_c – the thermal conductivity coefficients of the shell material and carbon. Accordingly, the effective thermal conductivity coefficient

$$\lambda_k = 1 \Big/ \left(\frac{m_g}{\lambda_g} + \frac{1 - m_g}{\lambda_{mc}} \right), \tag{3}$$

where λ_g – the thermal conductivity coefficient of the gas.

According to the additivity principle the effective specific heat capacity of the porous composite material of the SCF matrix will be ($\rho_g / \rho_k = 1$):

$$c_k = n_g \varepsilon_m \bar{\rho}_m c_m + \varepsilon_c \bar{\rho}_c c_c, \tag{4}$$

where $\bar{\rho}_m = \rho_m / \rho_k$, $\bar{\rho}_c = \rho_c / \rho_k$; c_m, c_c – the specific heat capacity of the microsphere shell material, carbon.

Let us further consider the material of the syntactic foam in general. We suppose that the filtration rate of the medium in the foam pores is small the possible heat transfer due to convection is also small. In addition, we shall not take into account the conductive heat transfer upon contact of the solid and gas (liquid) phases. As a result, the heat conductivity and radiant heat exchange between the more and less heated parts of the matrix are predominant in the SCF provided that the medium in the pores is diathermic [7 – 12].

The noted combination of two types of heat transfer leads to a specific behavior pattern of the SCF material thermal conductivity with temperature increase. In the case when the pores are filled with gas the thermal conductivity is small at low temperatures since the thermal resistance of the spaces between matrix parts is very large, the pores impede heat transfer. As the temperature T increases, the conductivity increases sharply (~ T^3). At high temperatures it becomes of the same order as the thermal conductivity of the matrix material in this case the medium behaves like a solid body whose thermal conductivity is close to the average thermal conductivity λ_k.

In order to take into account the mentioned peculiarities of the heat propagation in the SCF material we select an elementary volume in it and assume that it consists of matrix parts oriented in the direction of the heat flow ξ; parts located transversely to this direction and spaces between them. The pores filled with gas form a series circuit with thermal resistances $R_n = 1/4\sigma\varepsilon_n\gamma_n d_n T^3$ [7, 10], where σ – the Stefan-Boltzmann constant, ε_n – the correction for incomplete blackness of the pores, d_n – the equivalent pore size in the direction of heat propagation, T – the surface temperature of the matrix parts that is close to the average temperature of the SCF material, $\gamma_n = \gamma_n(m_n)$ – a pore shape factor that depends on the porosity m_n ($0 < \gamma_n < 1$).

As a result, after averaging by the elementary volume for the effective thermal conductivity coefficient λ_s, we obtain the formula:

$$\lambda_s = (1-\varphi_n)\lambda_k + \varphi_n / \left(R_k + \varphi_{sn}\left(R_n + R_k\right)\right). \tag{5}$$

Here, φ_n – the ratio of the voids area of the cross-section to the heat flow to the total cross-sectional area, φ_{sn} – the ratio of the voids length along the longitudinal line to the total length of the elementary volume in the direction of the heat flow, $R_k = 1/\lambda_k$.

Analyzing the dependence (5) we make sure that as the temperature of the SCF material increases, the thermal conductivity coefficient λ_c increases significantly tending to the theoretical limit $\lambda_{max} = \lambda_k\left(1-\varphi_n + \varphi_n/(1-\varphi_{sn})\right)$. It should be noted that formula (5) is similar to the Lob's formula for cellular bodies. The first term in (5) is due to the longitudinally located matrix parts, the second – the remaining components of the material.

If we consider the heat propagation in the porous body with through holes along the heat flow, then it is easy to see that

$$\lambda_s = \left(1-m_n\right)\lambda_k + m_n\left(\alpha_k + 4\sigma\varepsilon_n d_n T^3\right), \tag{6}$$

where α_k – the coefficient that takes into account the convective heat transfer.

Another more natural situation is that heat propagation in the SCF material is modeled using the solid body with through holes which are located chaotically with respect to the direction of the heat flow ξ. In this case, in (5) $\varphi_n = \varphi_{sn} = m_n$ the effective thermal conductivity coefficient

$$\lambda_s = \left(1-m_n\right)\lambda_k + m_n / \left(R_k + m_n\left(R_n - R_k\right)\right). \tag{7}$$

By definition the thermal diffusivity coefficient of the SCF material

$$a_s = \lambda_s / (\rho_s c_s).$$

Here the effective density ρ_s is approximately equal to

$$\rho_s \cong (1-m_n)\rho_k ; \tag{8}$$

effective heat capacity

$$c_s = m_n\bar{\rho}_n c_n + (1-m_n)\bar{\rho}_k c_k \cong (1-\mathrm{m}_n)\bar{\rho}_k c_k , \tag{9}$$

where c_n – the material heat capacity of the gaseous phase that is in the pores; $\bar{\rho}_n = \rho_n / \rho_s$, $\bar{\rho}_k = \rho_k / \rho_s$ – relative densities.

As it is known the rate of change of the material average temperature is described by the dependence:

$$c_s\rho_s\frac{\partial T}{\partial \tau} = \frac{d\lambda_s}{dT}\left(\nabla T \cdot \nabla T\right) + \lambda_s\nabla^2 T , \tag{10}$$

where τ – time.

In the case where $d\lambda_s / dT$ and the gradient ∇T are small equation (10) is simplified:

$$\partial T/\partial\tau = a_s\nabla^2 T , \tag{11}$$

Where according to (8) – (9)

$$a_s = a_k + \frac{\rho_k/\rho_s - 1}{\rho_s c_s R_k + c_s\left(\rho_k - \rho_s\right)R_n} , \quad a_k = \frac{\lambda_m}{\rho_k c_s} .$$

Further, let us suppose that along the surface $\xi = 0$ a heat flow of power q acts upon the selected elementary volume of the material. Assuming that the heat propagates in the ξ axis direction we estimate the heating of the material under consideration, for this we supplement the equation (11) with the initial and boundary conditions:

$$T(\tau = 0) = T_0 , \ -\lambda_s\,\partial T/\partial\xi\big|_{\xi=0} = q , \ T(\xi,\tau)\big|_{\xi\to\infty} \to T_0 ,$$

where T_0 – the initial temperature of the material.

It can be seen that the solution of problem (11), (12) will be:

$$T = T_0 + \left(2q/\lambda_s\right)\left(\left(a_s\tau/\pi\right)^{1/2}\exp\left(-\xi^2/4a_s\tau\right) - 0.5\xi^2 F\left(\xi/2\left(a_s\tau\right)^{1/2}\right)\right),$$

where F – the error function.

3. Conclusions

Thus, the relationships for evaluation of thermophysical properties of the specific material of the SCF are obtained taking into account the characteristics of the binding carbon, microsphere shells and the gas contained therein. The formulas for determining the thermal conductivity coefficient, specific heat capacity and thermal diffusivity that can be used at high temperatures are proposed. In addition, alternative dependencies are presented. A solution for the one-dimensional model problem on the heating of the SCF material which clearly illustrates the influence of thermophysical parameters on the temperature change of a given material over time is presented.

The work was carried out with the financial support of the Ministry of Education and Science of the Russian Federation within the framework of the Federal Target Program "Research and development in priority areas of development of the scientific and technological complex of Russia for 2014–2020" agreement No. 14.583.21.0057 of July 28, 2016 (Project ID RFMEF158316X0057)

References

[1] Fedyaev V L, Galimov E R, Samoylov V M, Danilov E A and Orlov M A 2016 *Production of thermally conductive carbon foams and their application in automobile transport* Proceedings of the International Scientific and technical conference "Innovative Engineering Technologies, Equipment and Materials – 2016" part 2 pp 211 – 214

[2] Michio I, Jieshan Q and Quangui G 2015 *Review Carbon foam: Preparation and application* Carbon vol 87 pp 128 –152

[3] Gallego N S and James W K 2000 Carbon foams for thermal management *International Seminar on Advanced Applications for Carbon Materials. Jeju Island, Korea. September 12-13*

[4] Klett J, Ott R and McMillan A 2000 Heat exchangers for heavy vehicles utilizing high thermal conductivity graphite foams *SAE Technical Paper 2000-01-2207*

[5] Jurchenko D and Kelley K 2007 Carbon foam liberates lead-acid batteries for emerging military missions *Defense Electronics*. Available at: http://www.rfdesign.com/

[6] Almajali M ant et al. 2010 *Mechanical properties of copper-coated carbon foams* Carbon vol 48 pp 1604–08

[7] Chudnovskiy A F 1954 *Thermal conductivity in dispersion media* (Moscow: Publishing house Gostekhizdat)

[8] Dul'nev G N and Novikov V N 1991 *Transport processes in inhomogeneous media* (Leningrad: publishing house Energoatomizdat)

[9] Pavlyukevich N V 2002 *Introduction to the theory of heat and mass transfer in porous media* (Minsk: Publishing house of Heat and mass transfer institute of the national academy of sciences of Belarus)

[10] Grinchuk P S 2014 *Contact heat conductivity under conditions of high-temperature heat transfer in fibrous heat-insulating materials* Journal of engineering physics and thermophysics vol 87 No 2 pp 466-72

[11] Kashapov N F, Nefed'yev E S and Fakhrutdinov I M 2008 *Filtering materials based on technical carbon obtained by electric arc plasma* "Vestnik" of Kazan State Technological University 6 pp 25-30

[12] Galimov E R, Fedyaev V L, Galimova N Ya, Takhaviev M S and Siraev A R 2017 *Mathematical modeling of processes occurring during deposition of sprayed particles of polymeric powder* Journal of Physics : Conf. Series vol 789 **012007**

IOP Conf. Series: Materials Science and Engineering **412** (2018) 012015 doi:10.1088/1757-899X/412/1/012015

Evaluation of interphase boundaries dynamics at the contact of melt with solid material

V L Fedyaev[1,2], E R Galimov[2], A V Belyaev[2], A V Dolgov[2] and V M Samoylov[3]

[1]IME - Subdivision of FIC KazanSC of RAS, Lobachevsky str. 2/3, Kazan, 420111 Tatarstan, Russia
[2]Kazan national research technical university named after A N Tupolev, K. Marx St. 10, Kazan, 420111 Tatarstan, Russia
[3]JSC NIIgrafit, 2 Elektrodnaya str., Moscow, 111524, Russia

engelgalimov@yandex.ru

Abstract. The results of mathematical modeling of the thermal processes that occur on contact of the melt with a layer of polymer material located on the surface of the hollow microspheres are presented. Dependencies that allow estimating the melting time of the given layer and other parameters of interest in the production of articles from syntactic materials were obtained.

1. Introduction
A set of hollow microspheres with the layer of the solid polymeric material on their surface is considered. It is assumed that the thickness of this layer is the same, the microspheres are tightly packed, at the initial time the temperature of the coated microspheres material is equal to Θ_0. For some time τ_n, the space between the spheres is filled with the melt of the same material as the coating material. The dynamics of the spheres heating and the movement of the interphase boundary starting with the time τ_n is estimated, provided that the time τ_n is small, at filling the voids the spheres' temperature remains equal to Θ_0, the average temperature of the melt is Θ_1. A periodicity cell including a coated microsphere and a melt layer also having a spherical shape with an effective radius R_e is sorted out. Figure 1 shows the corresponding calculation scheme. Here 1, 2, 3, 4 – the gas cavity, the microsphere shell, the polymer coating layer, the conditional layer of polymer melt; R_0, R_1, R_c –the gas cavity radius, the outer radius of the microsphere shell, the coating surface radius.

2. Body text
In the case of axial symmetry of thermal fields in the absence of internal sources – heat sinks, gas flow in the cavity, melt flow the non-stationary heat transfer in the elements of the considered system is described by the equations:

$$\frac{\partial \Theta_i}{\partial \tau} = a_i \left(\frac{\partial^2 \Theta_i}{\partial R^2} + \frac{2}{R}\frac{\partial \Theta_i}{\partial R} \right), \tag{1}$$

where Θ_i – the temperature of the substance in the i-th area (i = 1, 2, 3, 4), $a_i = \lambda_i / (\rho_i \cdot c_i)$ – temperature diffusivity coefficient of the material of the corresponding medium; ρ_i, λ_i, c_i – density, thermal conductivity coefficient, specific heat, τ – time.

Content from this work may be used under the terms of the Creative Commons Attribution 3.0 licence. Any further distribution of this work must maintain attribution to the author(s) and the title of the work, journal citation and DOI.
Published under licence by IOP Publishing Ltd

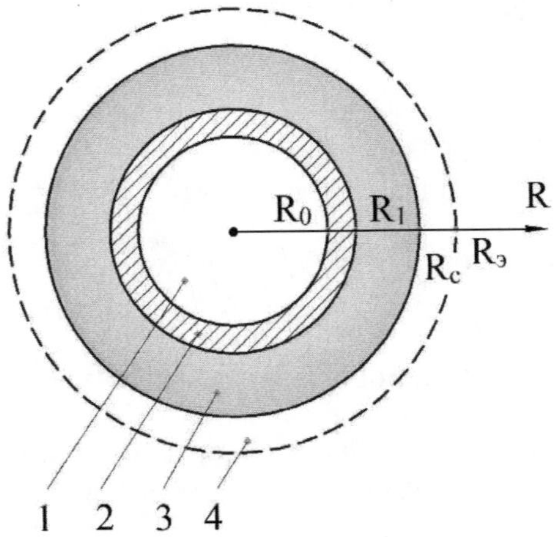

Figure 1. Spherical periodicity cell

At time $\tau = \tau_n$ the temperatures of the gas, the microsphere shell, and the coating

$$\Theta_1(R,\tau_n)=\Theta_2(R,\tau_n)=\Theta_3(R,\tau_n)=\Theta_0, \tag{2}$$

the melt layer temperature

$$\Theta_4(R,\tau_n)=\Theta_1. \tag{3}$$

The boundary conditions have the form:

$$R = 0:\ \partial\Theta_1 / \partial R = 0 \ \text{(symmetry condition)}; \tag{4}$$

$$R = R_0:\ \Theta_1=\Theta_2,\ \lambda_1 \frac{\partial\Theta_1}{\partial R} = \lambda_2 \frac{\partial\Theta_2}{\partial R} \ \text{(ideal thermal contact)}; \tag{5}$$

$$R = R_1:\ \Theta_2=\Theta_3,\ \lambda_2 \frac{\partial\Theta_2}{\partial R} = \lambda_3 \frac{\partial\Theta_3}{\partial R}. \tag{6}$$

Taking into account the fact that the coating and melt materials are the same (at a certain temperature $\rho_3 = \rho_4$, $\lambda_3 = \lambda_4$, $c_3 = c_4$) layers 3 and 4 will be considered as one layer with its thermophysical properties characterized by the parameters ρ, λ and c. Inside this layer, an interphase surface located between the solid material and its melt is marked out. The motion equation of this surface is written as:

$$R_* = R_*(\tau),\ R_*(\tau_n) = R_c;$$

with $R = R_*$ $\Theta = \Theta_*$,

$$\left(\lambda \frac{\partial\Theta}{\partial R}\right)_{R_*+0} = \left(\lambda \frac{\partial\Theta}{\partial R}\right)_{R_*-0} + \rho L v. \tag{7}$$

Here Θ_* – phase transition temperature, L – latent heat of melting, $v \equiv \dfrac{dR_*}{d\tau}$ – motion speed of the melt front (crystallization).

At the fictitious boundary $R = R_e$, we assume:

$$\frac{\partial\Theta}{\partial R} = 0, \tag{8}$$

which means that there is no heat exchange with the medium surrounding the selected periodicity cell.

Formulated in this way the thermo-conjugate problem (1) – (8) can be solved numerically. However, for engineering practice the analytical dependencies are useful. In order to obtain them, we simplify this task for this purpose along the radius-vector R we distinguish a prism including the shell

IOP Conf. Series: Materials Science and Engineering **412** (2018) 012015 doi:10.1088/1757-899X/412/1/012015

material of the microsphere, solid and molten polymer. In this case, the calculation scheme takes the form shown in figure 2.

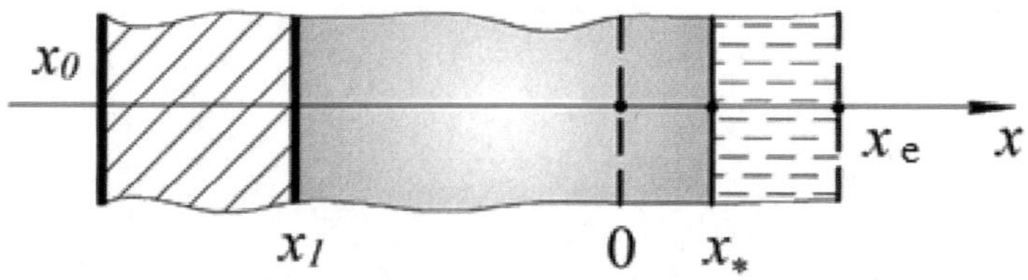

Figure 2. Simplified calculation scheme

In this scheme the origin of the x-coordinate directed across the layers is on the polymer layer surface, $x_* = x_*(\tau)$ – the motion equation of the interphase boundary (at the initial time point $\tau = 0$, $x_*(0) = 0$) the coordinates x_0, x_1 characterize the position of the shell surfaces of the hollow microsphere, x_e characterizes the thickness of the effective layer of the melt in the moment of its contact with the coating material of the microsphere.

In accordance with this scheme in the layer $x_0 \le x \le x_1$ occupied by the microsphere shell,

$$\frac{\partial \Theta_s}{\partial \tau} = a_s \frac{\partial^2 \Theta_s}{\partial x^2}; \tag{9}$$

in the polymeric material ($x_1 \le x \le x_e$)

$$\frac{\partial \Theta}{\partial \tau} = a \frac{\partial^2 \Theta}{\partial x^2}. \tag{10}$$

Here, $\Theta_s = \Theta_s(x,\tau)$, $\Theta = \Theta(x,\tau)$ – the temperature in the shell, polymeric material $a_s = \lambda_s / (\rho_s c_s)$ ($\rho_s \equiv \rho_2$, $\lambda_s \equiv \lambda_2$, $c_s \equiv c_2$), $a = \lambda / (\rho c)$.

At time $\tau = 0$

$$\Theta_s(x,0) = \Theta_0\ (x_0 \le x \le x_1),\ \Theta(x,0) = \Theta_0\ (x_1 < x < 0),$$
$$\Theta(x,0) = \Theta_1\ (0 < x < x_e). \tag{11}$$

Considering that the gas in the microsphere cavity conducts heat poorly, there is no convective heat transfer, at the boundary $x = x_0$ we write:

$$\frac{\partial \Theta_s}{\partial x} = 0. \tag{12}$$

Provided an ideal contact in the contact plane of the polymeric material and the microsphere shell

$$x = x_1:\ \Theta_s = \Theta,\ \lambda_s \frac{\partial \Theta_s}{\partial x} = \lambda \frac{\partial \Theta}{\partial x}. \tag{13}$$

At a movable interphase boundary (phase-transition front) $x_* = x_*(\tau)$:

$$\Theta(x_* - 0, \tau) = \Theta(x_* + 0, \tau) = \Theta_*,$$
$$\left(\lambda \frac{\partial \Theta}{\partial x}\right)_{x_*+0} = \left(\lambda \frac{\partial \Theta}{\partial x}\right)_{x_*-0} + \rho L v(\tau). \tag{14}$$

Finally, for $x = x_e$ in accordance with (8) we have

$$\partial \Theta / \partial x = 0. \tag{15}$$

The problem (9) – (15) can be slightly simplified if we take into account the fact that the microsphere shell thickness $h_s = x_1 - x_0$ is very small its material has a high thermal conductivity. As a result, considering the shell layer as a thermally thin body assuming that its average temperature $\overline{\Theta_s} : \Theta_0 + \alpha\tau$ where α is the average heating rate of a given layer at the boundary $x = x_1$, instead of (13) we obtain condition

IOP Conf. Series: Materials Science and Engineering **412** (2018) 012015 doi:10.1088/1757-899X/412/1/012015

$$\lambda \frac{\partial \Theta}{\partial x} : \alpha \rho_s c_s h_s = \beta_s . \tag{16}$$

If we further assume that the value $\beta_s = \beta_s(\tau)$ is small, the microsphere shell from scheme in figure 2 can be eliminated thereby converting the problem (9) – (15) to the problem of integrating the heat transfer equation (10) in a polymer material layer under the initial condition

$$\Theta(x,0) = \begin{cases} \Theta_0 (x_1 \le x < 0; \\ \Theta_1 (0 < x \le x_e. \end{cases} \tag{17}$$

and boundary conditions (16), (14) and (15).

The problem represented by the relations (10), (16), (17), (14) and (15), many similarities with the problems considered in [1, 2]. From the results of their numerical solution under the condition of thermal insulation at the ends of the computational domain it follows that the function $x_* = x_*(\tau)$ can approximately be written in the form:

$$x_* = x_*(\tau) = x_m - \alpha_m(\tau - \tau_m)^2 , \tag{18}$$

where x_m – the maximum displacement of the interphase boundary towards the melt due to its cooling on contact with solid material, τ_m – the time point when the boundary motion reverses it begins to move in the direction opposite to the $0x$ axis direction. Since at time $\tau = 0$ $x_* = 0$, the coefficient

$$\alpha_m = \frac{x_m}{\tau_m^2} .$$

Note that the value of x_m can be estimated by referring to the heat conservation equation of the layer under consideration during the time τ_m. In this case, it is necessary to take into account that heat losses through the surfaces $x = x_1$ and $x = x_e$ are absent. Approximately

$$x_m = 0.5c\left(h_c\left(\Theta_0 - \Theta_*\right) + 0.5x_e\left(\Theta_1 - \Theta_*\right)\right) / \left(L - 0.5c\left(\Theta_1 - \Theta_0\right)\right) \tag{19}$$

(h_c – the thickness of the polymer coating layer).

Further, we require that

$$x_m : 0.5\delta_m\sqrt{\tau_m} , \tag{20}$$

where δ_m – an additional parameter characterizing the velocity v_0 of the interphase boundary at the initial time,

$$\delta_m = v_0\sqrt{\tau_m} .$$

It is easy to make sure [3, 4] that in the region occupied by the melt ($x > x_*$),

$$\Theta = \Theta(\mathrm{x}, \tau) = \Theta_1 - C_1\left(1 + \Phi\left(\frac{x}{2\sqrt{a\tau}}\right)\right); \tag{21}$$

in the solid phase ($x < x_*$)

$$\Theta = \Theta(\mathrm{x}, \tau) = \Theta_0 + C_0\left(1 - \Phi\left(\frac{x}{2\sqrt{\tau a}}\right)\right). \tag{22}$$

Here C_0, C_1 are the integration constants found from the condition that the ambient temperature at the front of the phase transition of the layers to the left and right of it is equal to Θ_*; $\Phi(\mathrm{z}) = \frac{2}{\sqrt{\pi}}\int_0^{\mathrm{z}} \exp(-\mathrm{u}^2)\mathrm{du}$ – the probability integral.

Hence, if $x_* = 2\gamma\sqrt{a\tau}$, then $C_0 = (\Theta_* - \Theta_0)/(1 - \Phi(\gamma))$, $C_1 = (\Theta_1 - \Theta_*)/(1 + \Phi(\gamma))$.

It should be noted that the solutions (21) and (22) are valid for infinitely extended regions occupied by the melt and solid material. At the same time, they approximate the temperature field in the vicinity of the interphase boundary quite well. The parameter γ is determined from the heat balance equation on it:

$$\lambda C_1(\gamma)\exp(-\gamma^2) = \lambda C_0(\gamma)\exp(-\gamma^2) - \sqrt{\pi}\rho a L\gamma . \tag{23}$$

For small γ approximately $\Phi(\gamma) \approx \frac{2}{\sqrt{\pi}}\left(\gamma - \frac{\gamma^3}{3}\right)$, consequently the coefficients

$$C_0 \approx (\Theta_* - \Theta_0)(1 + \frac{2}{\sqrt{\pi}}(\gamma - \frac{\gamma^3}{3})),\ C_1 \approx (\Theta_1 - \Theta_*)(1 - \frac{2}{\sqrt{\pi}}(\gamma - \frac{\gamma^3}{3})) . \tag{24}$$

After substituting these values of C_0, C_1 into equation (23) as a result of a number of transformations to estimate the parameter γ, we obtain the formula:

$$\gamma = \sqrt{\pi}c(2\Theta_* - \Theta_1 - \Theta_0) / (\pi L + 2c(\Theta_0 - \Theta_1)) . \tag{25}$$

From this formula, it is clear that the moving velocity of the interphase boundary will decrease with a decrease of the specific heat and an increase of the latent heat of the material melting. Indeed, with a decrease of the heat capacity with other indices being constant the thermal diffusivity increases, and the heat dissipates faster. If the latent heat of melting increases, more energy is required to overcome intermolecular bonds.

Using the obtained solutions, we estimate the magnitude of the maximum displacement of the interphase boundary x_m. Since at $\tau = \tau_m$ the boundary velocity $v(\tau_m) = 0$ from (14) after substitution of the functions (21) and (22) into this relation, taking into account the dependence $\gamma_m = 0.5 x_m / \sqrt{a\tau_m}$ we find:

$$x_m = \sqrt{\pi a \tau_m}\left(2\Theta_* - \Theta_1 - \Theta_0\right) / \left(\Theta_0 - \Theta_1\right) . \tag{26}$$

Comparing formulas (20) and (26) we define the parameter δ_m:

$$\delta_m = 2\sqrt{\pi a}\left(2\Theta_* - \Theta_1 - \Theta_0\right) / \left(\Theta_0 - \Theta_1\right) . \tag{27}$$

Due to the fact that in accordance with the technological requirements the initial movement of the interphase boundary must be reversed in such a way that the polymer coating layer would melt we limit the value of x_m setting $x_m = 0.5\, x_e$.

Hence we obtain the estimate of the time τ_m:

$$\tau_m = \left(x_э / \delta_m\right)^2 .$$

At the time $\tau = 2\tau_m$ the interphase boundary will return to its original position ($x_* = 0$), and the coating material will begin to melt. According to formula (18) the interphase boundary will overcome the distance h_c in time $\tau_c = \tau_m\left(\sqrt{1 + \frac{h_c}{x_m}} - 1\right)$. Therefore, the total execution time of the operation $\tau_0 = 2\tau_m + \tau_c$.

3. Conclusions

Thus, having the thermophysical parameters of the contacting media, given the geometric dimensions, using the obtained dependencies it is possible to estimate the dynamics of the interphase boundary motion, determine the run-time of the technological operation, thereby minimizing the risks of the appearance of defects [5, 6].

The work was carried out with the financial support of the Ministry of Education and Science of the Russian Federation within the framework of the Federal Target Program "Research and development in priority areas of development of the scientific and technological complex of Russia for 2014–2020" agreement No. 14.583.21.0057 of July 28, 2016 (Project ID RFMEF158316X0057)

References

[1] Mazo A B, Fedyaev V L and Khismatullin N I 1989 *Calculation of the interphase boundary dynamics at contact melting of metals* Modeling of nonlinear processes in mechanics and heat engineering, Materials of the workshop of Kazan Physics and technology Institute of the Kazan branch of the USSR Academy of Sciences Kazan vol 24 pp 66-77

[2] Mazo A B 1996 *Mathematical modeling of processes of hot metal processing* (Kazan Math. Soc.) p 209
[3] Alishaev M G 1986 *Motion of the phase transition front after contact of the solid and liquid phases of the metal* Izvestiya Visshikh Uchebnykh Zavedenii. Chernaya Metallurgiya vol 9 pp 35-38
[4] Kozdoba L A 1975 *Methods for solving nonlinear heat conduction problems* (Moscow: Nauka)
[5] Kashapov N F and Luchkin A G 2014 *Application of low-temperature plasma for applying of reinforcing coatings on plastics* Izvestiya Visshikh Uchebnykh Zavedenii. Fizika vol 57 No 3-3 pp 160-63
[6] Kashapov N F, Gavrilova V A, Tukmakov A L and Tukmakov D A 2012 *Numerical simulation of the spraying process of the charged aerosol on the surface in an electrostatic field* "Vestnik KHTU im. A.N. Tupoleva" No 2 pp 177-182
[7] Adamson A 1979 *Physical chemistry of surfaces* (Moscow: Mir)
[8] Betchelor Dzh 1973 *Introduction to fluid dynamics* (Moscow: Mir)
[9] Galimov E R., Fedyaev V L, Galimova N Ya, Takhaviev M S and Siraev A R 2017 *Mathematical modeling of processes occurring during deposition of sprayed particles of polymeric powder* Journal of Physics : Conf. Series vol 789 **012007**
[10] Samoylov V M, Danilov E A, Galimov E R, Fedyaev V L, Galimova N Ya and M.A. Orlov 2017 *Production of thermally conductive carbon foams and their application in automobile* transport IOP Conf. Series: Materials Science and Engineering vol 240 **012062**

Influence of the additives of natural sulphur-organic compounds on thermal stability of oil fractions of arlan petroleum

L R Gainullina[1], V P Tutubalina[1], L S Sabitov [1,2]

[1]Kazan State Power Engineering University, 51 Krasnoselskaya street, Kazan, 420066, Russian Federation
[2] Kazan Federal University, 18 Kremlyovskaya street, Kazan, 420008, Russian Federation

gainullina7819@mail.ru

Abstract. The influence of additives of natural sulfur-organic compounds on the thermal stability of the oil fraction of Arlan petroleum in an electric field is studied.
It has been established that the introduction of additives of natural sulfur-organic compounds into the petroleum fraction leads to a decrease in the formed solid sediment and acid number, which indicates an increase in the dielectric characteristics of the oil and an increase in its thermal stability in the electric field.
The optimum amount of addition of natural sulfur-organic compounds in an amount of 0.5% reduces the amount of water-soluble acids formed, and, consequently, also the aging of the oil.

Introduction

Transformer oil of petroleum, widely used in power engineering as a heat sink and insulating medium, under the influence of high temperatures, electric field and molecular oxygen of air is subject to aging [1]. As a result of the aging of the oil its performance properties deteriorate [2, 3]. To improve the performance of the transformer oil, various types of additives are used.

The increase in the extraction and processing of sulfur and high-sulfur petroleum, set the task of using sulfur-organic compounds in industry and agriculture. Organo-compounds having unique properties can be widely used as various types of additives for transformer oils to improve their performance properties.

Experimental part

The purpose of this work is to study the effect of natural sulfur-organic compounds on the performance properties of transformer oil.

The object of the study was the oil fraction with Tkip = 300-400 ° C Arlan oil. Natural organosulfur compounds were obtained by adsorption separation of the oil fraction on silica gel ASK and aluminum oxide.

In the process of adsorption separation of the oil fraction on the adsorbents used, five fractions of organosulfur compounds were isolated. Physicochemical characteristics of organosulfur compounds isolated from the oil fraction of Arlan oil are given in Table 1.

Content from this work may be used under the terms of the Creative Commons Attribution 3.0 licence. Any further distribution of this work must maintain attribution to the author(s) and the title of the work, journal citation and DOI.
Published under licence by IOP Publishing Ltd

IOP Conf. Series: Materials Science and Engineering **412** (2018) 012016 doi:10.1088/1757-899X/412/1/012016

Table 1

Physico-chemical characteristics of sulphur-organic compounds oil fraction of Arlan petroleum

Indicator name	Fraction of sulphur-organic compounds				
	I	II	III	IV	V
Density ρ_4^{20}	1,0	1,0320	1,0310	1,1070	1,0827
Refractive index	1,5560	1,5740	1,5830	1,680	1,6270
Molecular mass	321	302	260	227	213
Sulphur content, % common sulphide	 11,3 10,9	 8,1 7,6	 7,4 6,2	 10,1 отс.	 7,9 отс.
Percentage of sulfide sulphur from total	96,5	93,8	83,8	-	-
The empirical formula	$C_nH_{2n-5}S$ n=21	$C_nH_{2n-8}S$ n=20	$C_nH_{2n-10}S$ n=17	$C_nH_{2n-15}S$ n=15	$C_nH_{2n-15}S$ n=14

From the experimental data given in Table 1 it follows that fractions I-III with a total sulphur content of 11.3% to 7.4% consist of 96.5% -83.8% of sulfides.

According to IR spectroscopy, the presence of an absorption band in the region of 1170 cm-1 indicates the presence of alkyl substituted mono-cycloalkanes. The absorption bands in the region of 775 cm^{-1} and 745 cm^{-1} indicate the presence of sulfides, respectively, with propyl and butyl radicals. The characteristic absorption bands for aromatic hydrocarbons were not detected by IR spectroscopy.

As a result of the investigation of fractions I-III by mass spectroscopy, it was found that the sulfides in these fractions are represented by saturated compounds, mainly thiamono- and thiabicycloalkanes in the absence of uncondensed bi- and tricyclic sulfides.

From Table 1 it follows that the organosulfur compounds of fractions IV and V do not contain sulfide sulfur in their composition. Fractions IV and V by the ratio of carbon to hydrogen can be represented by the empirical formula $C_{15}H_{15}$, which corresponds to thiophene derivatives with condensed benzene nucleus.

Using IR spectroscopy in IV and V fractions, absorption bands were observed in regions 2955, 2921 and 2873 cm^{-1}, as well as in the 1587, 1481 and 1280 cm^{-1} regions, indicating the presence of thiophenes and their derivatives.

In the UV spectra, an intense band of 240 nm is detected, which indicates the presence of polyalkyl-substituted thiophenes. PMR spectra in the range of 0.9 and 1.21 ppm. correspond to the protons of the saturated thiophene ring and its alkyl substituents, and the signals are 1.92 and 2.18 ppm. indicate the presence of methyl substituents thiophene ring.

Fractions IV and V are probably polyalkyl substituted thiophenes and cyclothiophenes with various substituents.

Mass spectroscopic analysis of organosulfur compounds of fractions I-V showed that in these fractions there are compounds corresponding to the empirical formulas shown in Table 1.

The bulk of fraction I is represented by saturated organosulfur compounds - sulfides, the empirical formula of which $C_nH_{2n-5}S$ indicates their cyclic structure.

In the fraction II, the compounds of $C_nH_{2n-8}S$ are represented by thiaindans. The remaining compounds represented by the general formulas $C_nH_{2n-10}S$ and $C_nH_{2n-15}S$ refer to alkyl-, cyclo- and bicyclobenzobenzothiophenes.

The fractions of organosulfur compounds I-V were used as an additive to the desulfurized oil fraction (sulfur content was 0.001%) to study the change in its performance properties.

Investigation of the operational properties of the desulfurized oil fraction in the presence and absence of additives of fractions I-V was studied in an apparatus described in the literature [4]. The amount of additives introduced into the desulfurized fraction varied within a wide range from 0.1% to 1.1% in 0.2% increments. The effect of the number of additives of fractions I-V was investigated by the amount of precipitate formed and the acid value, which were determined according to GOST 6370-83 and GOST 6307-75, respectively.

The obtained experimental data on the effect of additives of fractions I-V on the thermal stability of the desulfurized oil fraction determined in accordance with GOST 982-80 are shown in Fig. 1,2.

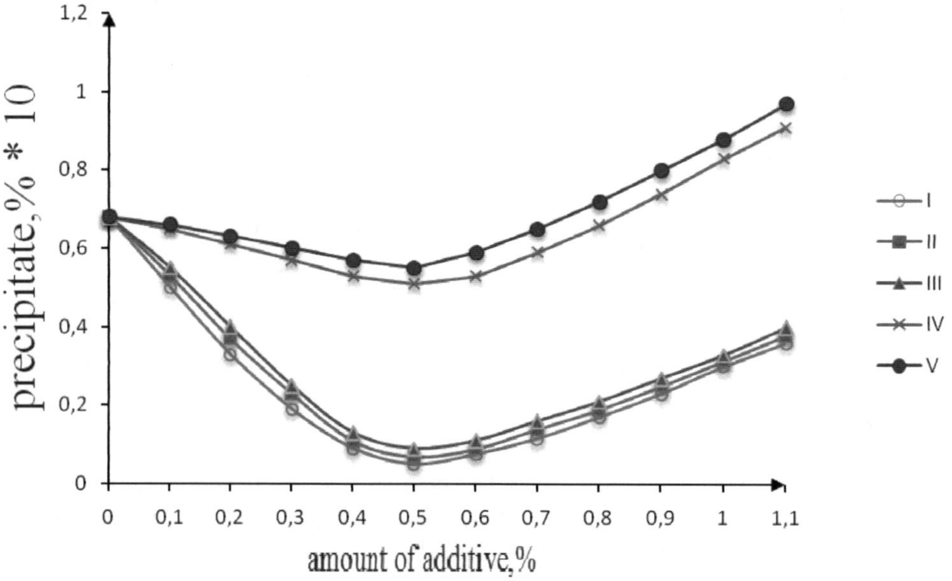

Fig.1. The effect of additives I-V on the amount of precipitate formed with aging of oil fraction

From the experimental data shown in Fig. 1, it can be seen that the additives I-III, consisting mainly of sulphides, refer to effective antioxidant additives, since in the presence of the latter from 0.1% to 0.5% the amount of precipitate formed decreases for all additives (I-III). The amount of solid precipitate formed in fractions I, II and III at an additive concentration of 0.5% was 0.005%, 0.007% and 0.009%, respectively. A further increase in the number of additives (I-III) to 1.1% contributes to the increase in the precipitate formed to 0.037%, 0.038% and 0.039%, respectively.

It also follows from figure 1 that the fractions of organosulfur compounds IV and V, which do not contain sulfide sulfur represented by benzothiophene derivatives, increase the thermal stability of the oil fraction by 1.58 times and 1.4 times, respectively. However, the greatest antioxidant effect at which a minimum amount of precipitate is formed, 0.058% and 0.06%, is achieved with the content of 0.5% additive IV and V in the oil fraction. Increase in the amount of additive of fractions IV and V from 0.5% to 1, 1% contributes to an increase in the amount of sediment to 0.13%, 0.15%. Consequently, the optimum amount of additive of organosulfur compounds of fractions IV and V is 0,5%.

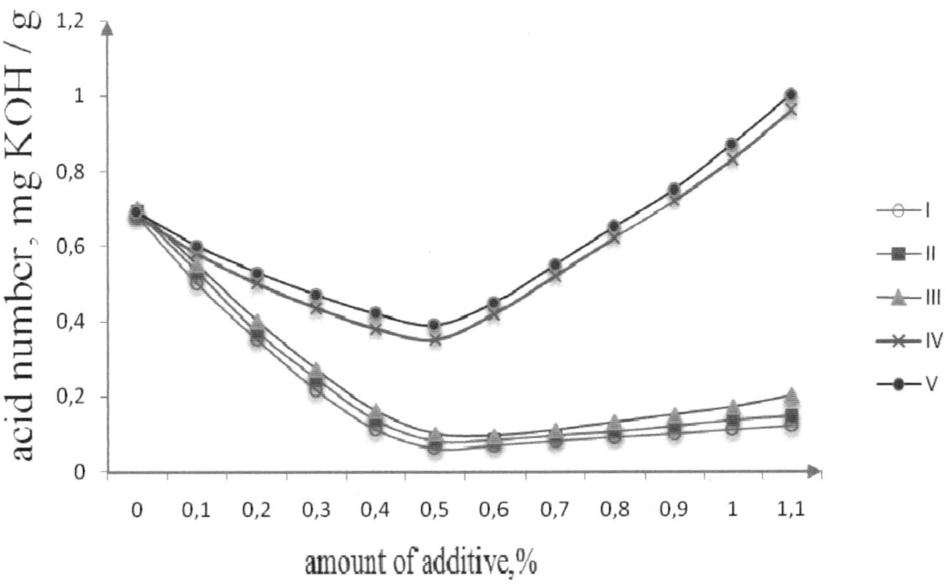

Fig. 2. The effect of the number of additives I-V on the acid number

It can be seen from the data in Fig. 2 that with an increase in the concentration of additives I-III from 0 to 0.5%, the acid number decreases from 0.068% to 0.006%, 0.008% and 0.01%, respectively, for all the investigated additives I-III. The increase in the concentration of additives I-III in the oil fraction from 0.5% to 1.1% is accompanied by an increase in the acid number, respectively, to 0.0105, 0.0146 and 0.02 mg of KOH / g fraction. This fact points to the inexpediency of increasing the concentration of additives I-III in the oil fraction.

The fractions of additives IV and V under identical experimental conditions in accordance with GOST 982-80 show inertness in the process of increasing the thermal stability of the oil fraction in comparison with the additives I-III. The acid number in the presence of additives IV and V in an amount of 0.5% decreases, respectively, to 0.034 and 0.038 mg KOH / g fraction. The course of curves I-III and IV-V is identical, which indicates the same nature of the effect of additives on the oxidation of oil hydrocarbons. The optimal amount of IV-V additives is also 0.5%.

Under the influence of an electric field, increased operating temperatures and structural materials such as metals, cellulose and varnishes, the transformer oil is aging. As a result of the aging of the oil, a precipitate forms and the acid number increases. Precipitation precipitates the aging of cellulose solid insulation and reduces the heat transfer rate, which worsens the insulating properties of the oil. The acid number refers to the important characteristics of the oil, since it indicates the presence of oxidation products of the hydrocarbons that make up the oil, and therefore its oxidation stability.

From the studies carried out it follows that the best antioxidant additives to the oil fraction are fractions I-III, consisting of oil sulphides. The additive fractions IV and V, which are derivatives of thiophene with condensed benzene nucleus, exhibit less antioxidant properties with respect to the oxidation of hydrocarbons in the oil fractions.

The formation of low molecular weight volatile acids during the aging of transformer oil leads to an increase in the oxidative conversion of its hydrocarbons, aging of solid insulation and corrosion of the metal. In this connection, the influence of additives of fractions I-III on the formation of volatile acids was studied. The study was carried out in accordance with GOST 982-80 in the presence of 0.5% of the additives of fractions I, II and III, respectively, to the desulfurized initial fraction of Arlan petroleum.

The results of the study are shown in Table 2.

IOP Conf. Series: Materials Science and Engineering **412** (2018) 012016 doi:10.1088/1757-899X/412/1/012016

Table 2

Effect of additives of sulphur-organic compounds on the formation of water-soluble acids

Oil fraction	Stability in accordance with GOST 982-80 for the content of water-soluble acids	
	Nonvolatile	volatile
Original oil fraction	0,0045	0,0059
Oil fraction + 0,5 % fraction I	0,0030	0,0019
Oil fraction + 0,5 % fraction II	0,0035	0,0022
Oil fraction + 0,5 % fraction III	0,0037	0,0026
Requirements according to GOST 982-80	0,005	0,005

From the data in Table 2 it follows that all the additives of fraction I-III contribute to reducing the amount of water-soluble acids formed.

Conclusions

1. The influence of natural sulphur-organic compounds on the increase of thermal stability of the oil fraction of Arlan petroleum was studied.
2. An optimum number of additives to be studied is found to be 0.5%, at which the thermal stability of the oil fraction increases.
3. It has been established that the investigated additives of fractions I-III, taken in an amount of 0.5%, reduce the amount of water-soluble acids formed, which increases the thermal stability of the oil fraction.
4. It has been established that the fractions of natural sulphur-organic compounds containing sulfide sulfur are the most effective additives to the oil fraction.

References

[1] Lipshtein R A, Shakhnovich M I 1983 *Transformer oil* (Moscow: Energoizdat) pp 41-173
[2] Vanin B V, Lvov Yu N, Lvov M Yu, Neklepaev B N, Antipov K M, Surba A S, Chichinsky M N 2001 *Electric stations* No 9
[3] Kharlampidi Kh E, Gaynullina L R, Tutubalina V P 2016 *Vestnik of the Technological University* T 19 No 6
[4] Vildanov R R, Tutubalina V P 2007 *Izvestiya vuzov. Problemy energetiki* No 1-2

IOP Conf. Series: Materials Science and Engineering **412** (2018) 012017 doi:10.1088/1757-899X/412/1/012017

Influence of structural-group composition of sulphur-organic compounds on water solubility in model mixtures

L R Gainullina[1], V P Tutubalina[1], L S Sabitov [1,2]

[1]Kazan State Power Engineering University, 51 Krasnoselskaya street, Kazan, 420066, Russian Federation

[2] Kazan Federal University, 18 Kremlyovskaya street, Kazan, 420008, Russian Federation

l.sabitov@bk.ru

Abstract. On model mixtures it is studied the solubility of water in transformer oil in the presence of suphur-organic compounds of the general formulas $C_nH_{2n-11}S$, $C_nH_{2n-7}S$, $C_nH_{2n-5}S$, from 0,6 to 3,0% concentrations. It is established that the greatest dissolving ability of water in model mixtures have suphur-organic compounds s of the general formula $C_nH_{2n-11}S$ due to the greatest recurrence they have.

Introduction

At present, the growth of the energy demand in modern electrical equipment, which is caused by the aging of the transformer park in the power system and the continuously increasing energy consumption, has increased. The world has a tendency that is characterized by a lack of production capacity, adequate to the needs of developing energy, which in the near future may lead to a shortage of electrical energy and an increase in the breakdown of electrical equipment.

Power systems make high demands on power transformers, which should provide high reliability in the process of their operation under the influence of high short-circuit currents, short-term and long-term overloads in operation at power facilities and modes associated with continuous voltage switching under load, providing greater reliability of power supply to consumers.

It should be noted that reliability and stability of power supply to industrial and civil facilities depends on the quality of transformer oil, which performs the role of liquid insulation in oil-filled transformers that produce the distribution of electrical energy in the power system. Modern monitoring of the state of liquid insulation (transformer oil) allows to detect incipient defects in electrical equipment that can lead to serious emergencies and expensive repairs [1,2].

The state of transformer oil in electrical equipment allows you to assess the speed of its aging and the performance of the equipment. The appearance of water in the oil helps reduce its electrical strength and the tangent of the dielectric loss angle, which determine the properties of the oil as a dielectric and relate to its diagnostic characteristics. Therefore, the analysis of the quality of the oil in use allows the timely detection of defects in electrical equipment at the stage of their occurrence. Maintenance of oil by its state is carried out directly in power transformers and increases reliability and the period of their operation [1, 2].

Content from this work may be used under the terms of the Creative Commons Attribution 3.0 licence. Any further distribution of this work must maintain attribution to the author(s) and the title of the work, journal citation and DOI.
Published under licence by IOP Publishing Ltd

The increase in the extraction and processing of sulphurous oils is associated with the production of petroleum products with a high content of sulphur-oragnic compounds, which accordingly aroused great interest in studying the effect of their structural group composition on the performance properties of petroleum products, in particular oil fractions, which are the raw materials for the production of transformer oil included in the composition its insulating structure [1, 2].

Experimental part

The purpose of the study was to study the effect of the structural-group composition of sulphur-oragnic compounds on the operational properties of the oil fraction of the Romashkino oil.

We find the complex of studies of the composition and properties of organosulfur compounds to reveal their effect on the performance characteristics of the oil fraction B.P. = 300-400 ° C, isolated from the oil of the Romashkino field. The oil fraction in its composition contains 1.68% of total sulfur, including 0.62% of sulfide sulfur.

The choice of the method for the isolation of sulphur-organic compounds from petroleum fractions is determined by their nature and the boiling point of the distillate [2].

To increase and differentiate the sulphur-organic compounds from the oil fraction of the Romashkino oil, a universal method of adsorption separation was used [2].

Adsorptive separation of sulphur-organic compounds was carried out in an experimental adsorber with a height of 1.2 m and a diameter of 0.1 m. As a adsorbent, broad-leaved fine-pored silica gel of the SHMS grade and aluminum oxide were used, which were located sequentially in the adsorber. To increase the activity of the adsorbents, the latter were calcined for 3 hours in a muffle furnace at a temperature of 150 ° C and 350 ° C, respectively. Adsorptive separation of the oil fraction was carried out at a temperature of 20 ° C. The quantitative ratio of silica gel to CMC: the oil fraction was 1: 2.5. The ratio of alumina: the oil fraction was also 1: 2.5. The eluents during the adsorptive separation of sulphur-organic compounds were petroleum ether B.P.= 40-70 ° C and benzene. Separation of sulphur-organic compounds was carried out according to the procedure of [2].

Potentiometric titration in the oil fraction of Romashkino oil elemental, hydrogen sulphide and mercaptan sulfur were not detected [3]. When the oil fraction was adsorbed through the combined adsorbent, a paraffin-naphthenic fraction was recovered in an amount of 46.83% of the general formula C_nH_{2n}(n=22), which freely flowed out of the adsorber.

The physicochemical characteristics of the adsorbed fractions of the adsorbent are shown in the table.

In accordance with the experimental data of the table, with increasing depth of the selection of the fraction, the concentration of sulfur in the latter increases, the hydrogen content in the molecule decreases with a simultaneous increase in the C: H ratio, which indicates an increase in the unsaturation of the selected fractions. The obtained empirical formulas of the table confirm this fact.

The table also shows the presence of a large number of aromatic fractions, which account for more than half of the initial fraction (53.17%). From the table it follows that as the depth of adsorbate collection increases, their density and refractive index increase.

Table 1
Physicochemical characteristics of adsorbed fractions

Fraction	Output, mass.%	Molecular mass	ρ_4^{20}	n_D^{20}	Element composition, %				C/H	Formula
					C	H	S	S_{sulf},mass. %		
Oil	100	304	0,91	1,5020	81,51	12,46	6,03	2,2	6,5	

Paraffin-naphthenic	46,83	308	0,83	1,4630	85,7	14,3	*No*	*No*	5,99	C_nH_{2n} (n=22)
Adsorbate 1	20,5	321	0,85	1,4876	78,5	12,1	9,4	3,4	6,49	$C_nH_{2n-5}S$(n=21)
Adsorbate 2	14,1	305	0,92	1,5230	78,7	10,82	10,48	3,0	7,27	$C_nH_{2n-7}S$(n=20)
Adsorbate 3	18,57	274	0,97	1,5570	78,8	9,49	11,71	2,3	8,3	$C_nH_{2n-11}S$(n=18)

Absorption spectra obtained in the ultraviolet region using the SF-4 spectrophotometer allowed us to approximately determine the distribution of aromatic hydrocarbons characterized by a different number of rings in fractions. Bicyclic compounds are found in the wavelength region 322-326 m. Tricyclic aromatic compounds are found at wavelengths of 377-385 m / s. The amount of tricyclic hydrocarbons in the fraction is 0.14%, which provides the oil fraction of Romashkino oil with increased resistance against oxidative polymerization of hydrocarbon fractions and consolidations, the flow of which promotes the intensive formation of a solid precipitate, a product of the oxidation of hydrocarbons and sulphur-organic compounds.

Sulfide sulfur in the adsorbed fractions was determined using the potentiometric titration method [3]. The content of total sulfur was determined in accordance with GOST 19121-73.

It is known [1, 2, 4] that the electrical strength, conductivity and dielectric losses in oil, characterizing it as a dielectric, depend on the concentration of water in the oil. In the process of operation of transformer oils in electrical equipment under the influence of an electric field of high tension and elevated operating temperatures, processes of thermochemical destruction of oil hydrocarbon molecules occur with the advent of active radicals, leading to the formation of condensation and water products.

The water present in the oil fraction in small amounts forms a true solution in it, which does not affect the value of the tangent of the dielectric loss angle. At the same time, when the water content in the oil fraction increases before the formation of the emulsion state, the numerical values of the tangent of the dielectric loss angle increase sharply.

The experimental data are shown in Fig. 1. The experiments were carried out in the temperature range 20-80 ° C. The tangent of the dielectric loss angle was determined in accordance with GOST 6581-75.

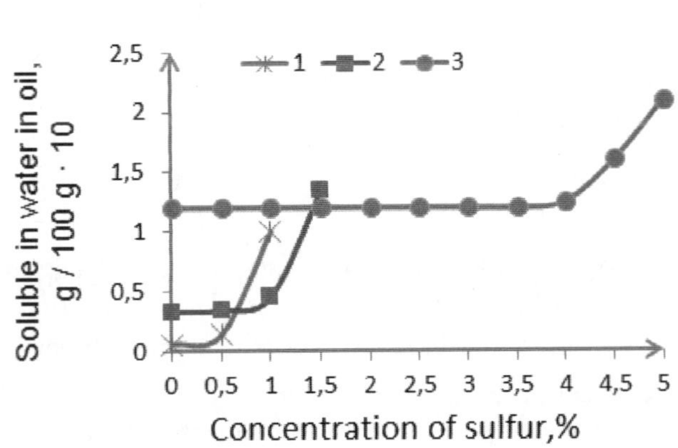

Figure 1. Effect of water concentration g / 100 g of oil fraction on the tangent of the angle of dielectric losses

1 – temperature 20 °C
2 – temperature 50 °C
3 – temperature 80 °C

According to the data in Figure 1, at temperatures of 20 ° C and 50 ° C, the emulsion state in the fraction passes at concentrations of 8.0 times and 2.8 times lower, respectively, than at 80 ° C.

The next series of experiments was carried out to study the effect of the structural-group composition of sulphur-organic compounds and their concentration on water solubility in the oil fraction of the Romashkino oil. The studies were carried out using model mixtures. Model mixtures were prepared using a hydrocracking transformer oil as containing a minimum amount of organosulfur compounds-0.01% based on total sulfur oil and sulphur-organic compounds represented by common empirical formulas $C_nH_{2n-11}S$, $C_nH_{2n-7}S$, $C_nH_{2n-5}S$, varying their mixture concentration from 0.5% to 3.0% in 0.5% increments. The characteristics of sulphur-organic compounds are presented in the table 1.

The effect of the above sulfur organic compounds on water solubility in model mixtures was studied at a temperature of 100°C in an electric field of 49kV/cm for 44 hours in an experimental setup described in the literature [2]. The concentration of water was determined in accordance with GOST 7822-75.

The experimental data obtained are shown in Fig. 2.

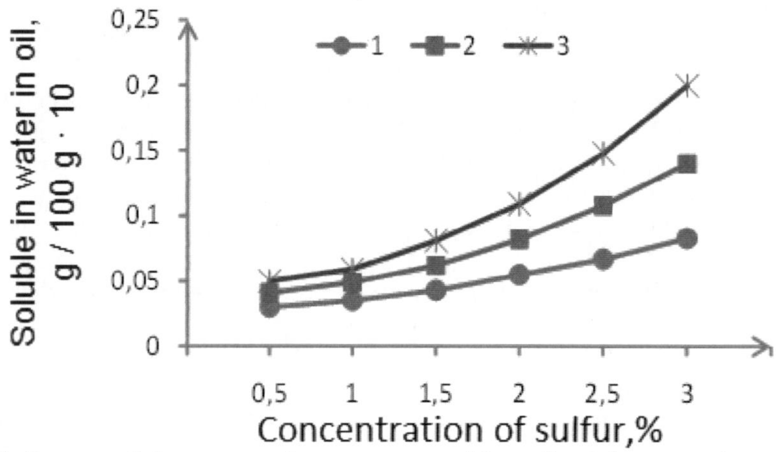

Figure 2. The influence of the structural-group composition of sulphur-organic compounds on the solubility of water in oil

1 – sulphur-organic compounds $C_nH_{2n-5}S$
2 – sulphur-organic compounds $C_nH_{2n-7}S$
3 – sulphur-organic compounds $C_nH_{2n-11}S$

Results and Discussion

Analysis of the data shown in Fig. 2, shows that the greatest dissolving capacity of water in the model mixture is possessed by sulphur-organic compounds of the general formula $C_nH_{2n-11}S$, since these compounds are characterized by the greatest cyclicity. Sulphur-organic compounds of general formula $C_nH_{2n-5}S$ as less cyclic are characterized by the lowest water solubility in hydrocarbons of the model mixture. A model mixture of hydrocarbons containing in their composition sulfur-organic compounds of the general formula $C_nH_{2n-7}S$ occupies an intermediate position between these mixtures.

An increase in the concentration of sulphur-organic compounds, regardless of their structural-group composition, contributes to an increase in water solubility in the model mixture. For example, when using sulphur-organic compounds of the general formula $C_nH_{2n-11}S$ taken at a concentration of 0.6%, 4.99% of water dissolves in the model mixture. The increase in the concentration of sulphur-organic compounds to 1%, 2% and 3% contributes to the increase in water solubility in the mixture, respectively, by 1.20 times, 2.0 times and 4.0 times. Similar data were obtained for mixtures containing sulphur-organic compounds of general formulas $C_nH_{2n-5}S$ and $C_nH_{2n-7}S$ (Fig. 1).

The decrease in the cyclicity of the molecule of sulphur-organic compounds corresponding to the general formula $C_nH_{2n-5}S$, as compared with sulphur-organic compounds of the general formula $C_nH_{2n-7}S$ and $C_nH_{2n-11}S$, leads to a decrease in water solubility in the model mixture. At a concentration in the model mixtures of 0.6%, 1.0%, 2.0% and 3.0% of the compound of the general formula $C_nH_{2n-11}S$, the water solubility in the model mixture is increased by 1.65 times, 1.95 times, 2.25 times and 2.4 times as compared to a model mixture containing sulphur-organic compounds at the same concentration of the general formula $C_nH_{2n-5}S$.

The sulphur-organic compounds of the general formula $C_nH_{2n-7}S$ occupy an intermediate position between sulphur-organic compounds of the formulas $C_nH_{2n-5}S$ and $C_nH_{2n-11}S$. These compounds are dissolved at concentrations of total sulfur equal to 0.6%, 1.0%, 2.0% and 3.0% respectively, more water in 1.30 times, 1.46 times, 1.58 times and 1.74 times than the sulphur-organic compounds of the empirical formula $C_nH_{2n-5}S$. In this regard, sulphur-organic compounds on the effectiveness of water dissolution in hydrocarbons of model mixtures can be arranged in the following order:

$C_nH_{2n-11}S > C_nH_{2n-7}S > C_nH_{2n-5}S$.

Conclusions.

1. For the preparation of model mixtures using sulphur-organic compounds of different structural group composition, the latter were obtained using the adsorption separation method.
2. The influence of the structural-group composition of sulphur-organic compounds on water solubility in model mixtures has been studied. The influence of water concentration and temperature on the magnitude of the dielectric loss tangent has been studied.
3. It has been established that the greatest dissolving capacity of water in a model mixture is possessed by sulphur-organic compounds of a general empirical formula $C_nH_{2n-11}S$, since these compounds are characterized by the greatest cyclicity. The lowest water solubility in hydrocarbons of a model mixture is characterized by sulphur-organic compounds of the general formula $C_nH_{2n-5}S$ as less cyclic.
4. Increasing the concentration of sulphur-organic compounds in the model mixture increases the solubility of water in the latter, regardless of their structural-group composition.

References

[1] Lipshtein R A, Shakhnovich M I 1983 *Transformer oil* (M.: Energoizdat) 296 p
[2] Vildanov R R, Tutubalina V P 2012 *Concentration of air in transformer oil* (LAMBERT Academic Publishing) 129 p
[3] Galpern G D, Girina G P, Lukyanitsa V G 1960 *Methods for analyzing organic compounds of petroleum* (Moscow: AN SSSR) 225 p
[4] Gainullina L R, Tutubalina V P, Kharlampidi H E 2016 *Bulletin of the Technological University* T19 No 14

Production technology of heat-resistant high-strength syntactic carbon foams for operation in extreme conditions

E R Galimov[1], E E Tukbaev[1], V L Fedyaev[1,5], N Ya Galimova[1], V M Samoylov[2], M A Orlov[3], A S Borodulin[3] and I R Mavleev[4]

[1]Kazan national research technical university named after A N Tupolev, K. Marx St. 10, Kazan, 420111 Tatarstan, Russia
[2]JSC NIIgrafit, 2 Elektrodnaya str., Moscow, 111524, Russia
[3]Bauman Moscow state technical university, Baumanskaya 2-ya str., building 1, house 5, Moscow, 105005 Russia
[4]Kazan Federal University, Syuyumbike street 10a, Naberezhnye Chelny, 423812 Tatarstan, Russia
[5]IME – Subdivision of FIC KazanSC of RAS, Lobachevsky str. 2/3, Kazan, 420111 Tatarstan, Russia

engelgalimov@yandex.ru

Abstract. Production technology of products based on the syntactic carbon foams designed to operate in extreme conditions is proposed. The components and regime parameters for the production of foams of an open-pore cellular structure with specified thermophysical and strength characteristics for manufacturing the large-sized products of complex shape are determined.

1. Introduction

Syntactic foams are composite materials consisting of a polymer, ceramic or metal matrix and hollow particles in the form of microspheres [1]. The term "syntactic" means the structure regularity when an elementary cell can be isolated the material structure can be described by its repetition [2-3].

Among the syntactic foam materials, carbon foams which have increased thermophysical and strength properties have a special place in terms of effectiveness and prospects of use [4].

The purpose of this work is to develop the production technology of syntactic carbon foams with an open-pore cellular structure for manufacturing the articles including large-sized ones with complex geometry.

2. Body text

Phenol-formaldehyde resin (PFR), solvent (acetone), and hollow glass microspheres were used as initial components. The preparation of the compositions was carried out by mixing the PFR, the solvent and the finished hollow glass microspheres at certain ratios. Manufacturing of the workpiece was carried out in shape-generating molding tools under pressure and elevated temperature with the simultaneous action of ultrasonic vibrations with a frequency of 15-18 kHz. Then, the workpiece was

Content from this work may be used under the terms of the Creative Commons Attribution 3.0 licence. Any further distribution of this work must maintain attribution to the author(s) and the title of the work, journal citation and DOI.
Published under licence by IOP Publishing Ltd

IOP Conf. Series: Materials Science and Engineering **412** (2018) 012018 doi:10.1088/1757-899X/412/1/012018

carbonized in an electrovacuum furnace with a removal of the gas phase, and the article was exposed to pyro-compaction.

Figure 1 shows the diagram of the processing line in which the articles are produced. The ground granules of PFR 70-80 μm in size are loaded in the supply tank 1 of the auger-type dispenser machine 2 for feeding in the mixing unit 3. The solvent from the container 4 is also supplied by the membrane dispensing pump 5 to the mixing unit 3 in which the binder material is prepared. Mixing of the binder with microspheres is carried out by feeding them from the supply tank 6 equipped with a pneumatic vibrator 7 through the heating device 8 into the mixer 9. The prepared binder material with viscosity required for mixing with the filler is supplied by a dispensing pump 10, and the mixing unit 9 has a chamber 11 with an ultrasonic generator built into it at the output flow part.

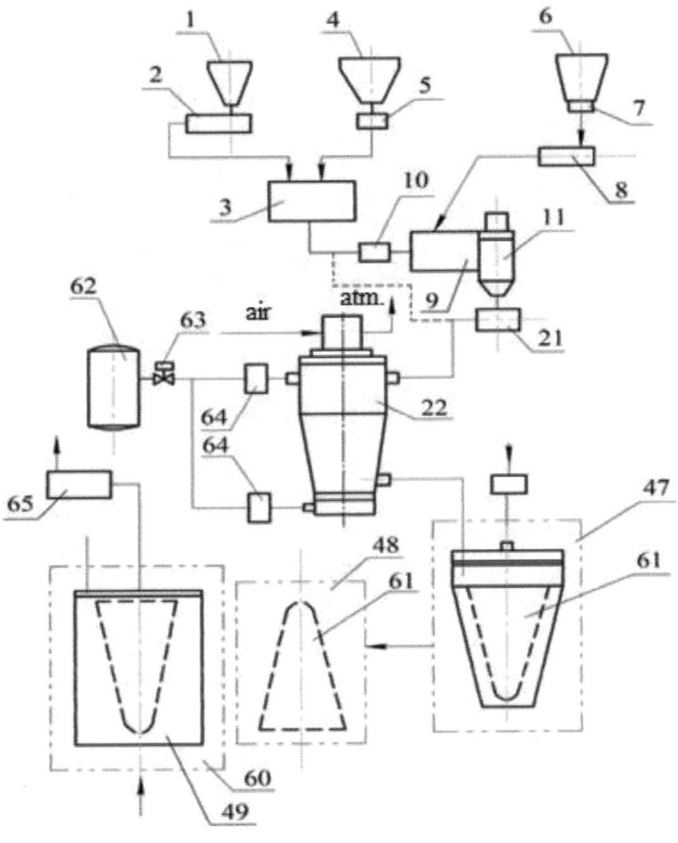

Figure 1. Diagram of the processing line

The mixing unit 9 (figure 2) of the binder material with the filler is a cylindrical body 12 inside which the core 13 with the Archimedes helical line 14is placed motionlessly and without tip clearance. The waveguide 15 of the ultrasonic generator 16 is located in the chamber 11 along the outflow axis of the mixed components therefrom. The binder material through the inlet opening 17 through the conical gap 18 passes the helical channel 19 into the chamber 11 acquiring a rotational movement. In the inlet part of the mixing unit 9 at the angle parallel to the inclination angle of helical line 14, the filler is supplied through the channel 20 in the direction opposite to the rotation of the binder composition. The filler is crushed and turbulized in the binder because of the different moving directions of the flows of binder and filler in the binder volume. The turbulization effect of the filler in the binder is enhanced by the additional influence of the ultrasonic vibrations primarily on the filler particles which have a density different from the density of the binder composition.

IOP Conf. Series: Materials Science and Engineering **412** (2018) 012018 doi:10.1088/1757-899X/412/1/012018

Due to the proportioning by volume of the binder fed to the special mixing unit of the components 9 the material viscosity up to the chamber 11 with the waveguide 5 is maintained within 10-15 Pa·s and after preliminary evacuation of the volatiles from the chamber 11 under the influence of the ultrasonic vibrations the viscosity increases to 18-20 Pa·s.

The resulting composition using the airless spraying device 21 is fed to the drying unit 22 (figure 3) of the shells of the microspheres encapsulated in the binder. The drying unit 22 is a cylindrical body 23 with a conical bottom in the lower part of which there is a pseudofluidization chamber 24 with a filter baffle 25 above which there is a suction pipe of the ejector device 26 for supplying the microspheres with the dried shell into the tooling of the product forming.

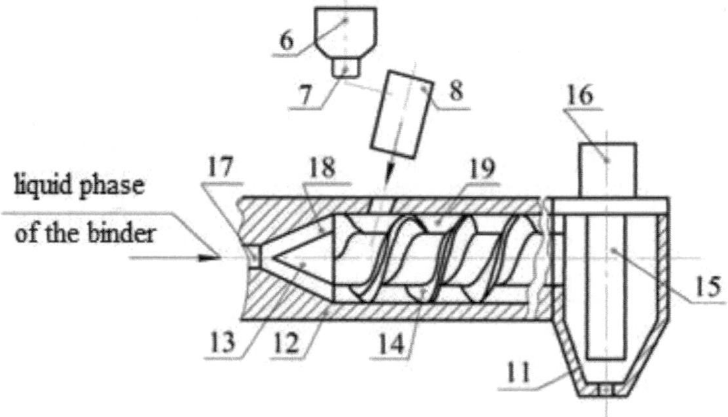

Figure 2. Mixing unit

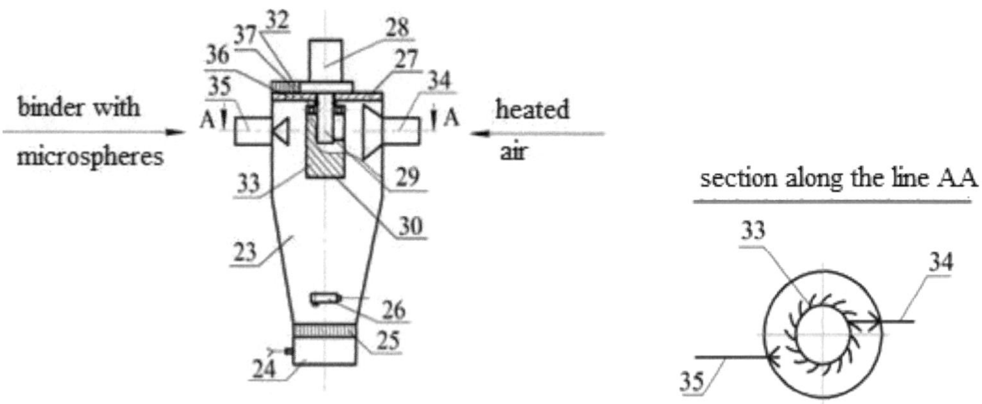

Figure 3. Drying unit

The pneumatic engine 28 is mounted in the upper part of the cylindrical body 23 closed by the lid 27, on its shaft 29 the cylinder shell 30 with the pneumatic vibrator 31 is fixed, and the compressed air is supplied to it through a central channel 32 on the shaft of the pneumatic engine 28. The blades 33 are positioned at an angle to the rotation axis on the outer surface of the shell 30. Under the lid 27 of the cylindrical body 23, the crevice nozzle 34 is positioned for supplying the heated air to the cylindrical shell surface and from the side opposite to it the nozzle 35 of the airless spraying device 21. The excess air from the drying unit 22 is removed through the channel 36 with the filter 37. A jet of viscous mass, sprayed by the airless spraying device, breaking into small particles falls on the surface of the rotating cylindrical shell with blades, meets with a stream of heated air, and falls down

IOP Conf. Series: Materials Science and Engineering **412** (2018) 012018 doi:10.1088/1757-899X/412/1/012018

along the blades. The microspheres that have fallen from the surface meet with the stream of heated air from the pseudofluidization chamber which is constantly in a bubbling state forming a dry shell on them. The drying unit operates in a cyclic mode for a definite volume of encapsulated microspheres. The air is supplied from the air heater 62 through the valve 63 and is controlled by the chokes 64.

Encapsulated in a casing made of binder material and dried fillers are transported by the ejector device 26 to the molding unit (figure 4) which is a split housing 38 with its inner cavity having an external configuration of the product being manufactured. The punch 39 is installed in the split housing 38, on the outer side it has a flexible silicone casing 40 which is tightened by a lid 41 to a split housing 38. The configuration of the punch 39 with the casing 40 is equidistant to the configuration of the inner surface of the split housing 39 with preservation of the channel 42 the size of which corresponds to the design thickness of the article 61. The grooves 43 connected with the central channel 44 are made from the outer surface of the punch 39, and the lid 41 is provided with openings 45 for air-filling of the encapsulated in the casing made of binder material fillers into the working volume 42 and the outlet channel 46 with filter cell (not shown) for the removal of compressed air at pneumatic transport.

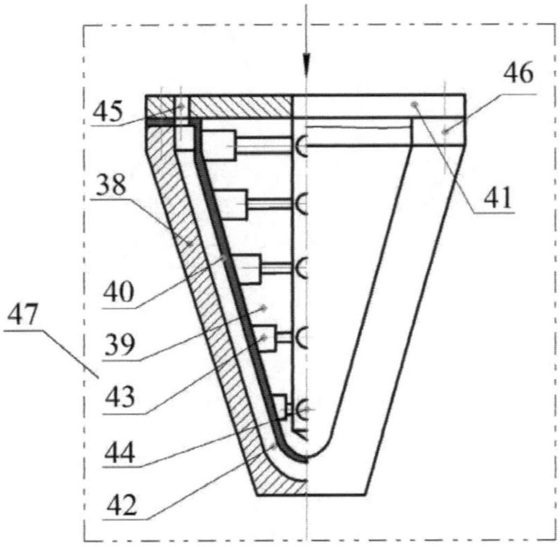

Figure 4. Molding unit

The preshaping of the workpiece is performed in the muffle furnace 47 at atmospheric pressure with a premolding of the silicone casing backfill at a compressed air pressure from 0.15 to 1.5 MPa through the central channel 44 and at temperatures up to 200-250 °C. After cooling of the formed workpiece, it is carbonized in the free state in an electrovacuum industrial furnace 48 at a vacuuming pressure of 10-80 mm Hg and temperature up to 900 °C with a rate of temperature elevation of 5 °C/min.

To perform the pyro-compaction of the outer working surface, the carbonized article 38 is loaded into the special tooling 49 (figure 5) its inner cavity is identical to the outer configuration of the pre-carbonized article.

The special tooling 49 is a set of rings 50 hermetically connected and tightened together by a stud. Each ring has an internal bore 51 and an opening 52 mated with the outer contour of the article. In this case, the series of openings 54 for supplying the gas phase are made on each ring's walls 53 sealing the product contour, and openings 54 are located in the horizontal plane and shifted by 180° relative to each other starting from the gas phase supply channel 55. The general sealing of the set of tooling rings is carried out by tightening them with top 56 and bottom 58 lids through the seal 57. For pyro-compacting, the tooling with the product 61 is loaded in the electrovacuum furnace 59. The pyro-compacting process of the pre-carbonized polymeric matrix is carried out by a known method of supplying the natural methane gas on the heated surface of the article from the outer side by infiltrating the gas in the presence of a pressure gradient. This method of depositing the pyrolytic carbon on the outer surface of the product with an open-pore cellular structure at a certain depth allows providing high-temperature properties of the outer layer while maintaining the strength and heat-shielding characteristics of the entire product. In this variant, the reactant gas passes through the porous structure in the radial direction to the internal cavity of the article. Vacuuming [5-12] of the internal cavity through the channel 59 is carried out by the vacuum station 65.

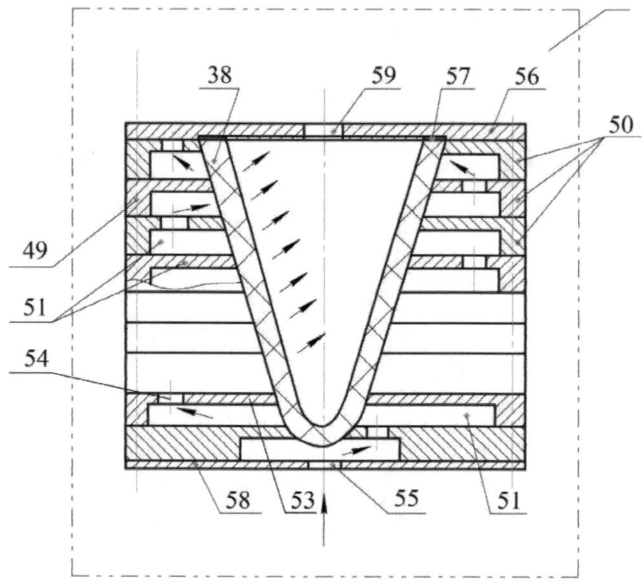

Figure 5. Tooling for product forming

3. Conclusions

Thus, the proposed technology ensures production of an article with a two-layer structure where the outer working layer compacted by pyrolytic carbon provides heat-shielding properties at operation up to temperatures of 3000 °C while maintaining the strength properties of the internal layers obtained by the carbonization of the carbon foam.

The work was carried out with the financial support of the Ministry of Education and Science of the Russian Federation within the framework of the Federal Target Program "Research and development in priority areas of development of the scientific and technological complex of Russia for 2014–2020" agreement No. 14.583.21.0057 of July 28, 2016 (Project ID RFMEF158316X0057).

References

[1] Tikhomirov V K 1975 *Foams. Theory and practice of their production and destruction* (Moscow: Khimiya Publ) pp 11-15

[2] Salosina M O 2016 *The thermal conditions of the solar probe taking into account the impact of high-speed dust particles* (Moscow: Proceedings of MAI) p 86

[3] Internet source: Firefly Energy Co. Available at: http://fireflyenergy.com/. (accessed 04.02.2018)

[4] Chen Y, Chen B, Shi X, Xu H, Shang W, Yuan Y and L Xiao 2008 *Preparation and electrochemical properties of pitch-based carbon foam as current collectors for lead acid batteries* Electrochimica Acta vol 53 pp 2245-49

[5] Kashapov N F, Luchkin G S and A G Luchkin 2010 *Vacuum technologies for the application of functional coatings* Bulletin of Kazan State Technological University number 2 pp 340-45

[6] Galimov E R, Shveyov A I, Shveyov I A, Lysanov D M and N Ya 2016 *Development of combined technologies and multifunctional equipment for application of polymer powder coatings* International Journal of Applied Engineering Research vol 11 No 3 pp 1583-89

[7] Samoylov V M, Danilov E A, Galimov E R, Fedyaev V L, Galimova N Ya and M A Orlov 2017 *Production of thermally conductive carbon foams and their application in automobile transport* IOP Conf. Series: Materials Science and Engineering vol 240 **012062**

[8] Galimov E R, Fedyaev V L, Galimova N Ya, Gimranov I R and M M Marat 2017 *Formation of polymeric powder coatings* IOP Conf. Series: Materials Science and Engineering vol 240 **012014**

[9] Galimov E R, Fedyaev V L, Galimova N Ya, Gimranov I R and A R Siraev 2017 *Dynamics of coalescence and spreading of liquid polymeric particles during coating formation* Journal of Physics: Conf. Series vol 789 **012006**

[10] Galimov E R, Fedyaev V L, Galimova N Ya, Takhaviev M S and A R Siraev 2017 *Mathematical modeling of processes occurring during deposition of sprayed particles of polymeric powder* Journal of Physics : Conf. Series vol 789 **012007**

[11] Garipov R M, Serova V N, Efremova A A and Zh Yu Gerkina 2015 *Production and properties of multilayer shrinkable polymer film* Bulletin of Kazan State Technological University No 3 pp 174-77

[12] Zagidullin A I, Garipov R M, Khasanov A I and A A Efremova 2014 *Effect of structure multilayer film on the barrier properties of a polymer film* Bulletin of Kazan State Technological University No 21 pp 151-53

IOP Conf. Series: Materials Science and Engineering **412** (2018) 012019 doi:10.1088/1757-899X/412/1/012019

Production and investigation of physical-mechanical properties of syntactic carbon foams

E R Galimov[1], V L Fedyaev[1,4], N Ya Galimova[1], V M Samoylov[2], E A Danilov [2], E I Timoshchuk [2], D V Ponomareva[2], A V Nakhodnova [2] and M A Orlov[3]

[1]Kazan national research technical university named after A N Tupolev, K. Marx St. 10, Kazan, 420111 Tatarstan, Russia
[2]JSC NIIgrafit, 2 Elektrodnaya str., Moscow, 111524, Russia
[3]Bauman Moscow state technical university, Baumanskaya 2-ya str., building 1, house 5, Moscow, 105005 Russia
[4]IME - Subdivision of FIC KazanSC of RAS, Lobachevsky str. 2/3, Kazan, 420111 Tatarstan, Russia
engelgalimov@yandex.ru

Abstract. The production technology of syntactic carbon foams based on the phenol-formaldehyde binders and microspheres was developed. The influence of the composition and regime technological parameters on the change of the physical-mechanical properties of foams was investigated.

1. Introduction

Syntactic carbon foam (SCF) is a perspective material of ordered structure possessing a number of unique properties: high specific physical-mechanical characteristics, adjustable coefficients of thermal conductivity, electrical conductivity, high heat and thermal resistance, resistance to hostile environment, adjustable specific surface and etc. [1-4]. The above operational properties of syntactic carbon foams make these materials extremely interesting for application in many fields of engineering.

The paper considers the production technology of SCF based on the phenol-formaldehyde resin (PFR) filled with phenolic microspheres. This technology includes mixing of the initial components, pressing the mixture, heat treatment of the pressed workpieces, their pyro-compaction and subsequent mechanical processing for making test specimens. Investigations of the physical-mechanical properties of the SCF were carried out on the prismatic samples with dimensions 10×10×20 mm. Microscopic studies were performed by the scanning electron microscopy method on fresh cleavages of destroyed samples using a Hitachi TM3000 (figure 1).

2. Body text

The hollow microspheres based on the PFR with a wall thickness of 1-2 μm and characteristic defects in the form of cavities were used as filler. According to the laser diffraction data (Fritsch Analysette 22 Compact), the average outer diameter of the microspheres was 34 μm.

The novolak type PFR brand SF-012A (СФ-012А) was the binding agent. The filler content in the SCF was varied between 33-85 wt. %. The mixing was carried out by a liquid-phase method in a three-necked glass reactor with a reflux condenser. The filler and the binder material were placed in a reactor a solvent (acetone) was added in an amount of 10 ml per 1 g of the binder. The mixture was heated with stirring to boiling, and then it was boiled under reflux for 90 minutes. From this

Content from this work may be used under the terms of the Creative Commons Attribution 3.0 licence. Any further distribution of this work must maintain attribution to the author(s) and the title of the work, journal citation and DOI.
Published under licence by IOP Publishing Ltd

IOP Conf. Series: Materials Science and Engineering **412** (2018) 012019 doi:10.1088/1757-899X/412/1/012019

composition, the solvent was removed by distillation under vacuum. The reaction mixture was transferred in portions of 200-400 ml to a 500-1000 ml round-bottomed flask and evaporated on a rotary evaporator IR-2M. Distillation conditions: residual pressure 36-39 kPa, temperature 40-45 °C.

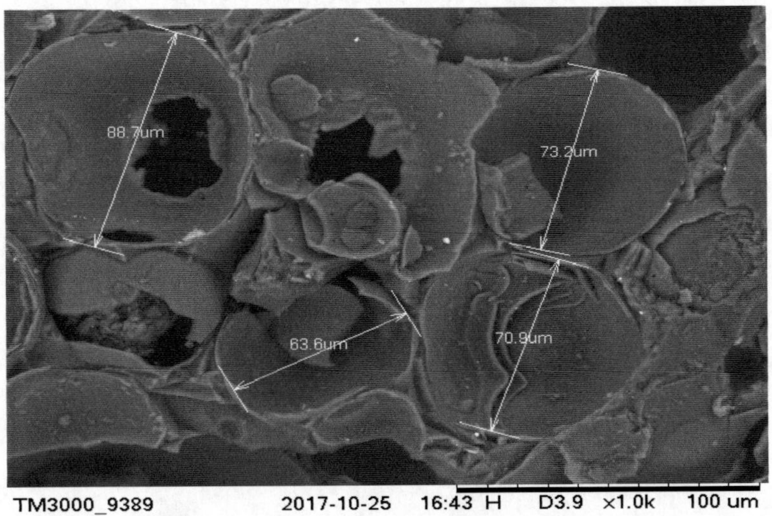

Figure 1. Characteristic linear dimensions of phenolic microspheres in SCF measured by scanning electron microscopy

The resulting bulk product was dried under vacuum using a rotary evaporator (residual pressure 1.3 kPa, temperature 110-120 °C) to obtain a granulated molding powder with a particle size of the order of 10 mm. The molding powder was crushed to a size of <500 μm and pressed 2-5 days after. Molding of workpieces was carried out using a hydraulic press PG-10 at room temperature in a mold. Figure 2 shows the dependence of the density of the pressed workpieces on the compacting pressure.

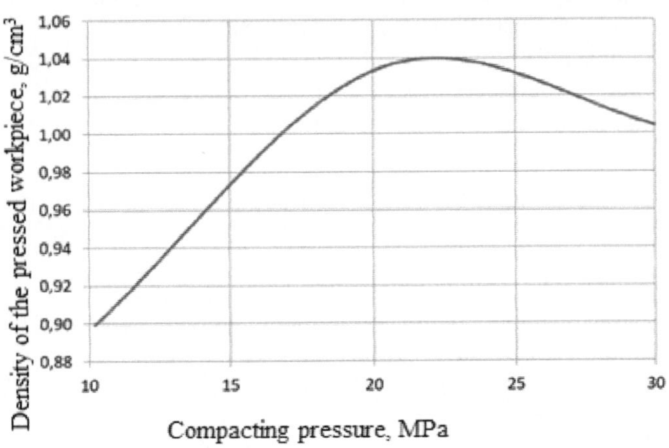

Figure 2. Dependence of the density of the pressed workpieces on the compacting pressure (mixture 50 wt. % PFR)

It can be seen that the density maximum of the "green" workpieces falls on the compacting pressure of about 21 MPa. Cracking of the samples is observed at a further increase of the compacting pressure. With this consideration in mind, pressing of samples was carried out at a pressure of 20.5-21.0 MPa, the rate of pressure increase and release of about 10 MPa/min and holding time at an end pressure of 3 min. Further, the workpieces were cured at a temperature of 150 °C with a holding time of 180 minutes in a vacuum oven. Subsequent carbonization was carried out in a laboratory muffle furnace the workpieces of "green" foams were wrapped in paper and placed in a steel container with graphite filling up. The rate of temperature elevation was 2.5 deg/min, the holding time at an end temperature 900 °C was 180 minutes. Carbonized workpieces were exposed to pyro-compaction at a temperature of 1050 °C and a methane pressure 1.1-1.4 kPa in an industrial furnace in several stages.

Figure 3 shows the diagram of the density change of the SCF workpieces during the processing. These data are typical for high-porosity samples of carbon materials: the workpiece has a high apparent density (about 1 g/cm^3) after pressing, during the curing process the workpiece expands and a partial loss of binder mass occurs this is accompanied by a drop of apparent density. At carbonization, the loss of mass because of the release of low-molecular volatile products at the thermal destruction of the binder dominates the shrinkage phenomena. Then, in the pyro-compaction process, a slow increase of mass occurs due to the deposited pyrolytic carbon.

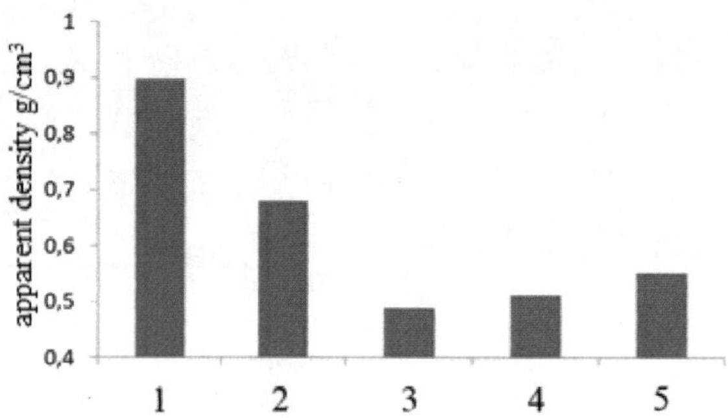

Figure 3. Change of the apparent density of the initial samples during processing: 1 – pressing, 2 – heat treatment at 150 °C, 3 – carbonization at 900 °C, 4 – pyro-compaction at 1050 °C for 20 hours, 5 – pyro-compaction at 1050 °C for 44 hours

Figure 4 illustrates the dependence of the samples apparent density on the PFR content in the initial molding powder after firing and pyro-compaction. It can be seen that an increase of the PFR content leads to an increase of the carbonized material density and this dependence is linear for large changes of PFR concentration. It follows that SCF density can be regulated in a wide range.

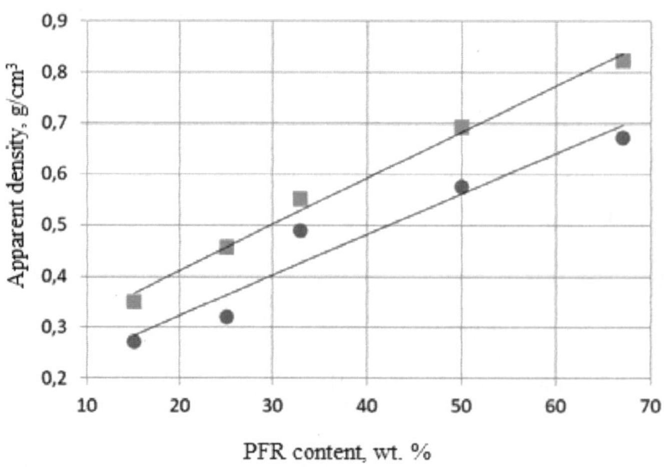

Figure 4. Dependence of the SCF apparent density based on the phenolic microspheres on the PFR content in the initial mixture for carbonized (●) and pyro-compacted (■) materials

Figure 5 shows, as an example, the change Dependencies of the SCF physical-mechanical properties on PFR content in the initial mixture.

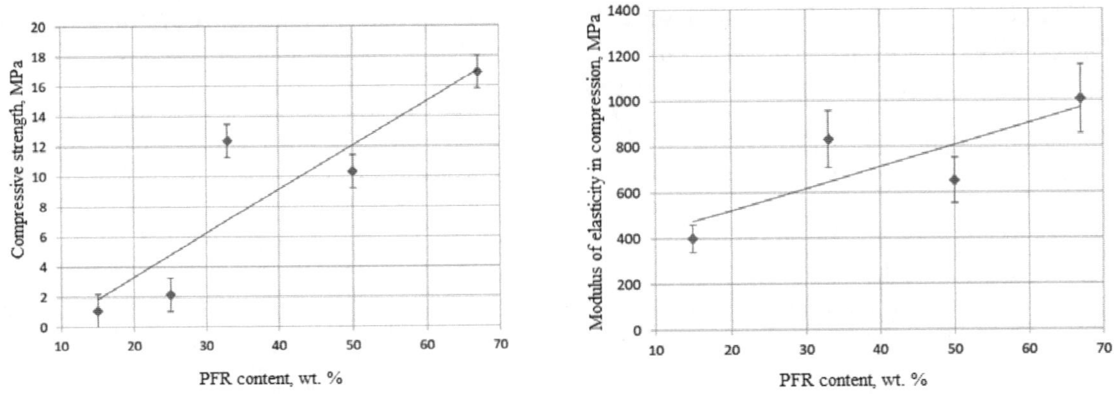

a) b)

Figure 5. Dependencies of compressive strength (a) and modulus of elasticity in compression (b) of SCF samples on PFR content in the initial mixture

3. Conclusions

It can be seen that compressive strength of SCF containing more than 33 wt. % of PFR (in the initial mixture) stably stands at a level above 10 MPa, the modulus of elasticity – higher than 500 MPa approximately linearly increasing with an increase of PFR content in the sample. The achieved level of properties allows us to speak about the possibility of using the received SCF as a structural material.

The work was carried out with the financial support of the Ministry of Education and Science of the Russian Federation within the framework of the Federal Target Program "Research and development in priority areas of development of the scientific and technological complex of Russia for 2014–2020" agreement No. 14.583.21.0057 of July 28, 2016 (Project ID RFMEF158316X0057)

References

[1] Tikhomirov V K 1975 *Foams. Theory and practice of their production and destruction* (Moscow: Khimiya) pp 11-15

[2] Shutov F A 1986 *Syntactic polymer foams* Advances in Polymer Science pp 63–123

[3] Inagaki M, Qiu J and Q Guo 2015 *Review Carbon foam: Preparation and application* Carbon Vol 87 pp128-152

[4] Anghelescu M S 2009 Integrated Engineering Thermal and Mechanical Analysis of Carbon Foam *Ph.D.* 122 p

[5] Kashapov N F, Luchkin G S and A G Luchkin 2010 *Vacuum technologies for the application of functional coatings* "Vestnik" of Kazan State Technological University No 2 pp 340-45

[6] Galimov E R, Shveyov A I, Shveyov I A, Lysanov D M and N Ya 2016 *Development of combined technologies and multifunctional equipment for application of polymer powder coatings* International Journal of Applied Engineering Research vol 11 No 3 pp 1583-89

[7] Samoylov V M, Danilov E A, Galimov E R, Fedyaev V L, Galimova N Ya and M.A. Orlov 2017 *Production of thermally conductive carbon foams and their application in automobile* transport IOP Conf. Series: Materials Science and Engineering vol 240 **012062**

[8] Galimov E R, Fedyaev V L, Galimova N Ya, Gimranov I R and M M Marat 2017 *Formation of polymeric powder coatings* IOP Conf. Series: Materials Science and Engineering vol 240 **012014**

[9] Galimov E R, Fedyaev V L, Galimova N Ya, Gimranov I R and A R 2017 *Dynamics of coalescence and spreading of liquid polymeric particles during coating formation* Journal of Physics: Conf. Series vol 789 **012006**

IOP Conf. Series: Materials Science and Engineering **412** (2018) 012019 doi:10.1088/1757-899X/412/1/012019

[10] Galimov E R., Fedyaev V L, Galimova N Ya, Takhaviev M S and Siraev A R 2017 *Mathematical modeling of processes occurring during deposition of sprayed particles of polymeric powder* Journal of Physics : Conf. Series vol 789 **012007**

Application of multi-agent system for control of parameters of precision stamping process of bevel gears

K N Gavarieva[1], L A Simonova[1], D L Pankratov[1], V G Shibakov[1] and R V Gavariev[2]

[1]Kazan Federal University, Naberezhnye Chelny Institute, 423812, Russia, Naberezhnye Chelny, Prospekt Syuyumbike 10A,
[2]Kazan national research technical university named after A.N. Tupolev-KAI,
Gavarievakn@mail.ru

Abstract. The article deals with the possibility of using an intelligent control system for the parameters of the technological process of precision stamping and mechanisms of crank hot-stamping presses.

1. Introduction

Modern metal-saving technologies are aimed at increasing the utilization factor of the material and obtaining high-quality products. Thus, the production of bevel gears (BG) corresponding to the 5-6 accuracy class is possible only by precision stamping. The products obtained this way are, by their shape and dimensions, extremely close to the finished part and have minimal allowances, or do not need further mechanical treatment at all [1].

To ensure high accuracy of parts manufacturing, it is necessary to eliminate as much as possible the negative impact of random errors affecting the technological process, which is achieved by automation [2].

As it is known, the most appropriate way to automate the process of precision stamping of bevel gears (BG) is to create and implement a multi-agent intelligent control system for process parameters (billet temperature) and the mechanisms of crank hot-stamping presses (Figure 1).

Content from this work may be used under the terms of the Creative Commons Attribution 3.0 licence. Any further distribution of this work must maintain attribution to the author(s) and the title of the work, journal citation and DOI.
Published under licence by IOP Publishing Ltd

IOP Conf. Series: Materials Science and Engineering **412** (2018) 012020 doi:10.1088/1757-899X/412/1/012020

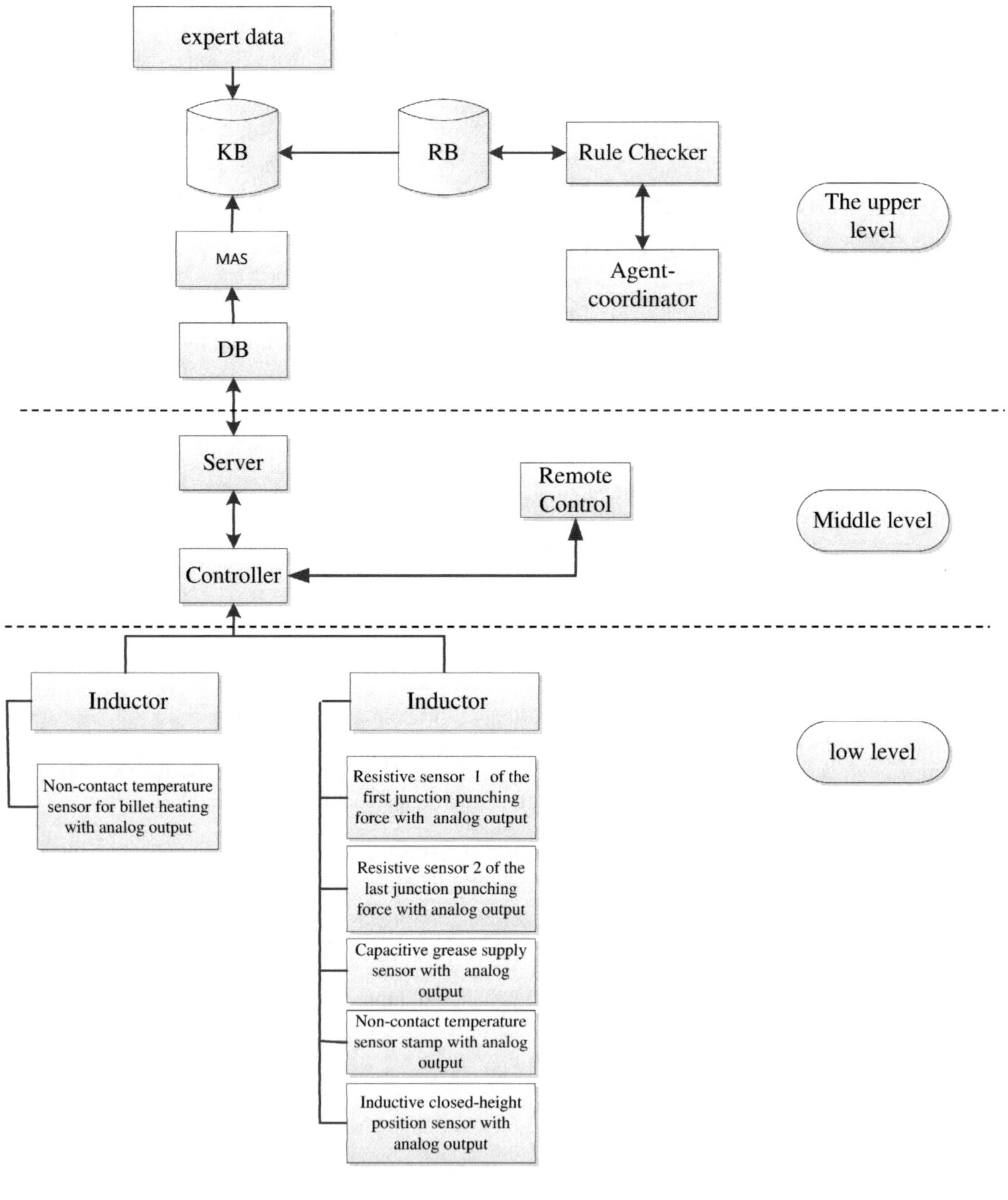

Figure 1. The structure of a multilevel control system for the precision stamping process of bevel gears BK-knowledge base, DB-database, BR-rule base, MAS-multiagent system.

2. The main body

The system is divided into 3 levels. The upper level is organizational, it is an intellectual superstructure of the system and is based on the knowledge of the organizational subsystem with the functions of reasoning, planning and decision-making in organizing the problem-solving process.

The middle level is coordination one. This level corresponds to the coordination subsystem, which is carried out through the interaction of the upper and lower levels of intellectualization. The task of this level is to plan the work of the lower level and the implementation of operational management.

Process information is displayed on the control panel display. The operator is necessary in the system in case of emergency situations, to debug the coating mode, as well as in the preparation and loading of blanks, and in unloading of processed products.

The lower level is executive. It is a hardware management system that solves a specific task with specific requirements for precision and functions. In order to obtain complete information about the process, sensors are added to the automatic control system. The information from the lower-level equipment is transferred to the remote in/out modules, and then to the controller. The system can be extended by adding measuring equipment and actuators.

This multi-level system as a basis has a knowledge base, which stores information about the parameters of the technological process of precision stamping of bevel gears (BG). The knowledge base is formed on the basis of practical experience, and formalized in the form of production rules, taking into account the rules of logical inference [3].

All new dependencies from the precedent base are analyzed in the block of checking rules for compliance with generally accepted ones. After adding new dependencies to the knowledge base, the system more accurately diagnoses the state of the installation nodes, which leads to a decrease in marriage.

On the basis of the presented structure of the technological process of precision stamping system of bevel gears (BG) we can represent agents of the process in the form of a single multi-agent system (MAS) (Figure 2).

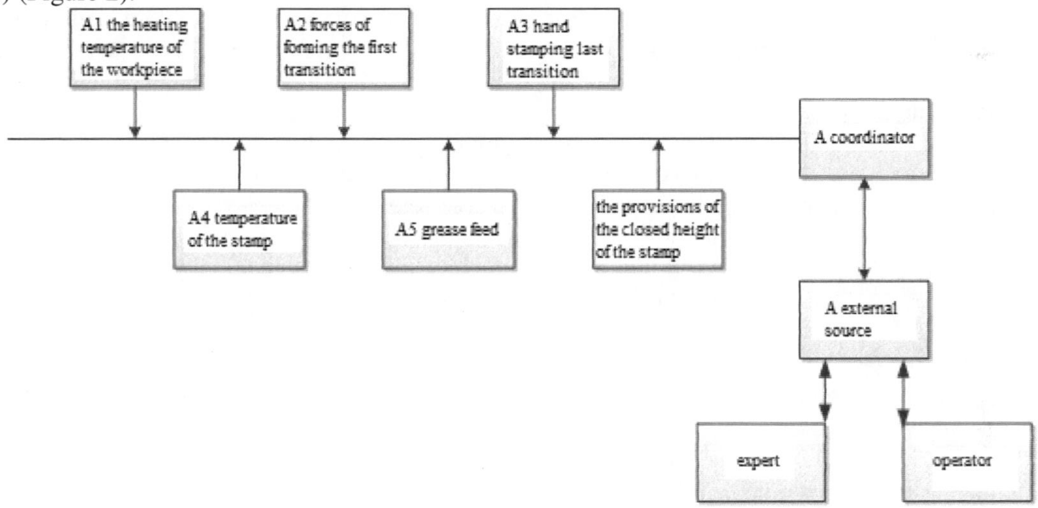

Figure 2. The structure of the MAS for a process of precision punching the bevel gears (BG)

In the presented system the whole spectrum of technological process tasks according to certain rules are distributed among all agents, each of which is a member of the group.

As it is indicated in papers [4-6], interaction of the agents entering into MAS with external environment is provided by the agent - external sources. Interaction between agents occurs through a coordinating agent, which is designed to coordinate the work of agents at all levels. It contains a repository of all the technological functions of the MAS and a knowledge base with rules for building the system.

The proposed control system generates control actions based on the knowledge base [7,8]. In the process of analyzing knowledge representation models, a product based on rules model was preferred. It has a high modularity of rules: when adding or changing a rule, everything that was created earlier, remains relevant. Ability to self-explain: the system explains the chain of rules that was used to obtain the output. In the course of the technological process, the precedent base and the knowledge base are developed.

To determine the production rules, according to papers [9,10], let us consider the fuzzy derivation by the example of the force-agent for stamping the first transition. The following signals from the sensors are available (Table 1).

Table 1. Examples of sensor signals

No	Signal	Denotation	Variant of origin	Denotation
1	The temperature of the workpiece	$t^{\circ}_{workpiece}$	Agent the heating temperature of the workpiece	A-1

For "workpiece temperature" the linguistic variable workpiece temperature is 1100-1200°C , where up to 1100 °C and its surroundings (0-50°C) is "low", the temperature is more than 1200°C is high temperature [11].

Therefore:

- if the temperature of the workpiece is low, then an increase in the yield stress is by 13% and there is an increase in the deformation force, consequently, an overload of the crank hot-stamping press (CHSP) appears and a press jam may occur;
- if the temperature of the workpiece is "high", then the plasticity of the metal decreases, which is the result of significant grain growth and subsequent burn of the metal (oxidation of grain boundaries)

3. Conclusion

Thus, the creation and implementation of a multi-agent intelligent system for controlling the parameters of the technological process (volume and temperature of the workpiece) and the mechanisms of crank hot-stamping presses makes it possible to increase operational, quality, and technical and economic indicators by reducing the number of errors in the development of the control system; reducing the time to identify and eliminate the causes of the malfunctioning of the system; reducing the probability of failure of expensive equipment.

References

[1] Berheev M M and Zalyaev I A 1988 *Fundamentals of Computer Aided Design Systems* (Ucheb. Posobie pod redakciej Yu. V. Kozhevnikova. — Kazan': izd. KGU) p 253

[2] Makarov I M, Lohin I M, Manko S V and Romanov M P 2006 *Artificial Intelligence and Intelligent Control Systems* (M.: Nauka) pp 214-17

[3] Morozov V K and Rogochev G N 2011 *Modeling of information and dynamic systems* (M.: Akademiya) pp 127-129

[4] Gavariev R V, Savin I A and Leushin I O 2016 *Impact of the functional coating on service durability of injection molds for the zinc alloys pressure casting* (Tsvetnye Metally) p 66–70 DOI: 10.17580/tsm.2016.01.11

[5] Gavariev R V, Savin I A and Leushin I O 2017 *Improvement of zinc castings surface quality by laminated protective coating* (Tsvetnye Metally) p 84-88 DOI 10.17580 / tsm.2017.05.13

[6] Gavariev R V and Savin I A 2017 *Improvement of surface quality of casting produced by casting under pressure* (Solid State Phenomena) p 988-993 DOI: 10.4028/www.scientific.net/SSP.265.988.

[7] Pankratov D L, Gavariev R V and Gavarieva K N 2016 *Influence of multilayer coatings on the operational stability of molds for injection molding* IOP Conference Series: Materials Science and Engineering Vol 134 No 1 **012031**

[8] Gavarieva K N, Simonova L A, Pankratov D L and Gavariev R V 2017 *Development of expert systems for modeling of technological process of pressure casting on the basis of artificial intelligence* IOP Conference Series: Materials Science and Engineering V 240 No 1 **012019**

[9] Shaparev A V and Savin I A 2017 Calculation of joint plastic deformation to form metal compound in cold condition (Solid State Phenomena) pp 313-18 DOI:10.4028/www.scientific.net/SSP.265.313
[10] Shaparev A V and Savin I A 2016 *Calculation of the amount of the reduction required for the formation of compound layers during cold rolling of bimetals* (Materials Science Forum) pp 328-333 DOI: 10.4028/www.scientific.net/MSF.870.328
[11] Safronov N N, Mingaleeva L B , Savin I A 2018 *Optimization of charge material composition in shs process with ferrosilide fabrication from gaseous wastes of metallurgical production* (Chernye Metally) pp 53-59

IOP Conf. Series: Materials Science and Engineering **412** (2018) 012021 doi:10.1088/1757-899X/412/1/012021

Stages of formation of scientific thinking of students in terms of engineering education at the university

I R Gilmanshin[1,2] and S I Gilmanshina[1]

[1]Kazan Federal University, 18 Kremlyovskaya street, Kazan, 420008, Russian Federation
[2] Kazan national research technical university named after A N Tupolev, K. Marx St. 10, Kazan, 420111 Tatarstan, Russia

is-er@yandex.ru

Abstract. The types of contradictions arising in the process of explaining science phenomena and processes in engineering education were considered. The authors proves the possibility and the stages of formation of scientific thinking of the students of the engineering directions at classical universities through the analysis of these contradictions.

Today, there is a wide use of intellectual capabilities of the modern educational environment. In this regard, there was an urgent need for the formation of scientific thinking among students of engineering and technical areas studying at the university.

The formation of scientific thinking is a long process of phased implementation of a number of interrelated intermediate goals, one of which is the transfer of mental activity of future engineers from the reproductive level to the level of creative activity. Such a transition is possible through familiarization with the analysis of contradictions and the formulation of problems, the promotion of hypotheses, their justification and proof in the explanation of natural phenomena and technological processes (for details see [1], [2], [3]. For the formation of scientific thinking, it is important in the course of engineering and technical education to continuously develop the methodology for solving experimental and computational-practical problems on the basis of interdisciplinary connections, applying methods of scientific knowledge with the gradual complication of the nature of tasks [4]. We believe that the emerging problematic situation as a structural component of creative thinking is a unit of the level of students' scientific thinking. This actualizes the development of a system of problem situations and tasks on the basis of cognitive contradictions.

In the process of explaining natural phenomena and technological processes, five types of contradictions can be naturally traced. This is a contradiction of the experiment (a contradiction between two or more experimental data); the contradiction between experience and theory (fact and generalization); the contradiction between old and new theories (or technologies); the contradiction between two coexisting unilateral theories (or technologies); the contradiction between objective and subjective factors in a particular system when explaining a natural science phenomenon, a

Content from this work may be used under the terms of the Creative Commons Attribution 3.0 licence. Any further distribution of this work must maintain attribution to the author(s) and the title of the work, journal citation and DOI.
Published under licence by IOP Publishing Ltd

IOP Conf. Series: Materials Science and Engineering **412** (2018) 012021 doi:10.1088/1757-899X/412/1/012021

technological process, or an environmental problem. Consider these contradictions in accordance with the classification of pedagogical problem situations, presented in [5].

The first type of contradiction is related to the fact that any scientific explanation is possible only on the basis of proven, firmly established facts. The contradiction in the field of educational experiment appears as a contradiction between inconsistent results obtained in the course of different experimental measurements of the same magnitude. The method of its resolution is to take into account the quality of the experiments from which the results were obtained, as well as the correct interpretation and evaluation of the results obtained (calculation of the measurement error). It should be remembered that in the experiment there are no effective criteria for eliminating contradictions. Only theory introduces an element of critical comprehension of results into the experiment. A university teacher should remember the duality of science. One side of knowledge is reflected in the determining role of experimental facts, the other side is connected with the heuristic role of the scientific theory. This duality of science should be constantly illustrated in the course of engineering and technical education with an emphasis on the role of contradictions.

Contradictions of the second and third types arise in university practice when considering theoretical material. It is known that in order to explain the totality of the experimental factual material, a hypothesis is always put forward, which can be the germ of a new theory (or new technology). However, there may be a contradiction between the inconsistency of the hypothesis of reality put forward. In this case, the cause of the contradiction is the inaccuracy of the hypothesis. As an example, we can recall the hypothesis of phlogiston (XVIII c), where hypothetical "fire matter" - phlogiston - was used to explain combustion.

A contradiction of the third type - between old and new (less and more complete) theories or technologies is removed if the positive content of the old theory (or technology) is transformed into a transformed theory or technology, and the supposed knowledge grows into a reliable one. The contradictions that arise in this process are connected with the general contradictions inherent in the movement of human thought from ignorance to knowledge and from less complete knowledge to a more complete one. A variant is possible when, under the imposition of certain conditions, the formulas of a new theory can turn into formulas that the old theory operated on. In the history of science, there are cases of establishing the truth of a new theory on this principle of correspondence. For example, there were several theories explaining the mechanism of the jump in the potential at the metal-solution interface, with its shortcomings.

Contradiction of the fourth type arises in connection with the separation of a single object in scientific knowledge into contradictory sides. To know the whole, as is known, it is necessary first to dismember, oppose one to another, and then reunify the unity. Thus, as a result of resolving the contradiction that existed between competing theories (radical theory and type theory, the first half of the nineteenth century), a new theory of the chemical structure of AM Butlerov emerged, incorporating the rational grains of both theories. However, the core of the last fundamental theory was a fundamentally new content of the concept "structure". A contradiction of the fifth type occurs in educational activity in the course of explaining a natural phenomenon or technological process. More precisely, due to the activity of the teacher, the contradiction between objective and subjective factors in the particular system under study must constantly attract the attention of students.

The types of contradictions considered are found in the conditions of engineering and technical education in the corresponding pedagogical problem situations. They can be classified for two reasons: a) according to the theory of step-by-step formation of mental actions - the definition of types of situations; b) by the nature (type) of the contradiction in the situation - the definition of the types of situations. Below is a classification of pedagogical problem situations, built on these two grounds.

The first type. Problem situations that arise when creating a motivation for the learning process and explaining the orienting basis of the action, in other words, before presenting theoretical material in lectures and in introductory conversations before the beginning of laboratory work. In these situations, there are contradictions between experience and theory (fact and generalization), between old and new (less and more complete) theories (or technologies), between two coexisting unilateral theories (or technologies), between objective and subjective factors in a particular system in a separate issue).

The second type. Problem situations arising during the formation of an action in a material or materialized form, otherwise, with the direct performance of laboratory work (performing actions with scientific equipment and training models). Here there is a contradiction between the inconsistent results obtained in the process of different experimental measurements of the same magnitude, between experience and theory (fact and generalization).

The third type. Problem situations in the formation of an action in the form of external speech (during classes when discussing the course and results of a laboratory experiment, in the form of oral and written speech while performing tests) in explaining the natural science phenomenon, the technological process, and solving problems. These are the contradictions of the experiment, between experience and theory (fact and generalization), between knowledge and application in practice.

The fourth type. Problem situations arising during the formation of an action in the form of internal speech (during compulsory independent extracurricular work of non-experimental nature). Contradiction between objective and subjective factors in a particular system (between knowledge and application in practice).

Thus, when forming students' scientific thinking, it is necessary to distinguish between the content (analysis of one of the problems of science or practical activity, for example, the problems of renewable energy [6], the utilization of industrial waste [7], the system analysis of the energy complex of enterprises [8]) and the methodical application of the method of problematic presentation of knowledge) of the side of the problem lecture.

Conclusion. The formation of scientific thinking in the conditions of engineering education is possible through the analysis of contradictions and the formulation of problems, the application of the method of hypotheses in solving complex educational and cognitive problems. This multifunctional process includes the steps: a) the formation of skills to analyze and generalize the experimental data, draw conclusions; b) it is logical to explain natural science concepts and technological processes; c) to formulate hypotheses, to give conjectures on the solution of the problem, to prove hypotheses theoretically and experimentally with the aim of explaining the natural science phenomenon or technological process; d) solve complex experimental and design engineering tasks on the basis of intersubject communications and carry out a critical self-assessment of the completed task.

References

[1] Gilmanshin I R, Gilmanshina S I 2016 *IOP Conf. Ser.: Mater. Sci. Eng.* 134 **012006**

[2] Gilmanshina S I, Khalikova F D 2016 *Teaching Gifted Adolescents in Terms of the Transforming Natural Sciences Education* Ifte 2016 - 2nd International Forum on Teacher Education Vol 12 pp 50-54

[3] Samigullina G S, Gilmanchina S I, Gaisin I T, Gilmanshin I R, Akchurina I R 2015 *Professional and Creative Development of Natural Geographic Course Teachers within the Process of Professional Retraining* International Education Studies 8(4)

[4] Gilmanshina S I, Gilmanshin I R, Sagitova R N, Galeeva A I 2016 *The Feature of Scientific Explanation in the Teaching of Chemistry in the Environment of New Information of School Students' Developmental Education* International Journal of Environmental and Science Education Vol 11 Is 4 pp 349-58

[5] Gilmanshina S I, Sagitova R N and et. 2015 *Professional Thinking Formation Features of Prospective Natural Science Teachers Relying on the Competence-Based Approach* Review of European Studies Vol 7 No 3 pp 341-49

[6] Gilmanshin I R, Kashapov N F, Gilmanshina S I, Galeeva A I 2016 *Landfill energy complex based on the renewable energy installations* IOP Conference Series: Materials Science and Engineering Vol 134 **012007**

[7] Azimov Yu I, Gilmanshin I R, Gilmanshina S I et al. 2016 *Modern technologies of waste utilization from industrial tire production* IOP Conference Series: Materials Science and Engineering, Vol 134 **012003**

[8] Kashapov N F, Gil'manshin I R, Konahina I A et al. 2014 *System analysis of the energy complex of engineering enterprise as a basic tool of effective energy management* IOP Conference Series: Materials Science and Engineering Vol 69 **012024**

IOP Conf. Series: Materials Science and Engineering **412** (2018) 012022 doi:10.1088/1757-899X/412/1/012022

Technology of e-learning in the university education

I R Gilmanshin[1,2], S I Gilmanshina[1]

[1]Kazan Federal University, 18 Kremlyovskaya street, Kazan, 420008, Russian Federation
[2] Kazan national research technical university named after A N Tupolev, K. Marx St. 10, Kazan, 420111 Tatarstan, Russia

is-er@yandex.ru

Abstract. Distinctive characteristics and disadvantages of electronic resources are distinguished. The structure and specificity of the developed author's practice-oriented electronic educational resources are considered. The application of e-learning technology as an active student-centered training for building individual educational trajectories in university education is substantiated.

In accordance with the world trends in the development of education, and university, including today, the actual technology is e-learning, based on the use of distance learning technologies in the form of electronic educational resources.

In Russia, the use of distance education technologies and e-learning in a network form is given increased attention at the state level. The priority project is the "Modern digital educational environment in the Russian Federation", whose passport was approved by the Presidium of the Council under the President of the Russian Federation on October 25, 2016.

Distance educational technologies are considered as a set of methods, forms and means of mediated, as a rule, through the Internet interaction of trainees with each other and with the teacher in the course of training.

The federal state standard of higher professional education provides for the active use of electronic educational resources in the course of training. Today, in the context of the new information and educational environment [1] of higher education, the relevance of electronic educational resources is unquestionable, since society needs professionals with professional thinking [2, 3] with emphasis on energy management [4] and information culture [5].

Under the electronic educational resource we will understand the educational resource presented in electronic digital form and including the structure, subject content and metadata.

Below are highlighted the distinctive characteristics of electronic educational resources, making them popular among learners and promising in application. This - accessibility, interactivity, convenience of information retrieval, openness for introducing new information and data, the possibility of studying the teaching material by students independently with the activation of cognitive activity, the possibility of studying various processes without the use of reagents and equipment.

The following indicators can be classified as essential shortcomings of electronic educational resources. Daily work on the Internet raises the issue of human information security. It is necessary

Content from this work may be used under the terms of the Creative Commons Attribution 3.0 licence. Any further distribution of this work must maintain attribution to the author(s) and the title of the work, journal citation and DOI.
Published under licence by IOP Publishing Ltd

to constantly update the technology in accordance with the requirements of the electronic resource. Simultaneously with the assimilation of the material of the electronic educational resource, it is necessary to remember the logical chain of judgments in the transition from one hyperlink to another, which can lead to a cognitive overload of the learner.

Nevertheless, today, electronic educational resources are being actively developed in many Russian universities. Similar resources of Kazan Federal University are available on the website www.edu-kpfu.ru, combining the previously existing on the basis of LMS MOODLE three sites: Zilant (http://zilant.kpfu.ru), Bars (http://bars.kpfu.ru).), Tulpar (http://tulpar.kpfu.ru).

At the department of chemical education of the Kazan Federal University work on the introduction of electronic educational resources has been going on since 2010. At present, the following electronic resources are used in the educational process, developed by the teachers of the department: "General theoretical bases of analytical chemistry. Qualitative Analysis "(winner of the contest of the Kazan Federal University in 2013)," Methods of solving problems in physical chemistry in school chemistry "," Chemistry Practices "," Theory of Chemistry Teaching "," Chemistry Techniques "," Didactic Games in Teaching Chemistry "and others [6]. In addition, an electronic resource has been developed for the Admission Committee to prepare for a unified state examination on general and inorganic chemistry. For the Privolzhsky interregional center of professional development of education workers, an electronic educational resource "Improvement of the professional competences of the teacher of chemistry in the conditions of the introduction of the GEF LLC" has been developed and is being used since 2016 at the courses of teachers' qualification improvement.

Structurally, all developed electronic resources include metadata, a work program of the discipline supported by the electronic educational resource, a glossary, sample questions for credit or examination, methodological instructions for the student and teacher. For each topic presented lecture and practice-oriented material for auditor and independent work, information resources, basic and additional literature, current and final control.

As an example, let's consider the specifics of the practice-oriented author's electronic educational resource "General theoretical bases of analytical chemistry. Qualitative analysis. "

The main difference of this resource is to contribute to the active study of the educational material, to develop the ability to think independently, to teach oneself thinking, relying on scientific chemical theories when solving problems and explaining natural phenomena. Otherwise, contribute to the formation of scientific thinking [7]. To do this, each topic is accompanied by a plan of presentation and methodological instructions, the organization of cognitive activity of students with the identification of a complex didactic purpose of studying the topic; lecture; information and cognitive unit, as well as a list of recommended educational literature. Given that historical facts are necessary for the general characterization of chemistry as a field of science and the development of a strong interest in it, every lecture provides information relevant to the history of chemistry. The content of the methodical block "Organization of cognitive activity of students" complements lectures. This relationship is focused on achieving the main goal of training - mastering the fundamental knowledge of the fundamentals of analytical chemistry.

In this electronic educational resource for each topic includes a lot of special tasks for training (typical tasks) and cognitive (with elements of creativity) character. The purpose of the training tasks is to form appropriate skills and abilities for students. Moreover, examples of solving all typical tasks from the self-control unit are given in the content of the relevant lectures. The goal of cognitive tasks is the formation of professional thinking (for more details see [2]). To carry out creative tasks, not only profound knowledge and skills are needed, a critical evaluation of the results obtained, the main thing is a clear awareness of the problem in a specific chemical task. At the same time, this electronic resource does not replace the task book.

In conclusion, we note that, according to the results of the questionnaire survey of students working with electronic educational resources on chemical and methodological disciplines for more than four years, their motivation and the quality of university education have been increased.

In general, the use of e-learning technology as an active student-centered learning is beneficial for building individual educational trajectories in university education.

References

[1] Gilmanshina S I, Khalikova F D 2016 *Teaching Gifted Adolescents in Terms of the Transforming Natural Sciences Education* Ifte 2016 - 2nd International Forum on Teacher Education Vol 12 pp 50-54

[2] Gilmanshina S I, Sagitova R N and et. 2015 *Professional Thinking Formation Features of Prospective Natural Science Teachers Relying on the Competence-Based Approach* Review of European Studies Vol 7 No 3 pp 341-49

[3] Samigullina G S, Gilmanchina S I, Gaisin I T, Gilmanshin I R, Akchurina I R 2015 *Professional and Creative Development of Natural Geographic Course Teachers within the Process of Professional Retraining* International Education Studies 8(4)

[4] Kashapov N F, Gil'manshin I R, Konahina I A et al. 2014 *System analysis of the energy complex of engineering enterprise as a basic tool of effective energy management* IOP Conference Series: Materials Science and Engineering Vol 69 **012024**

[5] Gilmanshin I R, Gilmanshina S I 2016 *The formation of students' engineering thinking as a way to create new techniques, technologies, materials* IOP Source: International Scientific-Technical Conference on Innovative Engineering Technologies, Equipment and Materials Vol 134 **012006**

[6] *Distance education of Kazan Federal University*. Retrieved 04/10/2017, from https://edu.kpfu.ru/course/index.php?categoryid=369.

[7] Gilmanshina S I, Gilmanshin I R, Sagitova R N, Galeeva A I 2016 *The Feature of Scientific Explanation in the Teaching of Chemistry in the Environment of New Information of School Students' Developmental Education* International Journal of Environmental and Science Education Vol 11 Is 4 pp 349-58

Formation of the future engineering chemical elite in lyceums for gifted adolescents at universities

S I Gilmanshina[1], F D Khalikova[1], Yu A Cherkashina[2], I R Gilmanshin[1,3]

[1] Kazan Federal University, 18 Kremlyovskaya street, Kazan, 420008, Russian Federation
[2]Kazan National Research Technological University, Kazan, Russian Federation
[3] Kazan national research technical university named after A N Tupolev, K. Marx St. 10, Kazan, 420111 Tatarstan, Russia

is-er@yandex.ru

Abstract. It sets out innovations in formation of the future chemical engineering elite in the lyceums for gifted teenagers created at universities. Such as Small Chemical Institute, professorial lectures, professorial schools, professorial mentoring, regular trips to the enterprises of petrochemical complex of Tatarstan, the scientific community lyceum students. The results of additional education in chemistry are presented in the form of convincing victories lyceum students at the various olympiads and conferences.

Tatarstan is one of the leading regions of Russia for oil production and refining. At the same time, there is a shortage of qualified engineering personnel in the industry, capable of solving complex problems of renewable energy [1], recycling industrial waste [2], system analysis of the energy complex of enterprises [3] and others. At the same time in the Republic of Tatarstan there are a lot of gifted children, capable of innovations and new scientific discoveries. Selection, education and development of gifted students in the process of general education is no less important task than their further educational trajectory in the university.

The above indicated was a prerequisite for the opening of the IT-Lyceum and the Lyceum. N.I. Lobachevsky at the Kazan (Volga region) Federal University (KFU) and the boarding school for gifted children. academician P.A. Kirpichnikova with in-depth study of chemistry at the Kazan National Research Technological University (KNITU).

IT-lyceum with KFU was opened on September 1, 2012. Training is conducted from the 7th grade. Individual selection on a competitive basis in three phases: distance testing (on-line), intramural stage (control work on the Russian language, mathematics, logic and core subjects), interview, taking into account students' portfolio. Lyceum is a structural subdivision of KFU, it is a resonant, unique educational institution not only in Russia, but also in the world. Teachers work professionally in modern Singapore and European methods, attaching students to analyze the contradictions and formulation of problems, explanation of natural phenomena (details in [4], [5]).

Despite the fact that the basic discipline of high school students - informatics, the guys have shown great interest in both the robotics and chemistry, after school is visited Small Chemical Institute at the Chemical Institute. A.M. Butlerov KFU (for more details on the forms of work with gifted adolescents in [6]).

Content from this work may be used under the terms of the Creative Commons Attribution 3.0 licence. Any further distribution of this work must maintain attribution to the author(s) and the title of the work, journal citation and DOI.
Published under licence by IOP Publishing Ltd

Small Chemical Institute - an innovative pilot project for gifted in the field of chemistry students, including four additional educational programs "Chemistry-8", "Chemistry-9", "Chemistry-10", "Chemistry-11" (each 72 hour program) was launched in Kazan (Volga region) Federal University in 2013. Enrollment is held on the advice of teachers of chemistry and the results of the entrance testing. Structurally, the Small Chemical Institute includes a club of young chemists, preparation for the passing of a unified state examination, olympiads, project activities of students at the departments of the Kh. A.M. Butlerov KFU (for details see [6]).

In the 2013/2014 school. 8 students of the IT-Lyceum attended the Small Chemical Institute, in 2014/2015 - 16, and in 2015/2016 school. 20 IT-lyceum students were involved in chemistry, in 2016/2017 - 26, and in 2017/2018 - 39 lyceum students. The number of students attending the Small Chemical Institute of Lyceum students correlates well with the positive dynamics of the number of prize-winners of various stages of the All-Russian Chemistry Olympiad (Table 1).

For example, at the regional stage of the All-Russian Olympiad in chemistry, respectively, in 2014. The winners were 2 people, in 2015 - already 4, 2016 9 IT-school students became the winners of the regional stage of the All-Russian Olympiad on chemistry, and in 2017 it was already 10 winners and one winner.

Table 1.
Correlation between attending the Small Institute of Chemistry students of IT-Lyceum and the number of winners / winners of various stages of the All-Russian Olympiad in Chemistry

Stages of the All-Russian Olympiad in Chemistry	Number of students awarded IT-Lyceum at KFU							
	2013/2014		2014/2015		2015/2016		2016/2017	
	Prize-Winners	Winners	Prize-Winners	Winners	Prize-Winners	Winners	Prize-Winners	Winners
Municipal	4	0	12	0	15	1	16	1
Regional	2	0	4	0	9	0	10	1
Final	0	0	1	0	3	1	2	1
The number of lyceum students attending the Small Chemical Institute at the Chemical Institute. A.M. Butlerova								
	2013/2014		2014/2015		2015/2016		2016/2017	
	8 students		16 students		20 students		26 students	

In general, for four years in the IT-Lyceum, a creative chemistry team has been created, consisting of prize-winners of the All-Russian Olympiad, the Volga Interregional Olympiad, the Interregional Subject Olympiads of the KFU, the All-Siberian Open Olympiad, the Nobel Prize-KNITU competition.

Considering the long-term perspective of the formation of the engineering chemical elite of the Republic of Tatarstan, an interest is the organization of engineering classes at the Lyceum named after. N.I. Lobachevsky at the KFU and the chemical lyceum-boarding school at the Kazan National Research Technological University (KNITU). More details on the Lyceum-boarding school. P.A. Kirpichnikov. Its main task is the preparation of talented students for various engineering directions of the KNITU, the formation of the fundamentals of engineering thinking among the lyceum students (for details see [7]). This will allow the republic as a whole and the chemical industry enterprises to receive in the future highly qualified industrial engineers.

The methodological basis for constructing the content of instruction in the Lyceum is a comprehensive and system-activity approach. The peculiarity of the additional education of the lyceum students is conditioned by such innovative forms of activity as the professorial lecture hall, professorial schools, professorial supervision, regular excursions to the enterprises of the oil and gas chemical complex of Tatarstan, the scientific community.

The teaching staff of the Technological University directly participates in educational activities. The first half of the day students are trained according to the program provided by the State Federal Educational Standard. After lunch, lessons continue in the format of professorial schools, lectures and research activities within the scientific community of lyceum students.

Professor's lecture halls are lectures by professors working in different directions. The classes cover and discuss with the lyceum students a variety of fields of science, the latest trends.

Professorial schools are a systematized (most often weekly) course of lectures given by a teacher with subjects with rigidly set ultimate goals. There are seven such schools in the Lyceum, within the framework of which the lyceum students are trained in the olympiads in chemistry, physics and biology. Since January 2015, a program has started, suggesting the attendance of professorial lectures by schoolchildren of other educational institutions in the city of Kazan and Zelenodolsk municipal district of Tatarstan.

Professorial supervision provides for the focal registration of a lyceum student (or a small group of lyceum students of 2-3 people) for a specific university professor to perform scientific work with subsequent protection at contests and conferences. The best lyceum students are given credit books, where professors evaluate their knowledge, which is an additional stimulus to successful educational activity.

Excursions to the enterprises of the oil and gas chemical complex of Tatarstan have a pronounced practice-oriented direction, where the lyceum students freely communicate with the leading experts in the engineering field of knowledge.

Preparation for olympiads on the natural-science profile (chemistry, biology, geography, physics) passes systematically throughout the year and there is a positive dynamics of winners and winners of the Olympiads.

For example, according to the results of the 2013/2014 academic year, there were 10 winners at the All-Russian Olympiad for schoolchildren in natural science subjects, 5 of them winners, in the 2014/2015 academic year there were 22 winners and 9 winners.

Thus, the use of innovative forms of supplementary education in chemistry, such as the Small Chemical Institute for the IT Lyceum at the Kazan (Volga region) Federal University, professorial lectures, schools, curatorship, regular excursions to the enterprises of the oil and gas chemical complex of Tatarstan, the scientific community of lyceum students in the Lyceum. P.A. Kirpichnikov at KNITU, effectively increasing the motivation to study chemistry, contribute to the achievement of high results by students in Olympiads and research conferences of students at various levels and, most importantly, the formation of a future engineering chemical elite.

References

[1] Gilmanshin I R, Kashapov N F, Gilmanshina S I, Galeeva A I 2016 *Landfill energy complex based on the renewable energy installations* IOP Conference Series: Materials Science and Engineering Vol 134 **012007**

[2] Azimov Yu I, Gilmanshin I R, Gilmanshina S I et al. 2016 *Modern technologies of waste utilization from industrial tire production* IOP Conference Series: Materials Science and Engineering, Vol 134 **012003**

[3] Kashapov N F, Gil'manshin I R, Konahina I A et al. 2014 *System analysis of the energy complex of engineering enterprise as a basic tool of effective energy management* IOP Conference Series: Materials Science and Engineering Vol 69 **012024**

[4] Gilmanshina S I, Gilmanshin I R, Sagitova R N, Galeeva A I 2016 *The Feature of Scientific Explanation in the Teaching of Chemistry in the Environment of New Information of School Students' Developmental Education* International Journal of Environmental and Science Education Vol 11 Is 4 pp 349-58

[5] Samigullina G S, Gilmanchina S I, Gaisin I T, Gilmanshin I R, Akchurina I R 2015 *Professional and Creative Development of Natural Geographic Course Teachers within the Process of Professional Retraining* International Education Studies 8(4)

[6] Gilmanshina S I, Khalikova F D 2016 *Teaching Gifted Adolescents in Terms of the Transforming Natural Sciences Education* Ifte 2016 - 2nd International Forum on Teacher Education Vol 12 pp 50-54

[7] Gilmanshin I R, Gilmanshina S I 2016 *The formation of students' engineering thinking as a way to create new techniques, technologies, materials* IOP Source: International Scientific-Technical Conference on Innovative Engineering Technologies, Equipment and Materials Vol 134 **012006**

IOP Conf. Series: Materials Science and Engineering **412** (2018) 012024 doi:10.1088/1757-899X/412/1/012024

Computer simulation of the torque distribution system of a hybrid heavy truck in conditions of slippage of one of the wheels

E O Gurjanova[1], D N Demyanov[2]

[1] Moscow Technological University, Vernadsky Avenue 78, Moscow, the Russian Federation
[2]Kazan Federal University, Naberezhnye Chelny Institute, 423812, Russia, Naberezhnye Chelny, Prospekt Syuyumbike 10A

demyanovdn@mail.ru

Abstract. Mathematical and computer models for the motion of a hybrid heavy truck in conditions of wheel slip have been developed. Algorithms for the redistribution of the torque between the wheels to provide controllability and stability of motion along the guided path have been proposed. The efficiency of the developed algorithms is confirmed by the results of computer simulation in the MATLAB / Simulink system.

1. Introduction

The use of hybrid transmissions is currently one of the most effective ways to improve the consumer properties of heavy vehicles. The appearance of cars equipped with AWD-system and the system "start-stop" can significantly improve the controllability of the car and increase its fuel efficiency [1]. Moreover, AWD system is more comfortable for the driver than ESC-system, as it does not involve the use of a brake. However, a significant disadvantage of the AWD system is its high complexity. Thus, Thus, the problem of developing control algorithms for AWD-system is very actual.

2. Mathematical modeling

Consider a four-wheeled car equipped with a hybrid transmission. The torque to the rear axle comes from the hybrid powerplant, the front axle is fitted with motor-wheels. The distribution of torque between the wheels of one axle is carried out by means of the limited slip differential [2].
To describe the motion of a heavy vehicle, we use the four-wheel model described in [3]:

$$\begin{cases} m\cdot\ddot{x} = (F_{xfl}+F_{xfr})\cdot\cos(\delta)+F_{xrl}+F_{xrr}-(F_{yfl}+F_{yfr})\cdot\sin(\delta)+m\cdot\dot{\psi}\cdot\dot{y}; \\ m\cdot\ddot{y} = F_{yrl}+F_{yrr}+(F_{xfl}+F_{xfr})\cdot\sin(\delta)+(F_{yfl}+F_{yfr})\cdot\cos(\delta)-m\cdot\dot{\psi}\cdot\dot{x}; \\ J_z\cdot\ddot{\psi} = l_f\cdot(F_{xfl}+F_{xfr})\cdot\sin(\delta)+l_f\cdot(F_{yfl}+F_{yfr})\cdot\cos(\delta)-l_r\cdot(F_{yrl}+F_{yrr})+ \\ \quad +\frac{l_w}{2}\cdot(F_{xfr}-F_{xfl})\cdot\cos(\delta)+\frac{l_w}{2}\cdot(F_{xrr}-F_{xrl})+\frac{l_w}{2}\cdot(F_{yfl}-F_{yfr})\cdot\sin(\delta). \end{cases}$$

(1)

Here is denoted: x, y – displacement in the longitudinal and transverse directions, respectively; ψ –

Content from this work may be used under the terms of the Creative Commons Attribution 3.0 licence. Any further distribution of this work must maintain attribution to the author(s) and the title of the work, journal citation and DOI.
Published under licence by IOP Publishing Ltd

yaw angle; F_{xxx} – the reaction force at the contact point of the wheel with the road, where the first index denotes longitudinal or transverse direction of force, the second index denotes the rear or the front axle, and the third index – the left or right side of the vehicle; m – vehicle weight; J_z – mass moment of inertia with regard to normal axle; δ – the steering angle of front wheels; l_f, l_r – the distance from the center of mass of the vehicle to the front and rear axle respectively; l_w – the wheel space.

The magnitude of the longitudinal and transverse reaction force can be determined by the formulas, described in the work [4]:

$$F_x = C_\sigma \cdot \frac{\sigma}{1+\sigma} \cdot f(\lambda)\ ;\ F_y = C_\alpha \cdot \frac{\tan(\alpha)}{1+\sigma} \cdot f(\lambda)\ . \tag{2}$$

Here: F_x, F_y – longitudinal and transverse component of the response reaction respectively; C_σ – normal tyre stiffness; C_α – lateral tyre stiffness; σ – the index of the wheel skidding; α – the slippage angle; $f(\lambda)$ – a function, describing the wheel parameters, numerical values of which is given in the work [4].

The slippage value for each of the wheels is determined by the formula:

$$\sigma = \frac{r \cdot \omega - \dot{x}}{r \cdot \omega}. \tag{3}$$

Here: r – rolling radius; ω – the angular velocity of the wheel.

The slippage coefficient is used to determine the condition of the wheel. Herewith, they distinguish two modes of a vehicle movement: the lead and the braking condition. In the lead condition the full slippage will take place at the slipping wheel and when the vehicle is stationary ($r = 0$), and at the braking condition – during the wheel slip movement ($r = \infty$) [5].

The slippage angles for the wheels of the front or rear axle are determined by the formulas [5]:

$$\alpha_f = \delta - \frac{\dot{y} + l_f \cdot \dot{\psi}}{\dot{x}}\ ;\ \alpha_r = \frac{\dot{y} - l_r \cdot \dot{\psi}}{\dot{x}}\ . \tag{4}$$

Here: α_f, α_r – the slippage angles for the wheels of the front and rear axle respectively.

The rotational rates of the wheels can be determined with the help of the equation of rotational motion dynamics:

$$\begin{cases} J_w \cdot \dot{\omega}_{fl} = T_{dfl} - T_{bfl} - r \cdot F_{xfl}\ ; \\ J_w \cdot \dot{\omega}_{fr} = T_{dfr} - T_{bfr} - r \cdot F_{xfr}\ ; \\ J_w \cdot \dot{\omega}_{rl} = T_{drl} - T_{brl} - r \cdot F_{xrl}\ ; \\ J_w \cdot \dot{\omega}_{rr} = T_{drr} - T_{brr} - r \cdot F_{xrr}\ . \end{cases} \tag{5}$$

It is indicated here: J_w – mass moment of inertia of the wheel; ω_{xx} – the rotational speed of the wheel, where the first index denotes the front or the rear axle, the second index – the left or the right wheel; T_{dxx}, T_{bxx} – torque and braking torque respectively, affecting the wheel, where the second index denotes the front or the rear axle, and the third index – the left or the right wheel.

The torque for the motor-wheels of the front axle comes from independent drives. The torques on the rear axle wheels are determined by the equations:

$$T_{dl} = \frac{T_d}{2} - q \cdot T_{clutch}\ ;\ T_{dr} = \frac{T_d}{2} + q \cdot T_{clutch}\ . \tag{6}$$

It is indicated here: T_{dl}, T_{dr} – the torque of the left and the right wheel respectively; T_d – general torque affecting the axle; q – the index of differential blocking; T_{clutch} – clutch torque.

The system of equations (1) – (6) enables to describe the motion of a heavy-duty vehicle taking into account the possible slippage of the wheels. By completing this system with the ratios to determine the coefficient of differential blocking and the torque of motor-wheels of the front axle, one can get a closed system of equations, which enables to simulate the motion of a heavy-duty vehicle.

3. Computer modeling

On the basis of the equations (1) – (6) in the MATLAB/Simulink system a computer model of the motion of the hybrid heavy truck in conditions of one of the wheels slippage was designed.

The following parameter values, close to the real characteristics of trucks of 'KAMAZ' family have been chosen: $m = 46000$ kg; $l_w = 2{,}04$ m; $l_f = 3$ m; $l_r = 4{,}27$ m; $r = 0{,}55$ m; $C_\sigma = 490000$ N/m; $C_\alpha = 431200$ N/m; $J_w = 13{,}5$ kg·m^2.

The system of torque distribution is modeled as a logical unit. Its input variables are the general torque, the speed of the vehicle, the state of charge and the voltage of the electric battery. The output variables are the value of the torque, affecting each wheel T_{dfr}, T_{dfl}, T_{drr}, T_{drl}.

The algorithm used to control the drives of the front axle motor-wheels is described in detail in [6]. The control of differential blocking is performed in the following way: the electronic control unit detects the moment when the wheel slips and redistributes the total torque to the other wheel.

Figure 1 shows the results of one of the computer experiments in which the movement of a hybrid truck with AWD system and without it was reproduced.

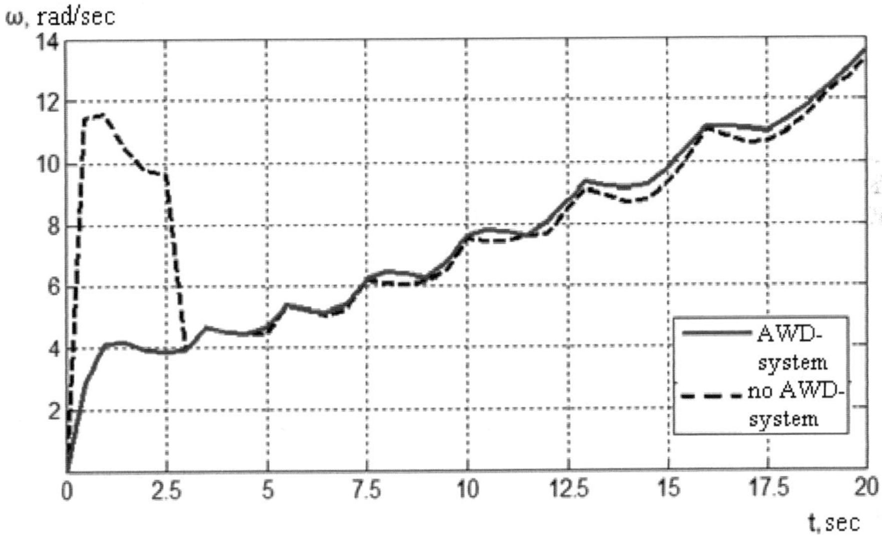

Figure 1 – The dependence of wheel rotating speed on time

The results of the simulation allow us to conclude that the use of the described control algorithms makes it possible to increase controllability and stability of a truck on different types of surfaces in the of wheel slippage.

4. Results

A mathematical and computer model for the motion of a hybrid heavy truck was developed in conditions of wheel slippage. We proposed algorithms for controlling motor-wheels and the pattern of change for the differential blocking coefficient, that provide improved controllability and stability. The results of the study can be used in practice in the development of active safety systems for hybrid trucks.

5. Acknowledgements

The work is executed at financial support of RFBR (grant № 17-08-00516).

References

[1] Qui L, Qian L, Zomorodi H, Pisu P 2017 *Global optimal energy management control strategies for connected four-wheel-drive hybrid electric vehicles* IET Intelligent Transport Systems

Vol 11 Iss 5 pp 264-272
[2] Andreev A F, Vantsevich V V, Lefarov A H 1987 *Differentials of wheeled vehicles* (M.: Machinostroenie) p 176
[3] Rajamani R 2011 *Vehicle dynamics and control* (Springer) p 500
[4] Dugoff H, Fancher P S, Segal L 1969 *Tire performance characteristics affecting vehicle response to steering and braking control inputs. Final Report* (US National Bureau of Standards) p 105
[5] Kiencke U, Nielsen L 2005 *Automotive Control Systems* (Springer) p 521
[6] Van E V, Gurjanova E O, Demyanov D N 2017 *Development of control strategies by transmission elements of a hybrid truck in the case of boundary conditions* XXVIII International innovation-oriented conference of young researchers and students (IIOCYRS - 2016) pp 219-222

Empirical heat transfer correlations of tube bundles under pulsating flows

A I Haibullina[1], L S Sabitov[1,2], A R Hayrullin[1], N H Zinnatullin[3]

[1]Kazan State Power Engineering University, 51 Krasnoselskaya street, Kazan, 420066, Russian Federation
[2]Kazan Federal University, 18 Kremlyovskaya street, Kazan, 420008, Russian Federation
[3]Kazan National Research Technological University, 68 Karla Marks, street, Kazan, 420015, Russian Federation

kharullin@yandex.ru

Abstract. In this paper, the influence of the parameters of pulsations on the heat transfer in tube bundles was considered. An analysis of the influence of the parameters of pulsations on heat transfer was carried out by empirical correlation. A new empirical correlation is proposed for calculation of external heat transfer in under of pulsating flow in tube bundles with different configurations. The maximum deviation of the empirical correlation with the data of the numerical experiment was 16.9%.

1. Introduction

Unsteady flows can be used to enhance the heat transfer of various heat exchange equipment. Therefore, many authors are engaged in the study of pulsating flows. Studies of heat transfer in pulsating flow are mainly devoted to individual pulsation regimes [1-4]. Therefore, for the frequently obtained data, it is not enough to obtain an empirical correlation, which can be used to calculate heat transfer in pulsating flows. Or empirical correlations are obtained for a narrow range of parameters [5]. Obtaining empirical dependence complicates a large number of designs of heat exchange equipment. Also, the complexity is complemented by the fact that the determining parameters of heat transfer for pulsating currents are greater than for steady-state heat transfer. However, in engineering practice, it is convenient to use the empirical correlation to calculate the heat transfer, which must be obtained for a given geometry and flow regime. To simplify the production of empirical correlation, one can use a numerical experiment instead of a real experiment. In the previous work [6], an external heat transfer in tube bundles of various configurations under pulsating currents was investigated by a numerical method. As a result, an empirical correlation was obtained for a wide range of parameters. However, the maximum deviation of the empirical correlation was ±35,7 %. The aim of this work is to obtain an empirical correlation with a smaller error for the data of the paper [6]. In addition, in this work, an analysis will be made of the empirical correlations obtained, which will allow us to estimate the influence of the regime parameters on the heat transfer for different bundles of tubes in the presence of flow pulsations.

2. Methods

Content from this work may be used under the terms of the Creative Commons Attribution 3.0 licence. Any further distribution of this work must maintain attribution to the author(s) and the title of the work, journal citation and DOI.
Published under licence by IOP Publishing Ltd

To find the empirical correlation was used regression analysis based on the method of least squares. In particular, multiple linear regression and its special case polynomial regression were used [7, 8].

3. Results and Discussion

In [6], an empirical correlation of heat transfer was obtained for pulsating flows in tube bundles (1).

$$\frac{Nu_p}{Nu_{st}} = 0{,}954\,\mathrm{Re}^{-0{,}201}\cdot \mathrm{Pr}^{-0{,}211}\cdot \beta^{0{,}184}\cdot Fo^{-0{,}230}\cdot \psi^{-0{,}053}\cdot \varphi^{0{,}085}\cdot s_1/D^{0{,}287} \qquad (1)$$

where Nu_{st} – Nusselt number for stationary flow; Nu_p – Nusselt number for pulsating flow; Re – Reynolds number; Pr – Prandtl number; β – relative amplitude of pulsations; Fo – Fourier number; ψ – pulsating ratio; φ – tube layout angle; s_1 –tube pitch, m; D – tube bundle, m. Equation (1) is obtained for the ranges $215 \le \mathrm{Pr} \le 363$, $0.25 \le \psi \le 0{,}5$, $100 \le \mathrm{Re} \le 1000$, $15 \le \beta \le 35$, $5.81\cdot10^{-4} \le Fo \le 14.53\cdot10^{-4}$. Coefficient of determination $R^2 = 0.906$, maximum error $\delta = \pm 35.7$ %. The definition of the parameters entering into equation (1) is given in [6].

To reduce the error of equation (1), geometric parameters of the tube bundles were removed φ and s_1/D (2). Thus, the place of one equation (1) yields 12 equations (Table 1).

$$\frac{Nu_p}{Nu_{st}} = A\,\mathrm{Re}^{m}\cdot \mathrm{Pr}^{n}\cdot \beta^{b}\cdot Fo^{c}\cdot \psi^{d} \qquad (2)$$

Table 1. The exponents of equation (2)

№	φ	s_1/D	A	m	n	b	c	d	R^2	Deviation δ. %	
										max	mean
Staggered tube bundle											
1	30°	1.25	1.020	–0.154	–0.203	0.198	–0.213	–0.092	0.910	±27.3	±4.7
2		1.50	1.673	–0.232	–0.214	0.199	–0.227	–0.082	0.944	±22.9	±5.3
3		1.75	1.800	–0.224	–0.215	0.181	–0.225	–0.069	0.949	±22.8	±4.8
4	45°	1.25	1.081	–0.170	–0.194	0.195	–0.210	–0.099	0.917	±30.5	±4.9
5		1.50	1.576	–0.249	–0.233	0.232	–0.252	–0.079	0.916	±29.0	±7.3
6		1.75	1.641	–0.230	–0.194	0.170	–0.237	–0.037	0.951	±36.3	±6.8
7	60°	1.25	1.101	–0.129	–0.211	0.193	–0.213	–0.018	0.903	±21.7	±4.6
8		1.50	1.638	–0.206	–0.224	0.192	–0.238	–0.022	0.937	±27.1	±5.1
9		1.75	1.957	–0.218	–0.193	0.142	–0.226	–0.006	0.959	±21.4	±3.9
In-line tube bundle											
10	90°	1.25	1.161	–0.179	–0.194	0.186	–0.228	–0.063	0.929	±21.4	±4.9
11		1.50	1.914	–0.213	–0.228	0.167	–0.234	–0.046	0.950	±21.8	±4.5
12		1.75	1.876	–0.212	–0.232	0.154	–0.255	–0.033	0.950	±22.0	±4.5

Consider the influence of numbers Re and Pr on Nu_p/Nu_{st} according to their degrees, depending on the configuration of the bundles of tubes. Since the degrees m and n are negative, then with increasing Re and Pr, Nu_p/Nu_{st} decreases for all φ и s_1/D. The values of the degree for Re in lie in the range $-0.129 \le m \le -0.249$. Minimal impact Re on Nu_p/Nu_{st} is observed in denser beams ($s_1/D = 1.25$,), m are in the range $-0.129 \le m \le -0.179$. When $s_1/D = 1.5$ and 1.75, m is in the range $-0.206 \le m \le -0.249$ .Influence of Pr on Nu_p/Nu_{st} less depends on φ and s_1/D (compare with the numbers Re), n is the range $-0.194 \le n \le -0.232$.

Influence β on Nu_p/Nu_{st} contrary to Re and Pr. If with the increase of Re and Pr there is a decrease in heat transfer, then with increasing β there is an increase in heat transfer for all φ and s_1/D, b is in the range $0.142 \le b \le 0.232$. When $\varphi = 60°$, $s_1/D = 1.75$ β has the minimal influence on Nu_p/Nu_{st} ($b = 0.142$). When $\varphi = 45°$, $s_1/D = 1.5$ β has the maximum influence on Nu_p/Nu_{st} ($b = 0.232$).

With increasing Fo, an increase in heat transfer occurs for all φ and s_1/D. Influence Fo on Nu_p/Nu_{st} less when $s_1/D = 1.25$, c is the range $-0.210 \le c \le -0.228$. When $s_1/D = 1.5$ and 1.75 ($-0.225 \le c \le -0.255$)

With increasing ψ there is a decrease in heat transfer for all φ and s_1/D. When $\varphi = 45°$, $s_1/D = 1.25$ ψ has the maximum influence on Nu_p/Nu_{st} ($d = -0.099$). ψ has almost no influence on heat transfer when $\varphi = 60°$, $s_1/D = 1.75$ ($d = -0.006$).

Table 1 shows that for some φ, s_1/D, the error δ has decreased to almost 20%. However, for some φ, s_1/D, the error δ remains about 30% or higher. To further reduce the maximum error δ, we remove ψ from equation (2), and the numbers Re, Pr, Fo and the relative amplitude β are grouped into the dimensionless complex $\beta/(\mathrm{Re}\cdot\mathrm{Pr}\cdot Fo)$ (3).

$$\frac{Nu_p}{Nu_{st}} = A\left(\frac{\beta}{\mathrm{Re}\cdot\mathrm{Pr}\cdot Fo}\right)^2 + m\left(\frac{\beta}{\mathrm{Re}\cdot\mathrm{Pr}\cdot Fo}\right) + n \quad (3)$$

Where (A, m and n) are given in Table. 2. As we can see from Table. 2. The maximum error δ decreased to 16.9%. The dimensionless complex equation (4) characterizes the ratio of the pulsation velocity to the stationary one, the increase of which leads to the growth of Nu_p/Nu_{st} irrespective of φ s_1/D and ψ..

$$\beta/(\mathrm{Re}\cdot\mathrm{Pr}\cdot Fo) = Af/u_{st} \quad (4)$$

Where A – pulsation amplitude, m; f – ripple frequency, Hz; u_{st} – speed in steady flow, m/s.

In Fig. 1-4 shows the curves constructed from equation (2) (solid line) and numerical experiment data for all φ and s_1/D and ψ.. According to Fig. 1-4 that equation (3) describes the data of a numerical experiment well. The coefficient of determination averaged $R^2 = 0.97$.

Table 2. The exponents of equation (3)

№	ψ	φ	s_1/D	A	m	n	R^2	Deviation δ. %	
								max	mean
1	0.25	30°	1.25	–0.0075	0.473	1.033	0.950	±14.7	±4.0
2			1.50	–0.1333	0.949	0.981	0.998	±3.4	±1.1
3			1.75	–0.1288	0.932	1.011	0.997	±4.4	±1.4
4		45°	1.25	–0.0279	0.544	1.004	0.978	±9.3	±2.8
5			1.50	–0.0893	0.974	0.958	0.993	±6.5	±1.9
6			1.75	–0.0540	0.863	1.000	0.995	±6.8	±1.4
7		60°	1.25	–0.0364	0.545	1.082	0.957	±14.3	±3.4
8			1.5	–0.0781	0.861	1.058	0.992	±8.1	±2.1
9			1.75	–0.1170	0.914	1.071	0.989	±5.6	±2.3
10		90°°	1.25	–0.0933	0.754	1.027	0.987	±8.4	±2.4
11			1.50	–0.1359	0.946	1.062	0.979	±11.2	±3.2
12			1.75	–0.1403	1.003	1.113	0.970	±10.5	±4.2
1	0.4	30°	1.25	–0.0221	0.532	0.976	0.956	±15.6	±4.1
2			1.50	–0.1510	1.017	0.928	0.997	±5.3	±1.5
3			1.75	–0.1422	0.983	0.966	0.997	±5.6	±1.8
4		45°	1.25	–0.0296	0.581	0.947	0.980	±10.4	±3.0
5			1.5	–0.1066	1.038	0.904	0.994	±6.2	±2.0

6			1.75	–0.0424	0.862	0.976	0.993	±7.3	±2.1
7		60°	1.25	–0.0400	0.543	1.083	0.950	±14.6	±3.7
8			1.50	–0.1251	0.937	1.040	0.990	±8.9	±2.4
9			1.75	–0.1621	0.997	1.058	0.980	±10.1	±3.2
10		90°°	1.25	–0.0867	0.738	0.999	0.980	±10.6	±3.0
11			1.50	–0.1643	1.007	1.027	0.976	±12.6	±3.2
12			1.75	–0.1765	1.076	1.082	0.977	±10.8	±3.5
1		30°	1.25	–0.0468	0.602	0.925	0.965	±14.6	±3.9
2			1.5	–0.1585	1.042	0.887	0.995	±6.8	±2.0
3			1.75	–0.1550	1.025	0.928	0.994	±6.5	±2.2
4		45°	1.25	–0.0612	0.674	0.891	0.986	±8.9	±2.7
5			1.50	–0.1370	1.113	0.861	0.996	±6.0	±1.8
6	0.5		1.75	–0.0707	0.928	0.951	0.990	±11.2	±2.9
7		60°	1.25	–0.0238	0.504	1.081	0.927	±16.2	±4.5
8			1.50	–0.1241	0.946	1.023	0.985	±10.8	±3.0
9			1.75	–0.1787	1.071	1.034	0.977	±10.4	±3.6
10		90°°	1.25	–0.0888	0.742	0.976	0.977	±16.9	±4.1
11			1.50	–0.1405	0.978	1.010	0.981	±11.0	±3.2
12			1.75	–0.1301	0.982	1.083	0.979	±10.6	±3.6

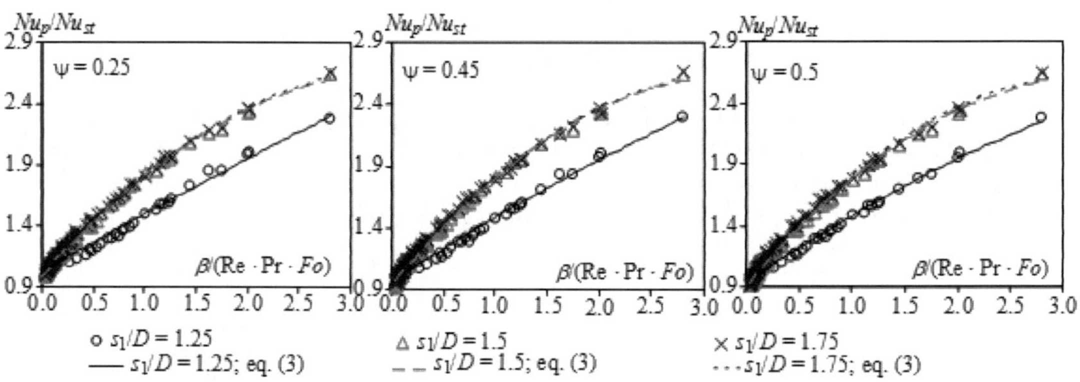

Fig. 1. Dependence of Nu_p/Nu_{st} from $\beta/(\mathrm{Re}\cdot\mathrm{Pr}\cdot Fo)$ for $\varphi = 30°$

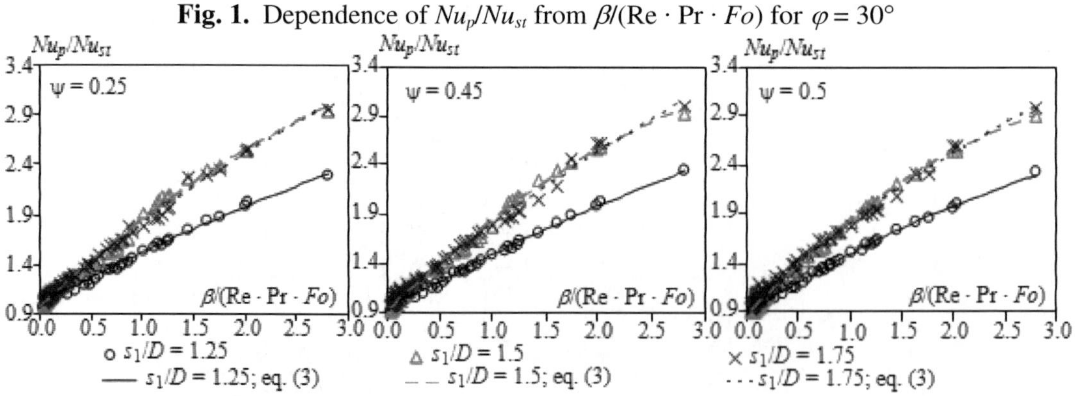

Fig. 2. Dependence of Nu_p/Nu_{st} from $\beta/(\mathrm{Re}\cdot\mathrm{Pr}\cdot Fo)$ for $\varphi = 45°$

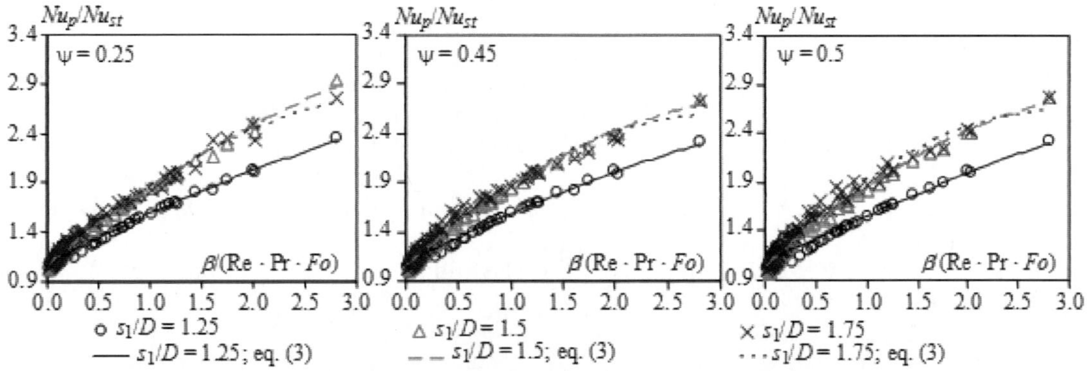

Fig. 3. Dependence of Nu_p/Nu_{st} from $\beta/(\mathrm{Re}\cdot\mathrm{Pr}\cdot Fo)$ for $\varphi = 60°$

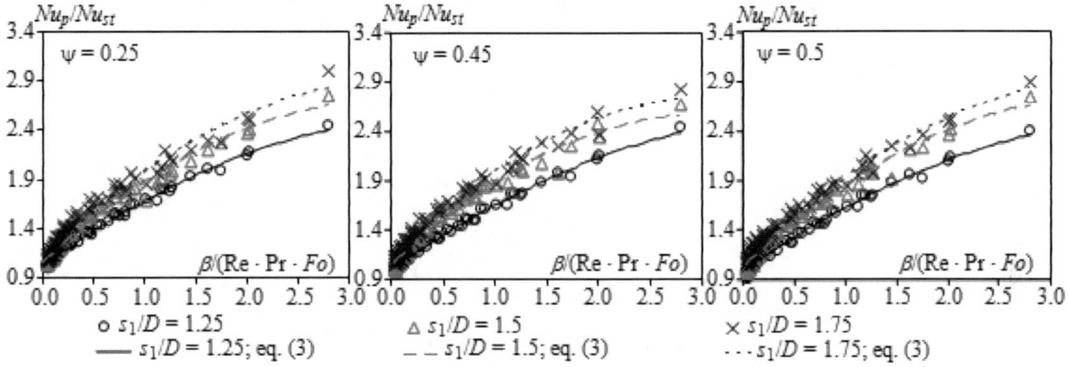

Fig. 4. Dependence of Nu_p/Nu_{st} from $\beta/(\mathrm{Re}\cdot\mathrm{Pr}\cdot Fo)$ for $\varphi = 90°$

Analysis of the obtained empirical correlation (2) made it possible to estimate the influence of the pulsation parameters on the heat exchange for different tube bundles in the presence of flow pulsations. Relative amplitude β has minimal effect on Nu_p/Nu_{st} when $\varphi = 60°$, $s_1/D = 1.75$, and maximum at $\varphi = 45°$, $s_1/D = 1.5$. The Fourier Fo number has a minimal effect on Nu_p/Nu_{st} when $\varphi = 45°$, $s_1/D = 1.25$, and the maximum at $\varphi = 90°$, $s_1/D = 1.75$. When $\varphi = 45°$, $s_1/D = 1.25$, the pulse ripple ψ has the maximum effect on Nu_p/Nu_{st} and has practically no effect on heat transfer when $\varphi = 60°$, $s_1/D = 1.75$.

3. Conclusion

As a result of the study, a new empirical correlation is proposed for calculating the external heat transfer in conditions of pulsating currents in tube bundles. The new empirical correlation (3) has a maximum error of ±16.9%, which is less than the maximum error of the equation presented in [6], which was ±35.7%. A new empirical correlation can be used to calculate heat transfer in pulsating flows in heat exchanging equipment at Reynolds numbers ≤ 1000. For example, shell and tube oil coolers, which are widely used in industry.

Analysis of the obtained empirical correlation (2) made it possible to estimate the influence of the pulsation parameters on the heat exchange for different tube bundles in the presence of flow pulsations. Relative amplitude β has minimal effect on Nu_p/Nu_{st} when $\varphi = 60°$, $s_1/D = 1.75$, and maximum at $\varphi = 45°$, $s_1/D = 1.5$. The Fourier Fo number has a minimal effect on Nu_p/Nu_{st} when $\varphi = 45°$, $s_1/D = 1.25$, and the maximum at $\varphi = 90°$, $s_1/D = 1.75$. . When $\varphi = 45°$, $s_1/D = 1.25$, the ripple ripple ψ has the maximum effect on Nu_p/Nu_{st} and has practically no effect on heat transfer when $\varphi = 60°$, $s_1/D = 1.75$.

The article is executed within the framework of the scientific project 18-79-10136 «Theoretical methods for modeling and developing energy-efficient import-substituting cleaners and deep processing of hydrocarbon raw materials at enterprises of the fuel and energy complex».

References

[1] Tae Ho Ji *et al* 2008 *Int. J. Heat Mass Transfer* **51** 1130–38
[2] Siddhanath V *at al* 2015 *Int. Research J. of Eng. and Tech.* **2** 487–92
[3] Khaibullina A, Khairullin A *et al* 2014 *The European Phys. J. Conf.* **76** 01004
[4] Ilyin V K, Sabitov L S, Haibullina A I and Hayrullin A R 2017 IOP *Conf. Ser.: Mater. Sci. Eng.* **240** 012026
[5] Ilyin V K, Sabitov L S, Haibullina A I and Hayrullin A R 2017 IOP *Conf. Ser.: Mater. Sci. Eng.* **240** 012025
[6] Ilyin V K, Sabitov L S, Haibullina A I and Hayrullin A R 2017 IOP *Conf. Ser.: Mater. Sci. Eng.* **240** 012027
[7] Rouaud M 2013 *Probability, Statistics and Estimation* p 60.
[8] Chiang C L *2003 Statistical methods of analysis* (World Scientific Publishing) p 617

IOP Conf. Series: Materials Science and Engineering **412** (2018) 012026 doi:10.1088/1757-899X/412/1/012026

Energy efficiency of pulsating flows at heat-transfer enhancement in a shell-and-tube water oil cooler

A I Haibullina[1], **L S Sabitov** [1,2], **A R Hayrullin**[1], **V K Ilyin** [1]

[1]Kazan State Power Engineering University, 51 Krasnoselskaya street, Kazan, 420066, Russian Federation
[2]Kazan Federal University, 18 Kremlyovskaya street, Kazan, 420008, Russian Federation

kharullin@yandex.ru

Abstract. In this paper the energy efficiency of pulsating flows during of enhancement heat transfer in a shell-and-tube water oil cooler was studied. A device for enhancement heat-transfer in an oil cooler based on a hydraulic system and a hydraulic cylinder is developed. A shell side flow of the oil cooler underwent pulsating. A Reynolds number of the oil flow based on the outside diameter of the tube was $Re_{oil} = 658$, a number of Prandtl $Pr_{oil} = 293$. It is established that with increasing frequency and relative amplitude of pulsations, an increase in heat transfer commensurate with increasing power consumption. It is shown that the energy efficiency of asymmetric pulsations is higher in comparison with the symmetric pulsations.

1. Introduction

To cool lubricating oil systems of various devices and installations (turbo generators, turbocompressors, hydraulic presses), oil coolers are used. Shell-and-tube water oil coolers were widely used. In shell and tube oil coolers, the oil is cooled by water. It is very often that oil is located outside the pipes (shell side), water inside the pipes (tube side). Therefore, the stable operation of lubrication systems of various devices and plants depends on the stable and uninterrupted operation of oil coolers. The oil temperature at the outlet of the oil cooler may increase due to a number of reasons. For example, in connection with an increase in the temperature of the cooling water or with the fouling of the heat transfer surface, which will lead to an increase in the flow of cooling water. [1].

In order to improve the efficiency of shell-and-tube heat exchangers, passive methods [2] of enhancement heat transfer are mainly used. At the same time, the existing passive methods of heat exchange intensification are difficult to apply for intensification of heat exchange of the existing heat exchange equipment, since they require a design change. However, using the active method of intensifying heat transfer (for example, the use of pulsating flows), it is possible to restore the heat transfer rate of the existing heat exchange equipment without parsing and replacing the structure.

The available pulsating flow studies are mainly devoted to non-stationary heat exchange in separate elements of heat exchange equipment. For example, unsteady heat transfer when flowing through a single cylinder or cylinder flow [3], tube bundles [4-6], different bodies (baffle bodies) [7] or flow in a pipe [8]. Despite the fact that positive results have been obtained in this direction, very few papers assess the energy efficiency of the application of flow pulsations using the example of industrial heat-exchange equipment.

Content from this work may be used under the terms of the Creative Commons Attribution 3.0 licence. Any further distribution of this work must maintain attribution to the author(s) and the title of the work, journal citation and DOI.
Published under licence by IOP Publishing Ltd

IOP Conf. Series: Materials Science and Engineering **412** (2018) 012026 doi:10.1088/1757-899X/412/1/012026

The purpose of this paper is to estimate the energy efficiency of the application of flow pulsations during the intensification of heat transfer in shell and tube oil coolers. Also, the development of a technical solution for the implementation of this method.

2. Methods

In Fig. 1 shows the pulsation machine, which can be used to intensify the heat transfer in the oil cooler.

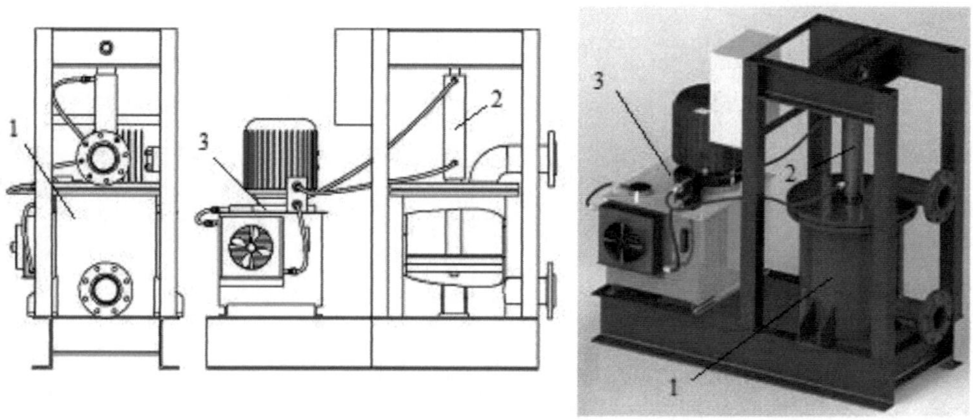

Fig. 1. Pulsation machine:
1 – pulsator; 2 – hydraulic cylinder; 3 –hydraulic system

Consider the principle of operation of the pulsation unit in conjunction with an oil cooler (Fig. 2). The pump 1 sucks the hydraulic fluid from the tank 9 and feeds it to the hydraulic cylinder 8 through the direction valve 7, then the hydraulic fluid returns to the tank through the radiator 6 and the filter 5. The hydraulic distributor serves to redirect the flow of the working fluid. Reciprocating motion of piston of hydraulic cylinder 8 occurs as a result of redirection of the flow of hydraulic fluid. The hydraulic cylinder is connected by a common piston rod with the piston of the pulsator, so the piston of the pulsator also performs a reciprocating motion (upwards).

Thus, the oil flow in the oil cooler reciprocates with a predetermined frequency f with relative amplitude β and pulsating ratio ψ. The required f and ψ are set using the hydraulic distributor. β is determined by the working path length of the pulsator piston.

The ripple frequency f is calculated by the following formula (1).

$$f = 1/T_p \text{ Hz} \tag{1}$$

where $T_p = T_1+T_2$ – ripple period, s (T_1 – pulsator piston up time, s T_2 – pulsator piston down time, s). The ripple pulsation ψ is found by the formula (2).

$$\psi = T_1/T_p \tag{2}$$

The relative amplitude of pulsations β is determined by the formula (3).

$$\beta = A/D \tag{3}$$

where D – outer diameter of tube bundle, m; $A = \langle \dot{V} \rangle / S \cdot T_1$ – the distance that the oil travels in the opposite direction in the tube bundle of the oil cooler, when the pulsator piston moves upward, m. Here S – average cross-sectional area in the bundle of oil cooler tubes, m^2; $\langle \dot{V} \rangle$ – average volume volumetric flow over time T_1, m^3/s calculated by formula (4).

$$\langle \dot{V} \rangle = \frac{1}{T_1} \int_0^{T_1} \dot{V}_t(t)dt \quad (4)$$

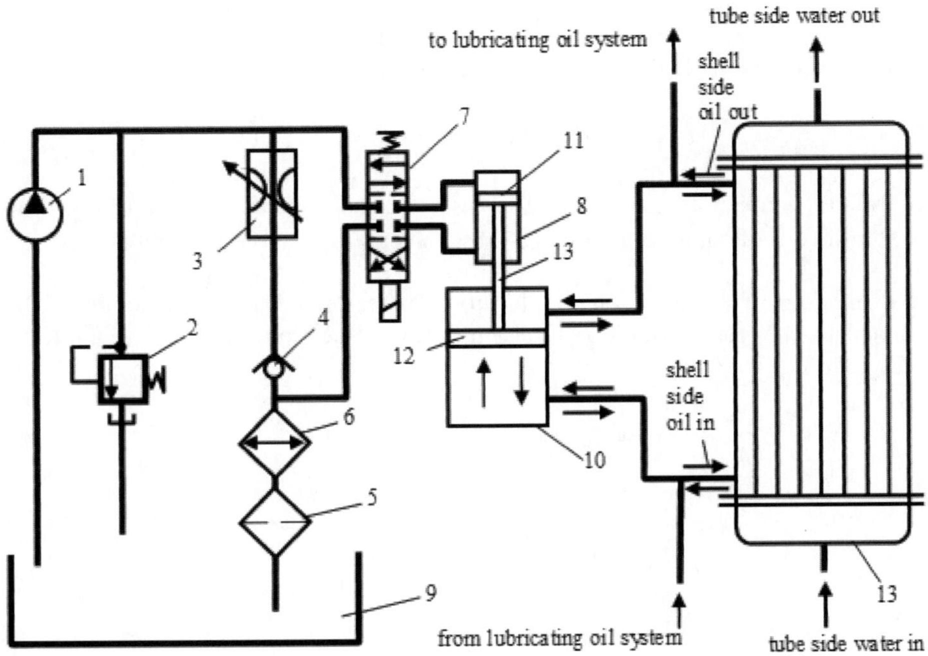

Fig. 2. The scheme of connection of the pulsating machine to the oil cooler:
1 – pump; 2 – relief valve; 3 – flow control valve; 4 – check valve; 5 – filter; 6 – air oil cooler; 7 – direction valve; 8 –hydraulic cylinder; 9 – reservoir; 10 – pulsator; 11 – hydraulic cylinder piston; 12 -pulsator piston; 13 -oil cooler

The energy efficiency of the oil cooler in the operation of a pulsating machine is calculated by the Kirpichev criterion (5) [9].

$$\frac{E_p}{E_{st}} = \frac{Q_p /(N_{pm} + N_p)}{Q_{st} / N_{st}} \quad (5)$$

where E_{st}– Kirpichev's criterion for the operation of an oil cooler in a stationary mode; E_p– Kirpichev's criterion in pulsation mode (when a pulsating machine is operating). N_{pm}– the power required to operate a pulsating machine, W; N_{st}, N_p – power of pumps spent on pumping a cooling and heating medium through a heat exchanger in a stationary and pulsating mode, W, Q_{st} – heat transfer rate of an oil cooler in a stationary mode, W; Q_p– heating capacity of oil cooler in pulsating mode, W.

The value of N_{pm} is calculated depending on the intensity of pulsations and the parameters of the oil cooler. To calculate N_{pm}, the mathematical model presented in [10] was used. N_{st}, N_p are calculated depending on the oil cooler parameters.

The heat output in the stationary and pulsating mode was calculated by formula (6).

$$Q = UA\Delta T_m \quad (6)$$

where A – heat transfer surface area, m^2; ΔT_m – mean temperature difference, °C; U – coefficient of overall heat transfer coefficient, W/(m^2·°C) calculated by formula (7).

$$U = \frac{1}{\frac{1}{h_1} + \frac{\delta}{k_{wall}} + \frac{1}{h_{2(p,st)}}} \tag{7}$$

where h_1 – heat transfer coefficient inside thetubes, W/(m^2·°C) is calculated by equation of Dittus-Bolter [11]; h_2 – coefficient of heat transfer outside of tubes, W/(m^2·°C) is calculated in dependence of the mode. For stationary mode $h_{2(st)}$ is calculated by formula (8).

$$h_1 = Nu_{st}\frac{k_{oil}}{D} = 0{,}354 \cdot \mathrm{Re}^{0,6}\,\mathrm{Pr}^{0,33}\,\varphi^{-0,1} \cdot (s_1 / D)^{-0,45} \cdot (\mu_{water} / \mu_{wall})^{0,14}\frac{k_{oil}}{D} \tag{8}$$

where Nu_{st}– Nusselt number for stationary flow; k_{oil}–thermal conductivity of oli, W/(m·°C); Re – Reynolds criterion; Pr – Prandtl criterion; φ – tube layout angle; s_1 – tube pitch, m; μ – dynamic viscosity, Pa · s.

During the operation of the pulsation unit, h2 (p) was determined by the equation (9, 10). Equation (9) was obtained in [12] for nonstationary heat transfer in the flow of oil in a bundle of tubes of different configurations.

$$\frac{Nu_p}{Nu_{st}} = 0{,}954\,\mathrm{Re}^{-0,201} \cdot \mathrm{Pr}^{-0,211} \cdot \beta^{0,184} \cdot Fo^{-0,230} \cdot \psi^{-0,053} \cdot \varphi^{0,085} \cdot s_1 / D^{0,287} \tag{9}$$

$$h_{2(p)} = h_{2(st)}\frac{Nu_p}{Nu_{st}} \tag{10}$$

where $Fo = a/(f \cdot D^2)$ – Fourier criterion, where a – thermal diffusivity of oil, m^2/s. Equation (9) is found out from the range $215 \le \mathrm{Pr} \le 363$, $0.25 \le \psi \le 0.5$, $100 \le \mathrm{Re} \le 1000$, $15 \le \beta \le 35$,. $5.81 \cdot 10^{-4} \le Fo \le 14.53 \cdot 10^{-4}$. The definition of the parameters entering into equation (9) is given in [12].

3. Results and discussion

Next, energy efficiency is considered when the oil cooler operates in a pulsating mode. The calculation was introduced for a range of parameters: volumetric flow rate of oil $\dot{V}_{oil}$ = 90 m^3/h, water $\dot{V}_{water}$ = 120 m^3/h, heat exchange area A = 63 m^2, Re_{water} = 41450, Re_{oil} = 653, Pr_{water} = 4,5, Pr_{oil} = 293, φ = 60°, s_1/D = 1.25, $0.2 \le f \le 0.5$ Hz, $0.25 \le \psi \le 0.5$,

In Fig. 3 shows the dependence of E_p/E_{st} on β.. According to Fig. 3 that the increase in the relative amplitude of ripple β does not have a significant effect on E_p/E_{st}. An increase in ψ leads to a decrease in E_p/E_{st} (Fig. 4). The ripple frequency f as well as β has no significant effect on E_p/E_{st} (Fig. 5). However, when β = 15, ψ = 0.5, E_p/E_{st} decreases from 0.88 to 0.79. When β= 15, ψ = 0.25, f = 0.312 Hz, the energy efficiency had a maximum value of E_p/E_{st} = 1.03. When β = 15, ψ= 0.25, f = 0.312 Hz, the energy efficiency had a minimum value of E_p/E_{st} = 0.73.

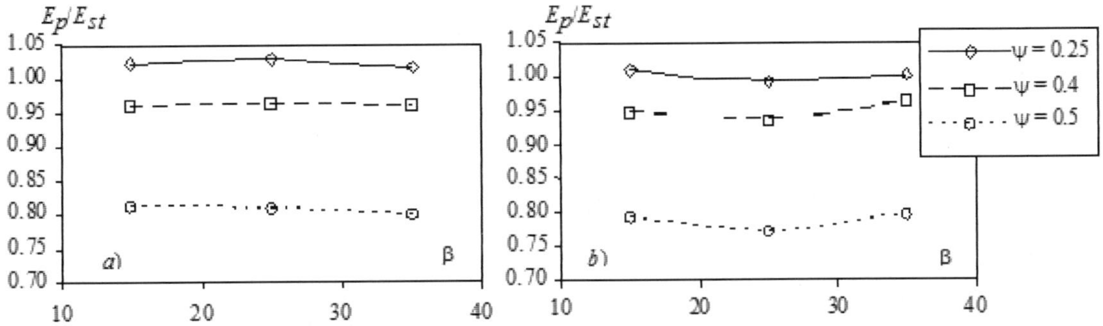

Fig. 3. Dependence E_p/E_{st} от β: *a)* f = 0,312 Hz; *b)* f = 0,5 Hz

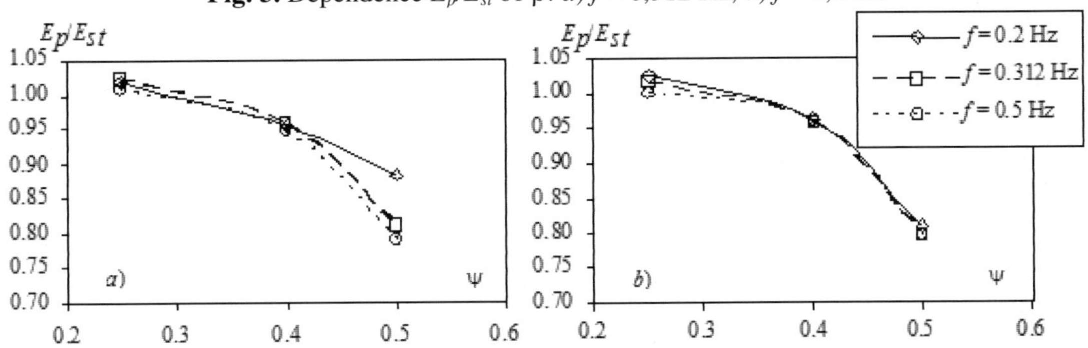

Fig. 4. Dependence E_p/E_{st} от ψ: *a)* β = 15; *b)* β = 35

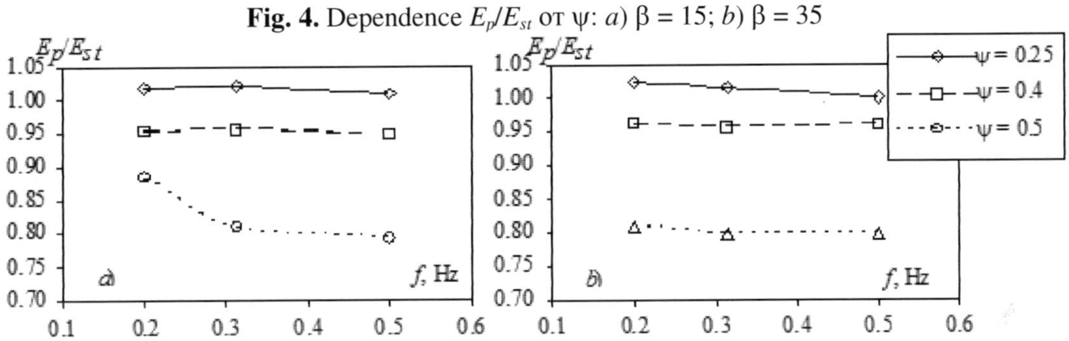

Fig. 5. Dependence E_p/E_{st} от f: *a)* β = 15; *b)* β = 35

4. Conclusion

The conducted research shows that flow pulsations can be used to intensify heat exchange in shell and tube oil coolers. To implement this method, you can use a pulsation machine. The dimensions of the pulsating machine and the power it consumes depend on the intensity of the pulsations and the parameters of the oil cooler. With an increase in the intensity of pulsations, it is possible to achieve a significant intensification of heat transfer. However, an increase in the intensity of pulsations leads to an increase in the power consumption. Therefore, it is necessary to find the optimum between the intensification of heat transfer and the increase in the power consumption. The conducted research shows that pulsation regimes with a minimum ripple ripple ψ have the highest energy efficiency.

Pulsation modes should be selected based on the maximum allowable pressure in the heat exchanger, since the higher the pulsation intensity, the higher the pressure jumps in the oil cooler.

The advantage of the proposed method is that the pulsating machine can be used to increase the heat exchange of existing oil coolers without changing their designs.

The article is executed within the framework of the scientific project 18-79-10136 «Theoretical methods for modeling and developing energy-efficient import-substituting cleaners and deep processing of hydrocarbon raw materials at enterprises of the fuel and energy complex».

IOP Conf. Series: Materials Science and Engineering **412** (2018) 012026 doi:10.1088/1757-899X/412/1/012026

References

[1] Brodova U M *et al* 2003 *Teploobmenniki energeticheskiy ustanovok* (Ekaterinburg: Socrates Press) p 965
[2] Popov I A *et al* 2009 *Fizicheskiye osnovy i promyshlennoye primeneniye intensifikatsii teploobmena.* (Kazan: Centre for Innovative Technologies) p 560
[3] Ashish S and EYK N 2017 *J. of Thermophysics and Heat Transfer* 1–13
[4] Ilyin V K, Sabitov L S, Haibullina A I and Hayrullin A R 2017 IOP *Conf. Ser.: Mater. Sci. Eng.* **240** 012025
[5] Ilyin V K, Sabitov L S, Haibullina A I and Hayrullin A R 2017 IOP *Conf. Ser.: Mater. Sci. Eng.* **240** 012026
[6] Khaibullina A, Khairullin A *et al* 2014 *The European Phys. J. Conf.* **76** 01004
[7] Kim S Y and Kang B H 1998 *Int. J. Heat Mass Transfer* **3** 625–34
[8] Li H 2007 *Int. Conf. on Power Eng.* 23–7
[9] Grigoriev B A *et al* 2017 *J. Phys.: Conf. Ser.* **891** 012151
[10] Hajbullina A I *et al* 2016 *Energy Problems*. **11-12** 56–67
[11] Dittus F W Boellter L M K 1930 *University of California, Berkeley, Publications* **2** 443
[12] Ilyin V K, Sabitov L S, Haibullina A I and Hayrullin A R 2017 *IOP Conf. Ser.: Mater. Sci. Eng.* **240** 012027

IOP Conf. Series: Materials Science and Engineering **412** (2018) 012027 doi:10.1088/1757-899X/412/1/012027

Thermal and hydraulic efficiency of the staggered tube bundle in pulsating flow

A I Haibullina[1], V K Ilyin[1], L S Sabitov[1,2], N H Zinnatullin[3], A R Hayrullin[1], A N Dolgova[1]

[1]Kazan State Power Engineering University, 51 Krasnoselskaya street, Kazan, 420066, Russian Federation

[2] Kazan Federal University, 18 Kremlyovskaya street, Kazan, 420008, Russian Federation

[3]Kazan National Research Technological University, 68 Karla Marks, street, Kazan, 420015, Russian Federation

haybullina.87@mail.ru

Abstract. Determination of the thermal and hydraulic efficiency η is one of the ways to assess the effectiveness of the method used to intensify heat transfer. In this paper, the hydraulic fluid efficiency of diamond-coated tube bundles is determined for a pulsating flow of a heat carrier. The Reynolds number was based on the outer diameter of the pipe and was Re = 100, the Prandtl number Pr ranged from 214 to 363. It was shown that an increase in the dimensionless amplitude β, frequency f, and pulsating ratio ψ leads to a decrease in thermal and hydraulic efficiency. The influence of the Prandtl Pr numbers on the thermohydraulic efficiency was also analyzed. It is shown that an increase in thermal and hydraulic efficiency is observed with increasing Pr.

1.Introduction

To date, there are many papers devoted to the study of heat transfer and hydrodynamics in bundles of pipes in order to improve the efficiency of heat exchange equipment.

In work [1], an experimental method was used to study heat transfer during transverse flow around a row of plates parallel to each other, which is similar to flow around a bundle of tubes. Air was used as the working fluid. The experiments were performed in the frequency range $10 \le f \le 50$ Hz, pulsation amplitudes of $13.33 \le A \le 15.35$ %, and Reynolds numbers of $250 \le \mathrm{Re} \le 2000$. The results show that both the frequency and the amplitude of the pulsating flow are influencing the heat transfer. The maximum intensification of heat transfer (enhancement heat transfer) by 136% was observed at Re = 253, f = 15 Hz, A = 13.35%.

In work [2], the intensification of heat transfer processes was experimentally studied in the artificial creation of a nonstationary flow regime by imposing low-frequency asymmetrical pulsations on the flow in the intertube space of a corridor bundle of pipes on the flow, and an assessment of the influence of such factors as a change in the frequency and ripple pulsation for the chosen range of Reynolds numbers ($0 \le f \le 0.8$ Hz; $0 \le \psi \le 1$; $20 \le \mathrm{Re} \le 100$). The authors show that the imposition of low-frequency pulsations on the liquid flow, with transverse flow around the cylinder located in the center of the corridor bundle of pipes, leads to an intensification of heat exchange in comparison with

Content from this work may be used under the terms of the Creative Commons Attribution 3.0 licence. Any further distribution of this work must maintain attribution to the author(s) and the title of the work, journal citation and DOI.
Published under licence by IOP Publishing Ltd

the steady-state current, on average, to 10%. At the same time, from the point of view of energy efficiency, the greatest effect is observed in pulsations with a degree of asymmetry $\psi = 0.25$, taking into account additional power inputs for pulsation generation.

In [3-5], an experimental method was used to study the heat transfer in a corridor bundle of tubes under conditions of low-frequency pulsations of the liquid flow. Flow pulsations were asymmetrical. As the working fluid, the water Pr = 5.5Re was used in the range $100 \leq \mathrm{Re} \leq 500$, the ripple frequency was $0{,}125 \leq f \leq 0{,}5$ Hz, the amplitude of the oscillation was $1{,}25 \leq A/D \leq 4{,}5$. The maximum increase in heat transfer by 90% was observed at Re = 500, $f = 0.5$ Hz, A/D = 4.5. It is shown that with increasing Re the heat loss decreases, and with increasing f and A / D the heat transfer increases. Also, the positive effect of using low-frequency asymmetric pulsations during heat exchange in a wavy channel was obtained in [6].

The efficiency of the heat exchange equipment depends on the values of the hydraulic resistance and the heat transfer coefficients of the heat exchange elements, which in turn affect the final metal capacity of the heat exchanger and the necessary power of the pumps required for pumping heat carriers. The author [7] studied the thermal-hydraulic efficiency of a discrete-rough channel in a pulsating flow. The evaluation of thermal hydraulic efficiency allows us to conclude that the use of superimposed pulsations in a discrete-rough channel is promising for intensification of heat transfer. The values *И* and η from the Strohal numbers Sh for a discrete-rough channel are given when pulsations are applied to the air flow. The maximum of the coefficient I = 1.33 for a discrete-rough channel is observed for Sh about 2.2 and the amplitude β = [0.4; 0.5]. It is noteworthy that the imposition of pulsations in a discrete-rough channel leads to a large increase in η. The lowest indices of various coefficients of thermal and hydraulic efficiency were obtained at Sh = 0.6, because at this value, the greatest increase in static pressure losses is observed. The obtained results open new possibilities for increasing the thermal and hydraulic efficiency of heat exchangers and cooling systems.

In the previous work [8] thermal and hydraulic efficiency of a corridor bundle of pipes was considered for pulsating flows at Pr = 5.5. In this paper, the thermal and hydraulic efficiency of a diamond beam in pulsating flows for data obtained in the course of numerical modeling in [9] was considered. The Prandett numbers were located in the range $214 \leq \mathrm{Pr} \leq 363$, the Reynolds baizing numbers on the outer diameter of the beam tube were Re = 100.

2. Methods

To estimate the thermal hydraulic efficiency of equation (1), the method proposed in [8] was used.

$$\eta = (Nu_p / Nu_{st}) / (\xi_p / \xi_{st}) \tag{1}$$

Where Nu_{st}, ξ_p – Nusselt number and hydraulic resistance in a steady flow, Nu_{st}, ξ_{st} – Average for the period of pulsations, Nusselt number and equivalent hydraulic resistance during pulsations of the flow. The parameter η in equation (1) is also called the Reynolds analogy factor (FAS), and the condition $\mathrm{Re}_p = \mathrm{Re}_{st}$.

3. Results and discussion

In Fig. 1-3 shows the dependencies Nu_p/Nu_{st}, ξ_p/ξ_{st} and η from Pr, β, f by which it can be seen that an increase in the regime parameters of pulsations β, f leads to a decrease in η, and an increase of Pr to an increase in η. With an increase in the number of Pr, a decrease occurs Nu_p/Nu_{st} и ξ_p/ξ_{st} (рис. 1). When the nature of the pulsations approaches the symmetric $\psi \to 0{,}5$ there is a decrease in the telpohydraulic efficiency η. It is shown that the increase β and f leads to an increase Nu_p/Nu_{st}, and an increase Pr and ψ to decrease Nu_p/Nu_{st} (Fig.1–3).

On Fig. 4–6 it is shown dependencies $\mathrm{Nu}_p/\mathrm{Nu}_{st}$, ξ_p/ξ_{st} and η от β/Fo from different numbers Pr and porosity, which shows that with an increase in the ratio β/Fo, there is an increase in heat transfer and hydraulic resistance in a non-stationary flow compared with a stationary one. Over the entire range β/Fo and Prrost numbers ξ_p/ξ_{st} exceeds the Nu_p/Nu_{st} (Fig. 6). Maximum values of thermal and hydraulic efficiency $\eta = 0{,}27$ (Fig.6 a) and $\eta = 0{,}40$ (Fig.6 b) are observed at a duty ratio $\psi = 0{,}25$ and minimum values of the ratio β/Fo.

On Fig. 7–9 it is shown dependencies Nu_p/Nu_{st}, ξ_p/ξ_{st} and η from $\beta/(Fo \cdot \mathrm{Pr})$ at different coefficients of porosity, which shows that with an increase in the ratio $\beta/(Fo \cdot \mathrm{Pr})$, there is an increase in heat transfer and hydraulic resistance. At the minimum values of the duty cycle $\psi = 0{,}25$ and relation $\beta/(Fo \cdot \mathrm{Pr}) = 28$ thermal hydraulic efficiency has a maximum value $\eta = 0{,}40$ (Fig.9).

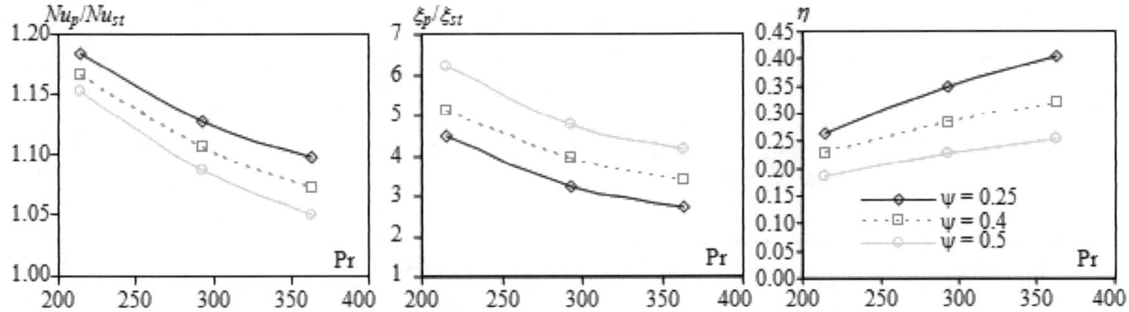

Fig. 1. Dependencies Nu_p/Nu_{st}, ξ_p/ξ_{st} and η from Pr at Re = 100; β = 15; f = 0.2, Hz

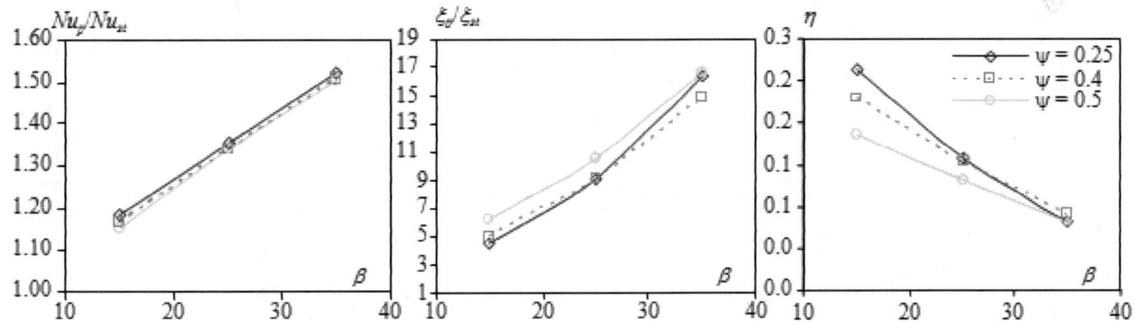

Fig. 2. Dependencies Nu_p/Nu_{st}, ξ_p/ξ_{st} and η from β at Re = 100; Pr = 215; f = 0.2, Hz

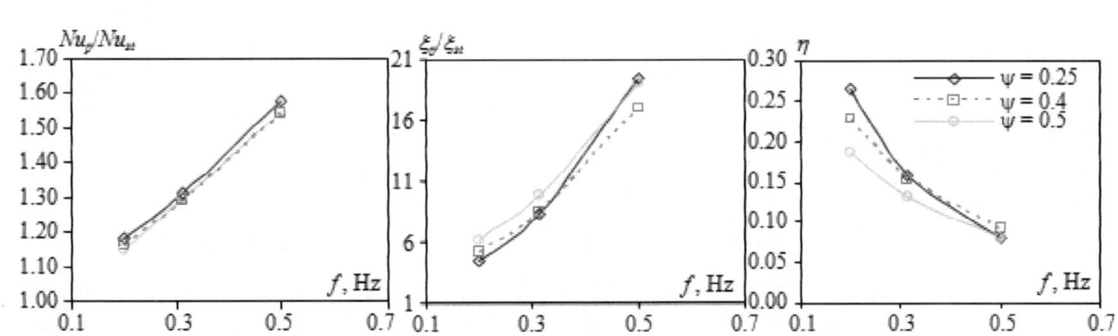

Fig. 3. Dependencies Nu_p/Nu_{st}, ξ_p/ξ_{st} and η from f at Re = 100; Pr = 215; β = 15

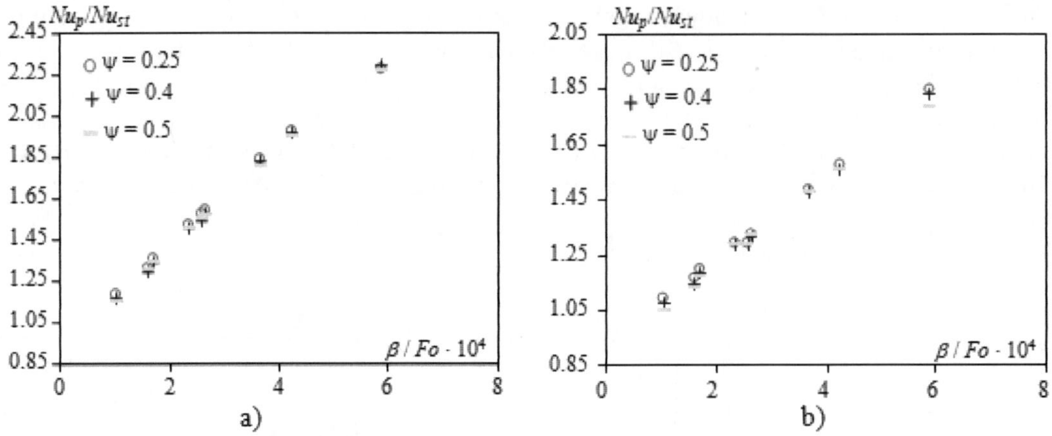

Fig. 4. Dependencies Nu_p/Nu_{st} from β/Fo at Re = 100: a) Pr = 215; b) Pr = 363

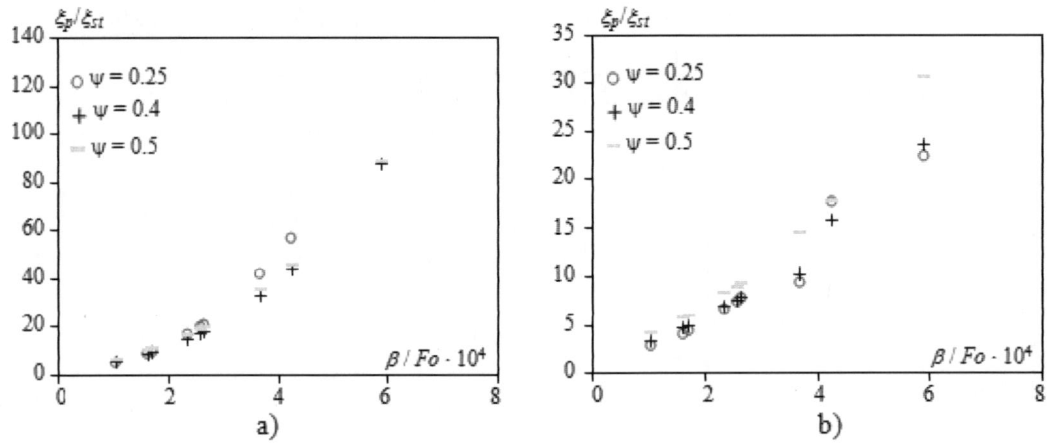

Fig. 5. Dependencies ξ_p/ξ_{st} from β/Fo at Re = 100: a) Pr = 215; b) Pr = 363

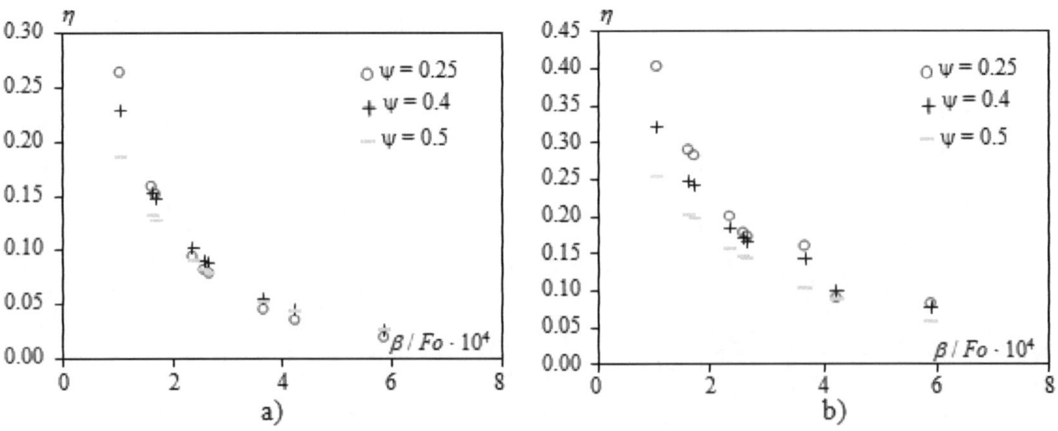

Fig. 6. Dependencies η from β/Fo at Re = 100: a) Pr = 215; b) Pr = 363

IOP Conf. Series: Materials Science and Engineering **412** (2018) 012027 doi:10.1088/1757-899X/412/1/012027

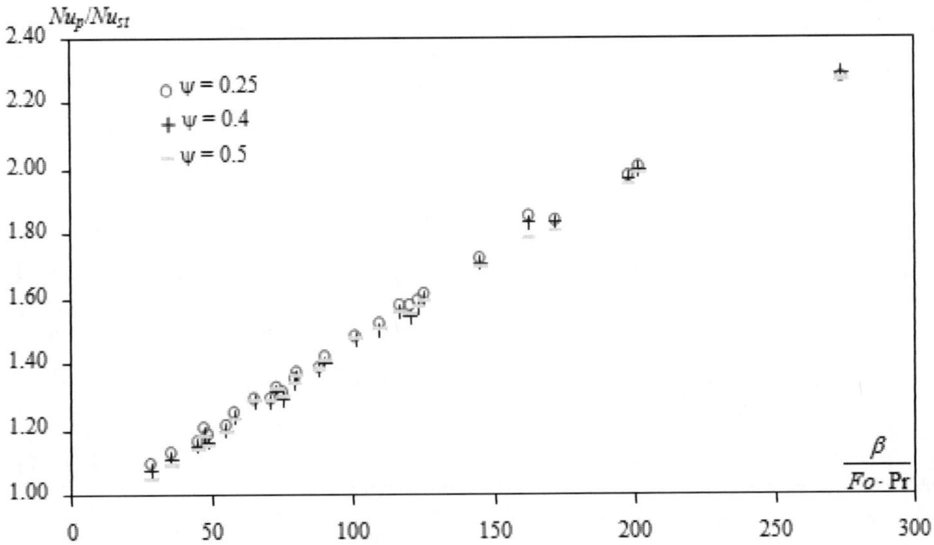

Fig. 7. Dependencies Nu_p/Nu_{st} from $\beta/(Fo \cdot \mathrm{Pr})$ at Re = 100

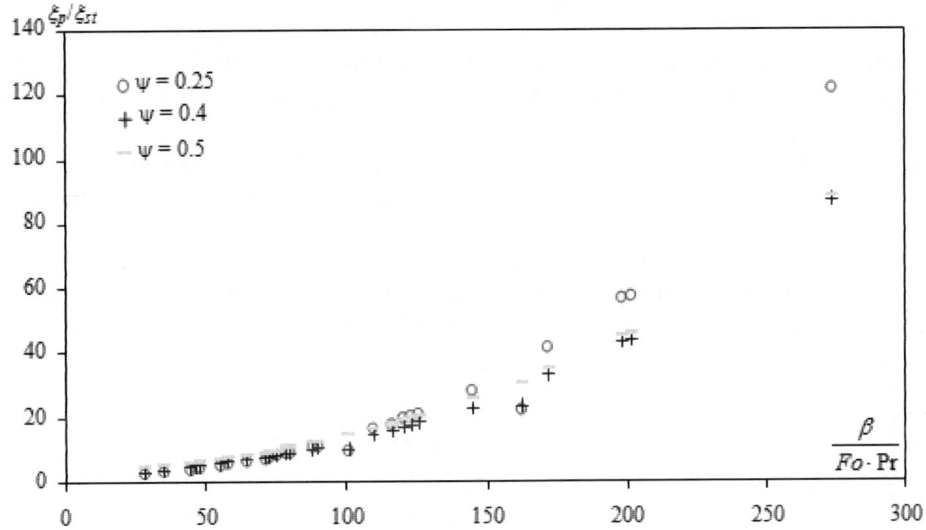

Fig. 8. Dependencies ξ_p/ξ_{st} from $\beta/(Fo \cdot \mathrm{Pr})$ at Re = 100

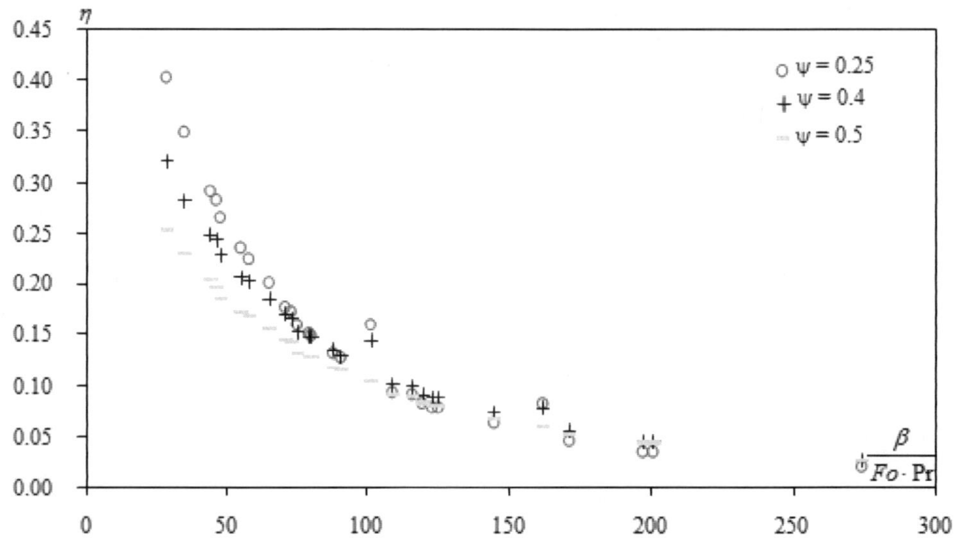

Fig. 9. Dependencies η from $\beta/(Fo \cdot \mathrm{Pr})$ at Re = 100

Conclusion

The hydraulic efficiency of a chess bundle of pipes is estimated at application of pulsations.

It is shown that as the Pr increases, an increase occurs in η, the maximum values of the thermal hydraulic efficiency $\eta = 0.40$ are observed at Pr = 363 and the ripple ripple $\psi = 0{,}25$.

As the frequency f and the dimensionless amplitude β of pulsations, a decrease in η occurs.

When the nature of the pulsations approaches a symmetrical $\psi \to 0{,}5$, the telopihydraulic efficiency decreases η.

The article is executed within the framework of the scientific project 18-79-10136 «Theoretical methods for modeling and developing energy-efficient import-substituting cleaners and deep processing of hydrocarbon raw materials at enterprises of the fuel and energy complex».

References

[1] Wantha C 2016 *Int. J. Heat Mass Transfer* 99141–148
[2] Musaeva D A, Sinyavin A A and Guriyanov A I 2012 *Modern Science: Researches, Ideas, Results, Technologies* 2 (10) 183–186
[3] Khaibullina A, Khairullin A *et al* 2014 *The European Phys. J. Conf.* **7601004**
[4] Khaibullina A I and Ilyin V K 2014 *Energy Problems* 1 pp 211–19
[5] Khaibullina A I, Khayrullin A R and Ilyin V K 2016*Energy Problems* 11 pp 1264–75
[6] Khaibullina A I *et al* 2015 *X-th International Youth Sci. Conf.: TinchurinReadings* 2 p 167
[7] Kolchin S A 2015 H*ydraulic resistance of discrete-rough channel with superimposed pulsations of flow*: dis. for the Candidate. tech. Sciences: 01.02.05, 01.04.14
[8] Ilyin V K, Sabitov L S, Haibullina A I and Hayrullin A R 2017 IOP *Conf. Ser.: Mater. Sci. Eng.* 240 **012025**
[9] Ilyin V K, Sabitov L S, Haibullina A I and Hayrullin A R 2017 IOP *Conf. Ser.: Mater. Sci. Eng.* 240 **012027**

IOP Conf. Series: Materials Science and Engineering **412** (2018) 012028 doi:10.1088/1757-899X/412/1/012028

Intellectual property as a factor of increasing innovation activity of economic entities

L Iksanova[1], **N Kashapov**[2]
[1] Kazan Innovative University named after V. G. Timirjasewa (IEML), 42 Moskovskaya str., Kazan, 420111, Russian Federation
[2] Kazan Federal University, 18 Kremlyovskaya street, Kazan, 420008, Russian Federation

liliyari@mail.ru

Abstract. Today, innovation is a powerful engine of social progress. Innovations determine economic growth in the all spheres of state activity through constant introduction of new technologies, products and services into production, which leads to reduction of dependence on imports of technologies, materials and consumer goods. Effectiveness development of enterprises in dynamically changing external and internal environment that requires constant innovation of production and management technologies depends on the existing intellectual property and the possibility of its generation in the foreseeable future. The intellectual property acts as a binding component of the scientific and industrial spheres, is a factor of increasing the innovative activity of the economic entities. The article proposes the principles of formation and development of the intellectual property of an economic entity, such as: efficiency, planning, dynamism, objectivity, balance, circulation of the intellectual property, synergy, rational transparency and minimization of transaction costs. Necessity of each principle of formation and development of the intellectual property of the economic subject is proved. On the basis of synthesis of approaches existing in scientific literature the author's definition of intellectual property of the economic subject in which presence of these principles is designated is offered. The article substantiates that the intellectual property has a direct impact on the activation of innovation processes due to the fact that intellectual property is the main "driver" scientific research, contributes to activation the processes of territorial and inter-sectoral transfer of innovation, contributes to the processes of industrial cooperation for create new innovative ideas and projects.

1. Introduction

Most scientists agree that innovation has become a major driver of economic and social development. So, according to J. Schumpeter economic dynamics is based on the spread of innovations in various spheres of economic life. The result of innovation is the impact on economic processes or directly on products [1].

The level of development of markets in industrialized countries no longer makes it possible to provide significant advantages of the company only due to material and financial factors. Individual advantages and leadership are increasingly becoming a consequence of the effective use of unique by its nature factors of intangible, intangible nature, including the receipt, processing, analysis, interpretation of information; forecasting, foresight, modeling of changes; the creation, use of commercialization of knowledge, experience and other intellectual products; the formation of an

Content from this work may be used under the terms of the Creative Commons Attribution 3.0 licence. Any further distribution of this work must maintain attribution to the author(s) and the title of the work, journal citation and DOI.
Published under licence by IOP Publishing Ltd

IOP Conf. Series: Materials Science and Engineering **412** (2018) 012028 doi:10.1088/1757-899X/412/1/012028

internal atmosphere of innovative susceptibility, business and emotional cooperation (partnership); the development of direct and communication channels with external audiences [2, p.268]. Thus, the success of work in the market and the solvency of economic entities depend on the existing intellectual property, which is an element of high-tech products for enterprises. In a highly competitive environment, only the company that uses its intellectual resources most productively wins.

The purpose of this article is to prove that intellectual property has a direct impact on the activation of innovation processes. In the process of the study, an abstract-logical method of research was used, including the following set of scientific methods: induction and deduction, analysis and synthesis, analogy, comparison.

2. Result of research

According to B. Santo, innovation is understood as a socio-technical and economic process, which ultimately leads to the creation of unique or the best technical properties of products or technologies, and if the innovation is focused on profit, its implementation brings additional income [3]. E. Hofmeister interprets the concept of "innovation" as a conscious, aimed at achieving a certain goal, the will of turning an invention into a process or method, or the skillful introduction of a product to the market [4]. Also, it should be noted the approach of P. F. Drucker, in whose opinion that innovation is a specific tool of entrepreneurship-an action that gives resources new opportunities to create wealth [5, p.39].

Over the past decade, the concept of "innovation" has come to a wider sense in the business world. Innovations are carried out in two directions. First of all, this concept includes updating or improving existing technologies and expanding the market of products. This direction is less expensive and therefore less risky. The second direction of innovation is associated with the creation of a completely new high-tech scientific and technical development. This direction is more expensive for financing, more risky for the purposes of commercialization and the introduction of development in production requires a longer time [6, p.6].

Of course, the focus on innovation helps to ensure the direction to achieve and maintain high profits. An economic entity with innovation that demands in the market will have a competitive advantage over other market participants. It is necessary understand that without innovation action impossible continue economic and cultural growth on the intensive path of development.

Implementation of innovations is the main factor of competitiveness of products of any country. Now innovations determine economic growth in all spheres of state activity by constant introduction of new technologies, products and services into production. The state plays a decisive role in the formation of the national innovation system. It defines the legal framework, creates infrastructure and ensures growth of sales the high-tech products of industrial enterprises [6, p.6].

Innovative action is a purposeful work of business entities for formation, testing, implementation results of research and development [7, p.29]. Economic entities prefer to invest in the creation of new technologies, increase of intangible assets, consolidation and monopoly of intellectual property rights, opening of new product markets.

Innovation activity is one of the most common characteristics of innovation, reflecting the degree of intensity, timeliness of action, ability to use advanced methods, technologies and potential for achieve scientific, technical and economic effect. The main factors of innovation activity of economic entities of various forms of ownership and economic activities are:

- ownership structure for production means and interest of owners in the long - term development and improvement of the activity of the business entity on the innovative basis;
- intensity of scientific and technological progress in the industry which this business entity belongs;
- availability of available financial resources for increase the level of innovation;
- staffing of the innovation process, its composition and structure;
- presence and pattern of use the intellectual property of the business entity and other factors.

The connecting component of the scientific and industrial spheres perform the intellectual property, thereby integrating science and material production. Intellectual property is the driving force of various industrial, technological, organizational changes, primarily of an innovative nature. A. A. Bovin and L. E. Cherednikova emphasize "intellectual property is the most important category of innovative management" [8].

It should be noted that the first mention of the intellectual property began to appear in scientific works at the end of the XVIII century in the form of the theory of natural law, the essence of which was that everything that was created by man in the form of material and creative objects was recognized as his property, which gave the possibility of exclusive right of disposal. Consider that the tradition of proprietary approach to copyright and patent law occurred in France and was embodied in the Patent law in 1791 [9, p.10], after which the countries that followed the French model of copyright began to recognize the personal non-property rights of authors. However, the idea of copyright as the "most sacred form of property" was embodied in the laws of some US States even earlier in 1789 [10, p.11].

The first international agreement in the field of the intellectual property is the Paris Convention for the protection of industrial property (1883), but the term "intellectual property" came into wide use in 1976, with the signing in Stockholm of the Convention establishing the world intellectual property organization, where the concept of the intellectual property includes both intellectual activity of man in the industrial, scientific, literary and artistic fields and legal protection of its results [10].

In the Russian legislation, the concept of the intellectual property is formulated in the article 1225 of the Civil code of the Russian Federation [11], and the Constitution of the Russian Federation establishes the need to protect intellectual property by law (art.44, 71) [12]. Providing services for the protection of the intellectual property, the state solves the problem of improving their quality, both in the domestic economic and foreign economic environment.

Depending on the sphere of application, there is a fairly significant range of views on the nature and content of the economic concept of "intellectual property". Thus, the legal approach for definition of the intellectual property considers intellectual property as a set of exclusive rights that allow to determine the scope of rights for created results of the intellectual action arising from the right holders. Protection of these rights allows commercial implementation of innovations [13]. According to I. A. Bliznets, "exactly objectively expressed the intellectual action result can participate in economic turnover, become a product, and function in the market. Such object should be protected by the state, society, law" [14, p. 4].

When using an economic approach aimed mainly at determining the effectiveness of the creation and use of the intellectual property, in this case, intellectual property is considered like the system of social relations about the results of intellectual work arising in the process of production, distribution, exchange and consumption.

As a basis for the formation and development of the knowledge economy, intellectual property has such specific features:

- it is not material in the traditional sense, what is its main difference from property in the traditional sense, although the forms that acquire intellectual assets can be embodied in physical (material) objects that have a certain economic value;
- generates income on condition fixing the intellectual property rights and in result does not allow public use;
- give the ability to create a sphere of influence in the market;
- main component of determining the market value of modern enterprises.

Basis of the synthesis of different approaches for the economic entity, we can offer the following meaningful definition: "intellectual property of an economic entity" is intangible assets of organization, and system of accumulated knowledge and competencies of staff which are used for purpose of entity functioning, give the right a benefit from their use and formed in accordance with such basic principles as efficiency, consistency, dynamism, objectivity, balance, synergy, rational transparency, minimization of transaction costs [13].

Let us consider the system for principle of developing the intellectual property for entities with different forms of ownership (table 1).

Table 1
The proposed principles of formatting and developing of intellectual property for an economic entity (based on the materials of their own research) [15, c.113]

Наименование	Содержание
1. Principle of efficiency	Imply obtained profit in result of the using the intellectual property of an economic entity will exceed the costs of its formatting and developing.
2. Principle of planning	Necessity in operating and future planning the need for intellectual property and its sources of funding for organization's divisions and business entity as a whole.
3. Principle of dynamism	Means that intellectual property of an economic entity have to continuous improvement with the trends of changes in the internal and external environment, especially the trends of scientific and technological progress, development of new competencies, rapid circulation the intellectual property elements between the divisions of the organization.
4. Principle of objectify	Formatting and developing the intellectual property of an economic entity should be based on scientifically based methods and models, with a minimum role of the subjective factor.
5. Principle of balance	There should not be a significant imbalance in developing the intellectual property between the individual units of the organization. This kind of imbalance will prevent the rapid and effective transfer of knowledge, technologies, ideas between the organization's units, and will hinder the implementation of joint projects for several departments of the economic entity.
6. Principle of circulation	There should be methods of transfer some types of the intangible assets between the structures of the economic entity, possibility of knowledge exchange and competencies of different units. This principle directly follows of principle of balance and determines the principle of synergy the intellectual property.
7. Principle of synergy	Imply need interaction the intellectual property of different structures for organization with the purpose to take additional economic effect. This principle may also involve the interaction of several types of the intellectual property in different structures of the organization, for example, when specialists of another structure are involved in the implementation of a modern information system at one of the divisions from a large organization on the project basis.
8. Principle of rational transparency	Means necessity providing maximum possible cost transparency for formatting of intellectual property and its management methods in the business entity for order to minimize possible misuse
9. Principle of minimization of transaction costs	The development of the intellectual property should be organized in such a way that in the future ensure minimization of all types of transaction costs arising in the course of transactions between the structures of the economic entity, and its transactions with external

	counterparties in the long term.

It should be accentuate that intellectual property contributes the expansion of the organization's capabilities, increases the level of business reputation, competitiveness, forms the intellectual potential of the enterprise, allows gaining a leading position in the market, stimulates and ensures progress, etc. According to O. V. Gurenkova, in the modern society advances in the area of science and technology began to determine the dynamics of economic growth of States, the level of competitiveness of products and services, the degree of national security and equal integration into the world economy [16 p.4]. Considering the main directions the intellectual property influence on the activation of innovative activity in the economy, we can distinguish the following.

Table 2

The main directions the intellectual property influence on the activation of innovative activity in the economy [17, c.22]

Specific properties and functions of the intellectual property	Directions influence on the activation of innovative activity
1. The intellectual property as the main factor of formatting knowledge economy	Knowledge of employees the institutions of higher education, research institutes, design offices, research departments of corporations and holdings are the main source of fundamental and applied research and development work, which, commercializing, transformed into innovation.
2. Organizational and economic function the intellectual property	The intellectual property is a factor that able to organize different types of capital the economic entity in order to ensure the maximum level of their productivity, including on an innovative basis.
3. Mobility the intellectual property	The intellectual property is a mobile resource. Specialists with knowledge freely move between enterprises of various industries and territories. There is a constant circulation of knowledge in the economy. Thus, the intellectual property helps to transfer of innovative technologies and solutions.
4. Scalability of the intellectual property	Using the intellectual property available for enable and accelerate innovation activity even for the small businesses.
5. Synergy of the intellectual property	The intellectual property contributes to the processes of cooperating of various enterprises, including economic clusters, which allows to implement joint innovative projects, to achieve a positive synergetic effect on the basis of joint generation ideas, developments, formatting of new innovative forms of economic integration.

3. Conclusions

Thus, the intellectual property is directly related with creative process and acting in the form of knowledge, skills of personnel or intangible assets, including licenses and patents for implementation of innovative technologies and methods of production, ensures the activation of research and development work. Not only Intellectual capital is formed and also the involvement of other people in the development of new ideas, technologies, original solutions. On the basis of joint work, including at the level of different kinds of state and municipal enterprises, the institutions of higher education, scientific organizations, special investment and innovation funds and other objects of the state innovation infrastructure, occur generation of ideas and their implementation is achieved, thus there is an innovative modernization, an environment is created that contributes to the formation of the intellectual property market.

In the case of an economic entity, the process of the intellectual property management is implemented through a set of functions, methods and organizational management structures in order to extract profit and increase the amount of capital as a result of the creation and use of intellectual property.

IOP Conf. Series: Materials Science and Engineering **412** (2018) 012028 doi:10.1088/1757-899X/412/1/012028

The obtained results of intellectual activity acquire a certain form and are the objects of legal protection. Protection of these rights allows for the commercial implementation of innovations. Exclusive rights for results of the intellectual action give their owner a competitive advantage until even more perfect object of the intellectual property is created. Therefore, the intellectual property should have the ability to continuously improve in accordance with the trends of changes in the factors of internal and external environments, which determines the principle of dynamism in the formation and further development of the intellectual property.

References

[1] Schumpeter J A 1982 *Theory of economic development* (Moscow, Progress) p 79
[2] Abdullin A L, Arslanov V A, Gazizov I S, Kashapov N F etc. 2014 *Intellectual property* (Kazan, "Fen" of Academy of Sciences of RT) p 268
[3] Santo B 1990 *Innovation as a means of economic development: Per. with Weng*. (Moscow, Progress) p. 296
[4] Hofmeister E 1981 Innovation stamen. WiVon the willingness of rum Technical walls (Berlin and Munchen) p 241
[5] Drucker P F 1992 *Innowacje a przedsiebiorczosc* (Warszawa) p 39.
[6] Abdullin A L, Arslanov V A, Gazizov I S, Kashapov N F etc. 2014 *Regional market of intellectual property. Part 1* (Kazan, Brig) p 6
[7] Paley T F 2011 *Innovation management. Ed. 2nd, revised. EXT* (Kazan, Foliant) p 29
[8] Bovin A A, Cherednikova L E 2014 *Intellectual property: economic aspec* (Moscow,INFRA) p.278
[9] Sergeev A P 1996 *Intellectual property Rights in the Russian Federation* (Moscow, Theis) p. 10
[10] Convention establishing the World Intellectual Property Organisation 1979 *Article 2* (Stockholm)
[11] The civil code of the Russian Federation (part four) from 18.12.2006 №230 –FZ (as amended on 01.07.2017) *Article 1225*
[12] The Constitution of the Russian Federation (adopted by popular vote 12.12.1993) *Articles 44, 71*
[13] Iksanova L R 2018 The principle of rational transparency for forming and developing the intellectual property. *Management of economic systems: electronic scientific journal* №2 (108) URL:http://www.uecs.ru/index.php?option=com_flexicontent&view=items&id=4790
[14] Bliznets I A 2001 *Intellectual property and the law. Theoretical question. Materials on copyright and related rights.* (Moscow) p 4
[15] Iksanova L R 2014 Strategy of developing the intellectual capital of a holding // *Aktual'niye problemy ekonomiki i prava* No 1 (29) p 113
[16] Gurenkova O V 2010 *Formation of a complex of intelligent products for logistics in the market of innovation* (Samara state University of Economics) p 4
[17] Iksanova L R 2014 *Features of intellectual capital influence on innovation activities* Akademisches Magazin Intellekt. Innovationen. *Investition* No 3 p 22

Formation of the diffusion layer in gas boring

V K Ilyin[1], L S Sabitov[1,2], A I Khaibullina[1], O V Ilyin[1]

[1]Kazan State Power Engineering University, 51 Krasnoselskaya street, Kazan, 420066, Russian Federation
[2] Kazan Federal University, 18 Kremlyovskaya street, Kazan, 420008, Russian Federation

l.sabitov@bk.ru

Abstract. Boriding in powder media is successfully used to increase the life of parts. To increase the depth of the diffusion layer and improve its properties, a method for activating the surface of a saturable part by means of electromechanical processing is proposed. The results on the structure of the resulting layer and its dimensions are presented.

Thermodiffusion boronization in powdered media can be successfully used to increase the life of parts operating under abrasive, thermal and chemical effects.

It is known from literature sources [1, 2, 3] that the durability of such a coating depends not only on its hardness, corrosion resistance, but also on the structure, phase composition, and also on the carrying capacity of the diffusion layer.

The task of increasing the bearing capacity of the boride layer is associated with an increase in the mechanical transmission of the layer and, as a rule, is solved by increasing the size of the diffusion layer or by thermal treatment.

To increase the diffusion layer and improve its properties, a saturating mixture containing boron carbide B_4C, introduce additionaly: Na_2BO_7, KF, Cr, Zn [2].

In such a mixture, the following reactions are possible:

$$Na_2B_4O_7 \xrightarrow{T} Na_2O + 2B_2O_3; \quad (1)$$

$$B_4C + 2O_2 \xrightarrow{T} B_2O_3 + CO + 2B_{aT}; \quad (2)$$

$$2Fe + 5B_2O_3 \xrightarrow{Fe} IOB_{aT} + 6O_2 + Fe_2O_3; \quad (3)$$

$$3Cr + 2KF + 2O_2 \xrightarrow{T+KF+O_2} CrF_2 + K_2O + Cr_2O_3; \quad (4)$$

$$3CrF_2 + 2Fe \xrightarrow{Fe} 2FeF_3 + Cr_{aT}; \quad (5)$$

$$Cr_2O_3 + 2Zn \xrightarrow{Zn} Zn_2O_3 + 2Cr_{aT}; \quad (6)$$

$$Cr_2O_3 + 3Zn \xrightarrow{Zn} 3ZnO_2 + 2Cr_{aT}. \quad (7)$$

Content from this work may be used under the terms of the Creative Commons Attribution 3.0 licence. Any further distribution of this work must maintain attribution to the author(s) and the title of the work, journal citation and DOI.
Published under licence by IOP Publishing Ltd

A large number of elements in the saturating mixture assumes its periodic correction. At the same time, the process does not exclude the formation of FeB on the surface of boride, which has lower mechanical properties and a higher index of brittleness as compared to Fe_2B boride, see Table. 1.

The formed atomic chromium diffuses into the part, forming intermetallic compounds, as well as deposits of pure chromium, which slows the diffusion of boron into the interior of the part.

As is known, zinc begins to evaporate at relatively low temperatures. At high temperatures it, like aluminum, can be a reducing agent. With insufficient amount of zinc in the saturating mixture, reactions (6) and (7) become impossible. The increased content of zinc in the mixture leads to its sintering and increase of brittleness of the boride layer. However, maintaining the optimum amount of zinc in the mixture is not possible, since the consumption of zinc depends not only on the temperature and the duration of the process, but also on the total surface area of the parts being processed, and on the presence of carbon in them.

Table 1. Comparative data of borides

№	Indicator name	Fe_2B	FeB
1	Microhardness H_{100} boride phases, GPa	13,8-16,5	17,5-21,0
2	Modulus of elasticity E, MPa	$3 \cdot 10^5$	$4 \cdot 10^5$
3	Poisson's ratio	0,18	0,15
4	The beginning of shearing during plastic deformation,%	3-4,5	1,5
5	The embrittlement factor (steel 45), $100 \cdot 10^3$	5,0	30

Thus, we see that the task of increasing the resource of parts due to the complication of the mixture and the introduction into it (Na_2BO_7), KF, Cr, Zn does not have a stable solution.

We also considered the mechanism of diffusion saturation of steel parts with boron from a powder of boron carbide (B_4C).

Under the influence of temperature, oxidation of boron proceeds according to reaction (2). Boron oxide B_2O_3 transports boron atoms to the saturating surface, where it is reduced by reaction (3). The atomic boron formed by the reactions (2) and (3) diffuses into the surface of the metal. In the diffusion of boron into the workpiece, an α-solution of boron in the iron is first formed, then, when the boron concentration reaches the limiting value, the fluctuation nuclei of the Fe_2B phase will begin to appear and grow.

However, such a sequence of phase formation is possible only if the rate of supply of active boron atoms to the surface is greater than or equal to the diffusion rate into the depth of the component $V_p \leq V_D$. In the case when the rate of formation and supply of atomic boron is much higher than the rate of its diffusion deep into the saturable part, an excessive amount of boron is rapidly formed in the surface layer, its concentration becomes sufficient to form at first, and then FeB, see Fig. 1. Layers with a high FeB phase content have increased brittleness, poorly machinable and have a low bearing capacity.

Obviously, in order for the coating to have better physico-mechanical and operational properties, it is necessary to obtain boride coatings containing the Fe_2B phase. For this, firstly, it is necessary to increase the diffusion of boron from the surface into the interior of the component, and second, to create the necessary activity of a saturating medium.

To realize the first point, we considered the mechanism of boron diffusion in iron. When evaluating the type of solid solutions formed by boron in γ-iron, most researchers [3, 4] consider that with γ-iron boron forms a solid solution of implantation.

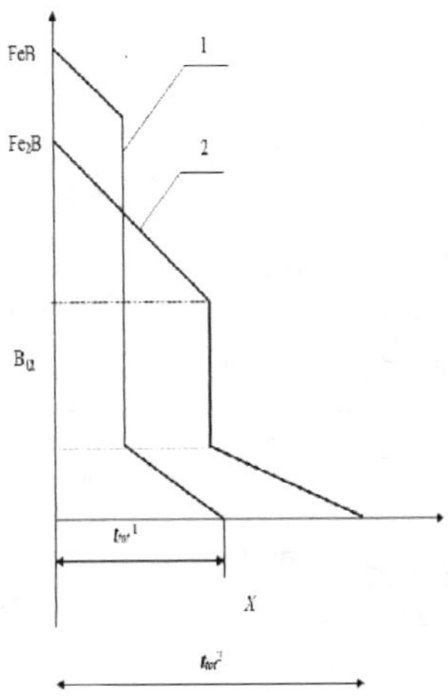

Fig. 1. Kinetics of boride layer growth: 1 – $V_{\Pi}>>V_D$; 2 – $V_{\Pi}\approx V_D$

With respect to α-iron, LG. Voroshin and L.S. Liakhovich in [3] suggested that boron atoms partially replace iron atoms and form a solid substitutional solution, and in part a solid solution of implantation.

Taking into account the mechanism of boron diffusion in iron, in order to increase the depth of diffusion into the interior of the part, the shape and mode of surface activation was chosen. Electromechanical processing (EMO) of the surface of the saturable part was used for these purposes.

To establish the optimum mode of activation of the surface of the parts, a series of experiments was conducted, during which it was established that the main parameters affecting the activation process are: the tool pressing force *F*, the current *I*, the running-in speed *V*, and the tool delivery relative to the workpiece *S*, that the most significant effect on the increase in diffusion thickness is exerted by the first two parameters.

In order to study the influence of the pressing force of the rolling tool and the current strength on the change in the linear dimensions of the parts to be restored and the thickness of the diffusion layer, a number of experiments were carried out. The maximum thickness of the boride layer was obtained after carrying out the activation under the following regimes: F = 400-450 H, I = 450-550 A; V =14m/min, S =0.1 mm/rev, respectively, for steels 45, XHG and 40X. The borating was carried out in powder B4C with the addition of 40% Al_2O_3 at a saturation temperature of 950 ° C and a holding time of 4 h Fig. 2.

An increase in the depth of boron diffusion due to the activation of the surface of the part contributes to the solution of the second problem - and to a decrease in its concentration on the surface, thereby reducing the probability of the formation of a fragile FeB phase.

The need for aluminum oxide in the process of chemical-thermal treatment is considered by many researchers as a measure to exclude sintering of the mixture. At the same time, the introduction of up to 40% of aluminum oxide into the saturating mixture also contributes to the stabilization of the formation of the Fe_2B phase on the surface of the part.

The latter occurs because the dimensions of the boron carbide particles are relatively small - therefore, the contact area of the mixture with the saturable piece is significant and the saturation process at the initial moment is active. After depletion of the particles of boron carbide contacting with the saturation surface, the process slows down, since the arrival of atomic boron from the reaction space to the surface of the saturable part is hampered by the low gas permeability of the boron carbide powder.

IOP Conf. Series: Materials Science and Engineering **412** (2018) 012029 doi:10.1088/1757-899X/412/1/012029

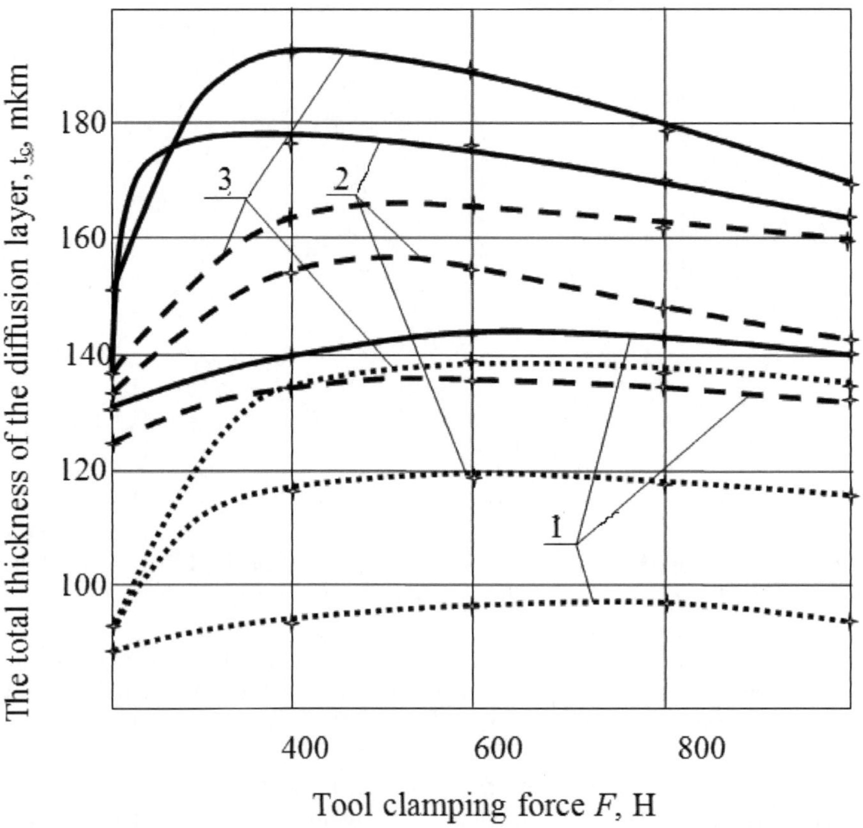

Fig. 2. Dependence of the thickness of the boride layer on the pre-activation regime: ——— steel 45; ········· steel HVG; — — — steel 40X.. EMO mode: 1–*I*=0 A; 2–*I*=1000 A; 3–*I*=500 A. *V*=14 m/min; *S*=0,1 mm/ob. The borating regime: *t*=950 °C; τ=4 hour

The introduction of Al_2O_3 reduces the area of direct contact of boron carbide particles with the saturating surface, thereby reducing the activity of the mixture at the initial moment, increasing its gas permeability, facilitating the connection of not only the boron carbide particles present in direct contact to the saturation process, but also the entire reaction space.

Investigation of the microstructure of the coatings showed that the diffusion coatings obtained in the borning with the preliminary EMO have a characteristic acicular structure. In Fig. 3 shows the microstructure of the diffusion coating on steel 45 and XBG after borning for 4 hours at 950 ° C in a mixture containing 60% boron carbide and 40% aluminum oxide. The most characteristic needle is observed in steel 45 (Fig. 3a). The tops of the needles are wedged into a matrix consisting of a solid solution of boron in iron·

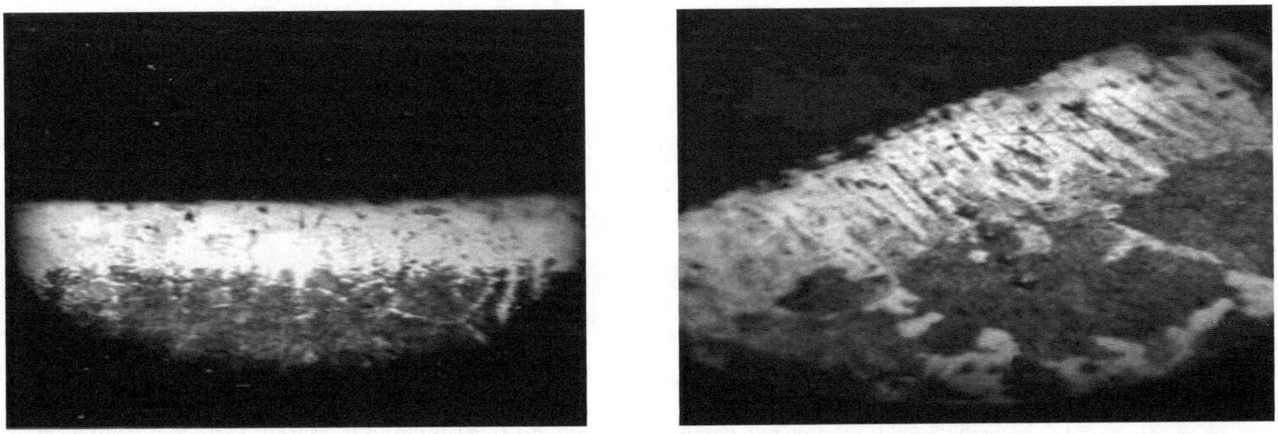

a) b)

Fig. 3. Microstructure of diffusion coatings on steels:
a) steel 45 b) steel HVG

On the HVG steel, the boride needles coalesce into a continuous coating. At the same time, their vertices are shortened, clearly delineating the lower boundary of the layer, under which, as the results of X-ray structural phase analysis have shown, cementite inclusions and a solid solution of boron in iron are found.

Thus, analyzing the results of the studies, we can draw the following conclusion: preliminary activation of the surface of the part by electromechanical processing allows to increase the depth of the diffusion layer, to reduce the presence of the FeB layer, to obtain a diffusion layer with better operational properties.

References

[1] Borisenok GV and others 1981 *Chemical-thermal treatment of metals and alloys: reference book* (Moscow: Metallurgy) p 424

[2] Ilyin V K 2004 *Restoration and strengthening of details of agricultural machinery by methods of diffusion saturation* Diss. the doctors. tech. sciences 297 p

[3] Voroshnin L G, Lyakhovich L S 1978 *Borating of steel* (Moscow: Metallurgy) p 240

[4] Ivanov S G, Guriev A M, Chernykh E V, Guriev M A, Ivanova T G, Garmaeva I A, Zobnev V V, Gong V 2014 *Thermodynamic modeling of reactions in a saturating medium during the diffusion boronization of steels* Fundamental problems of modern materials science Vol 11 No 1 pp 13-16

IOP Conf. Series: Materials Science and Engineering **412** (2018) 012030 doi:10.1088/1757-899X/412/1/012030

Mechanical filtration, based on elective concentration of particles, as an innovative method of water treatment

O R Karataev[1] and E S Karataeva[2]

[1]Federal State Budget Educational Institution of Higher Education «Kazan National Research Technological University», Kazan city, Karl Marx street, bld.68, Russia
[2] Kazan (Privolzhsky) Federal University, engineering institute, Kazan city, Saidashev street, bld.12, Russia

e.s_karataeva@mail.ru

Abstract. The paper describes a method for purification of waste water in a hydrocyclone by solid phase particles. To achieve more efficient cleaning, a technique is proposed for calculating the separation processes in hydrocyclones based on the equation for the radial motion of a particle associated with an air bubble.

Most natural and process fluids represent multiphase heterogeneous systems containing various inclusions and are characterized by nonlinear behavior in porous media. The effects observed in the filtration of heterogeneous systems are very diverse, which predetermines a qualitatively different form of the filtration laws for them [1].

The process of extracting suspended particles from wastewater into the field of centrifugal forces in the hydrocyclone allows to intensify the process and achieve an almost complete recovery of the particles of the solid phase by separating the particles of the smallest fractions [2].

The hydrocyclone (Fig. 1) consists of a cylindrical chamber, into which tangentially through the inlet branch *1*, installed in its upper part, at increased pressure (up to 0.8 MPa), production wastewater is fed, preliminarily saturated with dissolved gas (nitrogen), and of a conical part with a cone angle α. The suspension, that enters the hydrocyclone, drains to form a rotating film, down its walls, having a circumferential V_f, axial V_z and radial V_r velocity components [3]. When the pressure decreases to the atmospheric one, a supersaturation of the dissolved gas is created and the suspension «boils». The solid phase particles under the action of the centrifugal force move to the wall of the hydrocyclone body, and the gas bubbles under the action of Archimedes' pushing centripetal force move towards them to the surface of the film. In the collision of the solid phase particles with gas bubbles formation of flotocomplexes takes place, which carry out the solid phase particles onto the surface of the film into the foam layer, which is removed through the upper branch pipe *2*. Besides, the solid phase particles serve as direct centers for the formation of gas bubbles that are released when the pressure decreases, which leads to a significant increase in the kinetic coefficient of pressure flotation. The clarified suspension is removed from the device through the lower discharge pipe *3* [3].

Content from this work may be used under the terms of the Creative Commons Attribution 3.0 licence. Any further distribution of this work must maintain attribution to the author(s) and the title of the work, journal citation and DOI.
Published under licence by IOP Publishing Ltd

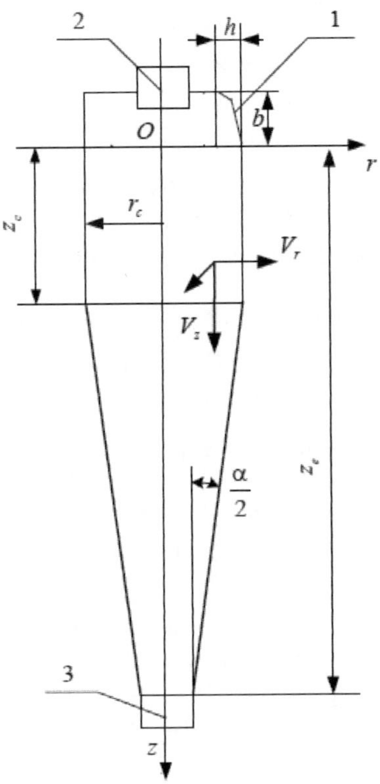

Fig. 1 - Scheme of wastewater treatment in a hydrocyclone under pressure

The simplicity of construction and the absence of moving parts are advantages that distinguish hydrocyclones from other centrifugal principle devices. The separation of heterogeneous systems in the devices of the hydrocyclone principle of operation can be intensified by using the flotation process [4].

We have proposed a technique for calculating separation processes in hydrocyclones based on the equation for the radial motion of a particle associated with an air bubble [5],

$$\mathrm{m}\frac{d^2r}{dt^2} = \pm m\frac{V_t^2}{r}\left(1-\frac{\rho_c}{\rho_f}\right) \mp \beta\left(V_r \pm \frac{dr}{dt}\right) \mp F, \tag{1}$$

where the upper sign under the terms refers to the case of the motion of the particle-bubble system to the wall of the device, and the lower one to the axis of the device; m is a particle mass; V_i is a tangential velocity component; V_r is a radial velocity component; ρ_c; ρ_f is a density of dispersion medium and dispersed phase, respectively; r is a current radius on which the bubble particle is located; β is a Stokes' resistance coefficient; F is Archimedes' force acting on an air bubble.

Taking into account the specific hydrodynamics of the hydrocyclones, after performing a series of transformations, we find the expression for the time function of the displacement

$$t=-0.5\left(\frac{m}{\beta}\right) ln\left|\frac{\frac{A'}{B^*}-r^2}{\frac{A'}{B^*}-r_1^2}\right| - 1.5\left(\frac{m}{\beta}\right) ln\left|\frac{r^2}{r_1^2}\frac{\frac{A'}{B^*}-r^2}{\frac{A'}{B^*}-r_1^2}\right| \mp ln\left|\frac{\frac{A'}{B^*}-r^2}{\frac{A'}{B^*}-r_1^2}\right| \tag{2}$$

$$B^* = \frac{Q_{\text{B}}}{2\pi h'}, \tag{3}$$

$$A' = \frac{\pi(A^*)^2}{6\beta}\left[d_T^3\rho_\Phi\left(1-\frac{\rho_-}{\rho_f}\right) - d_n^3\rho_c\right], \tag{4}$$

where d_T is diameter of a solid particle; diameter of an air bubble d_n can be determined in various ways, for example, through the critical value of the Weber number.

The condition for coupling a bubble to a particle

$$\pi\frac{d_n}{2}\sigma_{zh}sin\theta \geq \frac{\pi d_n^3}{6}\rho_c\frac{A^2}{r^3}+\frac{\pi d_T^3}{6}\rho_f\frac{A^2}{r^3}+\frac{\pi a^2}{4}\left(\frac{4\sigma_{zhg}}{d_n}-1.3\rho_c d_n\frac{A^2}{r^3}\right)+F_{mp} \quad (5)$$

where a diameter of the circle along which the bubble is attached to the surface of the particle; θ is a contact angle of wetting; σ_{zhg} is a surface tension on the liquid-gas section; F_{mp} is a force of viscous friction; A is a constant value for given geometric and mode parameters of the hydrocyclone operation.

Furthermore, a dependence was obtained to determine the basic size of the hydrocyclone - D:

$$D=\frac{2.56V_{vh}(tga)^{0.4}(d_T^3-nd_n^3)}{ndVtg(\frac{a}{2})} \quad (6)$$

where d = d_n when $d_n > d_T$ or d = d_T when $d_n < d_T$, in its turn, in this equation $d_r=d_{min}$ is a minimum diameter of solid particles (droplets), which should be separated in a cylindrical conical cyclone with a diameter of the cylindrical part *D;* V_{vh} is a speed of flow velocity in a feed nozzle of the hydrocyclone with a diameter d_{vh}; *a* is an angle of the conical body part; $n=\frac{\rho_c}{\rho_f-\rho_c}$; ν is a kinematic viscosity of the suspension (emulsion) flow.

Thus, the process of separation of heterogeneous liquid systems in the field of centrifugal forces in hydrocyclones provides high efficiency. This method is relevant for the treatment of waste water in the oil refining sector, since it is a resource-saving.

References

[1] Lagutkin M G and Pavlovsky G V 1999 Analysis of possibility of organizing the flotation process in a hydrocyclone *Thesis report of 12 International scien. conf. «Mathematical methods in engineering and technology»* vol 2 (Velikiy Novgorod) pp 214-215

[2] Masagutov D F, Pushnov A S and Kashapov N F 2014 Influence of geometric characteristics of regular structured attachments on hydraulic resistance and efficiency of heat and mass transfer processes *Chemical Technology* **9** vol 15 pp 564-571

[3] Yablonsky V O 2012 Choice of design and modes of operation of the hydrocyclone for wastewater treatment of PVC production by pressure flotation *Bulletin of Volgograd State Technical University* **5** vol 1 pp 68-70

[4] Mongait I L and Rodziller I D 1958 *Methods of wastewater treatment* (Moscow: State scientific and technical edit) p 251

[5] Karataev O R, Markina A A and Shamsutdinova Z R 2014 Wastewater treatment of petrochemical industry using crystallization and flotation methods *Bulletin of Kazan Technological University* vol 17 pp 351-355

IOP Conf. Series: Materials Science and Engineering **412** (2018) 012031 doi:10.1088/1757-899X/412/1/012031

Peculiarities of calculating the longitudinal mixing of adsorbent in a fixed layer with account of surface friction

O R Karataev[1] and E S Karataeva[2]

[1] Federal State Budget Educational Institution of Higher Education «Kazan National Research Technological University», Kazan city, Karl Marx street, bld.68, Russia
[2] Kazan (Privolzhsky) Federal University, engineering institute, Kazan city, S.Saidashev street, bld.12, Russia

e.s_karataeva@mail.ru

Abstract. The paper presents a theoretical one-dimensional model of the dynamics of adsorption in a fixed layer, taking into account the first-order chemical reaction. The general case is considered when the resistance to mass transfer is concentrated both in the external and internal phases.

1. Equilibrium in sorption processing

Sorption treatment of wastewater is most rational if they contain predominantly aromatic compounds, nonelectrolytes or weak electrolytes, dyes, unsaturated compounds or hydrophobic (for example, containing chlorine or nitro groups) aliphatic compounds. If only inorganic compounds are contained in water, as well as lower monohydric alcohols, this method is not applicable.

Between the amounts of the substance adsorbed by the sorbent and remaining in the solution, in dilute solutions a equilibrium occurs, that obeys the law of distribution, according to which both longitudinal mixing and the flow of substance are simultaneously taken into account due to a chemical reaction (a negative source of substance) [1, 2].

$$\frac{\partial C(x,\tau)}{\partial \tau} + \delta \frac{\partial C_a(x,\tau)}{\partial \tau} + W \frac{\partial C(x,\tau)}{\partial x} = D_L \frac{\partial^2 C(x,\tau)}{\partial x^2} - K_c C(x,\tau), \quad (1)$$

$$C_a(x,\tau) = AC^*(C_a) - B. \quad (2)$$

Here A and B are constants, $C_a(x,\tau)$ is a concentration of the adsorbed substance in the sorbent in the cross section x at a particular time τ ; $C(x,\tau)$ is a concentration of adsorbent in the flow at a distance x at a particular time τ ; W is a flow rate; K is a mass transfer coefficient; K_c is a reaction rate constant; $\delta = \frac{(1-\varepsilon)}{\varepsilon}$ is a coefficient; ε is a fraction of the free section of the adsorbent (a volume-constant porosity of the fixed layer); C^* is a concentration of the target component in the equilibrium flow with an average content of the adsorbent C_a in the layer; D_L is a coefficient of the longitudinal diffusion (mixing).

The first two terms on the left-hand side of equation (1) are the rate of change in the mass of the target component in the gaps between the particles and inside the particles, respectively. The third term corresponds to the increment in the mass of the target component due to convective transport

Content from this work may be used under the terms of the Creative Commons Attribution 3.0 licence. Any further distribution of this work must maintain attribution to the author(s) and the title of the work, journal citation and DOI.
Published under licence by IOP Publishing Ltd

with the flow. The first and second terms on the right-hand side (1) take into account the change in the mass of the target component due to molecular diffusion (longitudinal mixing) and the behavior of the chemical reaction, respectively.

In the initial section of the fixed layer at an arbitrary moment of time τ the concentration of the target component is constant and equal to c_0, since in (1) the second derivative of the concentration in time is presented, owing to the account of longitudinal mixing [3].

The driving force of the processes under consideration is the difference between the working (excess) pressure over the initial solution and the osmotic pressure of the solution. The most promising are reverse osmosis and ultrafiltration in local wastewater treatment systems with their small expenditures for concentrating and isolating the relatively valuable components and water purification. For example, purification of wastewater, generated during metalworking, etching, painting, plating from toxic substances such as cyanides, chromium, nickel, copper, zinc and others, which can also be regenerated [4].

2. Calculation of longitudinal mixing of adsorbent in a fixed layer with account of surface friction

Transformations (2) with due account allow to write down the following:

$$\frac{\partial C(x,\tau)}{\partial \tau}+W\frac{\partial C(x,\tau)}{\partial x}=D_L\frac{\partial^2 C(x,\tau)}{\partial x^2}-\delta K\left[C(x,\tau)-\frac{C_a(x,\tau)}{A}-\frac{B}{A}\right]-KC(x,\tau),$$
$$0\le x<\infty, 0<\tau<\infty, \tag{3}$$

$$\frac{\partial C_a(x,\tau)}{\partial \tau}=K\left[C(x,\tau)-\frac{C_a(x,\tau)}{A}-\frac{B}{A}\right],\ 0\le x\le\infty, 0\le\tau>\le\infty. \tag{4}$$

Unlike microfiltration and ultrafiltration, impurities trapped in the reverse osmosis process have dimensions at the level of molecules, associates, ions, acid residues.

The value of the flow passing through the reverse osmosis membrane is just as important as its selectivity with respect to various types of dissolved substance. One of the reasons for the formation of the surface layer is the special nature of material loading under friction. The main distinguishing feature of the friction process is the discrete, stochastic character of the frictional contact interaction [5]. This type of interaction for surfaces that come into contact for the first time is due to an initially imperfect surface relief. In the case of interaction of the used surfaces it is due to the presence of a microrelief (equilibrium roughness), formed on the surfaces in the process of friction [3]. In the proposed model, the material is represented by a set of N elements representing power centers and having the ability of translational motion. The viscous force of interaction between two elements (the concentration of substance C from the time of separation) is determined by the relative speed of motion of the elements and is directed opposite to the vector of this velocity [4].

$$F(x,s)\bullet\!\!-\!\!\circ\, C(x,\tau)=\left[C_0F(\tau)-B\delta\gamma\int_0^{\tau}F(t)F_1(\tau-t)dt\right]e^{\frac{Wx}{2a^2}}+$$
$$+B\delta\gamma\begin{cases}\dfrac{1}{\sqrt{a_2}}e^{-\frac{a_1\tau}{2}}\sin\sqrt{a_2}\tau \ \ when\ a_2>0,\\ \dfrac{1}{\sqrt{-a_2}}e^{-\frac{a_1\tau}{2}}sh\sqrt{-a_2}\tau\ \ when\ a_2<0.\end{cases}, \tag{5}$$

$$........\Phi(x,s) \bullet\!\!-\!\!\circ\, C_a(x,\tau) = \left\{ k\,C_0 \int_0^{\tau} e^{-\gamma(\tau-t)} F(t)dt - k\,B\,\delta\,\gamma \times \right.$$

$$\left. \int_0^{\tau} e^{-\gamma\xi} \left[\int_0^{\tau-\xi} F(t)F_1(\tau-t)dt \right] d\xi \right\} e^{\frac{Wx}{2a^2}} + \frac{k\,B\,\delta\,\gamma\,e^{-\gamma\tau}}{\left(\gamma^2 - a_1\gamma + a_0\right)} \times \quad (6)$$

For each material layer, you can define properties different from those of the other layers, so different pairs of layers can have different characteristics. This approach allows you to design a microelevation from several materials with different properties, thereby modeling the material with coatings.

The result of solving the temperature problem and, consequently, the behavior of the friction system as a whole is strongly influenced by the choice of the boundary conditions [5].

References

[1] Karataev O R and Ananikov S V 2013 Dynamics of adsorption taking into account the chemical reaction during cleaning of swimming pool drains Herald of Kazan Technological University **14** (Kazan: Kazan Technological University Press), pp 59-65

[2] Doetsch G 1958 Guide to practical application of Laplace transform (Moscow: GIFML) p 208

[3] Panteleev A V and Yakimova A S 2001 *Theory of functions of complex variable and operational calculus in examples and problems* (Moscow: High school) p 445

[4] Dwight T B 1964 Tables of integrals and other mathematical formulas (Moscow: Science) p 228

[5] Rubtsov V E 2004 Modeling of deformation and thermal processes in surface layer of elastoplastic material in friction: Candidate of physical and mathematical sciences Diss. (Tomsk) p 145

IOP Conf. Series: Materials Science and Engineering **412** (2018) 012032 doi:10.1088/1757-899X/412/1/012032

Modeling of dynamics of vapor compression cooling system

D L Karelin, A V Boldyrev, S V Boldyrev and A M Belousov

Naberezhnye Chelny Institute, Kazan Federal University, Naberezhnye Chelny, Republic of Tatarstan, Russian Federation

karelindl@mail.ru

Abstract. A mathematical model of cooling system with a vapor-liquid compression unit is presented. The modeling of dynamics of the parameters of the vapor compression system during the system start-up in the cooling mode was carried out. It is noted that with the accepted assumptions the evaporation and condensation temperatures stabilize fast enough: in the evaporator - in 0.5 s, in the condenser - in 2 s.

1. Introduction

The cooling system is one of the main elements responsible for stabilizing the thermal state of power units. As a result of intensive development of industries producing heat and electric engines for mobile machines in the direction of increasing specific power and environmental safety, the issue of increasing the efficiency of their cooling remains actual. Works of many authors [1-6] focus this issue offering various options for intensifying the process of heat exchange, in particular, by increasing the temperatures of the coolant. But this method is applicable not for all systems.

There are papers [7, 8] which authors propose to increase the efficiency of cooling systems by increasing heat transfer due to phase transitions of coolants in heat exchangers. In particular, the works [9, 10] propose a cooling method using vapor compression cooling systems (VCCS), combining the advantages of the temperature increasing and the phase transition of the coolant. Modeling of VCS parameters dynamics can be used to predict the duration of heating and cooling of power units, heat exchangers and equipment, as well as other specific time intervals for data collection which is important in the design of control units.

2. Materials and methods

In this paper, authors propose a mathematical model of VCCS (Figure 1), which includes models of compressor, heat exchangers (condenser and evaporator), expansion valve, and allows to calculate the transient characteristics of condensation and evaporation (boiling) temperatures.

The basic equations of the mathematical model of VCCS which were used to study the work of the cycle are presented below.

Content from this work may be used under the terms of the Creative Commons Attribution 3.0 licence. Any further distribution of this work must maintain attribution to the author(s) and the title of the work, journal citation and DOI.
Published under licence by IOP Publishing Ltd

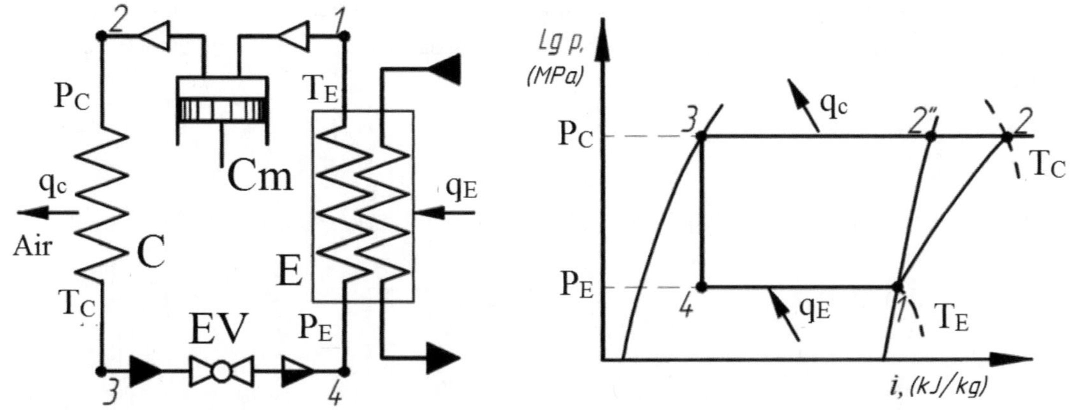

Fig. 1 Circulation of the working agent and the thermal cycle of the vapor compression cooling system (Cm – compressor, C – condenser, E – evaporator, EV – expansion valve)

1. Compressor

Specific work of adiabatic compression was determined by the equation:

$$l_a = P_1 \cdot v_1 \cdot \frac{k}{k-1} \cdot \left[\left(\frac{P_2}{P_1} \right)^{\frac{k-1}{k}} - 1 \right]. \qquad (1)$$

Specific indicator work of compressor

$$l_i = \frac{l_a}{\eta_a}, \qquad (2)$$

here η_a – the adiabatic efficiency for the rotary vane compressor was assumed to be 0.82 [11].

2. Condenser

Heat transfer coefficient of the working agent during condensation in round tubes [11]:

$$\alpha_c = 1{,}26 \cdot \frac{\lambda_c \cdot {\gamma'_c}^{0,1}}{(r_c \cdot v_c)^{0,5} \cdot {\gamma''_c}^{0,1} \cdot \sigma_c^{0,3}} \cdot q^{0,5} \cdot L^{0,35} \cdot \left(d_{c\_inner}\right)^{-0,25}. \qquad (3)$$

Heat transfer coefficient of air in the condenser which is a tubular heat exchanger with a staggered arrangement of finned tubes [12]:

$$\alpha_{c.air} = 0{,}23 \cdot \frac{\lambda_{c.air}}{b} \cdot \left(\frac{\upsilon_{f\,c.air} \cdot b}{\nu_{c.air}} \right)^{0,65} \cdot \left(\frac{b}{d_{c\_out}} \right)^{0,54} \cdot \left(\frac{h}{b} \right)^{0,14} \cdot \varepsilon_c \cdot \varepsilon_z. \qquad (4)$$

Differential energy equation of condenser (two-phase region):

$$G_c \cdot C_{p_c} \left(\frac{dT_c}{dt} \right) = G_c \cdot (i''_2 - i_3) - F_c \cdot \frac{1}{\frac{1}{\alpha_{c.air}} + \frac{\delta_{c.tube}}{\lambda_{c.tube}} + \frac{1}{\alpha_c}} \cdot \frac{(T_c - T_{air}') - (T_c - T_{air}'')}{\ln\left(\frac{(T_c - T_{air}')}{(T_c - T_{air}'')} \right)}. \qquad (5)$$

3. Evaporator

Heat transfer coefficient during boiling of working agent in round tubes [13]:

$$\alpha_e = 0{,}65 \cdot 10^5 \cdot \frac{\lambda'_e}{d_{e\_inner}} \cdot \left(\frac{\upsilon''_e \cdot d_{e\_inner}}{\nu''_e} \right)^{0,73} \cdot \left(\frac{\upsilon'_e \cdot d_{e\_inner}}{\nu'_e} \right)^{-0,73} \cdot \left(\frac{\nu'_e}{a'_e} \right)^{0,3} \cdot \left(\frac{L_e}{d_{e\_inner}} \right)^{-1,69}. \qquad (6)$$

Heat transfer coefficient of antifreeze in a multi-pass shell-and-tube heat exchanger-evaporator was determined from the similarity equation for the flow around a bundle of tubes with a staggered arrangement [11]

$$\alpha_{e.af} = 0{,}41 \cdot \frac{\lambda'_{e.af}}{d_{e.out}} \cdot \left(\frac{\upsilon_{e.af} \cdot d_{e.out}}{\nu_{e.af}} \right)^{0,6} \cdot \left(\frac{\nu_{e.af}}{a_{e.af}} \right)^{0,33} . \qquad (7)$$

Differential equation of energy equilibrium for the evaporator (two-phase region):

$$G_c \cdot C_{p_e} \left(\frac{dT_e}{dt} \right) = F_e \cdot \frac{1}{\frac{1}{\alpha_{e.af}} + \frac{\delta_{e.tube}}{\lambda_{e.tube}} + \frac{1}{\alpha_e}} \cdot \frac{(T'_{af} - T_e) - (T''_{af} - T_e)}{\ln\left(\frac{(T'_{af} - T_e)}{(T''_{af} - T_e)} \right)} - G_c \cdot (i_1 - i_4) .$$

(8)

Formulating of differential equations required following assumptions: iso-enthalpy of throttling process in the expansion valve; constancy of antifreeze and air temperatures; constant mass flow of coolants; equality of the temperatures of the inner surface of the walls of the evaporator tubes and the condenser at the boundary of the vapor skin to the temperatures Te, Tc of the processes on the diagram, respectively (Fig. 1), use of the heat transfer equation (6) for the slug-annular, annular flow regime.

3. Results and Discussion

As a result of the calculations carried out by equations (2) - (8), the transient characteristics of temperatures of the working agent during evaporation and condensation at the start-up of the system in cooling mode and ambient temperature of 25°C were obtained (Fig. 2).

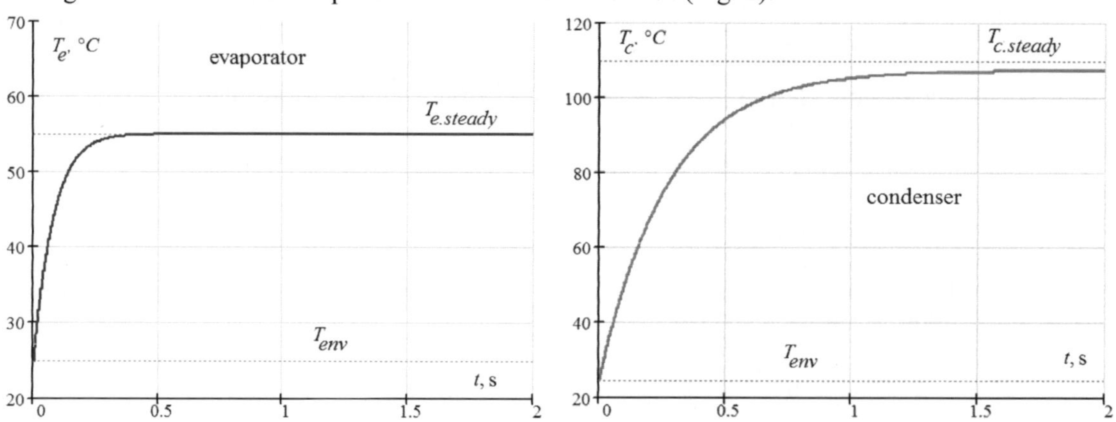

Fig. 2 The transient characteristics of temperatures of the working agent during evaporation and condensation at the start-up of the system in cooling mode

It was noted that, due to the assumptions in the mathematical model, the temperatures stabilize pretty fast (Fig. 2) due to the small values of the masses of the working agent m_e and m_c contained in evaporator and condenser, calculated by the standard method. And besides the evaporation temperature reaches a steady value within 0.5 s, and the condensation temperature within 2 s, that may be caused by significant difference in the heat capacities of air and antifreeze.

4. Conclusion

Further development of the mathematical model assumes taking into consideration: changes of temperature of antifreeze and air and mass flow of coolant; the formation of a vapor skin on the inner surfaces of tubes; the cost of heat for heating the heat exchangers themselves, the mathematical model of the compressor drive, and adjustment of the expansion valve.

References

[1] Livencev F L 1964 High-temperature cooling of internal combustion piston-engine (Moscow: Mashinostroenie) p 204

[2] Patrahalcev N N and Savastenko A A 2004 *Boosting of internal combustion engines* (Moscow: Legion-Avtodata) p 176

[3] Krivov V G, Sinatov S A, Kim F G and Ustinov N A 1986 Heatsink into the space behind the cooling jacket of the forced diesel engine during its high-temperature cooling *Engine construction* No 11 pp 5–11

[4] Kravchenko S A 1993 Diesel power plant of the mainline diesel locomotive on the basis of high-forced diesel locomotive engine with a heat recovery system and high-temperature cooling (St. Petersburg: CNIDI) p 24

[5] Mkrtumyan E A 1938 Cooling of diesel engine at elevated temperatures of cooling water *Proceedings of Red flag Moscow MMI named after N.E.Bauman* Vol 38-39/4 p 72

[6] Altynova N E 1984 Intensification of heat transfer in the cooling system of powerful electric converter devices (Novocherkassk: NPI) p 197

[7] Sklifus Ya K 2015 Reducing the energy consumption of the diesel engine cooling system by changing the functional scheme and the heat transfer method (Rostov-na-Donu: RGUPS) p 159

[8] Utilenko A I 2009 Principles of design of highly efficient cooling systems for electronic devices (Ryazan) p 419

[9] Karelin D L, Gureev V M and Mulyukin V L 2015 Simulation of a cooling system with a vapor-liquid compression unit *Bulletin of KSTU named after A.N. Tupolev* (Kazan: editorial board of KNRTU-KAI) No 5 pp 5–10

[10] Karelin D L 2017 Method for calculating the parameters of the thermodynamic cycle of a vapor compression cooling system *Proceedings of Academenergo* (Kazan) No 3 pp 23–31

[11] Rozenfeld L M and Tkachev A G 1960 *Refrigerating machines and devices* (Moscow: GITL) p 666

[12] Voronin G I and Dubrevskij E V 1973 *Effective heat exchangers* (Moscow: Mashinostroenie) p 96

[13] Avchuhov V V and Payuste B Ya 1986 *Tasks for heat and mass transfer processes* (Moscow: Energoatomizdat) p 144

IOP Conf. Series: Materials Science and Engineering **412** (2018) 012033 doi:10.1088/1757-899X/412/1/012033

Synthesis of the automatic control system with time-delays as in the case of internal combustion engine

V S Karimov

Kazan Federal University, Naberezhnye Chelny Institute, 423812, Russia, Naberezhnye Chelny, Prospekt Syuyumbike 10A, E-mail: kvs-chelny@mail.ru

Abstract. There was conducted the synthesis of the automatic control system with time-delays for multivariable plants in terms of input terminal and the state based on the system embedding technology. To reproduce the plant state vector the compensatory-observation scheme was used, compensating temporary time-delays in controlling. Illustrated by multivariable model of the internal combustion engine the law of control was realized.

Time-delays, concentrated in the ducting of a plant, has a negative impact on controlling processes and can lead not only to the deterioration of the quality of the controlling processes but to the stability of the control system as a whole. Consequently, it's vital to take them into account while designing an automation system. The problem of controlling by such like plants is complicated because of their multivariable nature.

The article offers a precise analytical solution of the task of the synthesis of the automatic control system with time-delays (ACS) for multivariable plants in terms of controlling and the state based on the system embedding technology [1, 2]. It is suggested that the phase plant state vector is unavailable for direct observation. The matrix transmittance function of the regulator or the states determining it is identified whereby the functioning of the control system will be described by the required matrix transfer function (MTF). The synthesis is conducted through the forced element of the closed-loop dynamic system. Illustrated by multivariable model of the internal combustion engine the law of control is realized, which compensates the input time-delays and the achievement of the desired effect in the control system.

Let the observed and the controlled line permanently installed plant with concentrated time-delays in controlling and state be represented as difference-difference equation.

$$\dot{x}(t)=\sum_{i=0}^{l}A_i x(t-\tau_i)+\sum_{j=0}^{r}B_j u(t-\theta_j),\quad y(t)=Cx(t), \tag{1}$$

where $\tau_0=0$, $0<\tau_1,\tau_2,...,\tau_l$ – the permanent time-delays with matrixes A_i, $i=0,1...,l$, $\theta_0=0$, $0<\theta_1$, θ_2, ..., θ_r – the permanent time-delays with matrixes B_j, $j=0,1,...,r$, $u(t)\in R^s$ – the

Content from this work may be used under the terms of the Creative Commons Attribution 3.0 licence. Any further distribution of this work must maintain attribution to the author(s) and the title of the work, journal citation and DOI.
Published under licence by IOP Publishing Ltd

vector of input variables, $y(t) \in R^m$ – the vector of output variables, $x(t) \in R^n$ – the phase plant state vector. The matrixes A_i have a size of $n \times n$, B_j – $n \times s$, C – $m \times n$.

The initial data are set taking into account the time-delays of signals in controlled plants (CO) – as a matter of form we will consider the negative time span $t < 0$, implying that the plant involved dynamic processes before the start time.

$$x(t) = \varphi_x(t),\ t_0 - \tau \le t \le t_0,$$

where τ – the maximum time-delay.

Let the compensatory-observation control scheme, calculating the current phase plant vector $x(t)$ be represented by the equations:

$$\dot{x}_м(t) = \sum_{i=0}^{l} A_i x_м(t-\tau_i) + \sum_{j=0}^{r} B_j u(t-\theta_j),\quad y_м(t) = LCx_м(t),$$

$$\dot{\hat{x}}(t) = \sum_{i=0}^{l} A_i \hat{x}(t) + Ly(t) - y_м(t) + \sum_{j=0}^{r} B_j u(t),\qquad (2)$$

where $\hat{x}(t)$ – the phase vector of the observer state, $x_м(t)$ – the phase vector of the plant model state, $y_м(t)$ – the vector of output of plant model, L – the matrix of the observer.

The analysis of the performance of the compensatory-observation scheme described by equations (2) showed that it enabled to completely compensate time-delays in controlling and reproduce phase plant $x(t)$ state vector.

The application of Laplace transformation to the equations (1) gives the operator form of the description of a plant:

$$px(p) = \sum_{i=0}^{l} A_i e^{-\tau_i p} x(p) + \sum_{j=0}^{r} B_j e^{-\theta_j p} u(p),\quad y(p) = Cx(p). \qquad (3)$$

Let the control law in the general case be described by matrix equation:

$$v(p) = K(p)\hat{x}(p) + u(p), \qquad (4)$$

where $K(p)$ – matrix transfer function of the size regulator $n \times s$, $v(p) \in R^s$ – the control vector at the input of the system.

Let the required performance of automatic control system (ACS) be set by matrix transfer function $E_y^v(p)$ from control input v to output y.

The task is: it's necessary to find MTF of the regulator $K(p)$ or states defining it for CO (1), where the functioning of ACS will be described by MTF $E_y^v(p)$.

Furthermore during the synthesis of the control system we will use the system embedding technology [1, 2]. Considering the equations (1), (2), (4) and the procedure performance of the system enclosing method the matrix in question (promatrix) of the task under consideration will have the following view:

$$\Omega(p)=\begin{bmatrix} (pI_n-\sum_{i=0}^{l}A_ie^{-\tau_ip}) & 0 & 0 & 0\ \ 0 & -\sum_{j=0}^{r}B_je^{-\theta_jp} & 0 \\ 0 & (pI_n-\sum_{i=0}^{l}A_ie^{-\tau_ip}) & 0 & 0\ \ 0 & -\sum_{j=0}^{r}B_je^{-\theta_jp} & 0 \\ 0 & 0 & (pI_n-\sum_{i=0}^{l}A_i)-L\,I_m & & -\sum_{j=0}^{r}B_j & 0 \\ -C & 0 & 0 & I_m\ \ 0 & 0 & 0 \\ 0 & -LC & 0 & 0\ \ I_m & 0 & 0 \\ 0 & 0 & K(p) & 0\ \ 0 & I_s & -I_s \\ 0 & 0 & 0 & 0\ \ 0 & 0 & I_s \end{bmatrix}.$$

The input matrixes α и β, used during the inputs systems have the following view: $\alpha=[0\ \ 0\ \ 0\ \ 0\ \ 0\ \ 0\ \ I_s]^{\mathrm{T}}$, $\beta=[0\ \ 0\ \ 0\ \ I_m\ \ 0\ \ 0\ \ 0]$ while $\omega=E_y^v(p)$, where ω – the image of synthesized system – the required transfer function of the system.

After the completion of the procedures of inputs technique – the consecutive matrixes factorization $\Omega=\Sigma\Xi$, $\alpha=\Sigma\delta$, $\beta=\pi\Xi$, $\omega=\pi\delta$, one can receive equations which should be satisfactory for MTF of the regulator *K*(*p*) during the synthesis in forced element $E_y^v(p)$ of the closed-loop dynamic system.

To solve matrix equations which appear as a result of application of inputs procedures it is possible to apply the device of matrix canonization [3].

The application of the system embedding technology during the synthesis in forced element produces the following equation as for the transmission matrix in question *K*(*p*):

$$C(pI_n-\sum_{i=0}^{l}A_ie^{-\tau_ip})^{-1}\sum_{j=0}^{r}B_je^{-\theta_jp}-E_y^v(p)=E_y^v(p)K(p)(pI_n-\sum_{i=0}^{l}A_i)^{-1}\sum_{j=0}^{r}B_j.$$

A great number of regulators in this case can be described by the formula:

$$\begin{aligned}\left\{K(p)\right\}_{\mu,\eta}&=(E_y^v(p))^{\%}[C(pI_n-\sum_{i=0}^{l}A_ie^{-\tau_ip})^{-1}\sum_{j=0}^{r}B_je^{-\theta_jp}-E_y^v(p)]\times\\&\times((pI_n-\sum_{i=0}^{l}A_i)^{-1}\sum_{j=0}^{r}B_j)^{\%}+\overline{E_y^v(p)}^{R}\mu(p)+\eta(p)\overline{(pI_n-\sum_{i=0}^{l}A_i)^{-1}\sum_{j=0}^{r}B_j)}^{L},\end{aligned}\tag{5}$$

where $\eta(p)$, $\mu(p)$ – arbitrary fractionary polynomial matrixes of the corresponding sizes. The terms of the equation solution (5), and consequently the existence of abundance of solutions (6) have the following view:

$$\begin{aligned}&\overline{E_y^v(p)}^{L}[C(pI_n-\sum_{i=0}^{l}A_ie^{-\tau_ip})^{-1}\sum_{j=0}^{r}B_je^{-\theta_jp}-E_y^v(p)]=0,\\&[C(pI_n-\sum_{i=0}^{l}A_ie^{-\tau_ip})^{-1}\sum_{j=0}^{r}B_je^{-\theta_jp}-E_y^v(p)]\overline{((pI_n-\sum_{i=0}^{l}A_i)^{-1}\sum_{j=0}^{r}B_j)}^{R}=0.\end{aligned}\tag{6}$$

Thus, we get the ratio (5) with the terms of existence of this solution (6), which allows to find a great number of regulators, satisfying the control law (4).

As in the case of synthesis of the control system of the internal combustion engine [4] the efficiency of the problem solution under consideration was exemplified.

References

[1] Asanov A Z 2007 *The system embedding technology and its applications: textbook* (Ufimsky State Aviation Technical University – Ufa: USATU) 227 p
[2] Bukhov V N 2006 *Inputs system. An alytical approach to the analysis and synthesis of matrix systems* (Kaluga: Scientific Literature Press by N.F. Bochkareva) 720 p
[3] Bukhov V N, Rjabchenko V N, Kosjanchukh V V, Zjabin E Y 2002 *Solution of linear matrix equations through idealization method* Kievsky University Vestnik. Edition: physico-mathematical sciences Is 1 pp 19–28
[4] Lino Guzzella, Christopher H Onder 2010 *Introduction to modeling and control of internal combustion engine systems (*Springer Science & Business Media, 2nd ed) 260 p

IOP Conf. Series: Materials Science and Engineering **412** (2018) 012034 doi:10.1088/1757-899X/412/1/012034

Influence of introduction of robotics on increase in efficiency of electrochemical production

N F Kashapov[1], **I I Khafizov**[1], **I G Nurullin**[1] **and Z B Sadykov** [1]

[1]Kazan Federal University, 18 Kremlyovskaya street, Kazan, 420008, Russian Federation

kashnail@gmail.com, khafizov@kpfu.ru, opros16@yandex.ru, zbsadykov@kpfu.ru

Abstract. The relevance of the study of the impact of the introduction of robotics to improve production efficiency is that it affects the level of development of domestic engineering. The article offers an economic justification of the use of the robot to automate the process of electrochemical machining of workpieces, in order to improve the quality of parts and improve production efficiency

The modern enterprises for increase in competitiveness and efficiency more and more attention is paid by production modernizations. An important role in this process is played by application of robotics.

The tendency of increase in the park of robots in modern industrial production is caused by a number of objective factors. As a rule, this increase in labor productivity at quality preservation of production at simultaneous reduction of expenses and a possibility of quick response to changes of objects of production and the consumer market [1].

Development of the direction of robotics has in recent years allowed to create effective decisions for production of small-scale and piece production with use of industrial robots. But at the moment, in Russia many enterprises, especially in mechanical engineering, use machines and technologies of semicentennial prescription which have already become outdated and don't show high efficiency [2].

As many scientists consider: the main push for rapid development of the domestic industry is introduction of new technologies. But unfortunately, the most part of hi-tech production are imported into Russia from the abroad as products of domestic production are noncompetitive or don't answer on these or those technical characteristics [3-7].

Robotics advantage - flexibility of application and a possibility of use in almost unlimited number of processes. Use of robots is possible, both separately, and in a complex. Feature of industrial robots – flexibility of change-over. For release of new products in most cases it is enough to enter other program. Such property is irreplaceable for the processing production as a half of volume of production becomes average or small parties.

If industrial lines aren't robotic, then on creation of products which are issued small volumes, only 5% of the general working hours will be spent. Other 95% will be occupied by setup of the machine, change of tools, loading and unloading of details, etc. If for production of the same party to use robotic

Content from this work may be used under the terms of the Creative Commons Attribution 3.0 licence. Any further distribution of this work must maintain attribution to the author(s) and the title of the work, journal citation and DOI.
Published under licence by IOP Publishing Ltd

production, then the ratio of time will be considerably increased towards productivity. Other positive effect from use of robots – economy of raw materials and materials.

Industrial robots have very high precision of positioning - 0,05-0,1 mm, and the repeatability allows to reach the necessary level of processing of a product at minimization of a manufacturing defect. The exception of a human factor in technological process considerably reduces percent of working mistakes, and use of robots results in constant repeatability during all industrial cycle [8].

In the sphere of processing of preparations less robots, than are involved in others, but it doesn't mean that their application is inefficient. Robotization allows to solve successfully many problems in the field of welding, test, cutting, control, etc. Use of automatic machines in production allows to move long time the heated heavy details at great speed and to work in hostile environment.

Application of robotization in the sphere of processing is justified in that case when on one robot several machines are necessary, and the automatic machine is programmed not only on installation of a detail, but also on her removal, transportation on the following stage, etc.

Having studied activity of several plants, we have revealed the following problem. At many enterprises when processing a surface of products use already outdated methods, manual human skills on sites with harmful working conditions. And communication with it at the exit turns out finished goods with the hung-up quantity of the rejected production, with high prime cost.

For change of a form, the sizes and receiving a surface with small roughness with high qualitative characteristics, it is possible to use electrochemical processing of metals.

Electrochemical metal working is one of ways of giving of a detail of a certain form, the size and also character of a covering of a blanket. In other words, such processing can alter a metal detail and cover her surface with special dusting by means of electrochemistry.

And also at electrochemical processings can be provided high precision [9].

For carrying out similar processing the highly specialized equipment - electrochemical machines is used. Because of low productivity such technological installations are created only under a certain process, but the similar equipment is irreplaceable. Such machines can have access to those parts of preparation of metal, where not to reach any in a different way metal working.

By unique these machines are done that after their impact on a detail, there is no need to use polishing machines or the grinding equipment which only the highly skilled personnel have to operate.

And also at these enterprises process of loading of the raw preparations for processing isn't automated: preparations are loaded into machines manually. The operator manually places preparation on the machine, slowing down thereby a production cycle and doing it to more expensive. And because of not exact laying of preparations in machines, in processing of details the high level of marriage appears.

In addition, introduction of robots instead of the person in these spheres positively affects health of the last as working conditions are connected with high temperatures and hostile environment.

For the solution of these problems we offer robotization of process with use of the ABB Robotics IRB 6620 LX robot. One robot can replace work of four people. The cost of the ABB Robotics IRB 6620 LX robot - $120 thousand; service life without repair of 10 years, with a condition of observance of rules of service. When studying opportunities on process automation, the choice I have fallen on this robot, because of his production characteristics and also because of the best ratio "price quality" [10].

Check of number of preparations at machines, their sorting according to a form and dimensions, control of the obtained data on compliance to the standard parameters specified in the drawing, etc. also belongs to automatic machine duties. That the robot could manage to perform necessary operations, it will have own transport network. And also development and creation of systems of automatic obtaining accuracy of processings and the systems of active control will help improvement of quality of finished products.

Robotization of process of loading of preparations on the electrochemical machine and their electrochemical processing will allow to increase accuracy, by stabilization of full processing, and

IOP Conf. Series: Materials Science and Engineering **412** (2018) 012034 doi:10.1088/1757-899X/412/1/012034

productivity due to reduction of the main and auxiliary time of processing (in the main time of subadjustments and preparatory and final time) [11].

Table 1 - Accounting of costs of quality before introduction of robotics

Main work	The costs of compliance with process		Costs due to non-compliance	
	The cost components	The value in rubles	The cost components	The value in rubles
Loading presets	The cost of loading raw workpieces for processing	32 hours*200 rubles= 6400 rubles	Losses due to the mismatch of materials, technology and tooling	600 rubles* 8 pieces = 4800 rubles
Quality management	The cost control of materials, processes and production, quality analysis, complaints handling	4 hours *150 rubles = 600 p.	Losses due to delay of operational quality information from the consumer. Losses due to failure of the evaluation criteria for acceptance	2000 rubles
Acceptance and storage	The cost of incoming inspection and storage	8 hours * 100 rubles = 800 rubles	Losses due to the mismatch between storage conditions	100 rubles *2 hours = 200 rubles
Process analysis and management	The cost of collecting, analyzing information, developing measures to improve the process	150 rubles *10 hours = 1500 rubles	Losses due to the lack of operational feedback necessary to transfer the process to a statistical managed state	2000 rubles
	Total costs of the compliance process	9300 rubles	The total cost due to imbalance	9000 rubles
	The full cost of the process		18300 rubles	

Table 2 - Accounting of costs of quality after introduction of robotics

Main work	The costs of compliance with process		Costs due to non-compliance	
	The cost components	The value in rubles	The cost components	The value in rubles
Loading presets	The costs of organizing work, the work load unprocessed workpieces for processing	7 hours*100 rubles= 700 rubles	Losses due to the mismatch of materials, technology and tooling	600 rubles * 3 pieces = 1800 rubles
Depreciation operation and maintenance	Depreciation and maintenance costs of the robot	7 hours * 700 rubles= 4900 rubles	Losses due to improper maintenance of the robot	100 rubles
Quality management	The cost control of materials, processes and production, quality analysis, complaints handling	2 hours *150 rubles = 300 rubles	Losses due to delay of operational quality information from the consumer. Losses due to failure of the evaluation criteria for acceptance.	1000 rubles
Acceptance and storage	The cost of incoming inspection and storage	8 hours * 100 rubles = 800 rubles	Losses due to the mismatch between storage conditions	100 rubles *2 hours = 200 rubles
Process analysis and management	The cost of collecting, analyzing information, developing measures to improve the process	150 rubles *5 hours = 750 rubles	Losses due to the lack of operational feedback necessary to transfer the process to a statistical managed state	1000 rubles
	Total costs of the compliance	7450 rubles	The total cost due to imbalance	4100

process			rubles
The full cost of the process		11550 rubles	

For assessment of expediency of introduction of the robot from quality, the decision to carry out accounting of costs of quality on process before introduction of robotics has been made. For accounting of costs of quality it is possible to use the POD method (prevention, assessment, defects), recommended for use by the ISO 9004 international standard which cornerstone the structure of costs of ensuring quality offered by A. Feygenbaum is.

Having carried out accounting of costs of quality, we compare costs of quality before introduction of robotics and after introduction to estimate expediency of introduction of robotization. On table 1 accounting of costs of quality on process before introduction of robotics (average values on branch are taken), and is given in table 2 - after introduction. Data are provided with calculation of costs of loading 1000 pieces of the raw preparations.

Apparently, from tables with accounting of costs of quality, introduction of robotics for this process it is expedient as allows to lower by 36,9% of costs of quality on process of electrochemical processing of products. After her introduction also decreases the expenses provided on mitigation of consequences of accidents, injuries and on certification of jobs. The consumed energy decreases: the robot can work in the conditions of smaller illumination. And also other advantages of use of the robot:

- Decrease in losses from marriage: introduction of the robot allows to reduce quantity of the rejected production from not exact installation of preparations on machines to 3 pieces on 1000 products;
- Minimization of expenses on a manpower: economy on the wages fund of 1600 thousand rubles a year;
- Because of economy of working space, reduction of time of a production cycle;
- Increase in labor productivity, for the account reduction of time of a production cycle and decrease in quantity of the rejected production;
- An exception of a human factor, and therefore – low coefficient of marriage;
- High efficiency of the machine. Lack of human weaknesses: work in the round-the-clock mode, with stable result.

By sight the simple solution of use of the ABB Robotics IRB 6620 LX robot for separate auxiliary process allows to reduce processing execution time significantly. In general, all this will lead to improvement of quality and decrease in product cost. Use of the robot is more ergonomic, and his arrangement in the top part of the conveyor to provide considerable economy of working space.

Introduction of robotics in production leads to improvement of quality of production at simultaneous reduction of expenses. As we think, application of this experience in the enterprises of mechanical engineering will give positive effect for growth of the general indicators of branch, considering that a situation in which it is.

Robotizations of production to reduce influence of a human factor and will allow to reach the best quality and quantitative index of products at the expense of the accuracy of work and resistance to influence of external factors. In present time, the production only computerized, with use of robotics, it is capable to guarantee quality daily to products.

Today a task of producers of products of mechanical engineering - to reconsider own strategy and to reveal possibilities of involvement of robotics at production and rendering services to take advantage of use of such decisions, for increase in the advantages on the markets representing production with the high qualitative characteristic and low prime cost.

References

[1] Khafizov I I and Nurullin I G 2017 *IOP Conf. Ser.: Mater. Sci. Eng.* **240** 012039
[2] Protsenko I and Ivanov B 2013 Industrial robots in modern production [Electronic resource] Mode of access: http://mirprom.ru/public/promyshlennye-roboty-v-sovremennom-

proizvodstve.html Date:19.02.2018

[3] Gil'Manshin I R and Kashapov N F 2014 *IOP Conf. Ser.: Mater. Sci. Eng.* **69** 012010

[4] Kashapov N F, Gil'Manshin I R and Konahina I A 2014 *IOP Conf. Ser.: Mater. Sci. Eng.* **69** 012024

[5] Fayrushin I I, Dautov I G and Kashapov N F 2017 *International Journal of Environmental Science and Technology* **14** (12) 2555–2560

[6] Gaisin A F, Son E E, Efimov A V, Gil'mutdinov A K and Kashapov N F 2017 *High Temperature* **55** (3) 457–460

[7] Kashapov L N, Kashapov N F, Kashapov R N and Denisov D G 2016 *J. Phys.: Conf. Ser.* **669** 012029

[8] 2018 Robotization of production: sphere of application, pros and cons of robots introduction [Electronic resource] *General Director Magazine* Mode of access: https://www.gd.ru/articles/9644-robotizatsiya-proizvodstva Date: 22.02.2018

[9] Khafizov I I 2013 The influence of the design of equipment and automation of electrical methods to improve the quality of processing *Bulletin of Kazan technological University* (Kazan: Kazan State Technical University) № 1 p 22-4

[10] 2016 Robotics IRB 6620LX [Electronic resource] Mode of access: http://www.abb.ru/product/seitp327/659aa8ac54906d1744257b9b005094b2.aspx#! Date:25.03.2018

[11] Polyakov Z I, Isakov V M, Isakov D V and Shamin Y V2006 Electrophysical and electrochemical methods of processing *Study guide for part-time students. Computer version* (Chelyabinsk: South Ural state University) p 89

[12] Nurullin I G and Khafizov I I 2016 The impact of innovation implementation and certification on the company's activities and reputation Modern technologies: topical issues, achievements and innovations: collection of articles of the winners of the IV international. science.- practical Conf. (Penza: ICSU " Science and education») p 44-6

[13] Akhmetov I D, Zakirova A R, Z B Sadykov and I I Khafizov 2017 *IOP Conf. Ser.: Mater. Sci. Eng.* **240** 012003

[14] Khafizov I I and Galimov A N 2017 *IOP Conf. Ser.: Mater. Sci. Eng.* **240** 012038

[15] Khafizov I I, Saveleva T N and Lyubtsov V S 2017 *IOP Conf. Ser.: Mater. Sci. Eng.* **240** 012040

[16] Khafizov I I 2017 *IOP Conf. Ser.: Mater. Sci. Eng.* **240** 012037

[17] Khafizov I I 2016 *IOP Conf. Ser.: Mater. Sci. Eng.* **134** 012014

[18] Devaev V M, Nikitina D V and Fadeev A Y 2016 *IOP Conf. Ser.: Mater. Sci. Eng.* **134** 012004

[19] Sirazetdinov R, Kamalov A, Nikitina D and Katsevman E 2016 *IOP Conf. Ser.: Mater. Sci. Eng.* **134** 012048

[20] Sirazetdinov R, Devaev V and Zakirzyanova G 2016 *IOP Conf. Ser.: Mater. Sci. Eng.* **134** 012049

[21] Sirazetdinov R T, Devaev V M, Nikitina D V, Fadeev A Y and Kamalov A R 2017 *IOP Conf. Ser.: Mater. Sci. Eng.* **240** 012065

IOP Conf. Series: Materials Science and Engineering **412** (2018) 012035 doi:10.1088/1757-899X/412/1/012035

The selection of energy-efficient main processing of the soil for sugar sorghum

N F Kashapov[1], M M Nafikov[1], I R Gilmanshin[1,3], A R Nigmatzyanov[2] and R F Saifutdinov[2]

[1] Kazan Federal University, 18 Kremlyovskaya street, Kazan, 420008, Russian Federation

[2] Tatarsky institute of retraining specialists agribusiness, Kazan,

[3] Kazan national research technical university named after A N Tupolev, K. Marx St. 10, Kazan, 420111 Tatarstan, Russia

Abstract. The implementation of any culture is the basic processing of soil are given great attention, as it largely affects the water-physical, biological and agricultural properties of the soil that ultimately determines yield.

Analyzing the prospects and contradictions of minimizing tillage, many scientists warn against the danger of simplifications and patterns. They note that the current campaign, in contrast to the former, the party-state, has a market-chinovnichny character. The initiative in it belongs to suppliers of pesticides and equipment. Instead of the state agrotechnological policy, based on the recommendations of scientists, the society is imposed a new style of market self-regulation of technological support of agriculture, which is a serious economic and environmental danger. Moreover, the advantages of the minimum and even zero tillage are advertised without serious indications of shortcomings that must be overcome by the system of agricultural enterprises.

Zero and close to it treatments are effective only in the conditions of plain relief, scarce water regime and relatively prosperous soils with respect to water permeability. The correct system of tillage is a rather knowledge-intensive task that requires professional training of specialists.

Thus, at the beginning of the century, the technology of tillage had almost two points of view: one proved the need to minimize it, and the other – to increase the intensity. Preference was given to the second direction, which entailed the intensification of treatment (including additional techniques, increasing the depth).

Field experiments on the study of basic tillage techniques were carried out in 2013-2015 according to the scheme:

1. Moldboard plowing to a depth of 22-24 cm, the plow PN-5-35;
2. The Maltsev plowing on 22-24 cm;
3. Processing the combined unit APK-6, 22-24 cm;
4. Treatment with anti-erosion cultivator KPE-3,8 20-22 cm.;
5. Treatment with a heavy cultivator KTS-10 20-22cm .

The results of the determination of contamination of crops which had been counted just before harvest showed that the lowest weed infestation was on the variant with a moldboard plow (16-24 PCs/m2).

Content from this work may be used under the terms of the Creative Commons Attribution 3.0 licence. Any further distribution of this work must maintain attribution to the author(s) and the title of the work, journal citation and DOI.
Published under licence by IOP Publishing Ltd

The highest contamination observed when conducting primary tillage KPE-3,8 (49-55 pieces/m2) and CBS-10 (49 to 56 pieces/m2). Among the weeds was dominated by oat grass and chicken millet (figure 1).

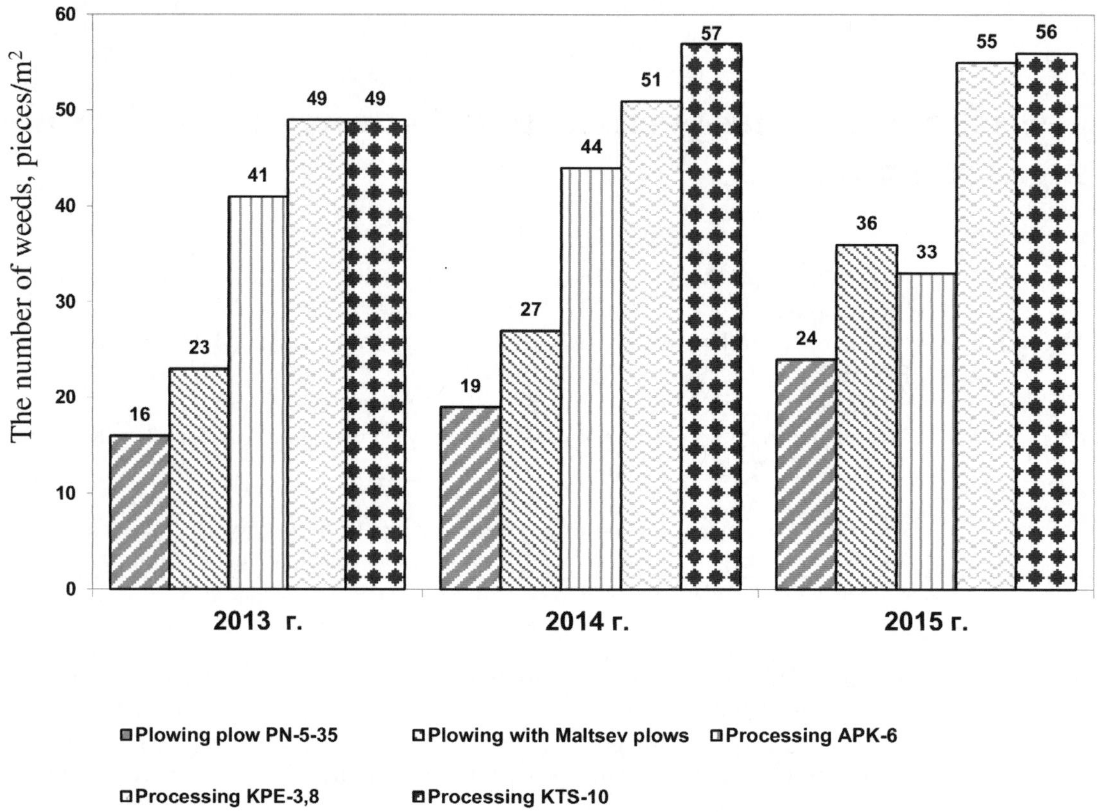

Figure 1-Clogged sorghum crops depending on the methods of basic tillage, PCs / m2

In our studies, the density of the addition of soil before sowing sorghum on the variants of soil tillage there was no significant difference, however, before the cleaning layer 10-30 cm soil was denser on the options in the processing of APK-6, KPE-3.8 and CCC-10 (table. 1).

Table 1 – the density of the composition of the soil depending on processing systems, g/cm3

Basic tillage	Before sowing				Before harvesting			
	0-10cm	10-20 cm	20-30 cm	0-30 cm	0-10 cm	10-20 cm	20-30 cm	0-30 cm
Plowing plow PN-5-35	1,06	1,12	1,14	1,11	1,08	1,15	1,17	1,13
Plowing with Maltsev plows	1,07	1,13	1,16	1,12	1,10	1,17	1.20	1,16
Processing APK-6	1,09	1,16	1.20	1,15	1,14	1,22	1,25	1,20
Processing KPE-3,8	1,08	1,14	1,18	1,13	1,12	1,20	1,23	1,18
Processing KTS-10	1,08	1,15	1,19	1,14	1,13	1,21	1,24	1,19

Analyses have shown that hardness of soil layers depended on systems of processing of the soil and degree of its moistening (table. 2). The data of table 2 show that in the layer 0-15 cm between the processing systems, this figure does not differ significantly, and at a depth of 20 cm and below it was higher in the plots of flat-cut loosening.

Table 2 – Hardness of soil in different systems of soil treatment, g/cm3, 2013-2015

Basic tillage	In the early growing season					Before harvesting				
	5 cm	10 cm	15 cm	20 cm	25 cm	5 cm	10 cm	15 cm	20 cm	25 cm
Plowing plow PN-5-35	2,9	5,2	14,1	17,1	30,2	12,1	18,5	27,3	32,1	33,8
Plowing with Maltsev plows	3,0	5,3	14,4	18,0	30,3	12,6	19,3	28,2	33,5	35,1
Processing APK-6	3,2	5,7	16,5	22,1	31,6	13,4	22,3	34,1	37,6	41,4
Processing KPE-3,8	3,0	5,4	15,6	20,1	30,5	12,9	21,2	31,8	36,3	38,2
Processing KTS-10	3,1	5,5	15,8	20,3	30,7	13,0	21,4	32,9	36,7	40,1

In experiments conducted systematic observations of the state of the soil nutrient regime. Three times during the growing season for three years on two layers of the arable layer in a two-fold replications carried out tests to determine the content of each of the major nutrients – NO3, P2O5, and K20 in the soil, we removed (15 definitions) the average for the 3 years in terms of layers of 0-15, 15-30 and 0-30 cm In the result of application of APK-6, KPE-3,8, KTS-10 sown pronounced differentiation of the parts of the arable layer fertility. In the layer 0-15 cm content of P2O5 on the dump treatment was 164 mg/ kg, in the processing of APK-6-176 and CCC-10-173 mg/ kg, K2O 179,186 and 184 mg / kg respectively. Exceeding version APK-6 P2O5-12, K20-15 mg.

Alkali-hydrolysis nitrogen by plowing contained 78 and for the soil treatment unit LDC-10 – 85 mg/kg soil. In the underlying layer (15-30 cm) opposite pattern. Plowing discovered 71 mg, and while tilling the CCC-10 – 55, P2O5 150 and 133, respectively, K2O – 161 and 150 mg/kg.

Consequently, the applied fertilizers, especially phosphoric and potassium ones at flat-cut loosening, are mainly retained in the upper (0-15 cm) layer and in the lower (15-30 cm) layer penetrate in much smaller quantities than is the case with plow treatment.

Further increase in feed production requires coordinated measures, the most important of which are: the creation of a biologically full-fledged and economically viable feed base, taking into account the achievements of intensification.

The greatest economic efficiency in the cultivation of sugar sorghum is obtained with the use of plowing. Net income on this option amounted to 1964 RUB/ha, profitability level of 64.3%, the energy efficiency ratio of 6.7 vs 383 - 1426 RUB./ha, 13.4 to 48.3 per cent and 4.4 to 5.7 obtained on other options.

In recent years, bioenergy assessment of agrotechnological operations, which is presented in table 3, has become increasingly important in the world practice.

Table 3-Energy assessment of basic tillage in sorghum cultivation

Basic tillage	Productivity of green mass, t/ ha	of Saved up energy, GJ / ha	of Spent energy, GJ / ha	Energy efficiency coefficient	BEP agro-ecosystems
Plowing plow PN-5-35	41,1	189,88	30,23	6,3	4,5
Plowing with Maltsev plows	35,9	165,86	29,11	5,7	2,6
Processing APK-6	26,5	122,43	28,01	4,4	1,1
Processing	31,4	145,07	27,91	5,2	1,7

KPE-3,8					

It allows to compare different technologies of production of agricultural products. The highest (6,7) in the experiment the coefficient of energy efficiency was obtained in the treatment of soil ploughs PN-5-35, the lowest (4,4) he was in the treatment of soil combined unit APK-6. Bioenergetic indicator of agroecosystem (BEP), these options amounted to respectively 4.5 and 1.1.

References

[1] Alabushev A V 1995 *Assessment of economic and energy efficiency of production of grain and sweet sorghum for intensive technology for the steppe zone of the Russian Federation* Proc Dokl Russian science - practice. Conf In the Volga nipti sorghum and maize pp 103-105
[2] Zhuchenko A A 1988 *Energy analysis in agricultureKishinev* pp 45-53
[3] Ziganshin A A, Sharifullin R L 1981 *Recommendations on programming yields of agricultural crops* Kazan 60 p
[4] Poponin A N 1998 *Evaluation of the energy efficiency of growing crops in the farming system: textbook* (M.: publishing house of ICCA) p 40
[5] Kashapov N F, Nafikov M M, Gazetdinov M X, Nafikova M M, Nigmatzyanov A R 2016 *IOP Conference Series: Materials Science and Engineering* 012013
[6] Nafikov M M, Yakushkin N M, Fomin V N, Kashapov N F, Nigmatzyanov R A 2016 *Sweet sorghum: technological and economic aspects of cultivation in the Republic of Tatarstan* (Textbook. Kazan: Publishing house "Snacks" IP Sizikov) p 40

IOP Conf. Series: Materials Science and Engineering **412** (2018) 012036 doi:10.1088/1757-899X/412/1/012036

Innovative technologies of grain sorghum processing

N F Kashapov[1], M M Nafikov[2], A R Nigmatzyanov[2], R A Mingazov[2]
[1] Kazan Federal University, 18 Kremlyovskaya street, Kazan, 420008, Russian Federation [2]Tatar Institute of agribusiness personnel retraining , Kazan, 420059, Russian Federation

arnig76@yandex.ru

Abstract. The article presents materials for the production of grain sorghum food products that allows you to completely eliminate from the diet foods containing gluten, vegetable oil from germ of grain sorghum, nanocellulose, tannin, amylopectinosis sorghum starch and glucose-fructose syrup.

In the Russian Federation sorghum although not the basis of the diet of the population, but flour from this culture is widely used in the diet of people suffering from allergies to gluten. Gluten intolerance (celiac disease), has recently become quite widespread in the world. The disease is provoked by the use of gluten - containing foods from wheat and rye. Flour from wheat and rye has improved baking properties, but hydrolysis of a new type of protein can form metabolites that have in certain cases, toxicity or allergenicity due to the lack of enzymes in the body for their digestion. One way to eliminate the symptoms of celiac disease is a diet, that is-complete elimination of gluten-containing products from the diet. Therefore, in some cases, such a substitute can be products using sorghum flour.

The dry sorghum contains up to 55% fat and serves as a raw material for vegetable oil production. It refers to edible vegetable oils, which by its physico-chemical parameters and acid composition is similar to corn [1,2]. Along with other vegetable oils, sorghum oil is used to prepare salads and other dishes.

In medicine tannins are used as a binding drug, as an antidote for poisoning with salts of lead and mercury, Antidiarrhoeal, hemostatic and antihemorroidal agent [3, 4]. Tannins can be used as a therapeutic and prophylactic agent to reduce the risk of heart disease and cancer. A large amount of glutamic acid in grain sorghum, as well as the presence of essential amino acids make waste cargotrailer production (and gluten extract) perspective raw material for production of glutamic acid, the amino acid gidrolizatov and pastes, the demand for which in the medical and food industries is increasing [5].

Frozen paste of amylopectinosis sorghum starch is rapidly and completely dissolves during thawing, almost no phenomenon of syneresis. This property is used in industrial freezing of food products. Amylopectinosis sorghum starch finds wide application in the textile industry, textile good dressing should be easy and removed from the finished fabric, and pastes amylopectinosis sorghum starch give a film with a high solubility and at the same time a high capacity dressing. This marked improvement in the quality of the fabric and reducing the amount of strand breaks. Ordinary sorghum starch is used as glue and dressing in the manufacture of paper, cardboard and various paper products, for the enrichment of bauxite ores. Dextrin obtained from amylopectin sorghum starch is better than dextrin from topioc. Pastes, dextrins from waxy sorghum transparent and sustainable, and their films possess high strength, flexibility and instantly re-dissolved. In addition, amylopectin starch is used in

Content from this work may be used under the terms of the Creative Commons Attribution 3.0 licence. Any further distribution of this work must maintain attribution to the author(s) and the title of the work, journal citation and DOI.
Published under licence by IOP Publishing Ltd

multi-color printing on fabric and paper. Starch is also used in mining, metallurgy and other industries [6].

Also extracted from grain sorghum tannins are used for tanning of leather and fur, preparation of inks, etching of textile fibers, for giving to various drinks tart and astringent taste and as a food dye.

As dyes in the food industry, isolated pigments from sorghum films are used.

The outer layer of sorghum grain contains wax. By extracting the envelopes of the grain sorghum with hexane to obtain 5-10% of crude wax to the weight of the shell. When applying for extraction of ethanol with crude to remove wax and pigments. The properties of this wax are close to the properties of Carnauba, used in the manufacture of furniture of the highest quality [7].

Sorghum is mostly a technical and fodder crop, the main products of processing are glucose-fructose syrups, sorghum starch and alcohol [8,9]. Sorghum starch is a versatile product free from any anti-nutrient components. Since sorghum endosperm often has a tint color, sorghum starch can also be slightly colored. The color of starch can also be enhanced by pigments of seed shells, which are released in the process of production solutions. In many cases, this is not essential, but white starch is preferred for the manufacture of some products. Therefore, the production of starch for food purposes, I prefer to use as raw material of sorghum varieties with light seed shells and white endosperm. By its physical and chemical properties (viscosity, gelatinization temperature, clear paste) starch sorghum similar to corn. It generally similar ratio of amylose and amylopectin, so the scope of use of these starches also coincide. Sorghum starch is used in the production of sauces, fillings for pies, puddings, desserts based on milk. It is used as an additive for baking bakeries and confectionery products in cases where it is necessary to weaken the action of gluten and give more softness and tenderness to the product while reducing the addition of sugar and fat (biscuit semi-finished product, ice cream cups, biscuits, bakery mixtures, etc.). This starch is used for technical purposes in the paper industry, in the textile and medical industry (figure 1).

IOP Conf. Series: Materials Science and Engineering **412** (2018) 012036 doi:10.1088/1757-899X/412/1/012036

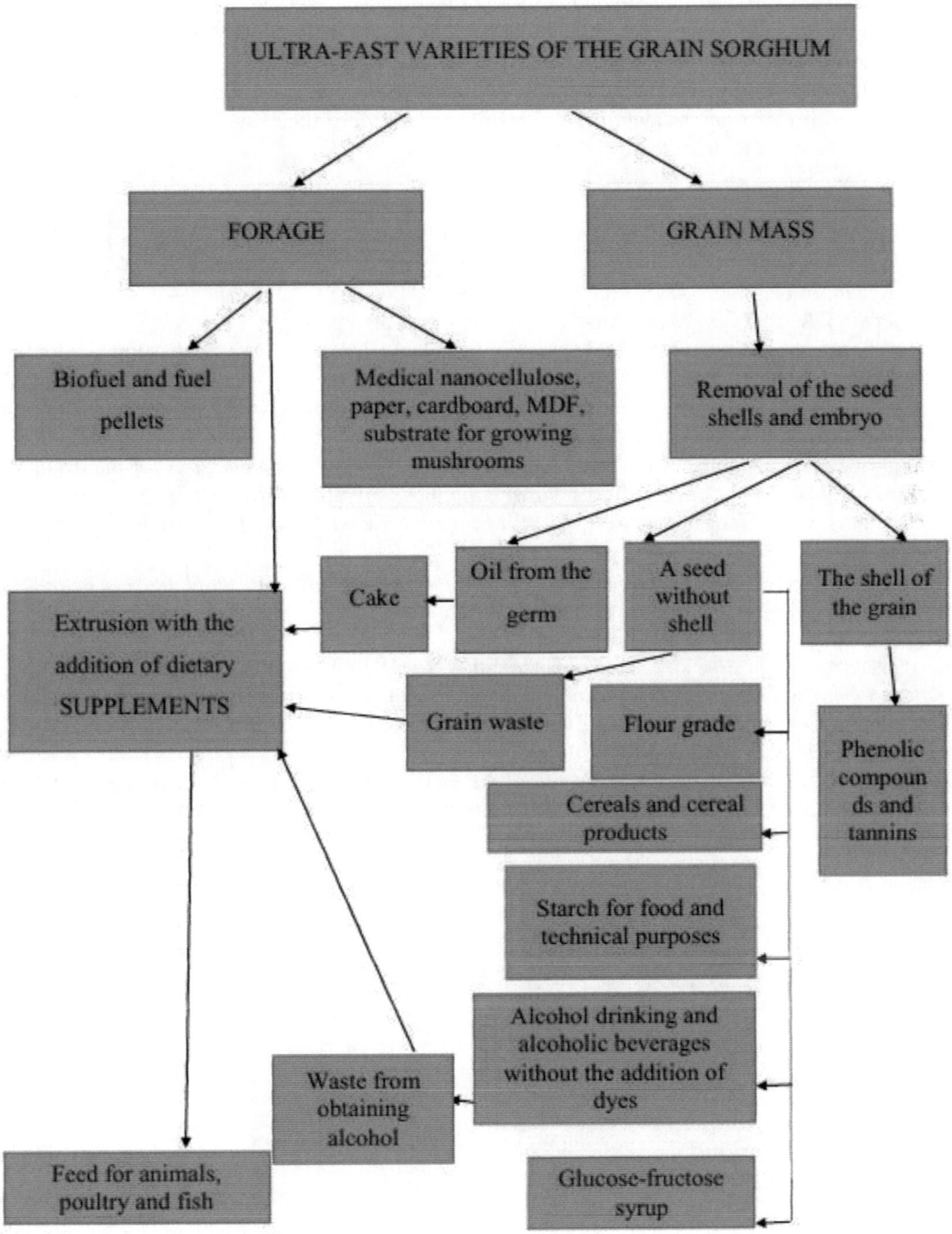

Figure 1 - Scheme of Cargo zero

The production process of Adobe Song Kerala Kerala is made from corn. Just like for baking starch, starch from Cargo is a good Serum for the production of shares substances. On its basis produce glucose-fructose crop (GFS) with a sedan is the ratio of the components. Schemes of production of GFS from song and corn Kerala are identical and include the following main stages: Dexter kerala a-Las - > glucose production (Lucille Share) —> artist glucose solution absorbent —" ISO glucose glucose.

Lower compared to corn cost of grain Cargo and higher content in it kerala provides the best performance of rolling for the production of GFS from Cargo. For the same reasons, production from Cargo Sparta is more rental than corn.

References

[1] Nafikov M M, Yakushkin N M, Fomin V N, Kashapov N F, Fomin D V, and Nigmatzyanov A R 2016 *Sugar sorghum: technological and economic aspects of cultivation in the Republic of Tatarstan* Kazan: Publishing house "ZnakC" IP Sizikov p 40

[2] SHepel' N A 1994 S*orghum* Volgograd: press Committeep 448

[3] Bol'shakov A Z, Bondarenko S M, Kadyrov S V and Klepko YU N 2008 Rostov-on-don: ZAO "Rostizdat" p 60

[4] Kadyrov S V, Fedotov V A and Bol'shakov A Z 2008 *Sorghum in the Central Chernozem region* Rostov - on-Don: ZAO "Rostizdat" p 80

[5] Alabushev A V, Alabusheva O I and Anipenko L N 2004 *Recommendations for the preparation of feed from sorghum and use in the diets of farm animals and poultry* Zernograd p 32

[6] Kovtunov V V, Gorpinichenko S I 2011 The main areas of use of grain sorghum grain farming No 6 (18) pp 28-32

[7] Amira C D 1992 *Small grains in monogastric and ruminant feed formula-tions: Prospects and problems. In: Utilisation of sorghum and millets* International Crops Research Institute for the Semi-Arid Tropics, Patancheru, India pp 183-90

[8] Kashapov N F, Nafikov M M, Gazetdinov M H and Nafikova M M 2015 *Waste-free production technology of alcohol from sugar sorghum* Materials of the International scientific and technical conference " Innovative engineering technologies, equipment and materials – 2015 "(MNTC" IMTOM–2015") p 288-91

[9] Kashapov N F, Nafikov M M, Gazetdinov M. H., Nafikova M M and Nigmatzyanov A R 2016 *Innovative technology for the production of ethanol from sweet sorghum* IOP conference Series: materials science and Engineering **012012**

IOP Conf. Series: Materials Science and Engineering **412** (2018) 012037 doi:10.1088/1757-899X/412/1/012037

Recycling of wastes of agricultural production and rural settlements to produce biogas and organic matter

N F Kashapov[1], M M Nafikov[1], I R Gilmanshin[1, 3], A R Nigmatzyanov[2]
[1] Kazan Federal University, 18 Kremlyovskaya street, Kazan, 420008, Russian Federation
[2] Tatarsky institute of retraining specialists agribusiness, Kazan, Russian Federation
[3] Kazan national research technical university named after A N Tupolev, K. Marx St. 10, Kazan, 420111 Tatarstan, Russia

Abstract.
In most regions of the Russian Federation today has a highly developed agriculture and processing enterprises, respectively with a high concentration of resources for production of biogas. On the Volga and southern districts account for a total of about 58% of the biogas potential. In the processing of all waste of agriculture and processing enterprises can fully supply gas to rural areas.

Every year, up to 14-15 billion tons of biomass are produced in the production of agricultural products on the territory of the Russian Federation. The energy of chemical bonds in the decomposition of this amount of biomass is equivalent to 8.1 billion tons of conventional fuel.

According to the research results Of the Institute of energy strategy of Russia, the total amount of organic waste of the agro-industrial complex of the Russian Federation in 2015 amounted to 225 million tons, including:

- livestock-58.3 million tons;
- crop production-147 million tons;
- processing industry APK 14 million tons (the sugar and canning factories, meat processing plants, alcohol Breweries, etc.).

The above data show that the greatest mass among organic wastes of agricultural production is occupied by wastes of plant growing (straw, stems, husk, tops, floors, etc.). Processing into biogas at the same time with the waste of livestock and poultry requires the development of a universal biogas technology and related equipment.

The analysis of national and world literature in the field of biogas technologies should begin with the coverage of the work carried out in the early 60s of the last century At the Institute of biochemistry. A. N. Bach of the USSR. It is these studies and their commercial incarnation was the starting point in creating a domestic industrial bio-energy and active fundamental research of the processes of biosynthesis of methane and biogasification [1,2,3,4,5].

The idea was developed and calculated in 1972-1973 and embodied in the project in 1979 This project was supported by the leadership of the Union of SSR and in 1980 included in the program of the USSR State Committee on science and technology.

Under this program, three large biogas plants were built between 1980 and 1990.

Development of market economy and emergence of new forms of ownership in agricultural production demanded development of the highly profitable technologies and the equipment working in any climatic zone and in any Russian remote place from the centralized power supply.

Content from this work may be used under the terms of the Creative Commons Attribution 3.0 licence. Any further distribution of this work must maintain attribution to the author(s) and the title of the work, journal citation and DOI.
Published under licence by IOP Publishing Ltd

Agricultural producers face the problem of manure utilization every year. Go a lot of money that is required for the organization of its export and warehousing.

It is known that manure is a very valuable fertilizer, and if there are several cows on the farm, then there are no problems with its use. Another thing when it comes to large farms with large livestock, where thousands of tons of biological material are formed per year. To make manure turned into high-quality fertilizer, we need equipped platforms, which are unnecessary costs. Therefore, many agricultural producers store it without observing the storage conditions, and then take it out and close up in the soil.

Biogas is a volatile substance without color and any smell, which contains up to 70% methane. According to its qualitative indicators, it approaches the traditional type of fuel – natural gas. Has a good calorific value of 1m3 biogas emit as much heat as is obtained by burning half a kilogram of coal. The formation of biogas we owe to anaerobic bacteria, which are actively working on the decomposition of organic matter, which is used as the manure of farm animals, poultry manure, waste processing plants, straw, etc. [9,10].

Table 1-Quantity and composition of combustible gas produced during processing of different types of biomass

Species biomass	Proportions of components,	the Specific biogas yield, m3/kg	methane,%
Chicken manure	100	0,311	59,8
Chicken manure and organic(straw, tops)	50/50	0,368	66,1
The cattle manure and straw	50/50	0,268	51,0
Pig manure and straw	50/50	0,473	65,0

Participate in the process of mesophilic bacteria. Their activity takes place at a temperature +30 – +40 degrees. Therefore, to activate the process it is necessary to create the most favorable conditions for their life. They should be similar to those in which microorganisms develop naturally, in the stomach of animals, where heat and anaerobic conditions are created. This is the two main conditions that contribute to the natural transformation of rotting manure mass into environmentally friendly fuel and valuable organic fertilizers [6].

To obtain biogas, you need a sealed reactor without air access, where the process of fermentation of manure and its decomposition into components will occur: methane (up to 70%); carbon dioxide (about 30%) and 1-2% - other gaseous substances (figure 1).

IOP Conf. Series: Materials Science and Engineering **412** (2018) 012037 doi:10.1088/1757-899X/412/1/012037

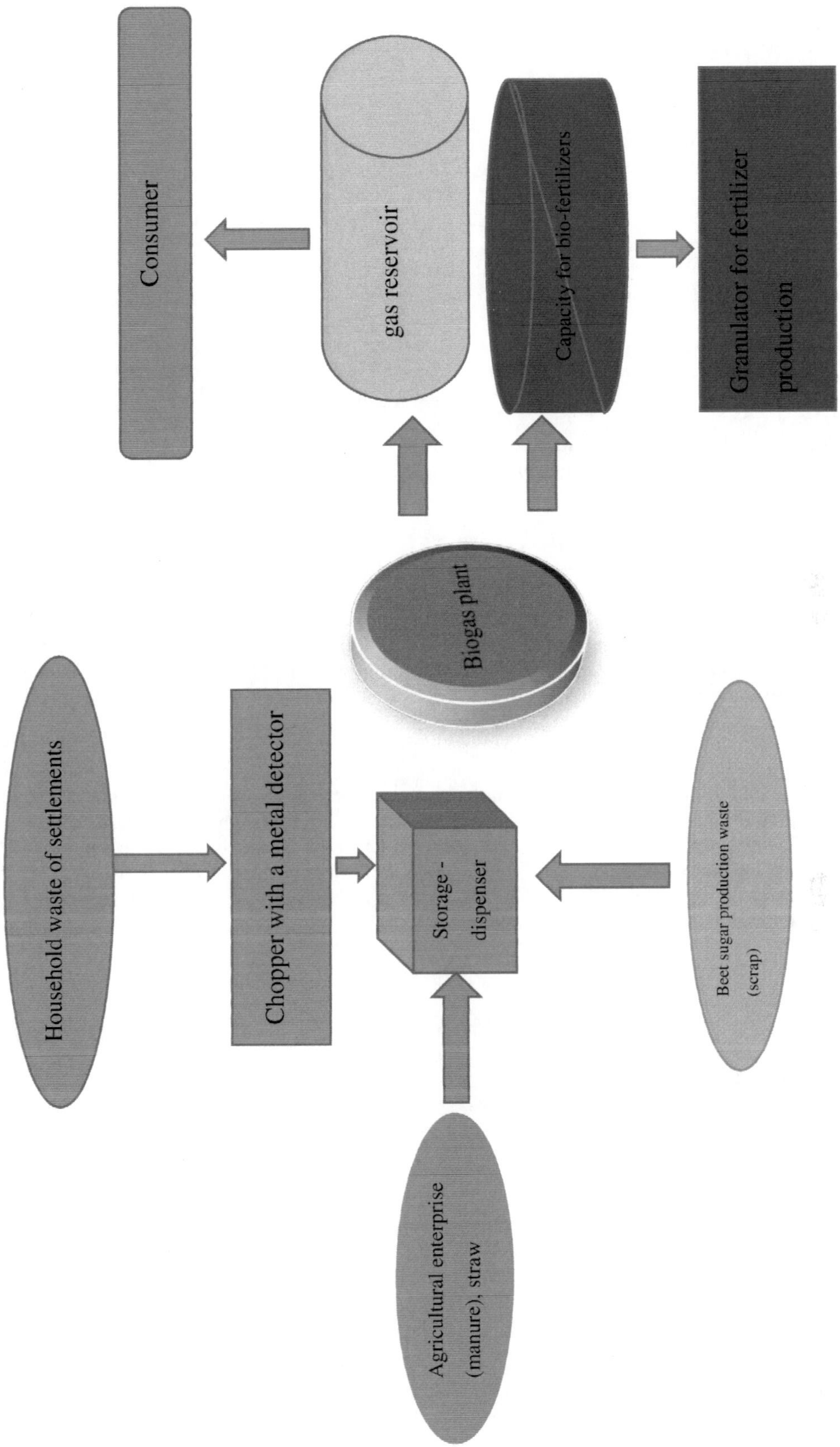

Figure 1. Scheme of biogas production from agricultural waste and household waste of rural settlements

The resulting gases rise to the upper part of the container, where they pumped then settles down and is calling Organic Dori, preserved in the result of processing CEN all the elements available in the nose of macro and micronutrients, dead pathogen, but also lost seeds son the life of the plant [7,8].

Calculations of the effectiveness of biogas and to enjoy the benefits of the use of biofuels an alternative, will help NS calculations. One bark weighing 500 kg produces about 35-40 kg of news per day. This amount will be enough to produce about 1.5 m^3. biogas, from which in turn it is possible to develop 3 kW/h of electricity.

If the farm contains 100 goals of cors is not difficult to calculate how many m^3 of biogas can be obtained at the output.

Sly Important is that for biofuel production it is necessary to use both one type of organic raw material and from a mixture of several components having a moisture content of 85-90%. Before the salad Breaker, it is necessary to remove from the mass to postpone the chemical simple, negatively affecting the process of the action of microorganisms.

Thus recycling of agro-industrial waste and sells pole sells will support the economy, will provide new jobs, the rational use of local waste and ensures maize landscapes of the environment.

References

[1] Baader V. *Biogas: theory and practice* 1982 (M: Colors) p 148

[2] Buffer C, Zanuda O 1985 *Biomass as an energy source: Per. English* (Moscow: Mir) p 368

[3] Boil D 1987 *Bioenergy: technology, thermodynamics, and costs* (M.:Agronomist)

[4] Vedenev A G, Men N 2006 *The construction of biogas plants* (Quick guide. Bishkek) p 28

[5] ZHirkov V, German A., Matveev Y and Ulanov M 2005 *Basics of construction of biogas plant for the processing of agricultural waste nearby*. "Ecomuseum", Karaganda, the Agency for renewable energy, Kiev p 17

[6] Mastepanov A M, Stepanov A D, Gorevalov S V and Belogor'ev A M. 2013 *Supply gas as a factor regionalization gas Rank* M: CI "Energy", - p 128

[7] Chetoshnikova L M 2010 *Power renewable energy: a training manual* (Chelyabinsk: Publishing house of the SUSU) p 69

[8] Sin A A 2002 *Biogas on the farmstead sells* (Barkley: Information Sisters company "Balakliya") p 68

[9] Gil'manshin I R, Kashapov N F and Galeeva A I 2016*Energetic salon of gas use as a way of forming a new model of tho management Problems and prospects of innovative development of the economy materials of the scientific forum* (scientific and practical XXI international conference).. - p. 239-242.

[10] Kashapov N F, Nafikov M M, Gazetdinov M X, Nafikova M M, Nigmatzyanov A R 2016 *IOP Conference Series: Materials Science and Engineering* 012013

IOP Conf. Series: Materials Science and Engineering **412** (2018) 012038 doi:10.1088/1757-899X/412/1/012038

About one approach to the assessment of technical equipment of agricultural enterprises in conditions of economy modernization

N F Kashapov[1], M M Nafikov[1], M Kh Gazetdinov[2], Sh M Gazetdinov[2] and A R Nigmatzyanov[3]
[1] Kazan Federal University, 18 Kremlyovskaya street, Kazan, 420008, Russian Federation
[2] Kazan state agrarian University Kazan, Russia
[3] Tatarsky institute of retraining specialists agribusiness, Kazan

Abstract. The problem of conformity of technical equipment and the number of employees in the agricultural enterprises arising in the process of modernization of the agricultural sector. As a means for evaluation of technical equipment are invited to use a special type of production functions, the distinctive feature of which is the simultaneous display of the micro - and macropaedia industry.

Introduction.

A modern agricultural production is a difficult organizationally-economic process functioning at his continuous providing capital goods, necessary for implementation of agrotechnological measures and operations at the production of goods and implementation of works, services. In this connection, the system of logistical support of production of agricultural goods in the agrarian sector of economy occupies a central place. An agricultural technique, along with the landed resource, makes basis of agrarian sector of economy, accordingly efficiency, the competitiveness of agricultural produce straight depends on material energywell-being and energyequipped of industry. In an agricultural production the system of logistical support with the resources engaged in her determines the more than half of all expenses [1].

The programs of modernisation of the economic system carried out in practice embrace her different elements. Modernisation affects quite different parties of life of country. At her realization an important value has correct combination of socio-economic, technological and political factors, because at liquidation of bottlenecks in one spheres, they involuntarily arise up in other. In our view, one of reasons such the phenomena consists in violation of principle of accordance, in underestimation of all limiting factors.

As a result, development of integration processes, addition of mainly intra-branch forms of associations, becomes the necessary condition of modernisation of agrarian sector of economy – major productive and processing concerns – by more difficult inter-branch and territorial forms, such as the apeak-integrated forming of holding type.

Exactly these forms of integration are specific for the agrarian sector of economy, and exactly to them today decision possibilities of native increase of efficiency and drawing on accomplishments of scientific and technical progress and innovations are related. However realization of these possibilities

Content from this work may be used under the terms of the Creative Commons Attribution 3.0 licence. Any further distribution of this work must maintain attribution to the author(s) and the title of the work, journal citation and DOI.
Published under licence by IOP Publishing Ltd

depends on the row of terms, including development of theory of co-operation of the integrated forming with the developing systems of the territorial and programmatic-having a special purpose planning and management. Only on this basis it is possible correctly to define a role and place of rural territories in realization of long-term tasks of socio-economic development of region and country. Necessity ecologically and the socially balanced passing to modernisation produces the row of requirements to forming of conception and strategy of development of rural territories. Term "rural territory" stops to be the simply general name for the intermediate levels of territorial structure of the socio-economic system or totality of the natural, social and technical phenomena on this territory [2, 3, 4].

Providing of steady development of agroindustrial complex in the conditions of entering of Russia to WTO maybe only on condition of his substantial technical and technological modernisation. Modernisation implies mastering and introduction of intensive technologies, new generations of agricultural technique [5, 6].

In obedience to strategy of machine-technological modernisation of agriculture of Russia on a period 2020 to next priority tasks are certain: to provide the height of the labour productivity no less what in 4 times, including by a way:

– technological rearmament of agriculture, promoting in 1,6-1.7 time the productivity of industries of plant-grower and stock-raising and attaining on this indicator of average world indexes;

– increases of gross production of agricultural goods in 1,9-2 times, using for this purpose, except intensive factors, present landed potential not engaged in a turn;

– technical retooling of industry the machines of new generation, allowing to provide the more than double increase of the industry average loading on the worker of machine-technological sphere [7].

Without technical and technological modernisation agriculture is not in strength substantially to promote production volumes, that especially topically in the conditions of the conducted politics of import substitution of agricultural produce. Thus a hardware of small forms of management on a mud flow is today a stumbling-stone.

Analysis and discussion of research results. In Republic of Tatarstan in 2016 agricultural organizations are purchase more than 5 thousand units of technique and equipment, to the amount of an about 6 milliard of rub.

At the same time from data of Tatagroleasing, a park of combines and tractors on 40% is outside the terms of exploitation, side by side with this, in obedience to data of Tatarstanstat, the quantity of these machines continues to grow short. In 2016 as compared to 2008 the park of tractors grew short on 30,6%, combine harvesters – on 33,8% (table 1).

However, as be obvious from a table, power powers per employee grew on 37,6%. It is constrained, first of all, by the not height of the productivity of the used technique, and with the high rates of decline of quantity busy in industry (in the investigated period – 38,6%). Material Energywell-being calculating on 1 ha of sowing area went down for 2008 - 2016 on 16,6%.

Very evident are next comparisons: in world practice there is a 21 tractor (in RT – 3,6 tractors) on 1000 ha of plough-land; combine harvesters on 1000 ha of sowing of grain-crops - 6,3 pcs. (in RT are 1,6 pcs.). Consequently, on the amount of tractors on 1000 ha of plough-land our republic falls behind from a world level more than in 5,8 time, and on the amount of combine harvesters - in 3,9 time.

Table 1 – Power powers are in agricultural organizations of Republic of Татарстан

	Years					
	2008	2010	2014	2015	2016	2016 in % by 2008
Power powers - all, thousand, hp.	4690	4296	3909	3885	3965	57,2
calculating on:						
– one worker, hp.	51,1	51,5	62,1	65,5	70,3	137,6
– 100 ha of sowing area, hp.	157,4	146,7	134,6	129,5	131,3	83,4

It is expected on: agriculture of Republic of Tatarstan: Statistical collection / Tatarstanstat. – Kazan, 2017. – 358 p.

On the estimations of specialists for providing of food safety of country, implementations of works in optimal agrotechnical terms it will be required to have material energywell-being calculating on 1 ha of sowing area – no less than 300 hp. (actually on Republic of Tatarstan in 2016 – 131,1 hp.), while in EC – 450-500 hp., the USA – 850 hp., Republic of Belarus – 500 hp. For Republic of Tatarstan optimal is a machine-tractor park including 55-60 thousand tractors taking into account mastering 3,4 million ha of plough-land, while in 2016 his quantity made only 10,8 thousand pcs. with middle power of 120-150 hp. Thus the amount of combine harvesters it is necessary to lead to 10 thousand pcs. (in 2016 are 2507 pcs.). Obviously, a native break is needed and not only quality but also quantitative.

Within the framework of Program of development of agriculture in Russia in a period from 2013 to 2020 subprogramme "Technical and technological modernisation of innovative development" is sent to creation of favourable economic environment assisting innovative development and bringing in of investments in industry, to the exit of Russia on leading positions in area of agricultural biotechnology. On realization of these aims from a federal budget it is pre-arranged to distinguish more than 23 milliards of roubles, from regional budgets is an about 12,5 milliard of roubles [7].

The resources of production come forward the basic factors of economic development of agrarian sector of economy in form material resources, labour and capital. Exactly they must be taken into account first of all in all set programs of development. There are two going near the decision of the indicated problem. The first is based on principle of distribution of each of resources on all agricultural enterprises, a deficit is here distributed and saved. The second supposes the selection of the most effective part of production, concentration of resources in her limits, removing to the same the problem of deficit for this part. Low effective productive units leave from functioning of the system [8, 9].

In connection with reduction of increase of labour resources in a country the problem of tying up of increase of units of technical equipments of production and their providing additional labour resources gets up especially sharply. For her decision many economists deem it wise the use of the second approach from afore-named.

In [10, 11, 12] it is reasonable, that, since the defined value, when the increase of material resources becomes surplus in relation to the resource of labour, maximum efficiency of his use goes down. In addition, dispersion of expensive, power-hungry new technique on the numerous enterprises of small and middle enterprise, with low removable office hours results in the decline of the real effect from application of new technique as compared to potential possibilities. Therefore at surplus of technical equipments of production it is necessary to accelerate their leaving due to pre-schedule liquidation of technique.

Such method of intensification of updating of technical equipped of production, though attended with losses in size of their remaining cost, does not need investment expenses and does not increase loading on an engineer and metallurgy. Leaving without a compensation supposes liquidation of the technical aggregates, not provided with labour resources, what the level of drawing on new technical and technological units will allow to promote. This is important, because the labour productivity on new technical and technological units is substantially higher, than on out-of-date.

Modernisation morally and physically out-of-date machine and tractor park over the pre-arranged leaving is difficult and responsible procedure. Before to begin her realization, it is necessary all-round to ground expediency of freeing of resources from the row of uneffective agricultural enterprises and use of them in more productive terms [13].

The preliminary estimation of technical equipped of machine and tractor park is required, to material well-being by his labour resources, estimation of productive possibilities of functioning agricultural enterprises. As means allowing to get the required estimations, it is suggested to use the special type of productive functions simultaneously representing micro- and macrodescription of objects. The construction of productive function of industry comes true on the basis of analysis of her structure, i.e. macrodescription of industry is base on her microdescription.

For the construction of branch productive function next assumptions are entered. It is assumed that a machine and tractor park executes the homogeneous types of works, services. And here an only limiting factor are labour resources, i.e. material resources do not limit a production within the limits of the distinguished funds.

Technological descriptions of technical and technological units are the productivity and norm of expenses of resource of labour on unit of the executed agrotechnological operation (λ). Accepted, that the norms of expenses of resources of other industries, and also productivity of aggregates, are identical for all agricultural enterprises. Then the technological structure of industry can be presented by means of closeness of distribution of h (t, λ) of stakes of the total productivity of technical and technological units of M (t) on different technologies λ in the moment of time of t. The function of h (t, λ) is certain at $\lambda \geq \nu$ (t), where ν (t) – the best of worked out and used in agriculture resource-saving technology that comes forward description of level of modernisation in the moment of time of t.

On the first stage of research we will be limited to consideration of permanent structure, i.e. we will not take into account appearance of new technologies because of modernisation. It means that ν (t) = const and h(t, λ) = ψ(λ), i.e. does not depend obviously on time. A productive function looks like

$$Y(t) = M(t)f(x), \quad x(t) = R(t) / M(t). \qquad (1)$$

Here the function of f(x) is related to distribution ψ(λ) by correlation

$$f(x) = \int_{\nu}^{\xi(x)} \psi(\lambda) d\lambda \ , \qquad (2)$$

where a top limit ξ(x) is determined by equalization

$$x = \int_{\nu}^{\xi} \lambda\psi(\lambda) d\lambda, \ x \leq x^* = \int_{\nu}^{\xi} \lambda\psi(\lambda) d\lambda. \qquad (3)$$

As f(x) is the function of distribution, correlations of f(0) = 0 are just, f(x*) = 1. The built productive function (1) in an obvious kind is determined by the technological structure of the industry set by distribution ψ(λ).

By means of an additional research it is possible to establish the properties of production function postulated usually: positivity of the first and negativity of the second of derivatives.

If the amount of workers is determined busy in the agricultural production of R(t) and productive potential of agricultural enterprise M(t), then correlations (1), (2), (3) describe productive potential of enterprise, that must be involved, to provide the level of production of agricultural goods, determined Y(t). In this case under the production of agricultural goods the volume of gross products not really got for a year is understood, and potentially attainable in the conditions of the set resources and technical and technological units.

It is known that in practice the level of loading of technical and technological units is subzero enough (on occasion at the level of 40%). Therefore actual distribution of busy workers on agricultural enterprises does not comport with distribution of technical and technological units on the norms of expenses of technological works, services. As a result we look after functioning of agricultural enterprises with the subzero labour productivity, and also shortage busy on agricultural enterprises with a high yield. To provide a maximal production of agricultural goods volume, it is required so to redistribute workers, first of all to provide technical and technological units with a burst performance.

Thus, the considered branch productive function gives an overhead estimation for the level of production of goods of agriculture as a function of present for her technical and technological units and busy on a production labour resources.

As follows from correlations (2) and (3), enterprises with the norms of expenses of labour higher ξ(x) must not participate in a productive process from the present deficit of agricultural workers. It is possible to suppose that technical and technological units and accordingly the productive technical equipments not involved in a productive process are subject to leaving first of all.

The described going near the construction of productive functions helps to ground the real possibilities of intensification of production due to the acceleration of leaving of ineffective technical

IOP Conf. Series: Materials Science and Engineering **412** (2018) 012038 doi:10.1088/1757-899X/412/1/012038

and technological units. In addition, he sets structural procedure on determining the amount and composition of leaving technical and technological units.

It is important to underline that the value of potential level of production of agricultural goods Y(t) always excels values productions that can be got at other variants of the use of technical and technological units.

Conclusion. Thus, as a major task of modernisation of economy the not increase of technical equipped of production is pulled out, and increase of their balanced with labour and material resources. The balance of factors of production is followed by delay of dynamics of updating of technical means of production as a result of elimination of low-productive and obsolete technical and technological units.

The considered approach of construction of productive functions of agriculture helps to define the most effective directions of the use of resources. A final decision about the rational concentration of technical and technological units on agricultural enterprises is accepted as a result of complex analysis of all factors and criteria of efficiency. In the conditions when the settling leaving of technical and technological units can not be in short space realized, offered approach can be used for forming of strategy of development of agrarian sector of economy.

References

[1] Trishkina L V, Grigorova E S 2011 *Analysis of co-operation of subjects of logistical support of enterprises APK* Announcer of Federal public educational institution of higher professional education the "Moscow state agroingenerniy university the name of V P Gorjchkina" V 6 (51) pp 27-30.
[2] *Politiques the rate of development territories rural*, ARTEVER, February 2014, p 4 http://www.grep.fr/projets/Artever/PDF/Artever France Rural.pdf.
[3] *Fact Sheet: die EU-Politik zur Forderung der Entwicklung des landlichen Raums* 2013-2020 // http://ec. Europa. Eu/agriculture/public/fact/rurdev2013.
[4] *Rural development in the Ruropean Union. Statistical and economic information* 2015 Directorate-General for Agriculture and Rural Development. REPORT
[5] Gazetdinov M Kh, Semicheva O S, Gazetdinov Sh M 2016 *Pre-conditions of forming of the territorial systems in the conditions of modernisation of economy* The Scientific messages of the Russian academy of enterprise V 48 pp 37-45
[6] Gerebin V M 2011 *Modernisation of economy and employment of population* are Questions of statistics V 10 pp 19-30
[7] Fisshin V N et al 2009 *Strategy of machine-technological modernisation of agriculture of Russia on a period to 2020* (M.: FGNU "Rosinformagrotex") p 80
[8] Gazetdinov M Kh, Akmaikin V M 2016 *Increase of efficiency of development of the productive program of enterprise in the conditions of market vagueness* (Kazan: publishing House "Brig") p 140
[9] Gazetdinov M Kh 2004 *Pre-conditions of stability of the economic systems in the changing terms of environment* Announcer of the Kazan state technical university the name of A.N. Tupolev V 1 pp76-78
[10] Kashapov N F, Nafikov M M, Gilmanshin I R, Gazetdinov M Kh, Nafikova M M and Nigmatzyanov A R 2017 *Energy-saving technologies of cultivation of sugar sorghum* IOP Conference Series: Materials Science and Engineering V 240 **012032**
[11] Kashapov N F, Nafikov M M, Gazetdinov M Kh, Nafikova M M and Nigmatzyanov A R 2017 *Comparative evaluation of different machines for seedbed for sorghum* IOP Conference Series: Materials Science and Engineering V 240 **012033**
[12] Kashapov N F, Nafikov M M, Gazetdinov M X, Nafikova M M, Nigmatzyanov A R 2016 *Justification of the choice of units for mains-noah soil cultivation of sweet sorghum and their effectiveness* IOP Conference Series: Materials Science and Engineering V 134 **012013**

IOP Conf. Series: Materials Science and Engineering **412** (2018) 012038 doi:10.1088/1757-899X/412/1/012038

[13] Kashapov N F, Nafikov M M, Gazetdinov M KH, Nafikova M M, Nigmatzyanov A R 2015 *Economic justification of the choice of machines and tools for primary tillage under sweet sorghum proceedings of the International scientific-technical conference* (MNTK "IMTOM–2015") pp 285-288

IOP Conf. Series: Materials Science and Engineering **412** (2018) 012039 doi:10.1088/1757-899X/412/1/012039

Mathematical modeling of the technological accuracy index deviation structure of the automobile parts

S V Kasjanov, D T Safarov

Kazan Federal University, Naberezhnye Chelny Institute, 423812, Russia, Naberezhnye Chelny, Prospekt Syuyumbike 10A

safarov-dt@mail.ru

Abstract. The technological components of the measured deviation of the accuracy index that arose during machining are ordered. Mathematical models have been developed to predict their magnitude during processing and also to find the causes of identified nonconformities in production, the components are ranked using the Pareto diagram.

1. Introduction

The world's leading automobile corporations have long since moved from a quality assurance based on alternative control to a constant reduction of quality indicators deviations value by regulating them. This approach may require the consideration of more factors.

Current International quality management standards require the development of systematic measurement of the most important (key) quality indicators, to build and analyze control charts [1], to apply methods for measurement systems of repeatability assessing and reproducibility. [2]. But identifying factors of instability and correcting planning takes time and reduces efficiency as soon as having processed. And there is no guarantee that the correction will reduce the deviation from the first attempt. Consequently, the effectiveness of management is lost.

2. Theoretical part

The proposed solution is shown on the example of the transition of longitudinal grinding (Fig. 1). In order to obtain the highest degree of adequacy of models, the process of forming in this paper is represented as the successive attachment of elementary areas to each other, caused by turning the workpiece by an angle $\Delta\varphi$ and axial movement of the tool by ΔS. To take into account all factors, the concept of a "pole" of the system is introduced. The blanking pole is the instantaneous position of the axis of rotation of the workpiece. The tool pole is the instantaneous position of the tool forming element [3-5].

Based on the definition of "products – the result of the process" (GOST R ISO 9000-08), the following basic principle is formulated: the measured deviation of the position coordinates of the elementary area of the treated surface from the given ideal value is a geometric sum of the technological components (procurement and tool) that arose in the preparation of the working stroke, in the processing of previous elementary sites and at the time of forming of this site.

The instantaneous radius of each site will be determined by the deviations from the ideal desired state of 2 poles: $\Delta\Pi_{3i}$ and $\Delta\Pi_{Иi}$. It is calculated in the operating technological coordinate system.

Content from this work may be used under the terms of the Creative Commons Attribution 3.0 licence. Any further distribution of this work must maintain attribution to the author(s) and the title of the work, journal citation and DOI.
Published under licence by IOP Publishing Ltd

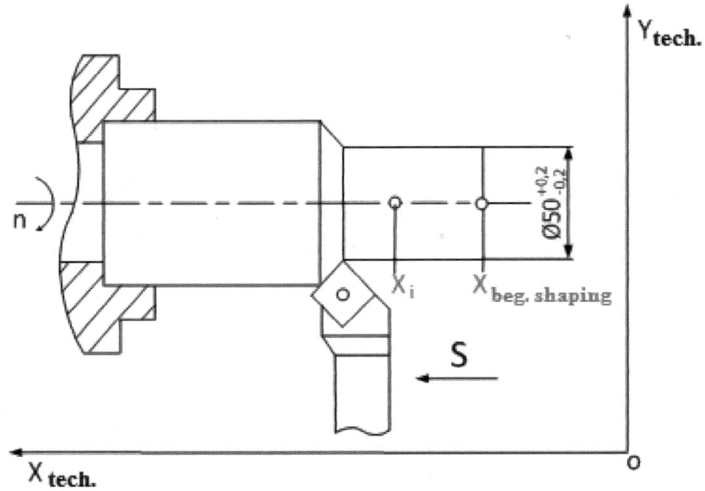

Figure 1. Technological scheme of the transition of external longitudinal grinding.

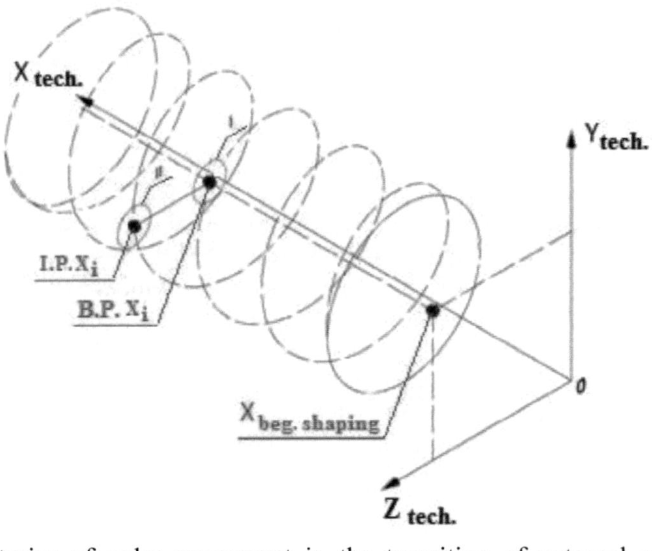

Figure 2. Ideal trajectories of poles movement in the transition of external cylindrical surface turning, I is the area of possible positions of the blank pole, II is the area of possible positions of the tool pole.

In the current X-coordinate of the position of the poles:

$$x_i = X_{pr.F.O.} - \sum_1^i \Delta\varphi_i * S \tag{1}$$

where $X_{pr.F.O.}$ – initial point of forming [mm.], $\Delta\varphi_i$ – the increment of the rotation angle [grade], S – submission amount [mm./turnover]

The actual distance of the elementary area from the instantaneous axis - R_i will be determined by the coordinates of the poles:

$$R_i = Y_{I.P.i} - Y_{ZP.i} \tag{2}$$

These coordinates of the poles are determined by geometrically summing of the coordinates of the point in their ideal position and technological components.

The Table shows a set of technological components that are relevant for this transition, in the order in which they occur.

General scheme of pole deviation factors

Table 1

№	Machine Forming Systems	Processes of the life cycle of the technological system				
		Equipment Characteristics	Setting up the task	Workpiece base	Figuration	Calculation point
1	Blanking branch	Position of the axis of rotation ΔY_{OCi}	---	The size of the allowance t_i and solidity HB_i	---	Deformation of the axis Δf_i
2	Tool Branch	Deprecation of the guide axis $\Delta Y_{Napr.i}$	Regulation of size $\Delta Y_{Nal.i}$	---	Accumulated dimensional wear Δh_{ri}	---

Thus:

The coordinates of the current position along the Y axis of the blanking pole:

$$Y_{Z.P.i} = Y_{ZPid.i} + \Delta Y_{OCi} + f_i \quad (3)$$

where $\Delta Y_{OCi} = \alpha * x_i$

The nonparallelism characteristic of the α-axis by the machine guide is determined by diagnosis.

The magnitude of the deformation of the axis at point i is calculated from known empirical relationships [2]. The coordinates of the position of the instrument pole are determined by:

$$Y_{I.P.i} = Y_{IPid.o} + \Delta Y_{Napr.i} + \Delta Y_{NAL} + \sum_1^{i-1} \Delta h_{ri} \quad (4)$$

$\Delta Y_{guide.}$ – the wear amount of the guide at the point

ΔY_{ADJ} – deviation of the coordinate of the adjustment dimension from the ideal value

Accumulated depreciation amount $\sum_1^{i-1} \Delta h_{r.i}$ is calculated from known empirical dimensions [6].

The accumulated instantaneous values R_i are added to the equation of the machined surface first in the machine coordinate system (2), and then it is converted to simulate the manufactured part in the measuring and construct coordinate systems.

The value of the identified components is ranked using the Pareto chart to plan the order of improvements.

3. Application of the developed modeling methodology

- To predict the deviation of key accuracy indicators during the development of production operations, including the use of virtual technologies.
- For automatic planning of the most effective corrective actions in case when an unacceptable amount of deviation is detected [6, 7, 8].
- To create a reliable information basis for effective preparation of production including innovative products, adding to the application of various methods of quality management [9].
- As a tool for comparative evaluation of the effectiveness of the methods used to control accuracy [5].
- As a basis for the effective diagnosis of machine system modules [10-12].

References

[1] Statistical process control (SPC). Reference Manual. Issued 1992, Second Edition, Issued July 2005. Copyrighht ©1992, ©1995, ©2005. Chrysler Corporation, Ford Motor Company, and General Motors Corporation

[2] Safarov D T, Fedorova K A and Ilyasova A I 2016 *Algorithms development of making special techniques in APQP manufacturing process of automotive components* IOP Conference Series:

IOP Conf. Series: Materials Science and Engineering **412** (2018) 012039 doi:10.1088/1757-899X/412/1/012039

Materials Science and Engineering V 134 Is 1 **012036**
[3] Kas'yanov S V, Safarov D T 2004 D*iagnostics of the technical condition of equipment and rigging in terms of technological accuracy* Automotive Industry No 5 pp 24-28
[4] Kas'yanov S V, Safarov D T 2007 *Formation of deviations in quality indicators for product development* Quality management methods No 2 pp 30-36
[5] Kas'yanov S V, Safarov D T 2007 *Productivity and effectiveness of methods for obtaining information about the quality of products* Methods of quality management No 10 pp 40-47
[6] Flohic J Le, Paccol F, Bouton N and Chanal H 2014 *A ModelBased Method for PID Tuning Applied to Serial Machine-Tool Innovations of Sustainable Production for Green Mobility* Energy-Efficient Technologies in Production Part 1 pp 797-812
[7] Jacobs H J, Jacob E and Kochan D 1981 *Spanungsoptimierung* (Veb Verlag Technik Berlin) p 280
[8] Kas'yanov S V, Kondrashov A G, Safarov D T 2017 *Regulation of Geometrical Parameters Deviations of Automotive Components Parts through Diagnostic Measurements Organization* International Conference on Industrial Engineering ICIE Vol 206 pp 1508-14
[9] Vasiliev V A, Odinokov S A, Borisova E V and Letuchev G M 2016 *Methods of quality management of innovation process* IEEE Conference on Quality Management, Transport and Information Security, Information Technologies pp 233-35
[10] Kas'yanov S, Safarov D T and Kondrashov A G *A method for diagnosing of the relative position and rigidity of tool rigs in boring operations in terms of the accuracy of the machined parts* The patent of the Russian Federation for invention No 2496611 Bulletin No 30
[11] Kas'yanov S, Safarov D T and Kondrashov A G *The automated complex for diagnostics of turning technological system on indicators of accuracy of processed details* The patent of the Russian Federation for useful model No 133040 bulletin No 28
[12] Kas'yanov S, Safarov D T and Kondrashov A G *Complex for diagnostics of the turning technological system in terms of the accuracy of the parts being processed* The patent of the Russian Federation for utility model No 133039 Bulletin No 28

Assessment of commercial feasibility of the device for the combined cutting of conducting materials by calculation of integrated koefficient

I I Khafizov[1] **Y Y Levshtanova** [1],

[1]Kazan Federal University, 18 Kremlyovskaya street, Kazan, 420008, Russian Federation

khafizov@kpfu.ru , inboxjulia@bk.ru

Abstract. Authors have considered the problem points connected with commercialization of university intellectual property items in the field of electrochemistry, namely the device of electrochemical cutting of conducting materials. Key criteria for evaluation of an intellectual property item are designated. The method of calculation of commercial appeal which is most acceptable for this development on the basis of the generalized integrated coefficient, theoretically is offered and its commercial value is almost proved.

Important indicator of efficiency of innovative development is her realization in the market, or production applicability as process of commercialization is based on the basis of those intellectual property items which have sufficient degree of commercial appeal to potential investors. If to address statistics, then it states not consolatory data regarding a ratio of the invested capital on carrying out scientific research, developments, ensuring protectability of OIC.

Authors were interested in a question of determination of "the commercial value" of scientific and technical development, namely the device for the combined cutting of conducting materials (the patent for the useful RUS 134097 model 4/23/2013) [5]. The useful model belongs to the field of mechanical engineering, is intended for receiving openings of any contour in conducting metals by means of electrochemical processing, the rights for which belong to the Kazan (Volga) Federal university.

As a rule, check of the knowledge-intensive product before execution of the offer to the customer/investor demands carrying out a complex of analytical researches - novelty and the inventive level, an opportunity and reliability of legal protection, the period of action and territorial accessory of the exclusive rights to the patent, the size of the market, the feasibility study and degree of commercial risk.

These indicators traditionally are considered in expert community as the most acceptable for evaluating a kommertsializuyemost of scientific and technical development. The three-level ball scale of estimation allows to determine the commercial value of an intellectual property item depending on the gained score. Undoubtedly, use of this method reserves also high degree of subjectivity. Advantage is that the method of ranging allows to estimate prospects of commercial realization at once of several OIC (further OIC) [1].

Content from this work may be used under the terms of the Creative Commons Attribution 3.0 licence. Any further distribution of this work must maintain attribution to the author(s) and the title of the work, journal citation and DOI.
Published under licence by IOP Publishing Ltd

Authors for evaluating useful model suggest to use the expensive method based on compilation of all financial investments actually on research implementation, registration of the rights and ensuring protectability. In order that assessment had complex character, a necessary condition is acceptance in attention of distinctive features of the device. First of all, it is the novelty level promoting improvement of competitiveness, scientific and technical importance and industrial applicability with a possibility of variable use of the offered device for the combined cutting of conducting materials as the last two requirements will allow to reach the most powerful economic effect.

For this reason, it is important to carry out assessment of "commercial appeal" by several techniques, and then to construct integrated model for generalization of all key indicators. In turn, creation of integrated model demands introduction of such concept as "correlation coefficients" (correction coefficients), the quantitative and qualitative characteristics of a subject of the analysis intended for definition of influence on commercial appeal. "Correlation coefficients" are offered to be calculated by means of ball assessment of qualitative and quantitative characteristics of the device.

Thus, assessment of commercial value of OIC by an expensive method by means of creation of integrated model it is feasible according to the following scheme:

$$CVz. = Int.z. \sum_{t}^{t-tk} 3,$$

where CVz. - level of commercial appeal of scientific and technical development;

Int.z. - the integrated coefficient received on the basis of generalization of "correlation coefficients" (correction coefficients) (tab. 1);

$\sum_{t}^{t-tk} 3$ – the generalized indicator characterizing financial investments on carrying out scientific research, compensation of developers, providing legal protection and the other incurred expenses for the entire period of research activity [2].

Table 1 - Coefficients of correlation of quantitative and qualitative characteristics of the device for the combined cutting of the conducting metals influencing her commercial appeal

Coefficient	Essence of coefficient	Admissible range of possible values of coefficients	Calculated values of coefficients (on the basis of expert assessment)
Kp1	Obsolescence of the device	K1 = 1 - (Tf/Tn),where Tf - the actual period of validity of the patent, years; Tn - nominal period of validity of the patent, years. from 0 to 1	0,5
Kr2	Physical wear of the device	from 0 to 1	0,7
Kr3	Level of complexity of the device	from 1 to 2	1,5
Kr4	Consumer value of the device	from 0 to 1,8	
Kr5	Level of the technical and economic importance	from 1 to 2	1,5
Kr6	Novelty of the device	from 0,4 to 1,2	
Kr7	Level of legal security of the device	from 0,6 to 1	1
Kr8	Level of prevalence of copyright of useful model	from 0,6 to 1	1
Kr9	Level of possible industrial use of useful model (readiness for introduction to economic circulation)	from 0,2 to 1	1

Kr10	Risk level of introduction of useful model to economic circulation	from 0,3 to 1	0,9
Kr11	Production volume	from 0,2 to 1	

The generalized integrated coefficient - is the work of values of quantitative and quality indicators (correlation coefficients) influencing commercial appeal of the device. Pays off depending on the applied assessment method.

We will explain that the expensive method at assessment of commercial potential of useful model takes into account basic correction coefficients - moral, physical wear, level of complexity and its relevance. The comparative method imposes more expanded requirements to assessment factors, than previous, and demands obligatory calculation of consumer value, level of the technical and economic importance, novelty and legal security, industrial applicability and that size of risk which the customer will face, having given preference to this useful model. Complexity of use of a method - receiving reliable information on the compared analog.

The profitable method as one of the most often used in practice, assumes establishment of relationship of cause and effect between unique features of technical solution and receiving the predicted profit on his use. Appraisers a number of additional factors, such as discount rate, planned character of settlement indicators therefore the subjectivity of assessment still remains at the high level is taken into consideration. Application of a profitable method isn't expedient concerning assessment of this useful model as the risk level of the potential customer at introduction of this scientific and technical development to economic circulation rather high, negatively is reflected in integrated coefficient. The risk is connected, first of all, with the high competition in the field of scientific research and inaccuracy of the predicted volume of future cash flows.

Thus, the generalized integrated coefficient for each method will be various [2]:

$Int.z. = Kr1*Kr2*Kr3;$

$Int.c. = Kr1*Kr2*Kr3*Kr4*Kr5*Kr6*Kr7*Kr9*Kr10;$

$Int.d. = K6*K8*K9*K10*K11.$

Settlement data on determination of commercial value "devices for the combined cutting of conducting materials are presented in the expensive way in tab. 2 [6]:

Table 2 - Settlement indicators of commercial appeal of the device

Item of expenditure	Percent to the total amount of expenses	General estimate
The financial investments necessary on idea realization, including:		64,84
1. Material inputs, one thousand rub: - purchase of accessories, materials;	7,3%	5,0
- expenses of working hours, people/hour; - cost of labor hour of one permanent member of staff, rub/hour; - the number of the employees involved in development		320 187 1
Expenses on ensuring protectability of development, including: 1. Registration of the rights for OIC, thousand rubles. 2. Obtaining patent for useful model 3. Maintenance of the patent for useful model within three years		1,0 1,8 1,2
Total amount of investments	86,9%	68,84
Obsolescence of scientific and technical development (Kr1)		1- (10/20) = 0,5

Physical wear of scientific and technical development (Kr2)		0,7
Level of complexity of technical development (K3)		1,5
Int. z. = Kr1*Kr2*Kr3 (in relation to an expensive method)		0,525
The generalized indicator of commercial appeal of technical development: ***CVz.***= Total amount of investments * ***Int. z.***, thousand rubles.		36,14

Advantage of an expensive method of calculation of commercial appeal before comparative is that unique features, namely physical and obsolescence, complexity of technical solution is considered that meets the main requirements of the carried-out assessment, increases her objectivity.

The mechanism of assessment of intellectual property items, namely useful model, by means of calculation of the generalized integrated coefficient allows to make reasonable corrections to quantitative and quality indicators and to influence result of calculation. Statement of the purpose and problems of evaluating, division of creative process into the interconnected stages and accounting of budget expenses on each of them, reliability and fully dimension of the provided information for calculation of total integrated coefficient, and, the main thing to check applicability of this approach to assessment of commercial potential of an intellectual property item. Having taken above-mentioned factors into account, it is possible to use a well-known way of generalization of results - the calculation of the average indicator based on introduction by an expert way of the weight coefficients which are most fully disclosing applicability and efficiency of the chosen assessment method for concrete technical solution.

Authors when calculating total weight coefficient takes into account the following factors - completeness and reliability of the provided information, compliance of the purpose of assessment.

The corrected commercial value of OIC according to an expensive method:

$$CDV\ z. = WCz. * CVz.$$

where CDV z. - the corrected development value;

WKz. - weight coefficient (authors have accepted - 0,15);

CVz. - commercial value of technical solution.

The coordinated result of assessment of useful model for receiving openings of any contour by means of electrochemical processing taking into account the corrected coefficient is equal in conducting materials to 30,719 thousand rubles. This technical solution reaches self-sufficiency, is with other things being equal capable to make profit during her realization. Use of an expensive method has allowed to carry out express assessment, to estimate the volume of own capital investments at development, to consider specific features, in general, in a complex to analyse industrial applicability and potential of realization.

Authors see a possibility of realization of economic potential of an invention in input him in economic circulation through licensing.

Sale of the rights to use OIC will allow to get profit. Having assumed that the license contract is signed for 3 years, the expected rate of payments for leasing payments will be not less than 4%, the expected output and, respectively, the price of a unit of production have every year a positive tendency of growth, we will receive the following results (table. 3):

Table 3 - Results of the predicted profitability from licensing of useful model (the RUS 134097 patent 4/23/2013)

Name of an indicator	Value of an indicator
Expected volume of product sales, piece.	18
The corrected cost of development, thousand rubles.	30,719
Expected rate of profitability of developers from product cost, %	10

Expected rate of percent on royalties, %	4
Expected sum of the license contract for a unit of production, thousand rubles.	33,79
The commercial value of licensing of an invention in three years, thousand rubles, CVl. = (0,04*(3*33,79+6*33,79+9*33,79))*3+33,79	106,78

The carried-out assessment has shown that the potential of this useful model isn't used by the owner, and, respectively, it is necessary to take measures for improvement of effective management of an intellectual property item for receiving profit on commercialization. And in general, as show statistical data probability of realization university to OIC doesn't exceed 10% of that set of perspective projects which are studied within intellectual activity. The difficulties which have arisen during assessment of the useful model considered by us can be caused by a variety of reasons, first, it has become clear that RID initially hasn't been focused on commercial realization therefore it is rather problematic to reveal commercial advantages; secondly, readiness for industrial applicability of the device from the point of view of developers and investors can be various (it is necessary to address the first point); thirdly, despite the revealed commercial value, often there is an inconsistency of actions between participants of process of management of intellectual property [3,4].

Therefore, before making the decision on commercialization of OIC, it is necessary to carry out express assessment, in view of key features of concrete development, to connect to work advisory council for carrying out the professional analysis of the actual technical condition of RID and determination of its commercial value for the local markets (the industrial enterprises to which the idea can be interesting and ready to invest own financial means). You shouldn't forget that perspective development is subject to creation of the innovative project, and it is possible even the innovative business direction therefore specification when evaluating is an important condition.

References

[1] Khafizov I I Stalemate 134097 Russian Federation, MPK V23N5/04. An electrode tool for the combined cutting of conducting materials No. 2013118839/02

[2] 2005 An innovative way of development for new Russia *Center of social and economic problems of federalism of Institute of economy of RAS* M.: Science 343 pages

[3] 2002 Methodical recommendations about determination of market value of intellectual property *Approve Ministry of property of the Russian Federation* No CK-4/2197

[4] Khafizov I I 2007 Low-waste fair division of scarce materials by the combined method. *The abstract of the thesis for a degree of Candidate of Technical Sciences* Public Educational Institution of Higher Professional Training VGTU publishing house p18

[5] Kozyrev A N and Makarov V L 2003 Estimation of cost of intangible assets and intellectual property M.: Interreklama 352 pages

[6] Lev B 2003 Intangible assets: management, measurement, the reporting *Lane from English L.I. Lopatnikov M.* Kvinto-Consulting 240 pages

[7] Akhmetov I D, Zakirova A R Sadykov Z B and Khafizov I I 2017 New electrode-tool for the combined kerf of electrically conductive materials *IOP Conference Series: Materials Science and Engineering* v 240 Is 1 **012003**

[8] Khafizov I I and Galimov A N 2017 IT-strategy and major aspects of quality management on the market of goods and services *IOP Conference Series: Materials Science and Engineering* v 240 Is 1 **012038**

[9] Khafizov I I, Saveleva T N and Lyubtsov V S 2017 Estimation of parameters of charge carriers in dielectric materials by CELIV method *IOP Conference Series: Materials Science and Engineering* v 240 Is 1 **012040**

[10] Khafizov I I and Nurullin I G 2017 Product quality as the main factor of increase of competitiveness (on the example of JSC «Kazan helicopter plant») *IOP Conference Series: Materials Science and Engineering* v 240 Is 1 **012039**

[11] Gil'manshin I R and Kashapov N F 2014 Energy service contracts in regional engineering center for small and medium businesses *IOP Conference Series: Materials Science and Engineering*
[12] Khafizov I I 2017 Ways of decrease in the material consumption in case of their separation by the combined methods *IOP Conference Series: Materials Science and Engineering* v 240 Is 1 **012037**
[13] Khafizov I I 2016 Economic efficiency and effectiveness of ways of separating materials electro diamond processing *IOP Conference Series: Materials Science and Engineering* V 134 No 1 **012014**
[14] Nurullin I G and Khafizov I I 2016 The impact of innovation implementation and certification on the company's activities and reputation *Modern technologies: topical issues, achievements and innovations: collection of articles of the winners of the IV international. science.-practical Conf.* (Penza: ICSU " Science and education») p 44-6

Estimation of charge carrier mobility in polymer thin-film structures

I I Khafizov[1], T N Saveleva[1]

[1] Kazan Federal University, 18 Kremlyovskaya street, Kazan, 420008, Russian Federation

khafizov@kpfu.ru tnsaveleva@stud.kpfu.ru

Abstract. In this article, the polymer polydiphenylenphthalide. The estimates of the mobility of charge carriers for FDP films of different thicknesses by the CELIV method are carried out. During the measurements it was concluded that the mobility value does not depend on the thickness of the polymer film within 100-800 nm.

Introduction

Polymers occupy one of the leading places among structural materials for mechanical engineering. Thus, the consumption of plastics in this industry becomes comparable (in units of volume) with the consumption of steel. Currently, there is a great interest in the study of new polymeric materials. One such material is polydiphenylenphthalide (PDF). It is a polymer of polyheteroarylenes class. He has the unique physical properties on the basis of which it can be used in the form of polymer films, binders for reinforced polymers, polymer adhesives in instrumentation, electronics, radio engineering; membranes, diaphragms in galvanic cells, batteries, electrolyzers, etc [1].

Polymers of this class were synthesized in the polycondensation group Of the Institute of organic chemistry of Ufa scientific center (UC) by electrophilic polycondensation [2].

Polydiphenylenphthalide and its properties

PDF is a dielectric. The parameters that determine the initial dielectric state are as follows:

- Band gap ~ 4.3 eV
- Electronic work function ~ 4.2 eV
- Electronic affinity of ~ 2 eV
- The first ionization potential of ~ 6.2 eV

Polyacrylonitrile have high heat and temperature resistance. The softening temperature of 3600 C. the Temperature of degradation (Tg – temperature of loss of 1% weight) in argon and in air is 4400 With [3]. The thermal stability of the phthalide cycle in polyarylenephthalides is determined by the nature of the groups that connect the phenyl nuclei in the main chain.

Content from this work may be used under the terms of the Creative Commons Attribution 3.0 licence. Any further distribution of this work must maintain attribution to the author(s) and the title of the work, journal citation and DOI.
Published under licence by IOP Publishing Ltd

IOP Conf. Series: Materials Science and Engineering **412** (2018) 012041 doi:10.1088/1757-899X/412/1/012041

The study of the winter hardiness of polyarylenephthalides showed that they have good resistance to aggressive media at elevated temperatures.

PDF has high film-forming properties. It is highly soluble in methylene chloride, chloroform, tetrachloroethane, dimethylformamide, cyclohexanone and other solvents. Watering from the solution, it is possible to obtain undirected films with tensile strength (800-900 kgf·mm2) with a relative elongation at rupture of 10-20% [4]. It is possible to allocate the following features of film formation of polymers of the PDF type. When using a solvent with a lower boiling point (chloroform) and applying films by centrifugation under the same conditions, a film of a greater thickness is formed than when using a solvent with a higher boiling point (cyclohexanon) [5].

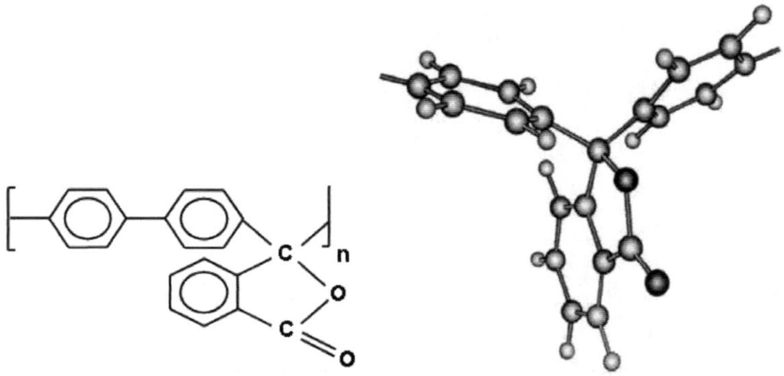

Figure. 1. Structural formula polydiphenylenphthalide and three-dimensional image

Morphology of thin polymer films (especially ultra-thin, obtained from solutions of 0.1% concentration) strongly depends on the nature of the substrate and its properties such as wettability, roughness and pretreatment conditions (cleaning, etc.).

Prototyping and research methods

The mobility of charge carriers is an important parameter determining the kinetic characteristics of charge carriers in the materials under study. To estimate the mobility in the films polydiphenylenphthalide was manufactured experimental samples of different thickness. The experimental sample was a thin-film structure of sandwich type indium oxide and tin / PDF / metal(Al). Film PDF received on the centrifuge. The thickness of the film varied depending on the speed of the centrifuge. Measurements were carried out on a specially assembled unit for the method CELIV [6].

The essence of the method is as follows: 2 successive pulses of linearly increasing voltage are supplied. The first impulse is the main one and the second one is the controlling one. The controlling impulse is necessary to evaluate the quality of contact blocking. As a result, we obtain transient current curves. The mobility of charge carriers is calculated by the formula

$$\mu = \frac{2d^2}{3At_{\max}^2\left[1+0.36\frac{\Delta j}{j(0)}\right]}$$

where Δ J is the current registered on the sample, J (0) is the displacement or capacitive current. It can be estimated either by the transient current curve, or by the following formula

$$j(t) = \frac{A}{d}\varepsilon\varepsilon_0 = j(0)$$

where A is the rate of rise of voltage, and d is the thickness of the sample.

Results and discussion

IOP Conf. Series: Materials Science and Engineering **412** (2018) 012041 doi:10.1088/1757-899X/412/1/012041

Figure 2 shows the results obtained on films with thickness of 800 nm and 170 nm at different rates of voltage increase. The insert shows the maximum voltage values.

The table shows the results obtained on the test sample. From the second column it can be seen that with the increase in the rate of voltage increase, the magnitude of mobility decreases. However, at a speed of 2700/mobility does not differ from the values of mobility obtained at a speed of 1500 In/sec. This is due primarily to the difficulty of estimating the time tmax of the curve of the transient current due to the weak signal and the extended high. At extreme points (near the maximum at 3000 V / s), the mobility value is comparable to the mobility measured at 2700 V/s and 1500 V/s.

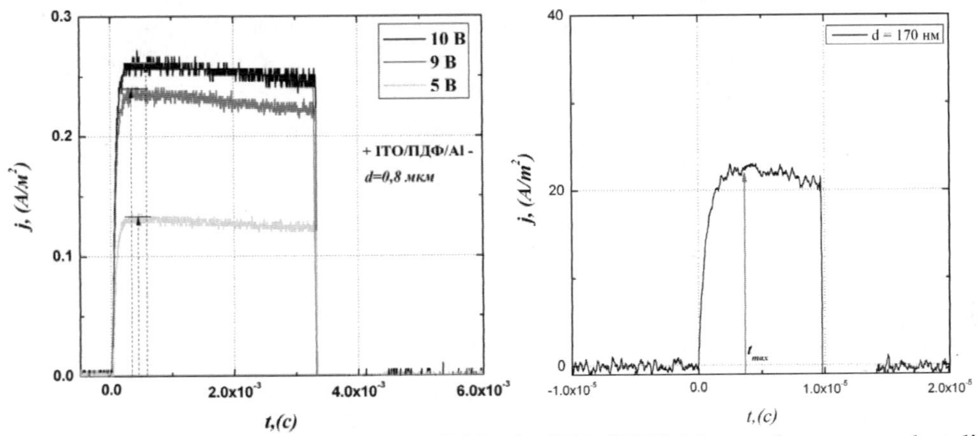

Figure. 2. Transient current curves FOR the ITO-PDF-Al sample measured at different voltage rise rates A=U/t.

d, мкм	*dU/dt, (B/c)*	*μ,(см²/Вс)*	t_{max},(мс)	*J(0),(A/м²)*	*ΔJ,(A/м²)*
0.8	3000	$3.8\cdot10^{-6}$	0.57	0.2	0.06
0.8	2700	$1.2\cdot10^{-5}$	0.34	0.1	0.06
0.8	1500	$1.2\cdot10^{-5}$	0.45	0.09	0.03
0.17	50000	$2\cdot10^{-5}$	4.1	15.6	6.38

Table 1. The results obtained from the transient current curves for

Measurements were made on the condition that at least one of the electrodes of the sample is completely (or partially) blocking. In particular, the lower transparent electrode ITO (indium tin oxide) was used as a blocking electrode for measuring mobility in this work. Therefore, due to the specific structure of the sample, the main charge carriers are holes.

Analyzing the obtained results, it can be concluded that the mobility value does not depend on the thickness of the polymer film within 100-800 nm.

References

[1] Lachinov A N and Vorobyova N V 2006 Electronics of thin layers of broad-band polymers-UFN Vol 176 No 12 pp 1249-66

[2] Zolotukhin M G, Kavardakov V, Salazkin S N and Rafikov S R 1984 *Some patterns of the synthesis poliatilenglikol homopolymerisation n-(3-chloro-3-phthalidyl)-biphenyl* (head of the. Coed) T 26 A No6 pp 1212-17

[3] Novoselov V I 1996 *The Interaction poliatilenglikol and their analogs with iodine* dis ... kand. chem. sciences. 02.00.06: protected 15.03.96 IOH UNTS RAS, Ufa p 135

[4] Salazkin S N, Zolotukhin M G, Kovardakov V, Dubrovina L V, Gladkova E A, Pavlova S S and Rafikov S R 1987 *Molecular-mass characteristics of polyarylene sulfide* Head of the coed A 29 No 7 pp 1431-36

[5] Wu C R, Johansson N, Lachinov A N, Stafstrom S, Kugler T, Rasmusson J and Salaneck W R 1994 *The chemical and electronic structure of the conjugated polymer poly (3.3-phthalidyliden-4.4-bihenylilene)* Synth. Metals V 67 pp 125-28

[6] Juska G, Arlauskas K, and Viliunas M 2000 *Extraction Current Transients: New Method of Study of Charge Transport in Microcrystalline Silicon* Phys. Rev. Letts V 84 p 4946

[7] I I Khafizov 2015 *Processing methods with imposing of electric field at low- waste division of materials* IOP Conf. Series: Materials Science and Engineering 86 **012013**

[8] Khafizov I I, Saveleva T N, Lyubtsov V S 2017 *Estimation of parameters of charge carriers in dielectric materials by CELIV method* IOP Conference Series: Materials Science and Engineering Vol 240 Is 1 **012040**

IOP Conf. Series: Materials Science and Engineering **412** (2018) 012042 doi:10.1088/1757-899X/412/1/012042

Problems of development of electrochemical production of Russia and possibility of their decision

I I Khafizov[1], I G Nurullin[1] and Z B Sadykov [1]

[1]Kazan Federal University, 18 Kremlyovskaya street, Kazan, 420008, Russian Federation

khafizov@kpfu.ru, opros16@yandex.ru, zbsadykov@kpfu.ru

Abstract. The relevance of the study of the current state of electrochemical production is that electrochemistry every year is becoming increasingly important in the domestic industry. The development of electrochemical methods of processing leads to an increase in both the processing capacity and the achievement of new technical effects, significantly increasing the strength, durability and other performance parameters of the parts.

Introduction.

The mechanical engineering around the world admits how the leading branch of production. The level of development of it of branch influences a condition of production capacity of the country, creates conditions for steady functioning of all branches of economy (transport and communication, agro-industrial complex, defensive and construction branches, etc.) and fillings by products of the market of consumer goods.

Depends on the level of development of this branch the most important specific indicators of gross domestic product of the state, labor productivity, level of ecological safety of the industry and, of course, defense capability of the country.

For today's time increase in growth of the competition is observed. The common goal on the present consists in creation of the main part of fixed capital of economy. In the course of investment and construction activity the equipment and machines of different function, devices and units, vehicles, knots turn into the active capital of economy of our state and by that form the production device. Thus, there is a reproduction process in national economy.

– it is possible to call the high level of wear of the fixed business assets one of the most serious problems of the Russian Federation. Average age of production oborudovaniye averages 20 years. More than 40% of oborudovaniye have age of 20 years, and at some enterprises even machines, nearly semicentennial prescription are used. Funds of many organizations of branch demand updating that is impossible without investment of big investments.

The high share of the outdated equipment in production doesn't give the chance of full use of production capacities the enterprises. In order that production of the domestic industry was competitive in the international markets, it is necessary to have the most modern of the equipment.

Electrochemical production gains escalating value in the industry every year. Construction of powerful power plants I have created favorable prospects for further development of electrochemical technology.

The modern electrochemical industry is one of the main consumers of electric energy. In this regard it is necessary to notice that many problems of power and electrochemistry were mutually

Content from this work may be used under the terms of the Creative Commons Attribution 3.0 licence. Any further distribution of this work must maintain attribution to the author(s) and the title of the work, journal citation and DOI.
Published under licence by IOP Publishing Ltd

IOP Conf. Series: Materials Science and Engineering **412** (2018) 012042 doi:10.1088/1757-899X/412/1/012042

connected. Development of chemical and especially electrochemical industry depends on the level of achievements in the field of power. At the same time in progress of power the important place is taken by electrochemical technology.

Main part.

Electrochemical methods develop in the industries by the advancing rates in comparison with chemical in connection with their obvious advantages. In electrochemical processes the equipment in many cases is simpler and more compact in comparison with alternative chemical processes, electrochemical methods of receiving a number of products are characterized by smaller number of production stages and operations, cheaper raw materials and bigger depth of his transformation, simultaneous formation of valuable products. These advantages often cause lower prime cost of the received production [1].

The technology of electrochemical processing has advantages in comparison with traditional methods of processing such as: a method of laser processing, laser cutting, use of milling and engraving machines, use of the multicoordinate high-speed processing centers, machining [2].

For example, uses of regularities of process of division by disk tools with calculated geometrical parameters with adaptive management on several coordinates, creation of new ways of management of the provision of various parts of the tool and calibration of a groove after division of preparation with the minimum allowance gives the chance of increase in accuracy of processing with the minimum losses of material at division. The exception of the subsequent operations on processing of places of division of details allows to accelerate several times processing, to considerably increase the accuracy and quality of details, to cut a consumption of scarce materials [3].

Advantages of electrochemical processing:

- the reached record indicators on roughness of a surface and accuracy of copying allow to exclude traditional finishing operations: fair grinding and polishing,
- lack of hangnails and a defective layer after electrochemical processing on the processed product surface;
- economy of working hours on production of products;
- economy of the electric power on production;
- electrochemical machines don't demand laying of special communications and is connected to the general exhaust system and the sewerage;
- electrochemical machines conform to requirements for noise influence and electromagnetic compatibility;
- electrochemical machines are rather compact (occupy no more than 4 sq.m.), are convenient in installation and transportation;
- practically all conducting materials at correctly developed technological processing and the picked-up electrolyte are exposed to electrochemical shaping;
- at serial production of products the prime cost, time of production of lot of products in two and more times decreases;
- at electrochemical processing, except the electric power, only water is spent [4].

In the conditions of the strained and developing competition, for saving the provision in the market, the enterprises of electrochemical production need to solve a set of different problems. For example, the problems connected with processes of management of quality of production, a conclusion of complex products with new consumer properties on the market, expansions of her range, etc. [5].

But it is only not enough to create a new product with excellent characteristics, it is very difficult to introduce it on the market. Such inventors as Thomas Edison and Nikola Teslo developed the inventions based on electricity use such as bulb or electric motor. But without due financial support and the thought-over marketing strategy of investors about such important, changed the world, innovations at that time nobody could learn [6].

At release of a ready-made product on the open market within the first two years close attention is necessary, continuous monitoring will allow to reveal and in time to correct all inaccuracies. It is possible to allocate the following factors which are the reasons of a failure of new products in the market:

- insufficient analysis of the market;
- defects of a product;
- lack of effective marketing actions;
- excessively high expenses;
- actions of competitors;
- a lack of support at removal of a product on the market;
- production problems.

Advance on the market of hybrid cars which distribution restrained not only the price, but also insufficient development at the first stage of infrastructure of their service can be an example [7].

Development of electrochemical methods of processing leads both processing productivities, and quality of details to increase. Besides, development of new ways of processing allows to reach new technical effects, for example, considerably to increase durability, wear resistance and to reach other operational parameters of details.

Dimensional electrochemical processing of metals will become a widespread method. It is very favorable to polish in every respect products and to remove roughnesses, especially from details of irregular shape or the tiny size, electrochemical way. In the same way the organic electrosynthesis allowing not only to receive clean products, but also to lower number of steps of reaction will develop more widely and more widely.

It is very important that electrochemical technology in all her versions - one of the cleanest and safe for Wednesday. Besides, electrochemical processes are capable to maintain ecological equilibrium. For example, the oxygen received as a by-product at electrolysis can be used for aeration of water in the rivers and lakes and to improve there biological conditions.

Many dangerous substances can be neutralized electrochemical influence, and to affect selectively, only the most dangerous toxic components, without spending the electric power for oxidation of other components. Thanks to it neutralization becomes especially economic. Purification of waters with electrodialysis, electroosmotic methods of drying of materials and other methods will promote protection of the environment too [8].

The main lack of electrochemical processes – high power consumption therefore power expenses make the main article of prime cost of products. Therefore decrease in energy consumption by improvement of technology and also rational, economic use of the electric power is especially important for electrochemical productions.

Conclusion.

At the moment, in the conditions of the cruel competition to the enterprises the problem of substantial increase of quality of different types of products, updating and expansion of her range according to modern requirements of the developing scientific and technical progress and also the increasing needs of the population lays down.

The high-quality transformations directed to change of technological way and with ensuring satisfaction of necessary requirements of production on the basis of innovative technologies, an optimum image the combining resource-saving and level of profitability modernizes the domestic industry and gives an impetus on active development.

Carrying out modernization of the industry is the urgent and objective need caused by the aspiration of the country to maintain long-term social and economic stability, to increase economic power, to strengthen positions in the international market and to ensure economic security.

References

[1] Klimovich K P and Odintsovo M A 2014 Problems and prospects of development of the machine-building complex of Russia in the conditions of market economy *Economic magazine* №4 Mode of access: http://cyberleninka.ru/article/n/problemy-i-perspektivy-

razvitiya-mashinostroitelnogo-kompleksa-rossii-v-usloviyah-rynochnoy-ekonomiki Date: 03.10.2017

[2] Temerlin D D and Gavrilov S V 2014 With Application of electrochemical processes in food industry *Bulletin of the Kazan technological University* No 23 Mode of access: https://cyberleninka.ru/article/n/primenenie-elektrohimicheskih-protsessov-v-pischevoy-promyshlennosti Date: 20.11.2017

[3] Khafizov I I 2013 *Precast electrode tool for electro-lubricating cutting* Innovative engineering technology, equipment and materials part 2 pp 114-7

[4] 2016 Electrochemical machines and electrochemical processing [Electronic resource] *Investment research and technological platform " second industrialization of Russia"* Mode of access: http://xn--80aaafltebbc3auk2aepkhr3ewjpa.xn--p1ai/elektrohimicheskie-stanki/ Date: 22.12.2017

[5] Mukhanov I V 2015 *Modernization of mechanical engineering of Russia in the conditions of economic sanctions: problems and solutions* The Economic journal Mode of access: http://cyberleninka.ru/article/n/modernizatsiya-mashinostroeniya-rossii-v-usloviyah-ekonomicheskih-sanktsiy-problemy-i-puti-resheniya Date: 13.01.2018

[6] Nurullin I G and Khafizov I I 2016 The impact of innovation implementation and certification on the company's activities and reputation *Modern technologies: topical issues, achievements and innovations: collection of articles of the winners of the IV international. science.- practical Conf.* (Penza: ICSU " Science and education») p 44-6

[7] Zadornova N A 2013 Innovative technologies in mechanical engineering *Izvestiya MGTU* No 1 Mode of access: http://cyberleninka.ru/article/n/innovatsionnye-tehnologii-v-mashinostroenii-1 Date: 15.02.2018

[8] 2016 Prospects of electrochemistry [Electronic resource] *Journal "Electricity in the world of chemistry"* (Moscow) Mode of access: http://electrohistory.ru/perspektivy-elektroximii/ Date:22.03.2018

[9] Akhmetov I D, Zakirova A R , Z B Sadykov and I I Khafizov 2017 New electrode-tool for the combined kerf of electrically conductive materials *IOP Conference Series: Materials Science and Engineering* V 240 Is 1 **012003**

[10] Khafizov I I and Galimov A N 2017 IT-strategy and major aspects of quality management on the market of goods and services *IOP Conference Series: Materials Science and Engineering* V 240 Is 1 **012038**

[11] Khafizov I I, Saveleva T N and Lyubtsov V S 2017 Estimation of parameters of charge carriers in dielectric materials by CELIV method *IOP Conference Series: Materials Science and Engineering* V 240 Is 1 **012040**

[12] Khafizov I I and Nurullin I G 2017 Product quality as the main factor of increase of competitiveness (on the example of JSC «Kazan helicopter plant») *IOP Conference Series: Materials Science and Engineering* V 240 Is 1 **012039**

[13] Gil'manshin I R and Kashapov N F 2014 Energy service contracts in regional engineering center for small and medium businesses *IOP Conference Series: Materials Science and Engineering*

[14] Khafizov I I 2017 Ways of decrease in the material consumption in case of their separation by the combined methods *IOP Conference Series: Materials Science and Engineering* V 240 Is 1 **012037**

[15] Khafizov I I 2016 Economic efficiency and effectiveness of ways of separating materials electro diamond processing *IOP Conference Series: Materials Science and Engineering* V 134 No 1 **012014**

[16] Khafizov I I 2015 Processing methods with imposing of electric field at low- waste division of materials *IOP Conf. Series: Materials Science and Engineering* V 86 **012013**

IOP Conf. Series: Materials Science and Engineering **412** (2018) 012043 doi:10.1088/1757-899X/412/1/012043

Research of boundary layers of molds, used in titanium casting

A I Khatmullina, E Yu Petrov and A M Lyakhovich

Kazan Federal University, 18 Kremlyovskaya street, Kazan, 420008, Russian Federation

alalam@mail.ru

Abstract. Boundary layers of the mold used for castings from titanium have been investigated by the method of electron microscopy with a system of probe microanalysis. Migration both components of mold and additives of titanium alloy is established to the surface layers of the form.

1. Introduction

Titanium is a metal that has a low specific gravity (titanium density is about 4.5g/cm3), but it has high specific strength. Pure titanium has high plastic characteristics. A sturdy, thin oxide film on the surface of titanium provides very high corrosion resistance. Titanium and its alloys are an important structural material for aviation, space, underwater technology, etc. The main problem of titanium alloys is the formation of a diffusion gas-saturated layer, which has a high hardness, reduced ductility, a cracking tendency, called the "alpha case". Many papers have been devoted to study of the alpha case, for example, [1-3]. Its formation occurs during casting, due to migration of mold elements and environment. The thickness of layer can reach several mm.

Objective: to research boundary layers of the mold used in the casting of titanium to identify the processes that occur during the formation of the alpha case.

2. Experimental

Mold composition MgO (83.2%), CaO (2.9%), SiO2 (1.9%), Na2O (SiO2) n (12%) was analyzed. The calculated content of chemical elements Mg: O: Si: Ca: Na (at.%) in the form is 41.9 : 50.3 : 2.7 : 1.0 : 4.1. The mold was sintered at a temperature of 1000 ± 50 ° C for 6 hours. Before sintering, casting mold was saturated with carbon dioxide to grasp its components. Then it's cooling for 10 hours in the furnace.

There were two types of samples: the initial casting mold (figure 1 left) and the form after contact with the titanium casting (figure 1 right). "EVO 50 XVP" (Carl Zeiss) scanning electron microscope equipped with the probe microanalysis system "INCA Energy-350" was used to study the relief, elemental composition of the surface and the cross section of the mold.

3. Results and discussion

Three surface regions (figure 2 left) and six regions of the cross-section were investigated in the initial form. (figure 2 right).

Content from this work may be used under the terms of the Creative Commons Attribution 3.0 licence. Any further distribution of this work must maintain attribution to the author(s) and the title of the work, journal citation and DOI.
Published under licence by IOP Publishing Ltd

IOP Conf. Series: Materials Science and Engineering **412** (2018) 012043 doi:10.1088/1757-899X/412/1/012043

It is established that in the sample there is a large number of pores located throughout the volume. The existence of regions with different structures was found on the mold surface. The relief of the

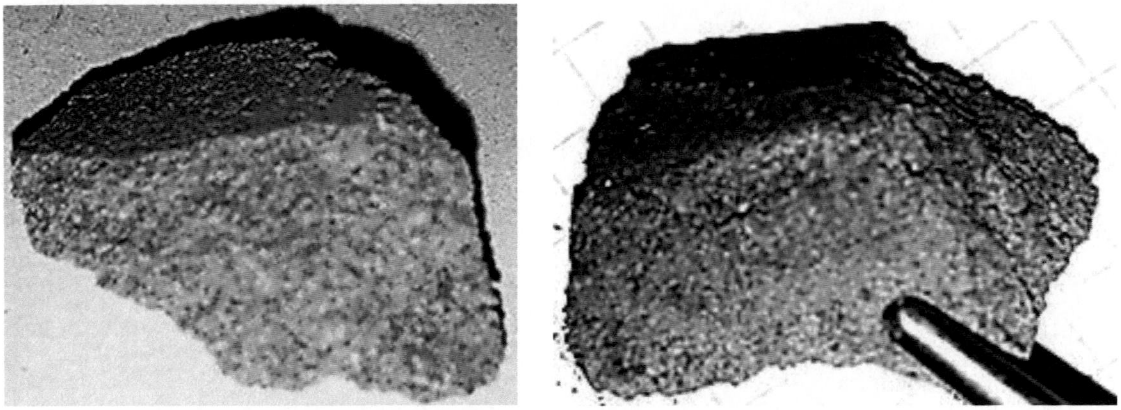

Figure 1. The samples of the original mold (left) and the form after contact with the casting (right).

cross-section is changing slightly. Investigation of regions of the surface with different structures showed that they differ in the elemental composition (table). The first section is characterized by an excess of the concentration of O and Ca with a reduced content of Mg compared with the calculated and average concentration. Probably, a significant increase in oxygen is associated with the formation of oxides with a higher degree of oxidation, and with formation of salts of carbonic acid. Carbonic acid is formed during the interaction of carbon dioxide with water present in the form.

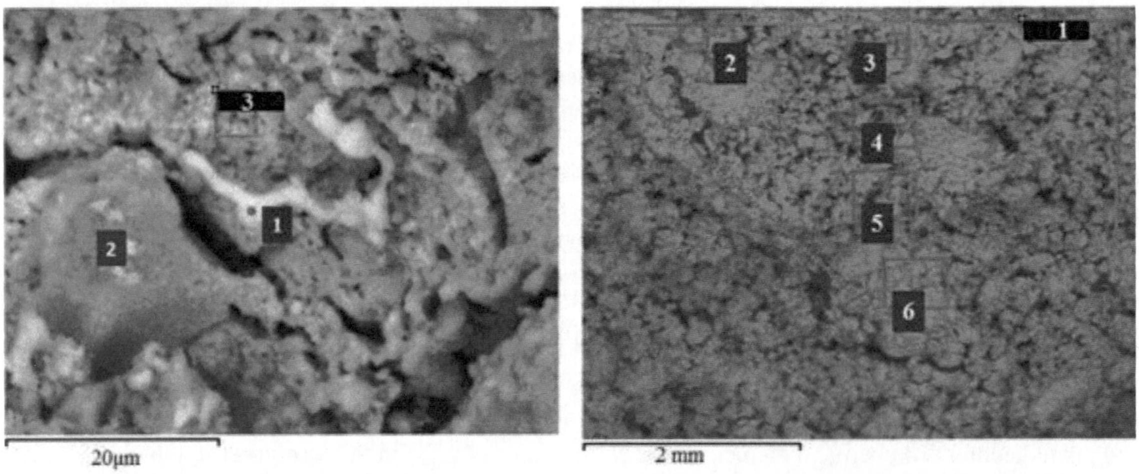

Figure 2. Image of the surface (left) and cross-section (right) of the mold before contact with the casting.

Table. Element composition (at.%) of the mold surface prior to contact with casting.

Element	regions			mean	
	1	2	3	surface	calculation
O	64.6	46.9	57.8	59.3	**50.4**
Na	3.0	5.1	4.0	3.5	-
Mg	18.2	35.5	25.6	26.6	**47.7**
Si	3.8	7.8	5.8	6.4	**0.8**
Ca	9.4	1.3	4.1	3.2	**1.2**

Fe	1.0	2.5	2.7	0.5	-
O*	**63.0**	**70.1**	**57.6**	**56.75**	-

In the second section, the Si, Na content is greater than in the first section; the third region is characterized by an intermediate content of the elements, between the first and second area.
A quantitative study of the elemental composition of the cross section showed that it is close to the average value of the surface composition and is practically the same at the depth examined (figure 3).

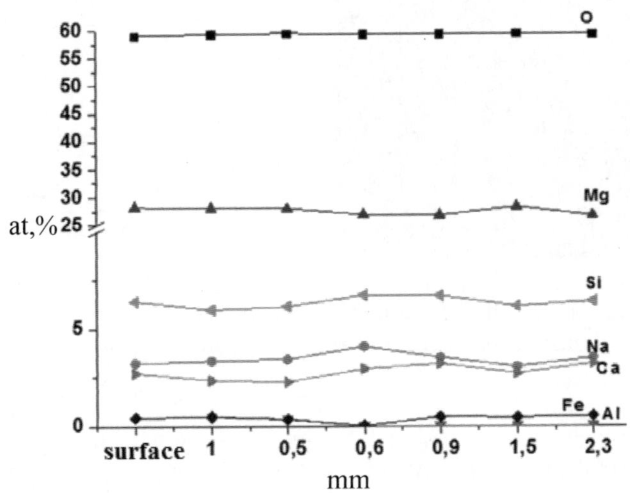

Figure 3. Change profile in the elemental composition of the mold.

Four regions of the surface (figure 4 left) and five sections of the cross section (figure 5 left) of form after contact with titanium casting were examined. The relief of the sample surface is uneven, there are depressions and elevations, for which a different elemental composition can be characterized (figure 4 right). This indicates that the mold is separated from the casting by layers located at different distances from the casting. The presence of titanium on the separation surface of the mold can be determined by the migration it from the casting.

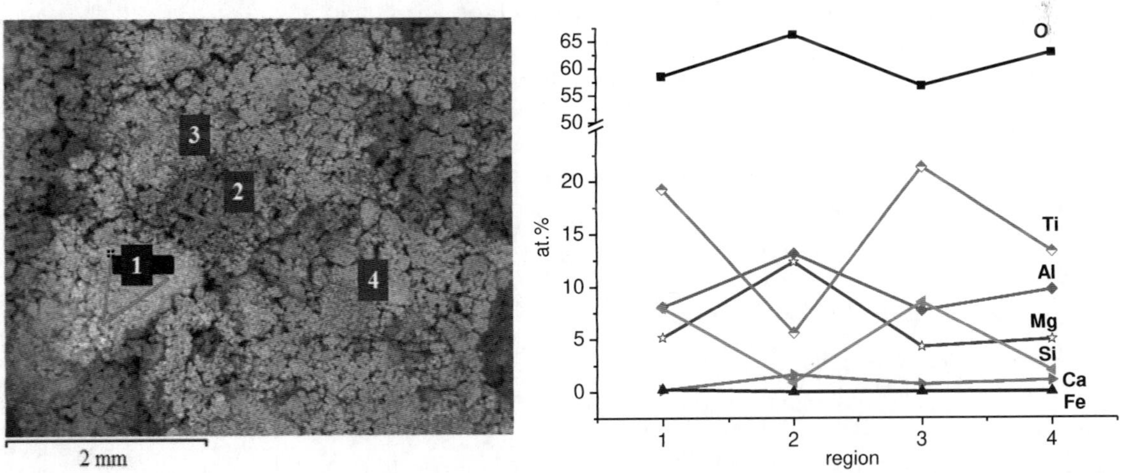

Figure 4. The relief image (left) and the elemental composition (right) of the investigated surface regions after contact with the casting

The cross section of the mold was obtained by its splitting. It has a rough structure. The distribution of elements along the cross section is the same and similar to the composition of the cross section of the mold without contact with the casting (figure 5right). It follows that the interfacial layer remaining on the surface of the casting is very thin; most of the interfacial layer remained on the casting.

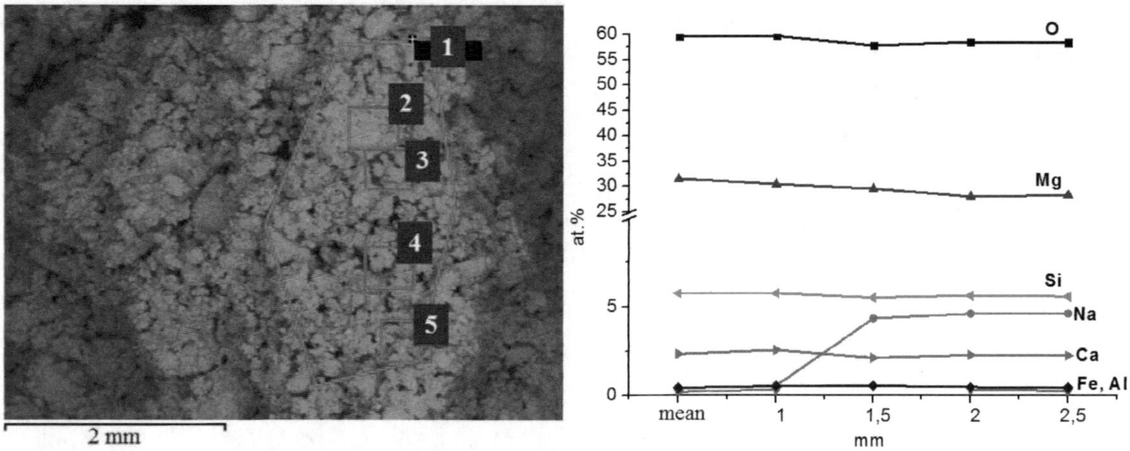

Figure 5. Image of the cross section (left) and the elemental composition (right) of the investigated regions after contact of the mold with the casting

4. Conclusions.

There is an increased content of O, Si, Ca and Na in comparison with the calculated values in the boundary layers of the initial mold. Perhaps this is due to the migration of these elements from the mold volume.

There are regions characterized by a different structure and element composition on the surface of the form.

After mold contact with the casting, we observe a migration of titanium to the surface layers of the mold.

Most of the interfacial layer remains on the casting, after peeling off mold from casting.

5. References

[1] Kanjer A, Optasanu V, Lavisse L, Marco de Lucas M, Dejardin S, Francois M, Berger P, Peyre P, Gorny C and Montesin T **2017** *Oxid Met* DOI 10.1007/s11085-016-9700-6
[2] Deshmukh V, Kadam R and Joshi S **2017** *Machining Science and Technology* **21**(1) 257
[3] Sung S, Kim Y **2005** *Materials Science and Engineering* A **405** 173

Acknowledgements
This work was funded by the subsidy allocated to Kazan Federal University for the state assignment in the sphere of scientific activities (1.1300.2017/4.6).

IOP Conf. Series: Materials Science and Engineering **412** (2018) 012044 doi:10.1088/1757-899X/412/1/012044

Profitability increase of pre-production and autocomponents issue project

R M Khisamutdinov[1], O I Yurasova[2], S Yu Yurasov[2], E A Ryabov[2]

[1]JSC «KAMAZ»

[2] Kazan Federal University, Naberezhnye Chelny Institute, 423812, Russia, Naberezhnye Chelny, Prospekt Syuyumbike 10A

sunnywind@list.ru, yurasova-olya@yandex.ru, docfile@yandex.ru, Evgeniy.ryabov@inbox.ru

Abstract. In modern conditions auto components domestic manufacturers should not only improve the quality of products to the world level, but also provide the greatest possible profit. As a result, a technique for developing modern technological documentation was proposed. In the framework of this methodology, requirements to modern technological documentation have been developed, factors of profitability decrease in the APQP project cycle have been identified, work grouping providing the best results for the APQP process stages has been grouped, technological information as the basis of enterprise management has been worked out, technological design and technical documentation processes have been developed, scheme of the process of technological design. In the paper, the key control characteristics of technological transitions are systematized, the principles of management of the preparation of production (technological design) based on technical documentation are formulated. The developed measures allow at the stage of technological design not only to obtain information for managing the auto component quality, but also to prevent potential losses, that is, to increase the profitability of the project.

Russia's accession to the WTO has led to the fact that auto components domestic manufacturers should not only improve the products quality to the world level, but also provide the greatest possible profit. One of the main conditions is to provide specialists and managers with the necessary set of technological information.

The introduction of quality management systems in the Russian automotive industry in accordance with GOST R ISO 16949-09 implies the transfer of as many works as possible to create information support for the early stages pre-production. It details the main provisions of the planning process, development, pre-production and production of the component (APQP-process). The purpose of the APQP-process is to ensure the output of the planned quality of mass-produced automotive components meeting the requirements and expectations of consumers.

The main principle is the execution of works by a multifunctional APQP-team of specialists. As the internal goal of the APQP process, the clear interaction of all its participants and ensuring timely implementation of the required stages of work at acceptable costs is indicated. The results of the APQP process, designed in an appropriate form, are included in the set of documents on the harmonization of automotive components.

The stages of the APQP process are shown in Fig.1

Content from this work may be used under the terms of the Creative Commons Attribution 3.0 licence. Any further distribution of this work must maintain attribution to the author(s) and the title of the work, journal citation and DOI.
Published under licence by IOP Publishing Ltd

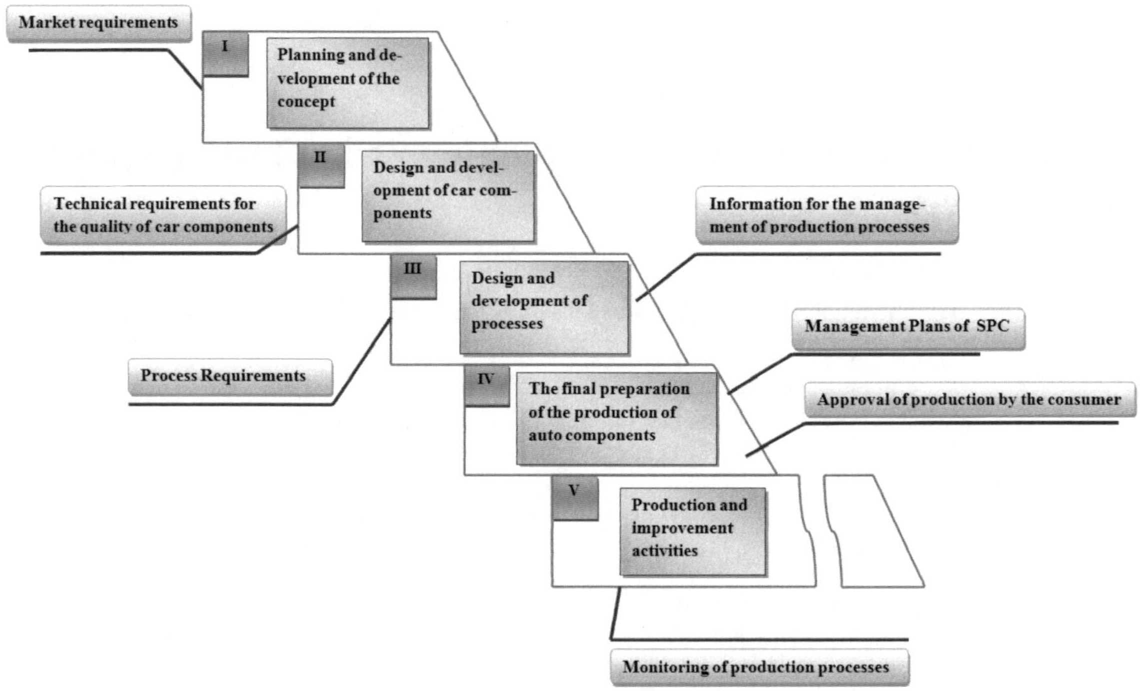

Fig.1 The content of the APQP process (GOST R 51814.6-05)

Activity on the production processes development is carried out at the third stage pre-production. This is not only a technological design with the formulation of standardized technological maps, but also the development of the life cycle processes of the projected commodity product, the processes of providing jobs with resources, the technology of collecting and storing associated products (waste, emissions, etc.), control technology (obtaining, analyzing, information processing, planning, reporting).

To organize management, each process should receive a full range of quantitative characteristics. These are the indicators of the quality of the product being created, the duration of implementation, the norms for the expenditure of all types of resources, and labor standards. These characteristics can be determined by the developer-technologist with the availability of appropriate software tools and electronic databases. Complex presentation of information about each operation provides undeniable advantages: not only are contradictions and omissions in information chains eliminated, but for the first time there is the opportunity to manage the system of production processes, which is the basis for long-term competitiveness.

At present almost all Russian suppliers of automotive components must certify their quality management systems according to the requirements of ISO / TS 16949-09. Establish a world-class output does not make sense unless it is possible to obtain as much profit as possible in each APQP project.

In the pre-production, miscalculations and omissions that can lead to serious consequences in the production process are unacceptable. Losses must be prevented at the production planning stage. It is necessary not only to eliminate, but also prevent further occurrence and development of losses.

The increase in the profitability of the project for the pre-production and production of auto components is an extremely difficult task for Russian machine builders. Let's analyze the possible economic results of the APQP project, the scheme is presented in Fig.

In the course of determining the market requirements for the product (Stage 1), in general, a detailed analysis of the legislation, the requirements of the prospective consumers, the capabilities of the competitors, as well as the supplier's ability to meet these requirements should be performed. At stage 2 (design), along with the development of documentation, it is required to manufacture and test prototypes to confirm the achievement of the planned quality requirements. If the designer does not provide them, loss of income is unavoidable, and if unreasonably overestimates - the cost increases.

Process development activities are carried out at the 3rd stage of pre-production. It is a mistake to think that this is only a technological design with the formulation of standardized technological maps. When developing processes in the course of technological design at stage 3, it becomes extremely important to lay the least expensive options for technological solutions. At the same time, it is not enough to develop them in the shortest possible time. It is equally important not to allow miscalculations and omissions in the content of operations, the selection of tooling and tools, the assignment of technological parameters, the choice of measuring instruments. Trying to save will lead to the fact that in serial production there will be a lot of problems - both at workplaces and when the products are delivered to the consumer.

The APQP project is aimed at ensuring that in the final pre-production (Stage 4) all works are performed with due quality, especially the installation of equipment, training of personnel, manufacturing or ordering of equipment. Many technologists and production managers are trying to "slip through" this procedure, not wanting to waste time confirming the correctness of the decisions taken. As a result, after the release of the auto component, problems may arise returning to the early stages pre-production.

In Figure 2, in the business project, the preparation of production was planned to receive a stable profit (dotted line), but the problem that arises requires time and financial resources to eliminate its causes [3].

Steps APQP can overlap, go parallel, refine. For example, suppliers, prices, equipment composition can change.

For effective management, you need the most complete information about the requirements for the process and its actual state. It can be concluded that the requirements of the standards are inconsistent in terms of optimizing the cost of the processes in the preparation of production, the complexity of determining planned losses, as well as the need to systematize and regularize the causes of potential losses at each stage of production planning.

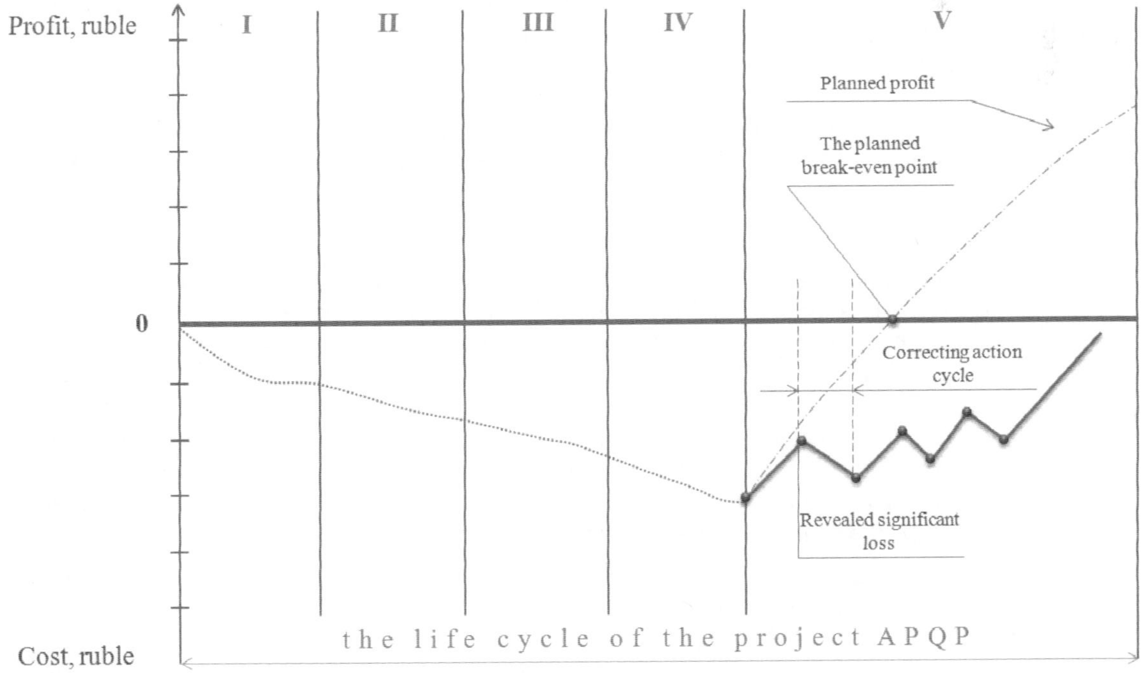

Fig. 2. Corollary of errors and omissions in the pre-production

To avoid losses in the early stages of the project is easier and cheaper than eliminating them after detection. For the manager, the profit planned from the project for the release of the new car component can be presented in simplified form as follows:

$$Profit_{cc} = \sum_{1}^{n} I - (Zpp + PPP) - Zz - (Sv + Pv) - (Zc + Pe + Ps), \quad (1.1)$$

where n - the number of positions of the supplied auto parts;
I - income from the sale of auto components;
Zpp - the cost of preparing their production;
PPP - losses in this training;
Zz - expenses for third-party purchases;
Sv - the cost of production of auto components;
Pv - losses in the process of their release;
Zs- expenses for their service;
Pe - losses in the course of their operation;
Ps - losses from sanctions by the consumer and supervisors [5].

The first condition of profitability is getting the maximum income from selling the car component to consumers. It is ensured by the unmistakable preparation of production in the shortest possible time, by maintaining the world-class quality of the product, by strictly observing the planned schedule of deliveries, by the impeccable organization of the service.

The second condition of profitability is the minimum cost of the product life cycle processes, as each of these processes will be repeated many times.

The third condition of profitability is the minimum of all losses, internal and external, with no sanctions at all. Losses must be prevented at the production planning stage.

Factors affecting the amount of income are mainly concentrated in the first three stages of the APQP process. The omission of any requirement or error in its importance can lead to a decrease in sales, which it is meaningless to search years after the launch of production. It is necessary to exclude the possibility of such errors occurring during the design work.

It is also important to prevent all the factors of profit reduction, classified in Table 1.

Tab. 1. Factors of decrease in profitability in the cycle of APQP-project

Stage APQP	Decrease in income	Increase in the planned cost of processes	Losses at the stage of manufacturing and delivery
1	- omission of the legislative requirement for the auto component; -Option of the customer's requirement for the auto component.	- Overestimation of the customer's requirements assessment; - Overstating the evaluation of the capabilities of competitors	- The omission of the legislative requirement for the safety of production; -Improving the legislative requirement for environmental friendliness.
2	Underestimation of technical requirements for the quality of the auto component	Overestimation of the technical requirements for the quality of the auto component	Errors in the contents of the requirements of the design documentation
3	- Errors in technical documentation;	-Expanding the requirements for	Errors in: -Technological

	-The designation of inadequate measuring instruments	equipment, the qualification of production personnel; - Neoptimality of planning; - Non-optimal organizational structure.	documentation; - Requirements to the quality of equipment; -Distribution of duties, authorities, responsibilities.
4	Poor installation of equipment	-	-No performance of requirements to the quality of equipment, personnel qualification; - Errors in working instructions
5	Undervaluation of current price	-	Non-fulfillment of requirements

The omission of legislative requirements for the safety and environmental friendliness of production will create the risk of obtaining sanctions from the supervisory authorities. Errors in the contents of design documentation requirements increase the probability of marriage. Errors in the technological documentation, the requirements for the quality of the tooling also lead to the marriage of products, cause a loss of time for correcting the indicated errors. Errors in the distribution of duties, authority, responsibility can cause any of the 7 losses, as well as the loss of time for correcting errors and omissions.

Failure to meet the requirements for tooling quality, staff qualifications, errors in working instructions can lead to losses for repairing defects, alteration. Negligence in the preparation of working instructions will lead to violations of technological discipline, a decrease in quality, and perhaps even injuries to workers.

To increase the efficiency of the enterprise, it is necessary to increase the effectiveness of management and minimize the cost of the process. To ensure minimum cost of the process, it is necessary to minimize labor costs and streamline labor intensity standards, to plan resource consumption, to apply the standard method of cost accounting and to calculate the cost of production. Thus, the cost of the process and the planned costs can be expressed by the following formulas:

$$Cost_{process} = \sum the\ costs\ of\ the\ plan + \sum Losses \quad , (1.2)$$

where $\sum$ the costs of the plan - the total costs of the process, rubles;
$\sum$ Losses - total losses on the process, rubles.

$$\sum the\ costs\ of\ the\ plan = Zupr + ZZAK + Zizg + Zbyt + Service + Zutil \quad , (1.3)$$

where Zupr - management costs, rubles;
ZZAK - the cost of obtaining and processing the order, rub .;
Zizg - the cost of manufacturing the order, rub .;
Zbyt - sales costs, rubles;
Service - the cost of maintenance, rubles;
Zutil - the cost of recycling, rubles.

For really effective management, which ensures a continuous improvement in the quality of products, the fullest information about both the requirements for the process and its actual state is needed. As a result, a set of requirements for interaction with consumers, the quality of the auto component and its components, and the characteristics of all processes of its life cycle should be developed. Each individual process should receive quantitative indicators of effectiveness and efficiency.

IOP Conf. Series: Materials Science and Engineering **412** (2018) 012044 doi:10.1088/1757-899X/412/1/012044

Thus, the complete technological documentation is the basis of enterprise management. For this purpose, a structural diagram of technological information is proposed as the basis of enterprise management (Fig. 3).

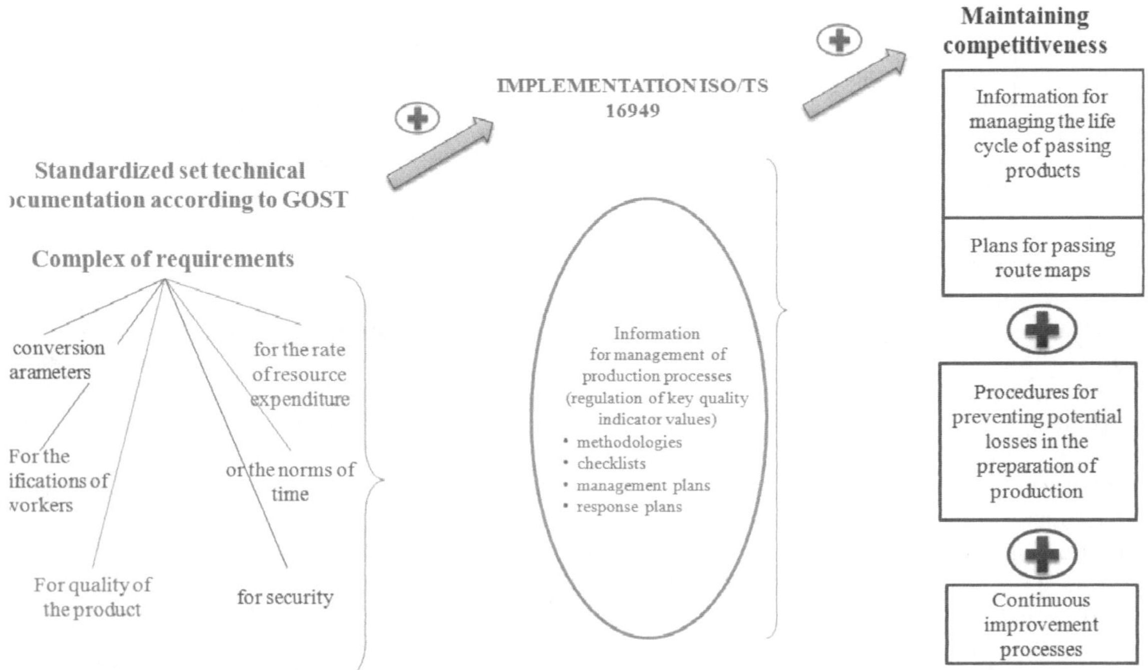

Fig.3 Technological information as the enterprise management basis

In the process of technological design, it is necessary to develop a full range of technologies for the activity of the enterprise associated with the new product and to issue the relevant documents. The volume of work on technological pre-production is increasing many times today. This increases the risk of errors and omissions in the content of different groups of documents.

But for this it is necessary to clearly define which documents are created at this or that stage of the APQP project, when exactly the requirements for their creation should be defined. The process of technological design must be strictly regulated.

Traditionally, the main content of technological design seems to be a two-level process: the development of routes and the development of separate operations as part of the routes. When obtaining a standardized set of technical documentation, this was quite enough.

However, for the preparation of modern production, which requires additional information, it is necessary to identify additional procedures in the design process related to the preparation of such information. This is the first condition.

The second mandatory condition is that it is necessary to embed technological design in the structure of the APQP process in order to ensure the formation and analysis of each document exactly when it is needed.

As a result, the following scheme of technological design is proposed and schematically shown in Fig. 4:

1) At stage 2 (design of the auto component), reconcile the content of its life cycle and create a documented flow route scheme so as not to leave any process uncontrolled.

This scheme is becoming an important tool for coordinating the work of diverse specialists in the APQP team.

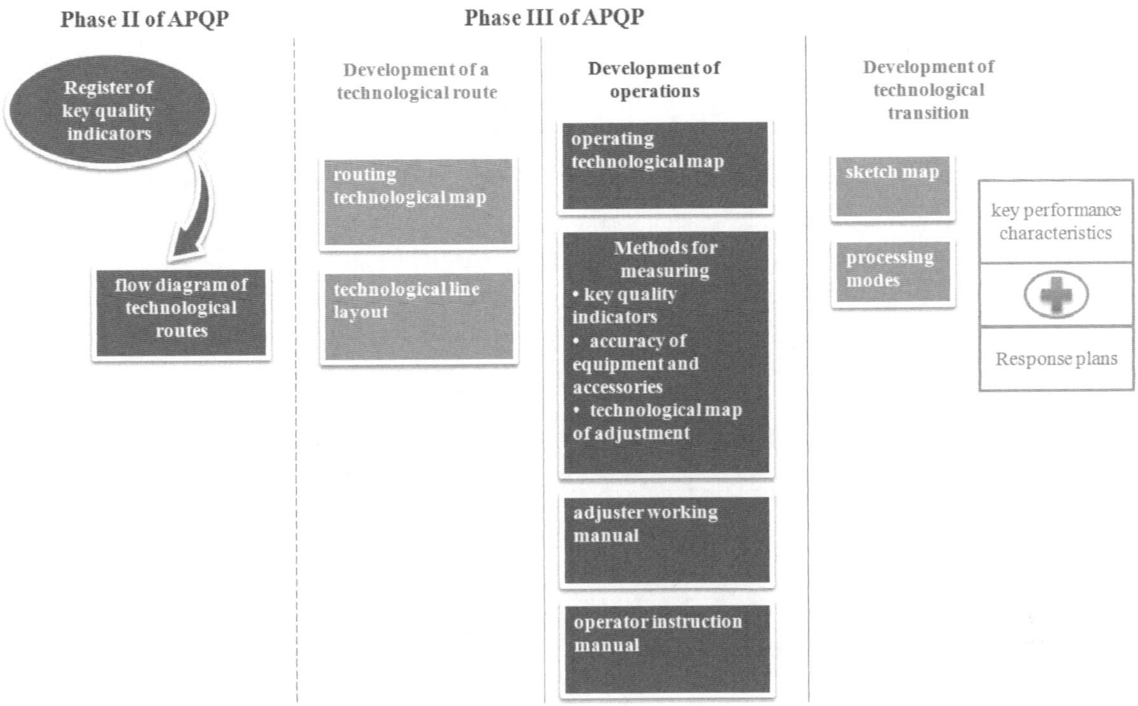

Fig.4 Technological design process scheme

2) In the 3rd stage of the APQP process "Process Development" in the cycle of the actual technological design during the procedure for analyzing the causes and consequences of potential defects (FMEA) [1], it is necessary to distinguish not 2, but 3 parts:

a) Routing design. It must be performed with additions to the routing technological map:

- Mandatory line layout (it provides planning of all transport operations);
- Methods for measuring the most important indicators of equipment accuracy.

In addition, the routing technological map also suggests the addition of those operations in which key quality indicators of the auto component are formed in order to automate the preparation of the process flow map as part of the PPAP set [1].

b) The design of technological operations, as usual, involves the definition of a sequence of transitions, but with a mandatory designation of those in which the key quality indicators are directly formed.

The following documents should be attached to the operational technological map:

- Methods for measuring each key quality indicator;
- Techniques for measuring the accuracy of the tooling;
- Technological setup card;
- Operator's manual;

Operational instruction of the operator.

c) The process of designing it has been separately carried out, since the management of quality indicators is carried out precisely in the transition. The modern technologist is obliged not to list the treatment modes, but to determine and rank those control characteristics that affect the values of the key quality indicators. This is a new condition for modern Russian technologists. For each key control characteristic, it is also necessary to indicate a response plan (the way it is regulated to maintain key quality indicators within specified limits). So, the process of technological design should become more transparent.

IOP Conf. Series: Materials Science and Engineering **412** (2018) 012044 doi:10.1088/1757-899X/412/1/012044

In the paper, key control characteristics are systematized for regulating the value of a key quality indicator and stabilizing production processes.
To do this, all factors that influence the result of processing in the transition are proposed, divided into 3 groups, as illustrated in Fig. 5.

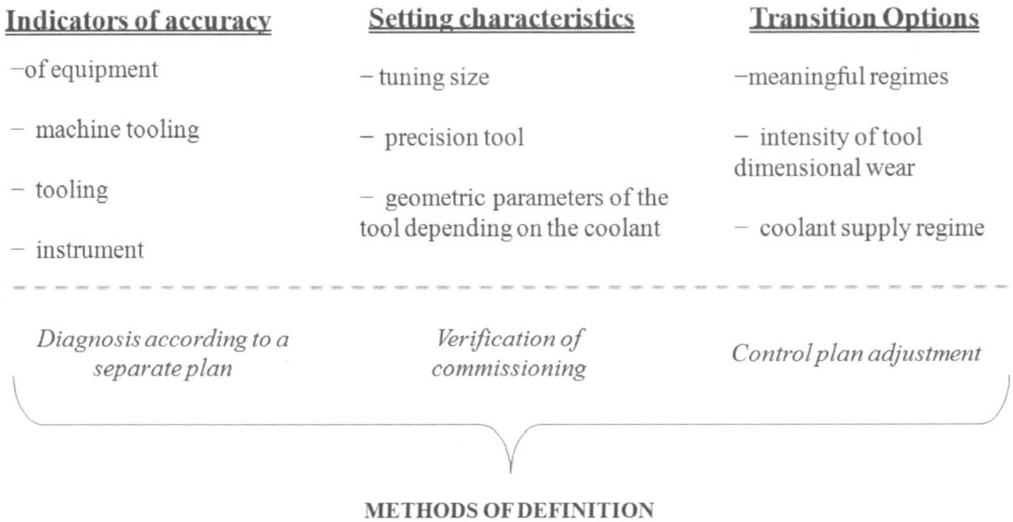

Fig.5. Systematization of key control characteristics of technological transitions

1 group - characteristics that act in long cycles of functioning of the workplace. This is the limiting accuracy of equipment and accessories. Their current values can be identified by diagnosis, and then used to predict the accuracy of processing.
2 group - setup characteristics, for example:
- adjusting size;
- the actual accuracy of the forming tool;
- geometric characteristics of the tool;
- Coolant characteristics.

Some of these characteristics can be determined during the setup process, and some - set by verification of adjustment (after processing and measuring the trial part).
3 group - the factors acting directly in the process of processing (these are cutting forces, temperatures, the intensity of the tool's dimensional wear, the coolant supply regime, etc.). Their contribution to the total deviation of the key indicator can be calculated.
Knowing this set of factors, it becomes much easier to plan and regulate the value of the key quality indicator of the auto component by known theoretical dependencies.
The developed documentation must be approved, its content verified. Verification is an affirmation that a document can be used as a criterion that is impossible without the implementation of production. For this purpose, a special document is created - the management plan for the installation part of the auto component.
After analyzing the results of processing the batch, it is possible to identify in the documentation various kinds of errors, or necessary improvements. Only after correction it is permissible to begin official coordination and subsequent approval of the set of documentation.
Procedures for approbation and application of documents are additionally ordered. For this purpose, a separate statistical management plan is developed (Fig. 6).

IOP Conf. Series: Materials Science and Engineering **412** (2018) 012044 doi:10.1088/1757-899X/412/1/012044

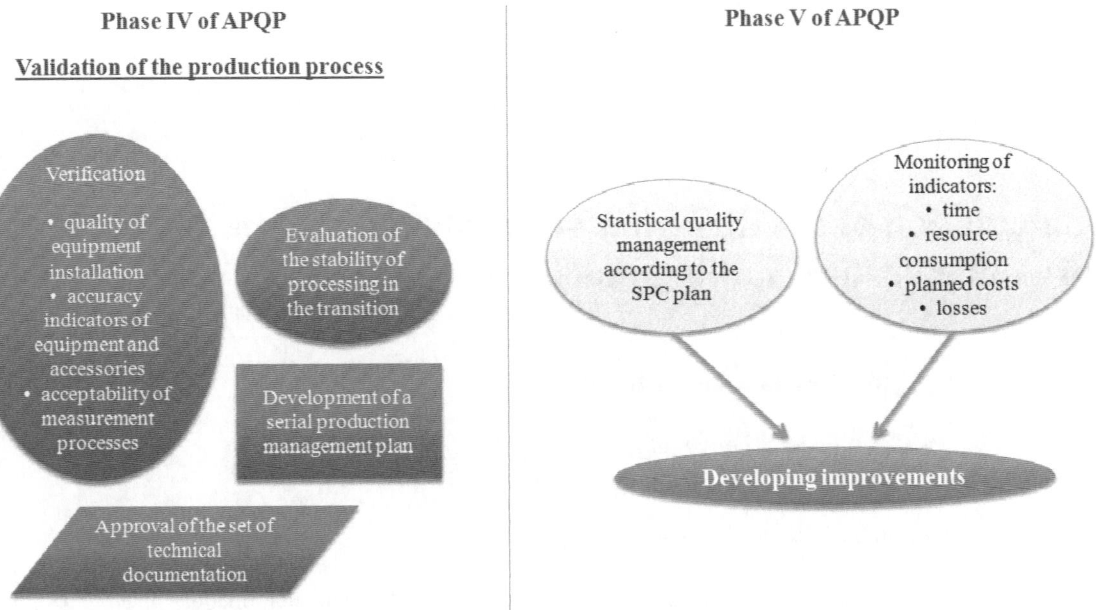

Fig.6 Management on the basis of technological documentation

Technological documents become an effective management tool. Thus, the set of technical documentation becomes more alive, close to the real processes. One of the advantages of the proposed information support is the ability to visually trace the causes of a potential error in the pre-production.
As a result, a technique for developing modern technological documentation was proposed. In the framework of this methodology, requirements for modern technological documentation have been developed, factors of profitability decrease in the APQP project cycle have been identified, work has been grouped to provide the best results for the APQP process stages (3,4,5 stages), technological information as the basis for enterprise management, developed processes of technological design and application of technical documentation, the scheme of the technological design process is presented. In the paper, the key control characteristics of technological transitions are systematized, the principles of management of the pre-production (technological design) based on technical documentation are formulated.
first paragraph after a heading is not indented (Bodytext style).

References

[1] 2005 GOST R 51814.6 - 2005. *Quality management systems. Quality management in the planning, development and automotive components pre-production.* - 70 p

[2] 2008 GOST R ISO 16949-09 *Quality management systems. Special requirements for the application of ISO 9001: 2008 in the automotive industry and organizations that produce the relevant spare parts.* - 51 p

[3] Grechishnikov VA, Kasyanov SV, Yurasova OI, Romanov V.B. 2016 Increase of competitiveness of the enterprise at a stage of manufacture autocomponents pre-production *the Bulletin of MGTU "STANKIN"* **2 (37)**

[4] Kasyanov SV, Yurasova OI 2015 Differential planning of a complex of quantitative indices of production processes of auto components at the stage of technological design *Socio-economic and technical systems: research, design, optimization. Publisher:" Kazan (Volga region) Federal University "* vol **2 (65)** pp 27-32

[5] Yurasova OI, Pautov G.A. 2015 Increase the efficiency of the machine-building enterprise by optimizing the cost of the process *Theory and practice of social development* **9** pp 53-55

IOP Conf. Series: Materials Science and Engineering **412** (2018) 012045 doi:10.1088/1757-899X/412/1/012045

Calculation of the direction of the strain axes in the working area of a vertical milling machine

R M Khusainov and A R Sabirov

Kazan Federal University, Naberezhnye Chelny Institute, 423812, Russia, Naberezhnye Chelny, Prospekt Syuyumbike 10A

rmh@inbox.ru

Abstract. The article studies the issues of static elastic deformations that appear during processing on milling type machines. The method of determining the principal axes of strain is observed.

1. Introduction

The requirements for increasing the accuracy of the parts produced assume the consideration of all factors that affect the processing accuracy [1,2]. Rigidity is considered one of the most important factors in the process of cutting. In actual processing conditions it is necessary to consider the rigidity of the entire technological system which includes the machine, the device, the cutting and the adapter tools, a workpiece. The traditional ways to increase the rigidity are to increase the rigidity of the details of the technological system and the rigidity of the joints between them. Meanwhile, the rigidity of the entire machine, that is the perception of cutting forces and the resistance to deformation from these forces, is affected not only by the intrinsic and contact rigidity of its elements, but also by the geometric characteristics of the layout of the technological system.

2. The strain axes

These characteristics can be estimated by the axes of rigidity or, in terms of the theory of the resistance of materials, the axes of the main strain. It is known [3] that at each point of the shank there are three mutually perpendicular directions along which linear deformations have extreme values. These directions are called the principal axes of strain, and linear deformations along the principal directions are the principal deformations. The most important are the axis of maximum strain in the direction of which the elastic deformations are least, and the axis of minimum strain, in the direction of which the elastic deformations are greatest. The importance of these axes is dictated by the following considerations [4]:

1. If the layout of the technological system is constructed so that the resultant of forces cutting is equal to or close to the axis of maximum strain, then the deformations will be minimal, therefore, the processing errors caused by elastic deformations are also minimized.
2. The strain axes of the main deformations have the property that if the resultant of forces passes along them, then this causes only linear deformations of the shank along these axes without causing angular deformations. This also minimizes elastic deformations.
3. Approximation of the resultant force of cutting to the strain axes increases the dynamic stability of the technological system.

Content from this work may be used under the terms of the Creative Commons Attribution 3.0 licence. Any further distribution of this work must maintain attribution to the author(s) and the title of the work, journal citation and DOI.
Published under licence by IOP Publishing Ltd

3. Calculation method for determining the strain axes

When investigating the rigidity of machines an experimental approach is generally used; it is carried out by means of a full-scale or computational experiment. The main method of the experiment is the "direction finding" of the position of the centre and the strain axes by a set of measurements when the angle of action of the loading force is changed. All these measurements take a lot of time, and in the first case they also need material costs. In this article we propose a calculated approach to determining the position of the strain axes. As we know [3] under any loading of the shank the linear and angular deformations of this body form a strain tensor:

$$T_\varepsilon = \begin{bmatrix} \varepsilon_{11} & \varepsilon_{12} & \varepsilon_{13} \\ \varepsilon_{21} & \varepsilon_{22} & \varepsilon_{23} \\ \varepsilon_{31} & \varepsilon_{32} & \varepsilon_{33} \end{bmatrix} \quad (1),$$

where ε_{ij} – linear and angular deformations.

The values of the principal deformations can be found from the solution of the cubic equation [3]:

$$\varepsilon^3 - I_1\varepsilon^2 + I_2\varepsilon + I_3 = 0 \quad (2),$$

where I_i – invariants of a strain tensor (1).

The components of the strain tensor (1) can be found experimentally under single loading or by finite element analysis of the technological system. In this study a three-dimensional model of a technological system was constructed. It included a carrier system of the machine, cutting and adaptive tools, a device, a workpiece. The layout of the technological system corresponded to the case of internal cavity milling [5]. By performing the calculation in the "Extended Simulation" module of Siemens NX linear and angular deformations were determined (Figure 1). By solving equation (2) the values of the main deformations were determined:

$$\varepsilon = \begin{bmatrix} -0{,}003 \\ 0{,}0002 \\ 0{,}002 \end{bmatrix} \quad (3).$$

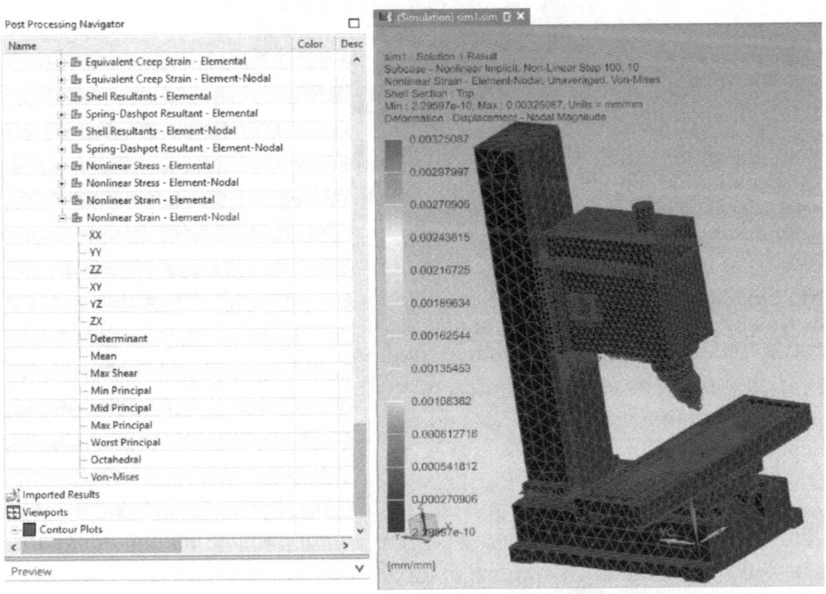

Fig. 1 – Deformations in the technological system

The component ε_1 corresponds to the principal greatest deformation, its direction is the axis of minimum strain; component ε_2 corresponds to the smallest principal deformation, its direction is the axis of maximum strain.

The directions of the principal axes of deformations can be determined by calculating the direction cosines l_i, m_i, n_i in the joint solution of the system of equations [6]:

$$\begin{cases} (\varepsilon_i - \varepsilon_{11})l_i - \varepsilon_{12}m_i - \varepsilon_{13}n_i = 0 \\ -\varepsilon_{12}l_i + (\varepsilon_i - \varepsilon_{22})m_i - \varepsilon_{23}n_i = 0 \\ -\varepsilon_{13}l_i + \varepsilon_{23}m_i - (\varepsilon_i - \varepsilon_{33})n_i = 0 \\ l_i^2 + m_i^2 + n_i^2 = 1 \end{cases} \quad (4).$$

According to the calculated values, the vectors forming the strain axes were constructed (Fig. 2, the cutting and back-up tools are formally not shown). One of the axes along which the deformations are greatest is the axis of minimum strain jmin, the second along which the deformations are the smallest is the axis of maximum strain jmax. It can be noted that the position of the main deformation axes corresponds to the results of the previously performed computational experiment.

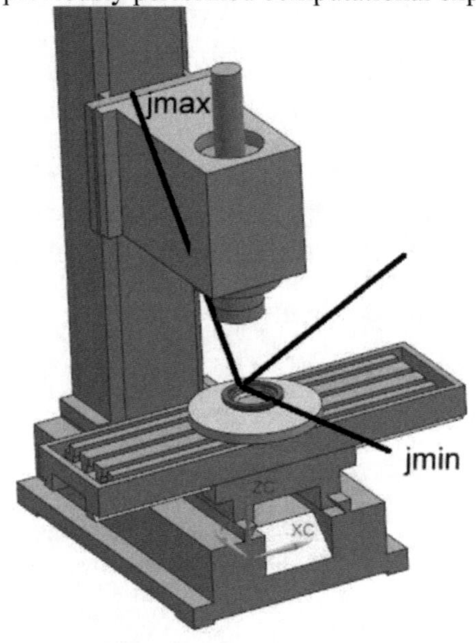

Fig. 2 – Axes of strain

4. Conclusion

Thus, the application of the proposed approach makes it possible to determine with minimum effort the position of the axes of strain. With the help of this result one can choose the cutting scheme or, conversely, change the layout of the technological system in such a way as to approximate the resultant forces of cutting to the axis of maximum strain, thereby reducing the elastic deformations and increasing the accuracy of processing.

References

[1] Irzaev G H 2015 *Model of control of products manufacturability in an industrial enterprise* Economics and management of systems control No 15 pp 50-57

[2] Govorkov A S, Akhatov R H 2011 *Analysis of the technological properties of aviation products based on the information image of the product* Review of Samara Scientific Center of the

Russian Academy of Sciences No 6-1 pp 285-92
[3] Kolesnikov K S, Aleksandrov D A, Astashev V K et all 1994 *Dynamics and strength of machines* Theory of mechanisms and machines
[4] Kudinov V A 1967 *Dynamics of machine tools* (Moscow: Mashinostroenie)
[5] Ryabov E A, Yurasov S Ju and Hisamutdinov R M 2016 *Internal contour processing by trochoidal milling with end mills* Innovative engineering technologies, equipment and materials – 2016 (ISTK "IETEM-2016") pp 150-154
[6] Grechishnikov V A, Petukhov Y E, Pivkin P M, Romanov V B, Ryabov E A and Yurasov S Yu and Yurasova O I 2017 *Trochoidalslotmilling* Russian Engineering Research Vol 37 No 9 pp 821-23
[7] Timoshenko S P and Gud'er Dzh 1975 *Theory of elasticity* (Moscow: Nauka)

IOP Conf. Series: Materials Science and Engineering **412** (2018) 012046 doi:10.1088/1757-899X/412/1/012046

Ways to Transfer Oil and Oil Products in Heating Conditions and Methods for Their Enhancement

E M Khusnutdinova[1], I A Konakhina[1], G R Khamidullina[1] and A O Mamonova[1]

[1]Kazan Federal University, 18 Kremlyovskaya street, Kazan, 420008, Russian Federation

rr-088@mail.ru

Abstract. This paper discusses the methods of oil transportation through pipelines. Particular attention is paid to the processes of complex rheological behavior of high-paraffin crude oils. Optimal methods applicable for transportation of heavy oil with the help of intensifiers are revealed.

Every year shows a significant increase in the volume of production of reserves difficult to recover in our country and elsewhere. These include heavy oil with a viscosity of 30 mPas-s and higher that contains much of paraffins and solidifies at relatively high temperatures. Reserves of such types of oil are about one trillion tons and exceed the volume of residual recoverable oil reserves of low and medium viscosity more than five times.

The oil deposit management is impossible without addressing the issue of how to convey the oil. Pipeline transportation of oil is the most economical way of delivering these natural energy resources to the place of their further processing. However, technological limitations related to the rheology of a high-viscosity liquid is the main obstacle to ordinary transportation by pipelines. To increase the pumping performance, viscous and highly viscous oils undergo preliminary treatment to increase fluid flow and reduce flow resistance of the pipeline. These include pumping using diluents, depressant additives; heat treatment of oil; preheating of oil and the use of an intensifier that operates as a flow swirler.

A number of trunk pipelines are operating in Russia for delivering viscous oil from the field to the place of further processing, where different methods of transportation are used.

The most accessible technology is the use of hydrocarbon diluents that, as a rule, are represented by gas condensate or low-viscosity oils. Mixing of highly viscous liquid being transported with diluent significantly influences its rheology, reduces not only its viscosity, but also the pour point of oil. This is primarily due to a decrease in paraffin content in the mixture, which happens through the dissolution of some of its parts with light fractions of the diluent. In addition, the asphalt-resin content in a low-viscosity liquid used as a solvent affects the growth of paraffin crystals and prevents the formation of its strong structural lattice.

The practiced method to enhance the rheological parameters of pipeline oil is the use of special chemical reagents (depressant additives). Depressors are high molecular weight oil-soluble substances capable to reduce oil pour point, viscosity and shear stress when introduced in small amounts into the oil with a high content of paraffin. Typical natural depressor are asphalt-resin substances contained in

Content from this work may be used under the terms of the Creative Commons Attribution 3.0 licence. Any further distribution of this work must maintain attribution to the author(s) and the title of the work, journal citation and DOI.
Published under licence by IOP Publishing Ltd

oil. Therefore, one of the ways to enhance the rheology of high-paraffin oil is the addition of products rich in these substances, e.g. tar, bitumen to it.

The transfer of heated high-viscosity oil, also called 'hot transfer', is currently the most common way of pipeline transport of these products. This method involves heating the oil either directly before its injection into the pipeline, or along the entire route.

In practice, the first option of the 'hot' pipelines is most applicable, where three types of stations are installed: pumping and heat stations: for heating and pumping the product, heat stations: only for heating, and pumping station: only for pumping the product [2]. The oil temperature can be increased both in tanks and in special furnaces of steam or fire type, respectively located at the point where the pipeline starts and at several points along its route.

Tanks receiving the crude oil are equipped with steam coil or sectional heating devices that allow maintaining the temperature of the transported products for pumping out by its booster pumps. These pump transferring oil through additional preheaters feed it to the main pumps and to the main pipeline afterwards [4].

However, while in motion, the oil cools down due to heat exchange with the environment. Consequently, it becomes necessary to periodically heat the high-viscosity liquid to prevent it from freezing and further free movement. Such additional points of heat treatment are available every 25-100 km along the entire route.

Then, the oil goes to the heated tanks of the final point.

In the second option, the pipeline through which the coolant is transferred serves as the heat exchanger. Usually, it goes parallel to the oil pipeline or placed on its heat insulated surface to reduce the heat losses of hot pipelines. The coolants are usually steam, hot water and other heated liquids [2].

"Hot transfer" became a frequent practice both in Russia and in other countries due to its simple and well-developed technology, but the problem remains in terms of reliability of designs [1].

Significant power consumption is associated with the need to overcome the forces of viscous friction. For this purpose, the oil heating is widely used so that it contributes to decrease their effective viscosity [5]. The heat treatment process consists in heating the oil to a temperature at which the solid paraffin hydrocarbons contained in the oil dissolve and in cooling at a predetermined rate in a certain state.

Enhancement of the oil rheology is driven by internal changes resulting from heat treatment. Natural cooling of oil releases paraffin crystals that are connected to each other so that they form a lattice, the strength of which depends on the paraffin content in the oil. While-in-treatment oil heating contributes to the complete dissolution of the solid paraffins contained demonstrating the paraffin crystallization upon further oil cooling. The success of this technology directly depends on the treatment temperatures, which are carefully adjusted for each oil product by way of trial and error. Experience has shown that oil cooling is best performed in a static state.

Oil rheology after heat treatment will be restored to the initial values as time goes. In order to avoid "freezing" the pipeline, it is necessary to know exactly the time the oil product properties recover, which are different for each of them. If the described parameters remain unchanged for a long time, such oil can be transferred through the oil line as a usual low-viscosity liquid [3].

A comparative analysis of transport technologies for high-viscosity oil shows that heat treatment is one of the most expensive types of oil preparation for transfer.

Currently, the matter of reducing flow resistance, energy consumption during transportation, reducing the weights and dimensions of heat exchangers remains a pressing problem in connection with the constantly growing volumes of high-viscosity oil transported through the main pipelines. One of the available solutions to this problem relates to the heat exchange intensification, which increases the energy performance of heat exchangers. This offers the opportunities through adjusting the shear rate in the flow to control the effective viscosity of oil and substantially reducing the flow resistance of the transfer lines [5]. When picking the heat exchange intensification methods, not only the efficiency of the surface itself should be taken into account, but also the versatility of the coolants, the fabricability of the surface, the processability of the heat exchanger, strength requirements. Of all the methods of intensifying heat transfer in pipes, the greatest attention is paid to the principle of screw swirling, both efficient and technologically feasible.

The issue of determining the regularities governing the hydrodynamic characteristics of the anomalously viscous oil flow in transfer lines and ducts is complex and requires systematization [6]. Despite the fact that this challenge is important from technology point of view, the researches have not yet been carried out sufficiently, and the presented analysis is aimed at solving the outstanding problem by means of the mathematical modeling apparatus.

References

[1] *Transportation of high-viscosity oils* 2016 (Youth Science Forum: Engineering and Mathematical Sciences: electr. sav. Paper on proceedings of the XXXV International Virtual Student Conference and Workshop) No 6(35)

[2] Aliyev R A, Belousov V D, Nemudrov A G, Yufin V A and Yakovlev Ye I 1998 *Pipeline transportation of oil and gas: college textbook* (Second edition, revised and enlarged – M.: Nedra) p 368

[3] Korshak A A and Shammazov A M 2002 *Fundamentals of Oil and Gas Engineering* (College textbook. Second edition, revised and corrected: — Ufa.: DesignPolygraphService LLC) p 544

[4] Sdvizhkov I V 2016 *Characteristics of transferring quick-hardening and high-viscosity oils and oil products through main oil lines* (Graduation thesis / National Research Tomsk Polytechnic University)

[5] Konakhina I A, Khamidullina G R and Khamidullina A F 2016 *Enhancement of Hydrodynamic Processes in Oil Pipelines Considering Rheologically Complex High-Viscosity Oils* IOP Conference Series: Materials Science and Engineering Vol 134 Is 1 **012019**

[6] Khusnutdinova E M, Konakhina I A and Khamidullina A F 2017 *Comparative Analysis of Thermohydraulic Performance of Enhanced Viscous Oil and Oil Product Heaters* IOP Conference Series: Materials Science and Engineering **012046**

Training transport specialists based on a contemporary view of self-organization of sophisticated developing systems

E M Khusnutdinova[1], G R Khamidullina[1], P P Pavlov[2], A E Aukhadeyev[2], R S Litvinenko[2] and A N Khusnutdinov[2]

[1]Kazan Federal University, 18 Kremlyovskaya street, Kazan, 420008, Russian Federation
[2]Kazan State Power Engineering University, 51 Krasnoselskaya Str., Kazan 420066

rr-088@mail.ru

Abstract. This paper discusses the problem of how to integrate natural-science, humanist and special technical knowledge in professional university-level training of transport specialists being a rapidly developing sophisticated system. This engages current science concepts of self-organization and self-development of sophisticated natural and man-made systems.

The urgency of the problem consists in the need to draw the attention of specialists of any field, first of all those who create sophisticated engineering systems that influence the human being and the environment, to the actively developing concepts of modern science about complex and often contradictory phenomena of self-organization and self-development.

The paper is aimed at the popularization among technical specialists of modern concepts of open, sophisticated, self-developing systems as exemplified by searching for innovative methods of studying and solving the problems of the transit system development in modern growing cities.

Content and methods.

The content of the paper includes teaching and scientific sources. 1. Textbooks on Current Concepts of Natural Science prepared in accordance with the State Educational Standard of Higher Professional Education for universities of different specialization. 2. Scientific monographs of Russian and foreign authors that substantiate and evolve insight into complex self-organizing systems. 3. Special sources of natural-science knowledge.

The methods are: the main method consists in a theoretical search for a model of studying and solving problems of managing the development of the transit system of modern cities, while the supplementary method discusses the meanings of new terms and definitions using didactic (derived from the Greek word didaktikos meaning apt to teach) schemes and drawings.

Scope.

Training of specialists in a current professional higher school sets up requirements to its teachers that emanate from conceptual, general scientific, methodological achievements of science based on the integration of various branches of knowledge. In each specialty, the general methodological concepts are detailed, they find their own particular theoretical and applied growth. This requires the teachers of higher school being widely erudite and profoundly competent, both are possible to achieve with great efforts.

Content from this work may be used under the terms of the Creative Commons Attribution 3.0 licence. Any further distribution of this work must maintain attribution to the author(s) and the title of the work, journal citation and DOI.
Published under licence by IOP Publishing Ltd

IOP Conf. Series: Materials Science and Engineering **412** (2018) 012047 doi:10.1088/1757-899X/412/1/012047

So, at the stage of training transport specialists, say representatives of technical disciplines, the concepts of the transit system as a complex developing system evolve, for the comprehension of which different knowledge is required and not only of technology itself, but also of the natural and humanist sciences.

Searching for a model for managing the development of the urban transit system (not to be confused with traffic management) suggests a heuristic idea, which includes four key theoretical positions:

1. Give consideration to the transportation means of modern cities as a socio-technical system based on the concepts of modern natural science about open complex systems having properties of self-organization and self-development like biological, living organisms. Emphasize the instability of the state of such systems, the unpredictability of development trends in the humanitarian context.

2. The development of the transit system including the humanitarian, social and technical, natural and cultural components is to be considered in the model of the formation and development of a complex natural and cultural system – Person-Personality with its main components: Activity and Consciousness.

3. Consider managing the evolution of the urban transit system in terms of the relationship between internal spontaneous processes and external influences by the systems hierarchically related and non-hierarchically linked. Consider Society, Culture, and Civilization as systems to which the transportation means relates hierarchically. Focus on the effect of these systems that meet common and prospective humanist interests as opposed to private and temporary interests.

4. Take a model of the structural and functional organization of the animal brain and the Human Brain being the material carrier of Consciousness, Intelligence and Reason as feasible for practical implementation of the model of managing the evoluition of the urban transit system. The human brain is a natural system, the most complex of all management systems known. Given that systems can be successfully recognized, if they are considered from the standpoint of models, analogies, metaphors of more complex systems, the biological structural-functional model of the human brain itself can serve for the development of a socio-technical model of conscious and intelligent management of the urban transit system to meet the humanist requirements.

Setup and functions of the human brain require special attention. There are examples from theoretical engineers who try to create man-made intelligence as well as to study and describe the brain in the most detailed way similarly to neurophysiologists [1]. A deep penetration into the notion of what the systems Person-Personality, Society, Culture, and Civilization mean. Appeal to the definitions Instinct, Consciousness, Reason is necessary for making them clear for application towards technical systems. This is because the transit system has intellectual and reasonable properties being a humanist phenomenon – a phenomenon of spiritual and material culture.

Modern science of the turn of the 20th and 21st centuries penetrated deeply into the nano-measurements of matter and into infinitely vast outer spaces. At the same time, it achieved considerable success in understanding how complex matter buildup arises and develops. There are representations of open spontaneous complex systems of physical and biological nature that offer self-organization and self-development [2, 3 and others]. Not only the world of natural nature, but also the world of culture - that created by man himself - is permeated with ideas about the unity of the emergence, organization and development of complex systems. They are formulated as modern concepts of natural science, which influenced the scientific worldview of society in the form of unity of natural and humanist cultures [4, 5, 6 and others].

Deep methodological importance is the notion of Chaos reigning in the world, and the Order; conditionality and non-conditionality (determinism and indeterminism); predictability and unpredictability of events and much more [7, 2 and others]. The main initiators of such a scientific worldview, called the evolutionary genetic synergetic one and has become a universally recognized paradigm, are: Ilya Prigogine, a Belgian physical chemist and Nobel laureate, and Hermann Haken, an outstanding German physicist and mathematician.

Evolutionary and genetic definitions, terms that belong to biological sciences for many centuries happened to be included in the methodological range of other sciences in the early 21st century, showing the similarity of the mechanisms of origin and development for complex systems and organisms of a different, non-biological nature. Now, from the biological, evolutionary and genetic positions, consideration can be given to such systems as the Person-Personality, Society, Culture, Civilization, Techniques and Technologies, socio-economic organizations, intellectual theoretical systems, material and spiritual culture in general. Modern science deeply substantiates this approach on both the theoretical and the empirical level investigating the real processes of development of non-biological complex systems. Specifically owing to this methodological basis, urban transit systems have not been considered yet. One of the areas of focus of the in-depth justification for the application of evolutionary and genetic methodology to use social systems is the idea of the development of local and world civilization as one of the most sophisticated cultural systems. In order to emphasize the methodological significance of this approach, it was called the science of the 21st century by its authors [8]. It formulates the definition of socioeconomic genetics as a new methodological concept of studying and solving problems of social, economic and deeply associated technological issues using the biological definitions.

One of the particularly complex systems is the technology created by the hands and brain of human being, which arose atop of the development of the infinitely complex hierarchy of systems [9 and others] of the Earth's material world.

The alliance of technology and science form an intricate system of an even higher hierarchical order that acquires properties characteristic of such systems called emergent - unexpected in nature (supernatural) and extremely high in their effect on nature, human and cultural environment. The scientific and technological progress, moreover – the scientific and technological revolution is fraught with emergent unexpectedness possibly resulting in the catastrophe, the destruction of the system of the highest order on the Earth, which is directly above technology and science – material and spiritual culture in general, civilization and biological life. Even now we can see and feel the intensifying inhuman effect of certain technical achievements, which has become the most important problem of modern society. However, common sense says that Homo sapiens must find a way to give a positive humane value towards the human being and nature to the emergence of the alliance of technology and science.

The city and its transportation means are a huge engineering system. It includes and is surrounded by systems different in nature: social, economic, energetic, and environmental. According to the concepts of modern science, such a sophisticated system can be compared with a living organism for a number of its properties: once appeared, it gets the ability to self-organize and self-develop, interact and compete with other systems [2 and others], and finally – intellectual properties. The engineering system makes these properties real with its own unique means, which must be disclosed, understood to be able to influence and manage them.

It is required to cover the dialectics of the relationship between spontaneous self-organization and self-development, i.e. self-management being an internal property of sophisticated systems, on the one hand, and external, necessary management of such systems, on the other. This requires a theoretical model of these relations that would satisfy the requirements of researching a complex object similarly to an object that is even more complex, capable to clear up the fullness of the content and processes of its development not reducing the insight into the object. In view of this, considering the modern unity of natural and humanist knowledge, the use of a natural, real and biological model – the human brain – is feasible in studying and solving the problems of complex engineering systems.

The brain of animals is the subject of researching by biologists, who in recent years have achieved considerable knowledge about its deep detailed structure at the level of molecular processes. The ideas of an integral and systemic functional structure of the Human Brain very important for the neurological theory and practice were obtained. The brain has been of interest for ages, and now with special strength and significance it is demonstrated by the engineering and technical community engaged in the development of man-made intelligence and the problems of its use in various spheres of human activity. The brain of the Person-Personality is constructed and functions in a particular way; it is developed as influenced by human activity in social conditions and its model is of fundamental importance for studying the problems of how to manage the development of a transit system that has a social and humanist component.

A broad historical-philosophical, general scientific and methodological and profound special-theoretical consideration of the foundations of the origin and development of technology is a condition for a specific theoretical definition and an empirical search for what is really existing and developing modern urban transportation means as an open sophisticated system.

First of all, the origin of technology, including transportation equipment, its genetics (from Greek genesis meaning origin) and its historical development – phylogenesis. As applied to technology, the terms genetics and phylogenesis are quoted, because they, as a rule, denote the phenomena of the biological world. Phylogenesis is a process of historical development of plant and animal organisms, their species, genera, etc. (Greek phylon, i.e. tribe). Phylogenesis is considered in unity and interdependence with the individual development of biological organisms – ontogenesis (from Greek ontos meaning being). If we consider technology as a method, technique, skill in obtaining, transporting and storing the most vital means, some of the first species of animals already possessed a certain technology and technical set (the terms are also quoted as traditionally not applicable to the animal world). This may be the basis for speaking metaphorically about the phylogenesis of technology – technogenesis.

The biological phylogenesis identifies the stage of the origin and development of Human as a species of animals gradually acquiring a special property – the Reason, the social and individual Consciousness. It is called anthropogenesis (Greek anthropos means human). Technogenesis and anthropogenesis have a close, organic interconnection, interdependence: Man creates a special kind of Technology – human technology, it becomes anthropogenic; Technology deeply transforms, improves the Man, creates Society equipped with technology – technical, technological society.

Three main periods are distinguished in anthropogenesis: the formation of Homo habilis took about a billion years; Homo erectus took several million years, finally, Homo sapiens took hundreds of millennia and continues now. Each period of anthropogenesis was accompanied by its instruments of life, e.g. the technology that the Man used.

For didactic purposes and visual perception of the idea of the genesis of technology, the chart is shown in Fig. 1.

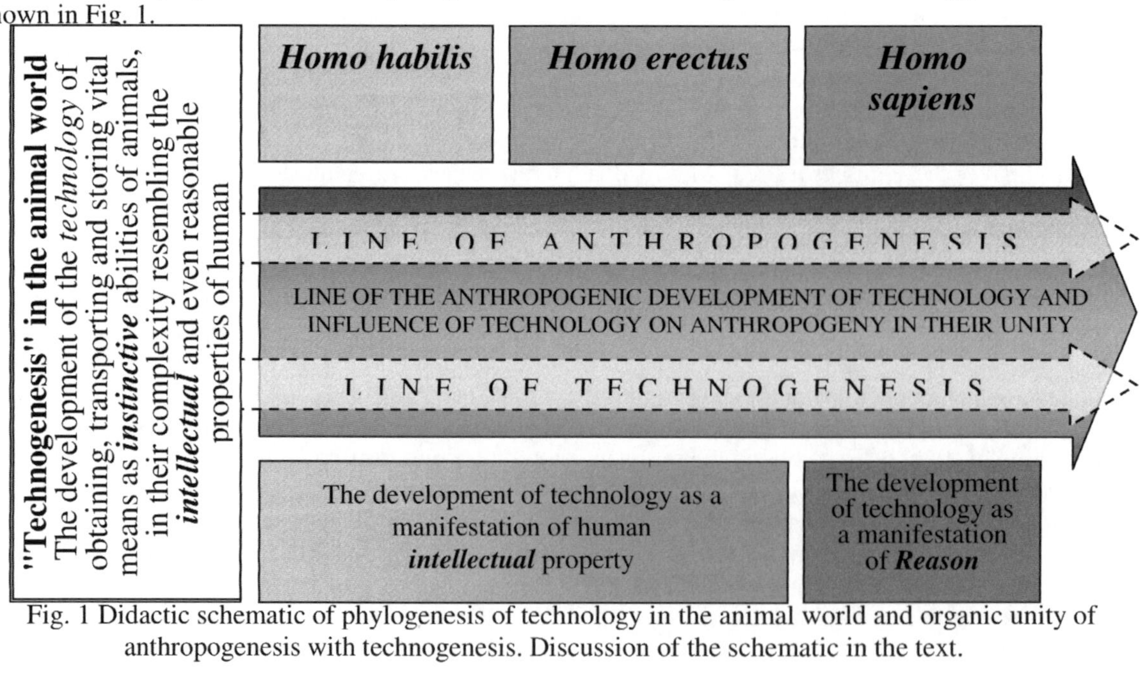

Fig. 1 Didactic schematic of phylogenesis of technology in the animal world and organic unity of anthropogenesis with technogenesis. Discussion of the schematic in the text.

Considering the transportation equipment itself, for sure, raises a question about its genesis as, at first, a relatively simple, then sophisticated, spontaneously self-developing system, as an industry of material and spiritual culture. We observe the simplest forms of transport technology already in animals: ants and beetles collectively transporting a straw or pushing huge objects; hedgehogs carry mushrooms on their prickles etc. From the standpoint of the modern science of sophisticated systems, they are truly the systems, since their organization ensures the achievement of a certain vitally important target. These animal forms of transport technology are instinctive (Latin instinctus means inspiration) associated with certain biological needs. Their manifestation is strictly fixed in the biological construction and is therefore predetermined, both in form and purpose.

The biological mechanisms on which the instinctive technical actions of animals are based are similar to the intellectual processes that ensure the actions and activities of human being. The theoretical and experimental approaches to studying the intelligence often intellectualize not only human being, but animals, and even artificial technological intellectual systems. Methodological boundaries need to be determined for comparing the intellectual processes of humans, animals and engineering devices. This will allow revealing the breadth, depth and, most importantly, the freedom and activity of the human intelligence in comparison with the instinctive intellectual properties of animals that are limited in freedom and with even more limited technological intelligence.

The definition of Intelligence in the ordinary worldview traditionally does not differ from the definitions Mind, Reason. Such a confusion of terms and definitions can be found in dictionaries, reference books and, unfortunately, in popular science encyclopedias. However, this is not the case. Clarification of terms and definitions of Instinct, Intelligence, Consciousness, and Reason is necessary because of their vagueness in the humanities and philosophy. In engineering sciences and practice, this becomes increasingly important since the properties of human being meant under these definitions are supposed to be possible, say in hardware – in mechanical man-made devices. Speculations on this topic are the beginning of the search for appropriate means and mechanisms for managing the spontaneous self-development of the transit system, similar to what the Human Brain is the mechanisms of Intelligence and Reason in technical terms.

Intelligence is a property of the animal body, which gives the opportunity to solve the problems of its life activity as an individual in a particular, sensually perceived, object environment. Another thing, the Reason is a very special type of realization of intellectual abilities, namely human ones. It relates not only to the solution of individual life problems, but also collective, social issues. Some collective actions of animals can be perceived as reasonable, but they are not realized by animals, they remain instinctive. A truly reasonable property of an individual Person-Personality arises and develops on the paths of the active entry of the individual into the system of knowledge accumulated by mankind – into the Public Consciousness (co-knowledge means joint knowledge) [10]. Individual mind is formed inside the Public Consciousness.

Consciousness, both Public and Individual, is the foundation, the material for the display of the properties of Intelligence and Reason. Consciousness is the Chaos of the vast knowledge that finds the Order in external practical, intellectual actions and internal, in the mind actions. Turning to the transit system as a living organism, moreover, as to the Person-Personality, that is, endowed with intelligence and reason, this thought can be expressed by the chart shown in Fig. 2.

Reasoning whether an anthropogenic transit system is similar to the transportation equipment of animals is not rhetorical. In modern outlooks about the spontaneous emergence of sophisticated and self-developing systems, the key question is their origin – genetics. In terms of technology, the definition of genetics (the term is borrowed from biology) is attached not to biological meaning, but rather to the analogy of the mechanisms of origin and development in humanist systems that are as complex and characteristic as the living organisms. As such a system, it is important to consider modern urban transit services and an engineering system of an even higher hierarchical order – a system that facilitates all the productive and non-productive needs of people as a single technological basis for social development [8], which is the City.

IOP Conf. Series: Materials Science and Engineering **412** (2018) 012047 doi:10.1088/1757-899X/412/1/012047

The beginning of genetics, the spontaneous self-organization and self-development of the man-made transit system created in human society, as well as in animals, is a vital necessity, first realized by people for their own purposes and forms. Developing in the public consciousness in the form of scientific and technical unity, this beginning acquires new ones and generates more and more perfect mechanisms, complicating the organization and enriching the properties of the transit system. At a high level of its development, the entire transit system generally becomes the genetic origin, the core of developing a system of a higher hierarchical level that ensures the development of the global technological system of society. In this process, the genetic origin of the self-development of the transit system serves as the basis for an even more complexly organized genetic origin of scientific and technological progress and social development.

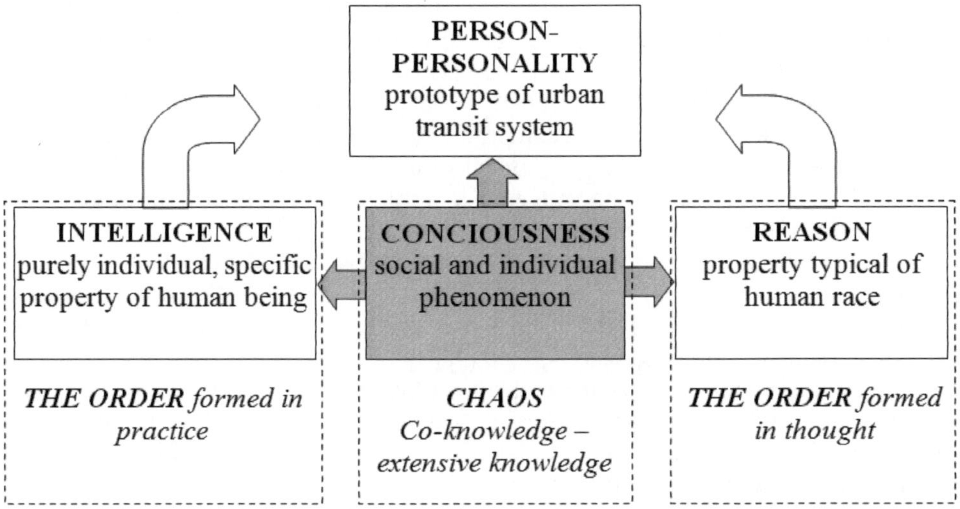

Fig. 2 A chart demonstrating the relationship between the phenomena of Consciousness, Intelligence and Reason, as a relationship of Chaos and Order as applied to the prototype of the transit system as a Person-Personality.

What is the beginning of the emergence and development of a sophisticated human transportation technology as a system – its genetic core – is the main object of search in the problem of the innovative model of the development management of the urban transit system. A productive idea may be to consider the genetic core of the transit system based on the ideas of socioeconomic genetics that have arisen and successfully developed in modern science, in which socioeconomic, including engineering structures are viewed in the likeness of biological, molecular and supermolecular genetic structures [8].

The transit system, at any stage of its development, as a man-made phenomenon, represents a communicative foundation that connects the realization of all production and non-production needs of society into a single unit, integrating it. It is made by people, society and creates society, is the main condition for its socio-economic and technological development. Therefore, according to biological terminology, it is the genetic core, the main condition determining the nature of social development. The central place in this core is taken by the idea of transportation technology, its properties, purpose, objectives, concentrated in the form of a set of certain engineering knowledge as formed in the public consciousness. Having such a fundamental importance for social development, the transportation technology focuses on other techniques and knowledge – the entirety of scientific and technical knowledge and practice, generally being the core of the relationship between science and technology. Transportation means as a sci-tech system, which takes the place of the genetic core of social development, is an incentive for scientific, technical, technological innovations and absorbs them into themselves.

IOP Conf. Series: Materials Science and Engineering **412** (2018) 012047 doi:10.1088/1757-899X/412/1/012047

Urban transportation means are a huge and complex engineering system, an industry of the material and spiritual culture of society. One of the first and most urgent challenges that facing researchers of such a system having decisive importance for people's vitality and ability to live, is the solution of the problem of the dialectics of the relationship between its internal properties of self-organization and self-development, on the one hand, and external target management, on the other.

The relationship between controllable and control systems, multiplying their tension, is influenced by the dialectics of internal mechanisms of self-organization and self-development existing in each of them, for example, such as, in technical terms, processes of dissipation mean loss of matter, information and energy [11, 3]. It is also, in terms of modern concepts of sophisticated physical and chemical, nuclear and molecular systems, the phenomenon of bifurcation is a split or even more splitting and unpredictability of development lines, a change of phase states and many others perceived by common sense as negative. However, these processes are dialectically related to indisputably positive phenomena that give the system rise to reorganize and route forward.

Managing the development of the transit system in its scientific and technical unity is one of the most pressing challenges of society related to the impact of transportation technology on health and livelihoods, on the cultural and ecological environment of people, and ultimately on the quality of life of our planet in general. These are the contradictions of emergence and bioethical problems of managing the development of the urban transit system.

The English word emergent means arising unexpectedly. The definition of emergence as applied to the rise of complex systems with new properties was previously regarded as an idealistic worldview concept that considers development as a jump-type process, where unexpectedly new higher qualities come into existence due to ideal forces (in the writings of S. Alexander and C. Lloyd Morgan.) [12]. Now, in the light of modern scientific views, this is a very real, materialistically explained property that belongs to sophisticated systems offering property of coherence of their structural and functional organization. This property is distinctly attributed to the alliance of modern science and technology in its influence on public life.

The emergence of a complex alliance of science and technology is characterized by the pretext of super – superefficiency, which is in tune with the properties of superconductivity, superstrength and similar those found in material coherent systems. In terms of the impact on people's lives, the sci-tech progress rises more and more steeply up the exponential curve leading to revolutionary changes in all spheres of life. It is necessary to foresee unexpected and possibly negative consequences that become ever-evident.

Positive consequences are important to envisage and to create conditions for their emergent growth. This requires knowledge of emergent and, at the same time, unpredictable in its development phenomena and processes in public culture; methods of forecasting, designing, managing them. The transportation sci-tech system taking the middle position in technological growth of society must necessarily be the focus of special humanist ethical requirements.

Bioethics is an interdisciplinary field of knowledge that covers a wide range of philosophical and ethical issues. The term bioethics was first applied to a special version of environmental ethics that considers the possibility of human survival in the conditions of anthropogenic civilization (the ethics of conservation of the human natural environment). However, later the meaning of the term changed due to the rapid growth of medicine, biological sciences and the use of high technologies in medicine. Alongside with this, the man-made problems of social development were exactly those which gave environmental and anthropological methodological significance to bioethics associated with the study and solution of problems of individual and public health, the vital activity of people in a natural and cultural, technically rich environment [13].

The environmental and anthropological methodology of modern science was reflected in the International Classification of Functioning, Disability and Health (ICF), developed under the auspices of the World Health Organization (WHO). It was published in 2001 and recommended by WHO as a technological tool for researching and solving practical problems of individual and public health. In the ICF, human health is considered as a potential for its vital activity, the context of which are, on the

IOP Conf. Series: Materials Science and Engineering **412** (2018) 012047 doi:10.1088/1757-899X/412/1/012047

one hand, the fundamental anthropological individual-typological and specific properties of human being, and the properties of the social and natural-ecological environment, on the other. The transit system is included in the ICF as one of the leading context factors of people's health and life.

The search for a model of managing the development of the urban transit system, should use the models of the setup and functions of the brain that are as possible in terms of complexity and corresponding to modern concepts of sophisticated systems. Such models, in continuation of the research by many biologists and technicians, have been developed from various perspectives by Russian scientists, physiologists Pyotr Kuzmich Anokhin, Nikolai Aleksandrovich Bernstein, German scientist, mathematician Hermann Haken, and American scientist Michael A. Arbib.

Having the properties of self-organization and self-development common to living organisms, a complex social and technical system can be managed by means of its specific social and technical means and tools. The controls are specific, similar, but certainly not identical to the biological – the brain controls, which are required to be defined from the position of a higher-level model, in order to be able to use them to manage the development of the system.

The model of the brain of highly organized animals according to Anokhin mainly reflects its functional properties in the vein of the mathematical cybernetics of N. Wiener as a science of proper control in the sense of imposing line of behavior on a complex system, technical or otherwise, based on certain principles and means of their implementation placed in the system itself. The model introduces physiological, psychological definitions consonant with the cybernetic definitions (specific biological and technical terms are highlighted in italics and a bright font for clarity): about afferent and efferent flows that provide direct and inverse links between the internal elements of the system with the external environment in which the behavioral act takes place in the form of a reflex ring. One of the properties that significantly differentiate the brain from non-biological systems is the extemporary organization of the processes of self-management by the behavior of a living organism, called a functional system by Pyotr K. Anokhin. This is an urgent organization of behavior that depends on situations and changing internal biological motives of behavior in animals or socio-psychological – in humans. The importance is given to the introduction of the definition of afferent synthesis being an analysis and generalization of signals – sensations entering the brain based on the specific and individual experience, which is the mechanism for obtaining the information – the core definition of cybernetics. Information accumulates as memory – the biological (psychological) definition actively used in technology. In terms of biological definitions, the memory lies at the heart of the animal manifestation of the property called intelligence.

Particular interest in the model of the brain's work on the principle of the formation of extemporary functional systems is shown by the mechanisms of biological and even psychosocial nature, such as the motive, that introduces ideas about the independent choice by the living organism of behavioral acts and the integral line of one's behavior. i.e. self-organization and self-management. A biological and psychological metaphorical notion of motive in technology is not found. Unfortunately, Anokhin's model presents the brain setup, as well as in cybernetics, as a black box, that is, a specific histological setup of the brain is not given a certain value. The functional system is not tied to the specific structures of the brain. The brain is not specified as a material carrier of the functional system. This is a significant drawback of the model to apply it as a tool for analyzing how a structurally functional system for managing the development of the transit system is being built. Other, more significant, shortcomings go on top of it. The model focuses on the properties of the animal's brain, i.e. on the processes that can be defined as intelligent. These processes are important, but in regard to the transit system, processes called reasonable and the functional structure of the human brain are of interest.

The model of the brain by Nikolai A. Bernstein extends and deepens the Anokhin's ideas. First of all, it is human brain to a greater degree meaning that it is the bearer of the mind, not only of the intelligence typical of all animals at some point. In the social systems to which transportation means belong, Reason is the ability to reflect the collective experience and knowledge of people acquiring a socially determined quality, called the Personality [14]. Personality is a complex, self-developing system governed in its formation and development by influences of the social environment, characterized by special social needs that distinguish it from animals, ideas about the necessary future [15], imagination and creativity.

Fundamentally important in the Bernstein's model is an appeal to the material structures of the brain in their connection with certain functions in contrast to the notion of the brain as a black box. It formulated a complex anatomical and functional organization of the brain. It is represented as a solid sophisticated system, each individual material (neural) element of which being very complex systems are in dynamic hierarchical relationships; they are functionally coherent (a primarily technical definition), i.e. consistent, unified. Thanks to the dynamic and, at the same time, coherent relations of the anatomical and functional units, the brain, showing activity in the processes of management, self-management and self-organization, can create functional systems (according to Pyotr K. Anokhin) that manifest emergence (philosophical, but with a more rigorous scientific meaning) in unexpected, unusually high ability to solve problems of vital activity of an organism.

The model of the brain by Hermann Haken is a contemporary view of how the brain created by a mathematician is set up and functions. It discusses the brain from a purely functional external side assuming that its real construction, as in cybernetics, does not matter much, moreover, owing to the principles of behavioral science that takes their behavior as the main thing in the study of complex living systems, but not the material structure. According to these principles, the material structure can be any, and in such systems as the brain – so complex that methodologically, in other words essentially, it is not possible to determine anatomical and functional links in exercising complex behavioral acts. Mathematical studies conducted by Hermann Haken state that the behavior of the animal body under conditions that force changing the activity upward or downward undergoes drastic changes from one form of organization to another, essentially different ones. At the same time, it is not possible to determine which areas of the brain do this, since the brain functions as an integral whole. The brain creates the organizational form of behavior as a function of the external conditions of behavior and internal conditions associated with the capabilities of the executive apparatus. It is impossible to foresee what it will look like amid infinite complexity and unexpected conditions, but the brain creates the necessary forms of behavior, for example, as Hermann Haken writes, a centipede with damaged legs suddenly organizes a completely unexpected and effective form of movement necessary to solve the problem of movement at any given speed. The insight into the setup and function of the brain set out by Hermann Haken more than half a century after the work of Nikolai A. Bernstein is of interest because it explains the brain from new, synergetic methodological, positions about complex systems. However, they lack the approach of need to take the brain as a concrete material carrier of functions, including the Human Mind.

Michael A. Arbib compared the brain with the computer, not begging the difference between the living brain and a computer though being a quite complex device. In this comparison, like Hermann Haken, Michael A. Arbib refers to the monstrously complex setup of the brain and, if possible, as a representative of theoretical cybernetics, describes this anatomical and physiological complexity. As a result of his comparisons, he comes to the conclusion that it is nevertheless necessary to know not only the functional, but also the structural organization of the brain in order to improve the computer and create more complex intelligent systems. In his book, Michael A. Arbib notes this deficiency in the description of the brain and computer of other authors. He insists on the need to strengthen the systemic, and somatotopic approach to the setup of the computer. The word somatotopic has a purely biological meaning: soma is a Greek word that means body; topos is a Greek word for location. Thus, it is referred to the need to search not only functional analogies, but also structural-functional, somatotopographic ones. Michael A. Arbib regards the brain as the carrier of intellectual processes, hence the brain of animals and human as an animal. The brain is not regarded as a carrier of Consciousness and Reason.

From a brief survey of existing animal and human brain models that can be used to study the problems of managing the development of transportation means as a sophisticated self-organizing and self-developing socio-engineering system, the question arises of using knowledge about the Human Brain, the material carrier of Reason, as a model for managing the development of socio-engineering urban transit system. Before the technical fellowship there is a need to utilize not only biological, but also humanist, psychological definitions in all their complexity and it becomes necessary to compare these definitions with the real structure of the urban transit system and the mechanisms for managing its development.

Referring to the human brain as a model for managing the development of the transit system requires abstracting away from many of its biological functions, concentrating, first of all, on the mechanisms and processes by which intelligent human motor behavior is organized and managed, which is a very complex self-organizing and self-developing system.

"The study of human behavior is an immense area of research. We mean under the behavior a variety of phenomena, from simple movements to highly complex behavior, such as communicating with other people or its actions in certain situations," writes Hermann Haken, the ideologist of the synergetic scientific worldview, " ... that's why I chose coordination of movements" [7]. We also select coordination of movements as coordination, structurally functional arrangement of transportation, as a technological system that serves industrial and non-productive public communication processes, requiring the movement of human and material flows.

The complexity of the motor behavior of a person is related to a number of quite certain conditions in which the human life and activity occur. From the force of gravity, in which the person must find the orientation of movements, to infinitely diverse and complex ideas that the person must realize in motion. The transit system as composed of different transportation equipment created by people, and which has become sophisticated and self-developing, is assumed to be less complex than the system that represents the motor behavior of a person. The mechanisms and processes of how to organize and manage the development of these two systems may be similar, but, naturally, are not identical. In the case of human motor behavior, this is due to the setup and function of human brain, and in the case of the transit system – to a certain socio-technical organization.

To study and solve the problems of managing the development of the transit system, obviously, there are all the methodological grounds for using the theoretical model of setting up and controlling the motor behavior of the person from the brain, as well as purely practical reasons, because transportation means are a system of moving, relocating equipment and technologies. Such a model should be a concept of the structure and functions of the human brain, but not animal one, conceptually formulated by Nikolai A. Bernstein. In the very difficult and long history of the development of concept of the brain – this is the first fairly complete idea of the Human Brain being the material carrier not only of the Intelligence, within certain limits inherent in animals, but of Reason peculiar only to human being.

The human motor function is a multilevel biodynamic tissue. It is called like that by Nikolai A. Bernstein because of the complexity of its structure, similar to the structure of the most complex anatomical structures of the body. The motion behavior as composed of a variety of movements has a macrostructure, consisting of five levels of movement consistently built on top of each other as the complexity of tasks rises in various spheres of life. Spheres of human activity in which the construction of movements is carried out are areas of immediate objectives and remote goals of movement, incentives and requirements for the organization of movements, to their spatial, temporal, force and, finally, to semantic characteristics, when movements realize intellectual and reasonable functions. This transition of the construction of movements from the solution of purely physical, mechanical challenges to problems of an intellectual, semantic and reasonable nature determines the preference to consider the human brain model described by Nikolai A. Bernstein.

Now let us talk about the levels of the construction of movements different in meaning. They are Latin lettered by Nikolai A. Bernstein.

Level A. The construction of movements in the sphere being the field of gravity. It is required to imagine the importance of knowing this part of the brain model for modeling the processes of controlling the transit system similarly to it. For the movements of the human body, probably there are some basic requirements for the transit system from some forces like gravity. In the transit system, this can be assumed as its organization (construction) in global conditions of functioning, such as natural and man-made energy resources, landscape and cultural-original conditions of local civilizational [8] of development, socio-economic basis in general.

Level B. Building the movements in structural and functional organization of one's own body. This leads to the idea of a similar existence of historically formed internal ties in transportation means that must be regarded and used in the management of transportation development as a self-organizing and self-developing system. This is obviously the manifestation of those effective acquisitions that have developed over the entire history of the formation and development of a technical, namely, transportation culture.

Level C. Building the motions in the coordinates of body and its parts relating with the objects of the surrounding environment. There is an idea of a special level of development management of the transit system – a level that ensures its openness towards the external technological environment of the city, entry into the system of a higher hierarchical, subordinate level. In the transit system, it is its organization in synergy, interaction with other non-transportation, production and non-production technologies of the city.

Level D. This is the construction of movements in the sphere of really existing semantic connections between surrounding objects. This raises an assumption about such a level of development management of the transit system. In a similar way, it is headed not to adaptation to the urban technological environment, as at the previous level, but to active participation in the development of the entire technological system of the city. In the transit system, this level, obviously, can be represented as the central link of its organizational and technological innovation development in the global system of production and non-production technologies of the city.

Level E. Building the movements in the sphere of semantic relationships between objects – movements that are perceived ideally, in thought, followed by real implementation in life. It is the highest level of human brain functioning that shows itself as the carrier of Reason, individual and social consciousness. The Level E is a prototype for managing the development of the transit system as a complex socio-technical, humanist system. It contains the potentials of the development of the transit system based on its integration into the international, outer space of sci-tech progress.

A todat technical specialist is well aware of how a computer as a tool for controlling a wide variety of processes is set up (see Figure 3).

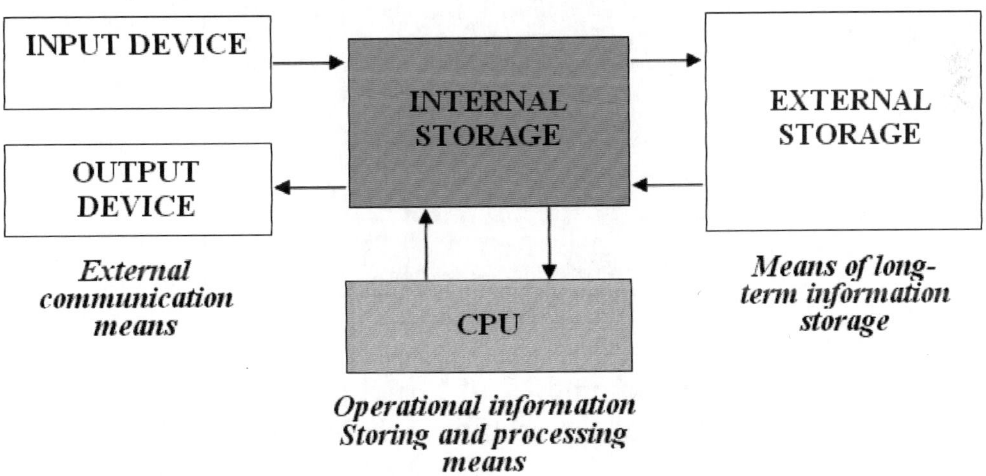

Fig. 3 Computer connection schematic.

This schematic should be combined with the setup of the human brain (see Figure 4). This schematic characterizes the construction of a real socio-technical organization for managing the development of the transit system. It is constructed amid brain, computer and social-humanist metaphors and demonstrates the basic elements of management organization and their relationship.

IOP Conf. Series: Materials Science and Engineering **412** (2018) 012047 doi:10.1088/1757-899X/412/1/012047

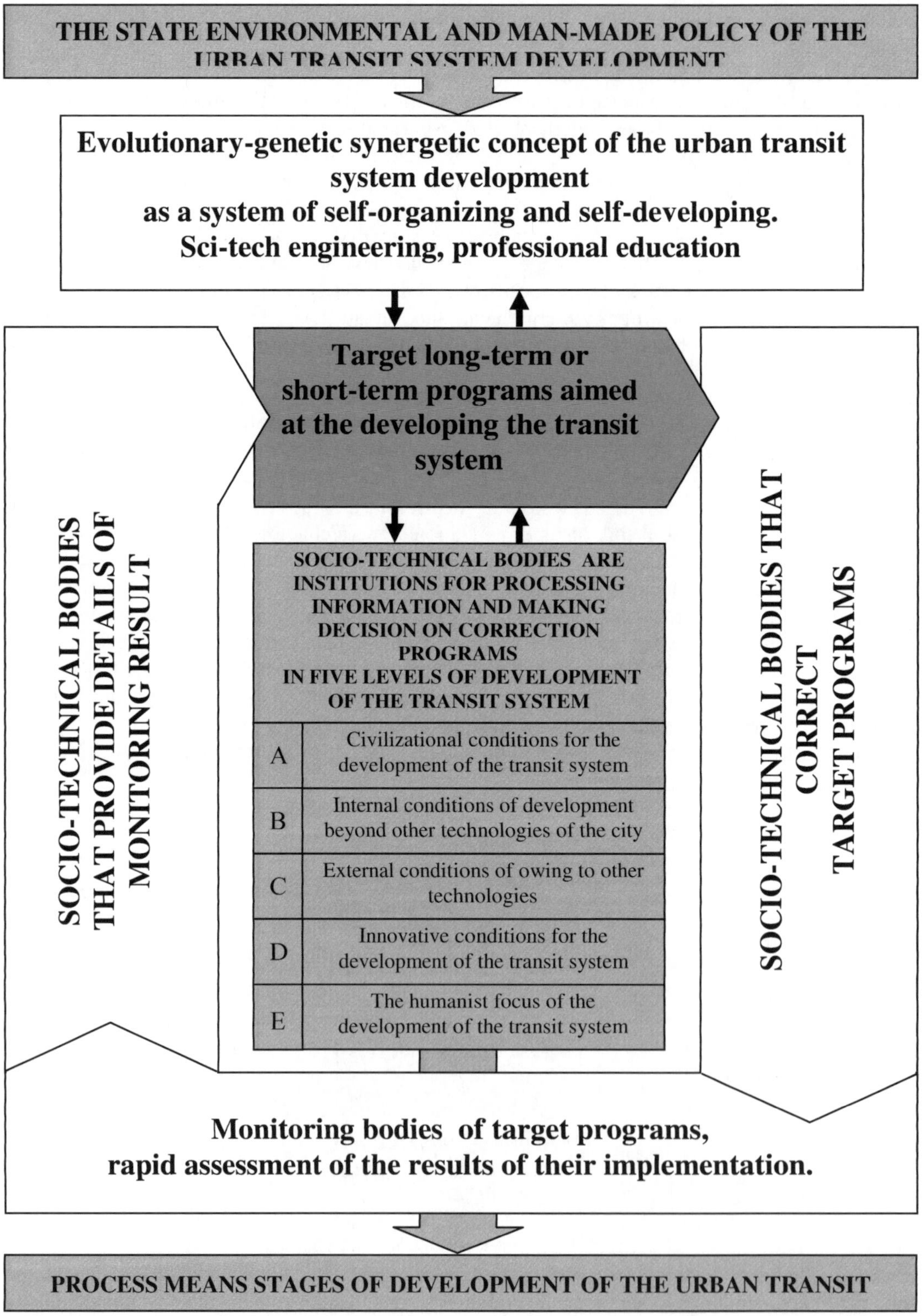

Fig. 4 A general chart showing how the management of the urban transit system development is organized.

The actual implementation of management according to this schematic will require a practical, empirical study: observations, experience and experiment. Then, theoretical analysis and generalization is necessary.

The institutions as really existing social bodies are the objectives of this empirical and scientific approach. The control activity of these social bodies can be compared with the intellectual and rational functions of the Person-Personality. Along with this, the study is aimed at the organization of management based on the principles of the setup of the human brain.

The chart shown in Fig. 4 with a didactic, search theoretical meaning has a clearly expressed applied say practical value with its theoretical justification.

Conclusions.

The use of modern ideas about the self-organization and self-development of complex systems of natural and man-made nature that emerged in the late 20th century and advancing in the 21st century reveals broad perspectives in understanding the problems of the contradictory development of the transit system of modern cities. To engage teachers and students of a professional technical higher school in this process, it is expedient to build it looking into a practical problem, such as searching for an innovative model for studying and solving the problems of the development of the transit system of modern cities. To simplify complex cognitive processes it is practicable to use didactic means, widely applicable in heuristic research. Thanks to the visual clarity which in a certain sense gives a visual idea of the logic of thought, they condition a particular sense perception of the content, structure of heuristic thought, and are its sensual tissue. This is the subject material for the rapid entry into and involvement in the research and discussion of the difficult scientific problem of specialists of different fields of knowledge as a fascinating, entertaining and productive didactic game. The game is an essential property of a person, a way the person communicates with the eventual [14 and others].

References

[1] Arbib M A 2004 *Metaphorical brain: Trans. from English* (M.: Editorial URSS) p 304

[2] Prigogine I, Stengers I 1986 *Order out of Chaos: Man's New Dialogue with Nature: Trans. from English* (M.: Progress) p 432

[3] Haken H.1980 Synergetics: Trans. from English. Ph.D in Physics and Mathematics V I Yemelyanov (M.: Mir)p 423

[4] Gorbachev V V and Bezdenezhnykh. V M 2004 *Concepts of modern science* (M.: The Economist) p 446

[5] Ruzavin G I 2007 *Concepts of modern science* (M., Units) p 287

[6] Khoroshavina S G 2005 *Concepts of modern science* (Rostov-on-Don: Phoenix) p 480

[7] Haken H 2001 *The principle of the brain: The synergetic approach to the activity of the brain, behavior and cognitive performance* (M .: Perce) p 351

[8] Kuzyk B N , Yakovets Y 2006 *Civilizations: Theory, History, Dialogue and the Future: in 2 volumes (*M .: Institute for Economic Strategies) p 768

[9] Khakimov E M 2007 *The dialectic of hierarchy and non-hierarchy in philosophy and scientific knowledge* (Kazan: Published by Feng Academy of Science)p 288

[10] Auhadeeva L A 2006 *Communicative culture of the future teacher as an object of systematic research* (Kazan: Publishing house of KSU) p 164

[11] Korolkov B P 2011 *Thermodynamic fundamentals of self–organization: monograph* (Irkutsk IrGUPS) p 120

[12] Prokhorov I M 1985 *The Soviet encyclopaedic dictionary* (M.: Sov. Encyclopedia) p 1632

[13] Auhadeev E I, Ilyazov R G, Karpov A M, Iksanov H W 2011 *Modern ecological and anthropological methodology of the study and solution of health problems: Book of reports*

Proceedings of the international conference dedicated to the 25th anniversary of the Chernobyl disaster p 296
[14] Leontiev A N 1975 *Activity, consciousness, personality* (M.: Politizdat) p 304
[15] Bernstein N A 1990 *The physiology of movements and activity* (M.: Nauka) p 496

Intensification of Oil and Oil Product Heaters by Means of Auger Inserts

E M Khusnutdinova[1], I A Konakhina[1], G R Khamidullina[1] and Y K Vachagina[2]

[1] Kazan Federal University, 18 Kremlyovskaya street, Kazan, 420008, Russian Federation
[2]Kazan Scientific Center of the Russian Academy of Sciences, the Federal State Budgetary Institution of Science
2/31 Lobachevskogo Str., P.O. box 261, Kazan, 420111

rr-088@mail.ru

Abstract. This paper discusses the approach of how to intensify the heat exchange processes under laminar flows of high-viscosity non-Newtonian fluids with the use of auger inserts. This approach of intensification refers to the methods of swirling the flow, which creates a rotational motion of the fluid throughout the entire duct cross-section being most relevant to the high-viscosity fluids.

One of the most important challenges that the companies of the country's energy sector face is the search for new ways of rational use of energy resources. The most promising way to reduce the weights and overall dimensions of heat exchange equipment is the heat exchange intensification. A description of the various types of intensifiers is given, in particular, in the works of Yu. G. Nazmeyev, A. Ye. Bergls, O. V. Mitrofanova, G. A. Dreitzer and other authors [1-3].

With regard to the flow of single-phase coolants, rough and developed surfaces are used, either by finning or creating cavities on the surface; swirling the flow with spiral fins; auger devices; swirlers installed at the duct inlet; mixing gas bubbles with the gas flow, and solids or liquid drops to the gas flow; rotation or vibration of the heat exchange surface; pulsation of the coolant flow; electrostatic fields influencing the flow; suction of the flow from the boundary layer; jet systems. The performance of these methods varies usually having an increase of heat transfer up to 2-3 times, while the energy costs are significantly different for various methods of intensification [1].

The laminar-flow region is a promising scope where the design and analysis of methods for increasing the performance of heat exchange equipment is quite a challenge of the day. Laminar flows are characterized by low heat transfer coefficients. The flow velocity and the temperature of the coolant vary over the entire duct height so that the thermal resistance occurs in the entire boundary layer, which can take the entire cross-section of the pipe or duct. Therefore, the small roughness on the heat exchange surface does not work for intensifying heat transfer in laminar flows [2]. Taking these features into account, it is possible to identify the approach of convective heat exchange intensification with swirling the flow by auger inserts [3]. Swirling the flow by means of a pipe-inserted auger considerably increases the energy efficiency of heat exchange versus the non-swirling flow.

Content from this work may be used under the terms of the Creative Commons Attribution 3.0 licence. Any further distribution of this work must maintain attribution to the author(s) and the title of the work, journal citation and DOI.
Published under licence by IOP Publishing Ltd

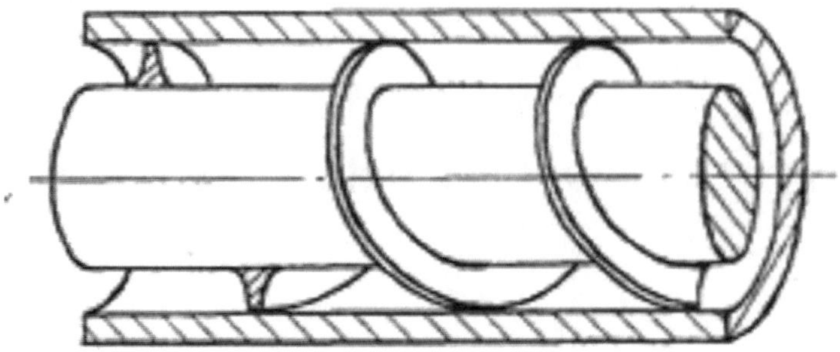

Fig. 1 Swirler - the auger insert in a round tube

In recent years, the rapid development of numerical methods and computer technology has made it possible to perform calculations for the swirling flow analysis to a high degree of detailing and a comparative performance analysis of various intensifier designs.

The auger inserts with multiple screw augmentation are expected to increase the intensification effect due to a strongest impact on the flow. The study of this effect requires the development of adequate full-scale mathematical models and methods of thermal hydraulics calculation.

The study of heat exchange intensification during the flow of heavy oil and oil products using the multi-screw auger inserts sets the following challenges:

1. Designing a mathematical model of hydrodynamics and heat exchange in the non-isothermal flow of a non-Newtonian fluid in a three-screw auger insert duct.
2. Making the Comsol Multiphysics software fit for addressing the challenge.
3. Numerical and experimental studies of non-isothermal flow of heavy oil in auger insert ducts for various geometric and duty parameters.
4. Analysis of the vorticity fields, the temperature and velocity profile following the results of numerical studies.
5. Evaluation of how the the auger fin angle, the screw pitch and the annular space have an effect on the vorticity fields, the temperature and velocity profiles and the thermal and hydraulic efficiency of the heat exchange equipment.
6. Summarizing and analyzing the results of numerical and experimental studies.
7. Providing recommendations on the use of auger inserts for heating oil and oil products at pumping stations.

The main effect on the thermohydraulic characteristics of the intensified ducts [4] is a decrease in the ducts' resistance to flow (internal flow) and the power expended for the transportation of oil and oil products, all other things being equal.

References

[1] Kadyirov A I, Khalitova G R and Vachagina Ye K 2016 *The hydrodynamic structure of a viscous flow in ducts with intensifiers as screw inserts* (Alademenergo papers. Publishing House: Research Center for Energy Problems of the Kazan Scientific Center of the Russian Academy of Sciences (Kazan) pp 58-74

[2] Popov I A, Makhyanov Kh M and Gureyev V M 2009 *Physics and industrial application of heat exchange intensification: Heat exchange intensification (*monograph: Center for Innovative Technology) p 560

[3] Nazmeyev Yu G 1988 *Heat exchange in laminar flow in discrete-rough ducts* (M.: Energoatomizdat) p 376

[4] Khusnutdinova E M, Konakhina I A and Khamidullina A F 2017 *Comparative Analysis of Thermohydraulic Performance of Enhanced Viscous Oil and Oil Product Heaters* IOP Conference Series: Materials Science and Engineering **012046**

IOP Conf. Series: Materials Science and Engineering **412** (2018) 012049 doi:10.1088/1757-899X/412/1/012049

Investigation of nanometallokeramic composite coatings obtained by vibro-arc surfacing

A V Kolomeichenko[1], N V Titov[1], Yu A Kuznetsov[1], L V Kalashnikova[1], O O Bagrintsev[1], S N Sharifullin[2,3]

[1] Orel State Agrarian University named after N.V. Parakhin, Russia
[2]Kazan Federal University, Russia
[3]Kazan state agrarian University, Russia

Saidchist@mail.ru

Abstract. The paper presents a description of the method of hardening the working bodies of agricultural machines, working in conditions of abrasive wear. The method includes vibro-arc surfacing of nanometallokeramic composite powder materials with simultaneous thermal diffusion hardening. The optimal composition and concentration of components of the material, ensuring the best physical and mechanical properties and resources of the hardened working bodies tillage machines.

1. Introduction

In all countries of the world the investigations on obtaining highly effective metal ceramic hard alloys that are able to increase the resource of working organs of tillage tools into several times are being carried out. But the first results of foreign metal ceramic plates application for hardening of plow shares running on sandy and loamy soils did not provide their wear resistance increase because of higher brittleness of plates [1-2].

In this connection nowadays the works on creation of new metal ceramic materials having high wear resistance and impact elasticity in the conditions of intensive abrasive wear at considerable static and dynamic loadings are continued. Now at the result of the complex of the carried out scientific investigations nanometal ceramic composite material (NMCM) allowing solving the set problem is elaborated. It consists of steel matrix (deposite powder), with incorporated aluminium oxides Al_2O_3 or silicone SiO_2, boron carbide B_4C, pounded to nanodimentional condition, and also alloying elements: boron, nitrogen and aluminium. Boron is included into powder composite content in the form of borax $Na_2B_4O_7$, nitrogen is included into nitric acid sodium [3]. Aluminium is used for ferrum deoxidation and formation ceramic phase by means of its conversion into aluminium oxides Al_2O_3 and at the following fusion into corundum. The latter in the mineralogical table by hardness occupies the 9th place it is very close to diamond hardness. Matrix is shockproof, consolidate and composite carcass NMCM. That is why steel filler powders with high hardness and abrasive resistance, for example PG – 10N - 01, PR - N70X17C3P4, PG - SR4 can be used reasonably as a matrix material.

2. Methods.

To obtain wear resistant hardening coating from the suggested NMCM we developed the short-circuited arc surfacing method with application of graphite electrode. At its simultaneous usage

Content from this work may be used under the terms of the Creative Commons Attribution 3.0 licence. Any further distribution of this work must maintain attribution to the author(s) and the title of the work, journal citation and DOI.
Published under licence by IOP Publishing Ltd

IOP Conf. Series: Materials Science and Engineering **412** (2018) 012049 doi:10.1088/1757-899X/412/1/012049

together with surfacing thermodiffusion hardening of the working organ surface by means of boron, nitrogen and carbon (boronnitrocementation) occurs. At depositing carbon is emitted at the account of sublimation of graphite electrode. Arc vibration of electrode provides the obtaining stronger and denser coating for the sake of the working organ material fusion and mixture of ceramic and alloying components NMCM [4]. More than that at short-circuited arc surfacing usage not so strong heat embedding as some other surfacing types usage takes place. Before surfacing NMCM in the paste form is applied on the working organ surface.

Preliminary done investigations stated that every component of NMCM considerably influences on the effectiveness of the working organs hardening. In this connection the researches on the influence of the NMCM components on hardness, microstructure and wear resistance of the hardened surfaces were carried out. To carry out the researches several pastes (table 1) were prepared. The pastes were prepared by mechanical mixture of the components some of which were broken to nanosized condition in a centrifugal mill in advance. As a bounding material 20% water solution of sodium silicate solute Na_2SiO_3 was used. After preparation the paste was applied on a sample surface prepared from steel 65Г. The given material selection is provided that it is used for production of the majority of working organs of tillage tools. The thickness of the applied paste layer was approximately 2,5…3,0 mm. Then the paste was dried to hardening. At the temperature of 90…95°C hardening time does not exceed 8…10 min.

Table 1 – Chemical composition of metal ceramic pastes

№ paste	Chemical composition of paste	Mass ratio of components in paste, %		
		№1	№2	№3
1	Steel powder PG – 10N - 01	15	15	13
	Borax $Na_2B_4O_7$	25	15	10
	Boron Carbide B_4C	45	55	65
	Nitric acid sodium $NaNO_3$	4	4	4
	Silicon oxide SiO_2	5	5	5
	Aluminium powder Al	6	6	3
		№4	№5	
2	Steel powder PG – 10N - 01	50	10	
	Boron Carbide B_4C	50	70	
	Borax $Na_2B_4O_7$		9	
	Cryolite Na_3AlF		5	
	Aluminium powder Al		6	
		№6		
3	Steel powder PG – 10N - 01	20		
	Boron Carbide B_4C	63		
	Borax $Na_2B_4O_7$	12		
	Aluminium powder Al	5		

Short-circuited arc surfacing of the samples was done on device VDGU-2, which is developed and produced in Federal agro engineering centre of all-Russian institute of mechanization. The device includes inverter thyristor current source of type MASTER 162 for 200…250A, operating console and vibrator with graphite electrode of diameter 6…10 mm fixed in it. Surfacing was done at the burning of direct arc in the following modes: current rate I=70…80A, voltage U=60B, vibration frequency of graphite electrode – 100…110 vibrations per second.

The hardness of the deposited layer and the hardened substrate of the sample were defined with computer microhardness tester KMT-1 according to Vickers method at loading F=1H and exposure

IOP Conf. Series: Materials Science and Engineering **412** (2018) 012049 doi:10.1088/1757-899X/412/1/012049

time t=15 c. At the same time 32 measurements were done: 16 – by the depth of the deposited layer, 16 – by по depth of the hardened substrate. Measurement of imprints was done with video device connected to a personal computer, using special software, with statistical processing and possible automated analysis of image data according to the standards of hardness measurement.

Relative wear resistance of the hardened samples was defined with device ИМ-01 – construction of All-Russian Research Institute of Agricultural Engineering under the name of V.P. Goryachkin (JSCo «ARIAE»). Duration of the tests was 30 min. Every test was done with three replications. As abrasive material high silica sand sizing 0,16…0,32 mm was used. Average contact pressure in the friction zone was maintained 0,33 МПа. Wear rate was defined by sample mass decrease by means of its weighing on balance VLR-200 with precision 0,1 mg before and after tests.

3. The results of the research.

The results of the conducted studies showed that the highest hardness of both the deposited layer and the strengthened substrate is a sample treated with the use of NFCM with a mass ratio of components in the paste: powder of steel PG-10N-01 - 20%, boron carbide B_4C - 63%, borax $Na_2B_4O_7$ - 12% and aluminum powder Al - 5%. Analyzing the compositions of the pastes that were used to harden the specimens, we can conclude that the determining role in obtaining the highest hardness is played by boron carbide B_4C and boron contained in the $Na_2B_4O_7$ borax. As a result of the melting of the paste, the formation of atomic nitrogen also occurs, which together with the carbon formed as a result of sublimation of the graphite electrode during arc combustion diffuses into the hardened substrate, forming a solid solution that increases its hardness.

The resource of working organs being hardened with the short-circuited arc surfacing method and working in the abrasive wear conditions considerably depends not only on hardness, but on the condition of the border of deposit layer and substrate, that is from the effect of substrate material fusion and its mixing with paste fusion [3]. The most complete idea about the deposit layer and substrate structure is presented with microstructure analysis.

The carried out researches allow stating that the microstructure of the hardened samples consists of 3 zones in spite of the content of applied pastes (Figure 1).

Substrate (zone 1) has evident phase changes to boundary line due to diffusion of the elements incorporated into deposited paste and carbon content. The deposited layer consists of 2 zones – transitional and basic hardened. Transitional zone is alloy of melted surface layer of substrate and paste material. Basic hardened zone is the hardest and consists of steel matrix, holding the formed ferrous carbides FeC, Fe_2C_3, borides FeB and Fe_2B and ceramic phases – ferrous spinel, boron carbide and corundum. Junction line of the deposited layer into the substrate is more obviously detected on the photographs of microstructure at small amplification.

Metallographic tests also showed that the structure of the deposited metal ceramic coating is heterophase and it is represented as space-distributed crystal phases, which form multiphased crystal solid body. Structure heterophase of coating is obvious at amplification 50x, where light phase is steel matrix, that keeps the formed carbides, borides and ceramic phases – dark light inclusions. At the result of phase transformations in the parent phase separate areas or crystals of new thermodynamically more stable phases appear. They grow, interact and as the result form heterophase structure [5].

IOP Conf. Series: Materials Science and Engineering **412** (2018) 012049 doi:10.1088/1757-899X/412/1/012049

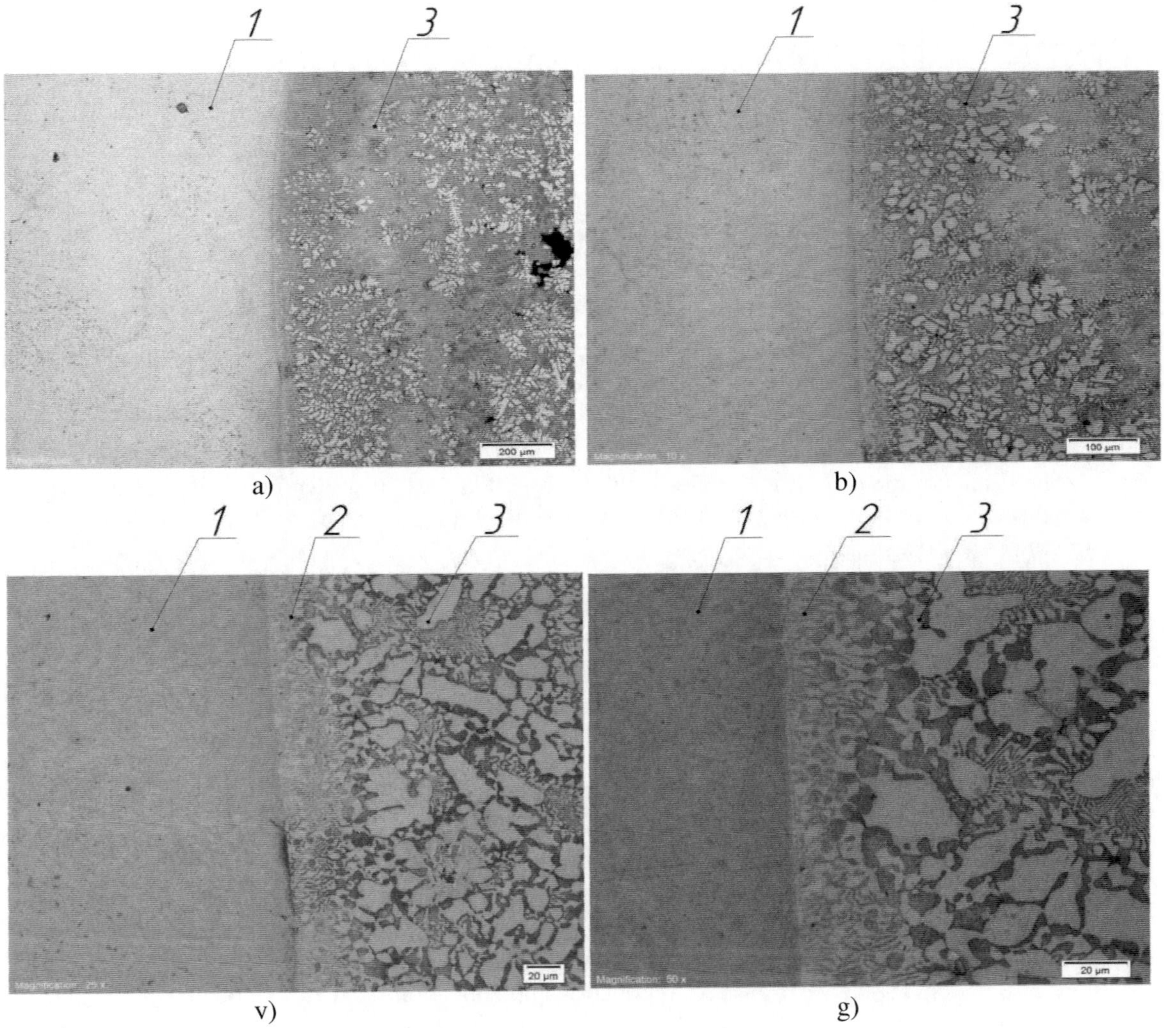

Figure 1 - Microstructure of the sample treated with the above paste, with an increase in 5x (a), 10x (b), 25x (v), 50x (g). Zones: 1 - the substrate; 2 - transient; 3 - basic strengthened

Different phases have different hardness. This is an explanation of some irregularity at hardness measurement by deepness of the deposited layer.

The results of the comparative tests on wear resistance of the hardened and unhardened samples are presented in Table 2.

Table 2 – Results of the comparative tests on wear resistance of the hardened samples

Sample number	Sample mass		Mass change	Relative wear resistance
	before test	after test		
standard	47,3400	47,3078	0,0322	1
1	41,5440	41,5350	0,0090	3,58

2	41,9614	41,9432	0,0182	1,76
3	51,5975	51,5823	0,0152	2,12
4	41,0270	41,0160	0,0110	2,93
5	42,5786	42,5647	0,0139	2,32
6	43,8791	43,8708	0,0083	3,88

Note: as a master-sample the sample from steel 65G was taken, it endured hardening and abatement at modes being used at tillage tools production

Analyzing the obtained data it is obvious that the maximum wear resistance possess samples No. 1 and No. 6. It is in 3,58 and 3,88 times correspondingly higher than hardened steel 65G possesses taken as transfer standard.

4. Conclusion.

The results of the conducted studies made it possible to establish that hardening, containing boron carbide B_4C and borax $Na_2B_4O_7$, is the most optimal for all parameters (hardness, homogeneity of structure, wear resistance).

References.

[1] Litovchenko N N, Kulikov V N, Titov N V 2013 *Vibro-arc surfacing with a graphite electrode of nanometallic ceramic composites* Welding production No 2 pp 51-53

[2] Titov N V, Vinogradov V V, Slobodchikov D A 2016 *Import-substituting technology for strengthening lancet paws of tillage machines* Innovations in Agriculture No 1 (16) pp 138-41

[3] Titov N V, Kolomeichenko A V, Logachev V N, Kravchenko I N, Litovchenko N N 2015 *Investigation of the hardness and wear resistance of vibro-arc surfacing using cermet materials* Welding International V 29 No 9 pp 737-39

[4] Titov N V 2015 *Increase of wear resistance of working surfaces of lancet paws of soil-cultivating machines by carbibifrodug hardening* Engineering and equipment for the village No 11 pp 38-41

[5] Titov N V 2012 *Hardening of the working organs of machines operated in an abrasive medium* Collected materials of the International Scientific and Practical Conference "Modern problems and solutions to them in science, transport, production and education - 2012", T II pp 46-48

Target Energy Monitoring Method for the Analysis of Production Companies and Facilities of Housing and Communal Services

I A Konakhina, G R Khamidullina, M A Kuzmina and E M Khusnutdinova

[1]Kazan Federal University, 18 Kremlyovskaya street, Kazan, 420008, Russian Federation

rr-088@mail.ru

Abstract: This paper discusses the aspects of target energy monitoring structures at the production companies and facilities of housing and communal services.

Energy conservation is one of the priorities of the strategic development of the Russian Federation in the coming decades. The success of our country to join the most developed economies depends on how successfully this challenge will be addressed and directly depends on the well-being of the population and the efficiency of domestic industry [1-5]. This problem is particularly relevant to the housing and communal sector, since about 70% of all energy resources in the energy mix of the country are consumed by this sector.

The technological expansion and equipment upgrade is impossible without economic incentives. If introduction does not yield a profit, it cannot be recognized as effective [4]. In the conditions of stringent financing, when all the risks of introducing methods for improving technologies and technical equipment fall entirely on the introducing organization, and there are no financial mechanisms for state backing, the solution may lie in the formation of a comprehensive mechanism for managing energy and resource conservation, which includes management tools and measures to make the processes happen. The key method that allows to create such a mechanism and keep it in health order is the target energy monitoring (TEM).

The TEM method requires a clear organizational structure charged with the information acquiring, processing, analyzing and decision-making. Thus, the international practice allocates such a structure to a separate unit (most often the energy management department is a special management body subordinate to one of the directors of the undertaking and independent of the chief power engineer) [1]. As experience shows, the introduction of energy management makes it possible to reduce energy costs by 10-12% only through effective accounting, without additional costs for technology upgrading.

In our country, energy management as a separate structure is normally missing, but the TEM method is widely used for analyzing energy efficiency. Specifically, it facilitates rapid response to cases of high-energy consumption [3].

The international standard ISO 50001: 2011 Energy management system – Requirements with guidance for use serves to address the problems related to energy conservation management. The provisions of the standard contain the basis to produce TEM systems at various levels of management,

Content from this work may be used under the terms of the Creative Commons Attribution 3.0 licence. Any further distribution of this work must maintain attribution to the author(s) and the title of the work, journal citation and DOI.
Published under licence by IOP Publishing Ltd

from individual organizations and industrial undertakings, to large municipal and production associations and entire regions (Fig. 1).

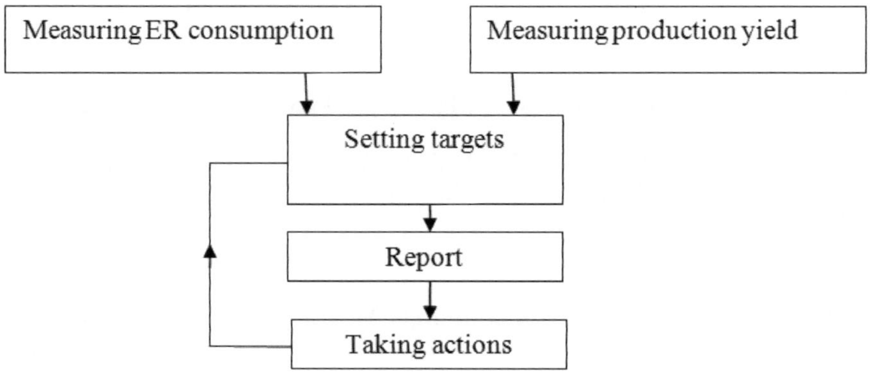

Fig. 1 TEM system chart for industrial undertaking [2]
ER – energy resource

The standard is based on the PDCA methods (Plan-Do-Check-Act), its key elements are shown in Fig. The primary data on the energy consumption level and production output are used to set targets. The report on the achievement (or not achievement) of the targets is used to develop measures to change the targets towards further improvement of the facility in terms of energy efficiency. It produces a feedback of corrective actions to be taken. The system of continuous improvement functions like that.

Thus, the target energy monitoring is one of the four basic elements of the energy efficiency management system (Fig. 2).

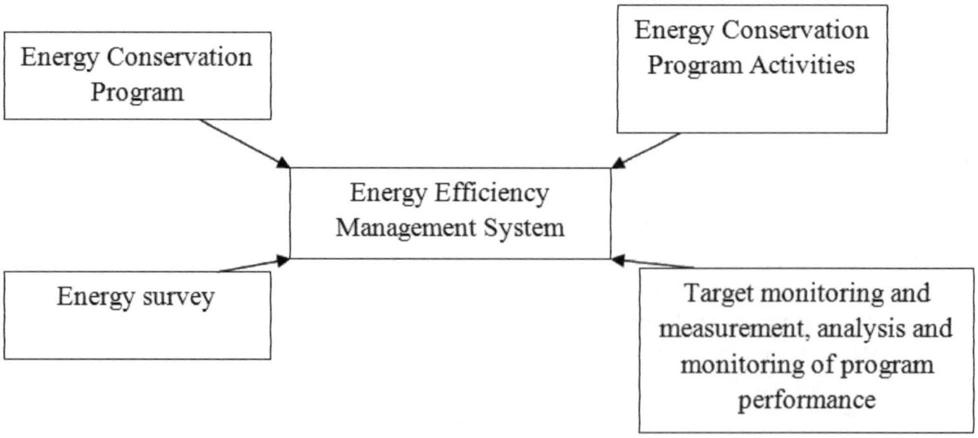

Fig. 2 TEM within
the energy efficiency management system [5]

The energy efficiency management system implies a target energy conservation program to be produced for designing the program activities. The activities happen in the conditions of continuous target monitoring, analysis of resulting data and monitoring of how the program is being introduced.

Periodic survey in the form of energy audit provides details on the achievement (or not achievement) of the targets with reference to which the "feedback" is formed (Fig. 1).

The TEM methods for industrial undertakings as well as housing and communal services includes:

1) Energy consumption monitoring that uses local meters installed at energy consumption metering units (energy-metering units (EMU)).

2) Setting the target function of energy consumption for each EMU based on the analysis of monitoring details. For undertakings, this function also includes the production output monitoring and the correlation between energy consumption and production output.

3) Producing a periodic reporting system.

4) Creation of task team in each unit, which will use monitoring data to design and introduce measures to improve the energy efficiency of their units.

5) Producing the "feedback" mechanism at all levels of the organizational structure of the undertaking or the housing and communal service facility to improve motivation and incentives for employees to save energy [2].

References

[1] Belyayev A S, Bubenok Ye K and Mukhin N V 2011 *The method of target energy monitoring and its enhancement for the analysis of the operation of an undertaking that produces several types of products* (Energy Safety and Energy Economy) No 2 pp 10-14

[2] Hyde D and Loskutov A V 1998 *Target energy monitoring in the energy management system* (Industrial Power Engineering) No 4

[3] Gonin V N and Badmazhapova Zh E 2014 *Enhancement of mechanisms for implementing energy-saving measures in the system of housing and communal services in the region* (Bulletin of ZabGU) No 6, pp 97-103

[4] Feraru G S 2010 *Organizational and economic approaches to energy saving in municipal organizations (as exemplified by housing and communal service facilities)* (Belgorod State University Scientific Bulletin. Series: Economics. Informatics) No 19-1 (90) pp 73-79

[5] Fedoskina L A and Abramov Ye I 2014 *Methodological aspects of forming the energy management system in the organization* (National Interests: Priorities and Safety) No 22, pp 23-30

Application of electric erosion machining for the restoration of splined surfaces

A T Kulakov[1], E P Barylnikova[1], F L Nazarov[2], S Yu Kovalenko[3] and K V Gribkov[3]

[1] Kazan Federal University, Naberezhnye Chelny Institute, 423812, Russia, Naberezhnye Chelny, Prospekt Syuyumbike 10A
[2] Engine Plant of PC KAMAZ, Russia, 423800, Russian Federation, Naberezhnye Chelny, Avtozavodsky Ave., 2
[3] Orenburg State University, Russia, 460000, Russian Federation,. Orenburg, Pobedy Avenue, 13, Senior lecturer of the Department of Motor Transport,

alttrak09@mail.ru

Abstact. The paper presents an analysis of the possibility of using electric erosive machining when restoring splined surfaces on specialized equipment.

It is well known that spline joints are used to fit such parts as gear wheels, pulleys, bushings etc. on the crank and in comparison with keyed joints they have a number of advantages, in particular less bearing stress at the spline boundaries and higher shaft strength with dynamic and variable loads. At the same time the experience in the operation of road transport shows that wear of slotted joints occurs frequently.

The restoration of splined surfaces is a complex and time-consuming process and includes as a rule such operations as surfacing, normalization, turning (broaching - in case of hole reconstruction), milling, heat treatment and grinding, which makes the recovery process economically unprofitable, for which reason the details are scrapped, and instead of them new ones [1,2,3] are installed during the repair. One of such details are carriages of synchronizers of IV-V transmissions of KAMAZ gear change box. These carriages are characterized by the wear of internal splined surfaces in conjunction with the secondary shaft of the box, accompanied by extrusion and cutting of a part of the metal from the side surface of the spline (Figure 1).

When such spline wear is restored existing technological processes [4,5] imply, as it was described above, surfacing with further machining and heat-strengthening, which is extremely unprofitable against the backdrop of the cost of the part, and given that separate synchronizer carriages are not supplied, this makes the whole unit is not repairable. With the cost of a new synchronizer assembly of about 4500 rubles, its replacement increases the cost of repair and operation of the vehicle as a whole.

Content from this work may be used under the terms of the Creative Commons Attribution 3.0 licence. Any further distribution of this work must maintain attribution to the author(s) and the title of the work, journal citation and DOI.
Published under licence by IOP Publishing Ltd

Fig. 1. Wear of internal splines of the synchronizer carriage of IV-V transmissions of KAMAZ gear change box

At the same time electric erosion machining is distinguished out of the many existing methods of processing parts due to its principle consisting in the ejection of particles from the metal surface by an electric discharge pulse. When copying the shape of an electrode or its cross-section the workpiece element in its shape is the inverse of the working surface of the tool. This operation is called broaching. There are methods of direct and reverse copying. With direct copying the tool is located above the workpiece, and with reverse copying it is below it. In process of electric erosion machining, the electrode is introduced into the part, ensuring the copying of the electrode [6,7]. This method is simple in execution and is widely used in industry - to produce holes of various shapes, shaped cavities, profile grooves and grooves in hard alloy parts, for tool hardening, for electric copying, grinding, cutting, etc.

The use of an electric erosion machining is associated with such basic advantages as:

- high accuracy and quality of processing;
- the use of cheaper, easily processed materials for manufacturing of the tool;
- processing of very hard materials;
- the ability to handle any current-conducting metals and alloys regardless of their physical properties.

At the moment, various electric erosion copying and broaching machines of both domestic and foreign production with manual control or with numerical program control are widely used in industry. Depending on the model the machine's productivity varies from 70 to 1100 mm^3 / min, the surface cleanliness is on average R_z = 80 ... 320 μm for roughing work and Rz = 20 ... 40 μm for finishing work, the surface roughness parameter at finishing regimes R_a = 0, 08 ... 3.2 μm with a processing accuracy of 0.003 to 0.04 mm. As it can be seen from the parameters of the resulting surface in most cases the parts do not need further processing.

Taking into account the peculiarities of the technology of electric erosion machining, its application is possible when restoring the slit surfaces of the synchronizer carriages.

Both in production and in the used recovery methods, after the machining of the carriage a heat-hardening step is performed [4,5]; it helps to create a surface with a high hardness of HRC 55, which is impossible initially due to the complication of machining. The electric erosion machining allows to process metals with any hardness of the surface, which simplifies the technology of spline joint restoration leading to surfacing of worn splines with a wire of high hardness, for example, 51 HFA (hardness is HRC 50-55), and further electric erosion machining of welded surfaces.

Orenburg State University together with the Naberezhnye Chelny Institute (branch) of KFU has developed and patented a method for restoring the surfaces of the synchronizers carriage with an electric erosive machining [8], the scheme of which is shown in Fig. 2.

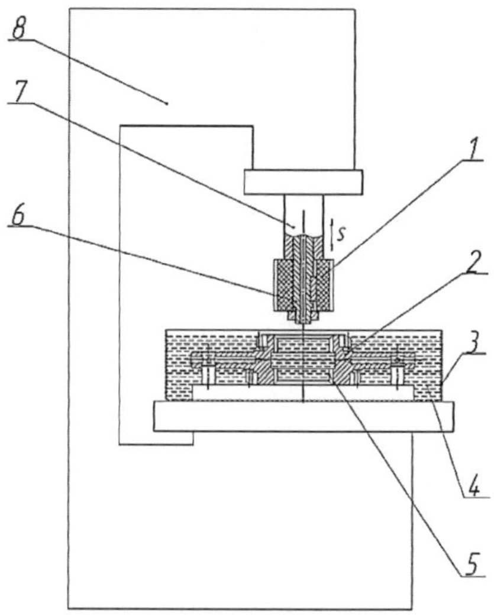

Fig. 2. Scheme of restoration process using the electric erosion machining of the worn out surfaces of synchronizer carriages:
1 - a template graphite electrode tool; 2 - the synchronizer carriage; 3 - bath; 4 - cooling liquid; 5 - weld layer of high hardness; 6 - electrode holder; 7 - the spindle of the machine providing the translational movement of the template electrode-tool; 8 – electric erosion machine tool

It is necessary to produce a template graphite electrode tool 1, for example graphite NK1, with a surface in the form of a copy of the interface surface of the response part (spline surfaces of the carriage), which is installed in the spindle 7 of the electric erosion machining, for example 4A611C. On the worn surface of the slots of the synchronizer bracket (piece No. 142.1701151), surfacing with 51 HFA wire or similar in carbon dioxide is performed, after which the part is immersed in a bath 3 filled with a cooling liquid (low-molecular hydrocarbon fluids of different viscosity) 4. In accordance with the characteristics of the welded high-hard material 5, the operating modes of the machine 8 are adjusted to provide the required dimensions, shape and surface cleanliness. In the process of electric erosion machining, current flows to the template electrode tool through the electrode holder. The processing of the welded high-hard material occurs with the translational motion of the template electrode tool, which provides the spindle of the machine. Thus, the surface of the teeth or slots of the synchronizer carriage is restored to the nominal size, desired shape and cleanliness of the surface of the finished part.

The technical result of the proposed methods is to provide the possibility of multiple restorations of the surface of the teeth or slots of the synchronizer carriage without machining to the nominal size, desired shape and surface cleanliness.

The proposed method of restoring the carriages of synchronizers of KAMAZ gearboxes using electric erosion machining is a promising and less resource-intensive method in comparison with the existing recovery methods used. With minor improvements (selection of operating modes of the plant and manufacturing of the necessary graphite electrodes), this method is applicable for the restoration of any parts having spline joints.

References:

[1] Kulakov A T, Denisov A S, Makushin A A 2013 *Features of design, operation, maintenance and repair of power units of trucks:: textbook* (M.: Infra-Inzhenerija) p 448

[2] Kulakov A T 2007 *Increase of reliability of autotractor diesel engines by improving the process of lubrication, cleaning and technology of repair of basic elements*:diss. ... d-r of technical sciences : 05.20.03 p 558

[3] Kulakov A T, Denisov A S 2007 *Maintenance of reliability of autotractor engines* (Saratov: Sarat. state techn. University) p 422

[4] Azamatov R A, Dazhin V G, Kulakov A T, Modin A I 2004 *Restoration of car parts of KamAZ: textbook (*Naberezhnye Chelny: KAMAZ) p 215

[5] Azamatov R A, Denisov A S, Kulakov A T, Kurdin P G 2007 *Restoration of the details of the power unit of KAMAZ 740.11-240 (Euro-1): textbook (*Naberezhnye Chelny: KamAZ) p 306

[6] *Restoring splines in holes* [Electronic resource]. Access mode: http://www.szkv.ru/vosstanovlenie-shlitsev-v-otverstiyah.html - 04.07.2017

[7] Molodyk N V, Zenkin A S 1989 *Restoration of machine parts* (M.: Mashinostroenie) p 480

[8] Kulakov A T, Denisov A S, Kalimullin R F, Kovalenko S Ju, Gribkov K V 2015 *Method for restoring a worn out surface of the slots of the synchronizer carriage* Pat. 2622551 of Russian Federation, MPK B 23 P 6/00, B 23 H 9/12

Development of a multistage gearbox for a new generation of heavy-duty trucks

I R Mavleev, V V Voloshko, I I Salakhov and I F Shaykhutdinov

Kazan Federal University, Kazan, Russian Federation

mirkampi@mail.ru

Abstract. The design of a multistage transmission for heavy trucks (patent No. 2508486) designed to implement the national program of import substitution in transport engineering in Russia. The main technical results implemented in the design of multi-stage gearbox.

1. Introduction

The national policy of import substitution in transport engineering in Russia began to be implemented at the very beginning of the XXI century. And in 2014, when economic sanctions were imposed, the policy of import substitution accelerated markedly and acquired the status of state importance. Large automakers develop their development programs, focusing on the import substitution program. Thus, the program of strategic development of PJSC "KAMAZ" has defined the main task of bringing to the market KAMAZ trucks of a new generation with fundamentally different technical characteristics and consumer properties. Multistage gearboxes will be used in the transmission of this generation of trucks.

Obviously, the designers of multi-stage gearboxes face the following tasks:

- creation of a reliable and rigid gearbox design with smaller dimensions and weight;
- increase the speed and smoothness of the transmission range of the divider without breaking the power flow and without turning off the clutch;
- providing the ability to test separate nodes transmission (differential divider, a main reducer and differential dual), which increases the quality of the overall assembly of the transmission conditions of mass production.

2. Development of a multistage gearbox

Now at Kazan Federal University the design of a multistage transmission with a differential divider and a demultiplikator in which the following technical results are realized is developed and patented:

The old sample divider, formed by a gear drive with a synchronizer, is replaced by a differential divider. Differential divider provides a significant reduction in the axial dimensions of the gearbox, the weight of the secondary and intermediate shafts. Also, the differential divider provides smooth control of the transmission ratio, which follows the design, which causes a reduction in the weight of the gearbox.

The use of a differential divider in the design of the gearbox allows you to switch the range of the divider without switching off the clutch smoothly and without breaking the power flow, which reduces wear and tear of friction clutch parts, increasing the power factor and increasing the fuel efficiency of

Content from this work may be used under the terms of the Creative Commons Attribution 3.0 licence. Any further distribution of this work must maintain attribution to the author(s) and the title of the work, journal citation and DOI.
Published under licence by IOP Publishing Ltd

the engine. At the moment of switching the divider range, which occur four times more often than the gear shift of the main gearbox, the engine does not go into partial load modes.

2. An additional series of toothed gears reverse eliminated by the application of a differential auxiliary gearbox with an integrated reverse gear. The differential demultiplikator with integrated rear transmission additionally reduces the axial dimensions and weight of the secondary and intermediate shafts. The use of an integrated rear-end differential demultiplikator allows the full power of the engine to be transmitted when the vehicle is in operation at different reversing speeds, which is particularly important for special-purpose vehicles.

3. The rigidity of the shafts of the main gear is increased by reducing their axial dimensions, therefore, their moments of inertia are reduced. This contributes to a faster alignment of the angular velocities of the rotating masses, reducing the load on the synchronizers, therefore, less wear blocking rings synchronizers when switching gears and faster gear shifting.

4. Increasing the reliability of the gearbox, since the technological process of its assembly can be carried out after preliminary bench tests of its individual components, namely, the differential divider, the main reducer and the differential demultiplikator, which in turn creates the prerequisites for automating the process of the general assembly of the gearbox.

The design of the multistage gearbox is presented in fig.1.

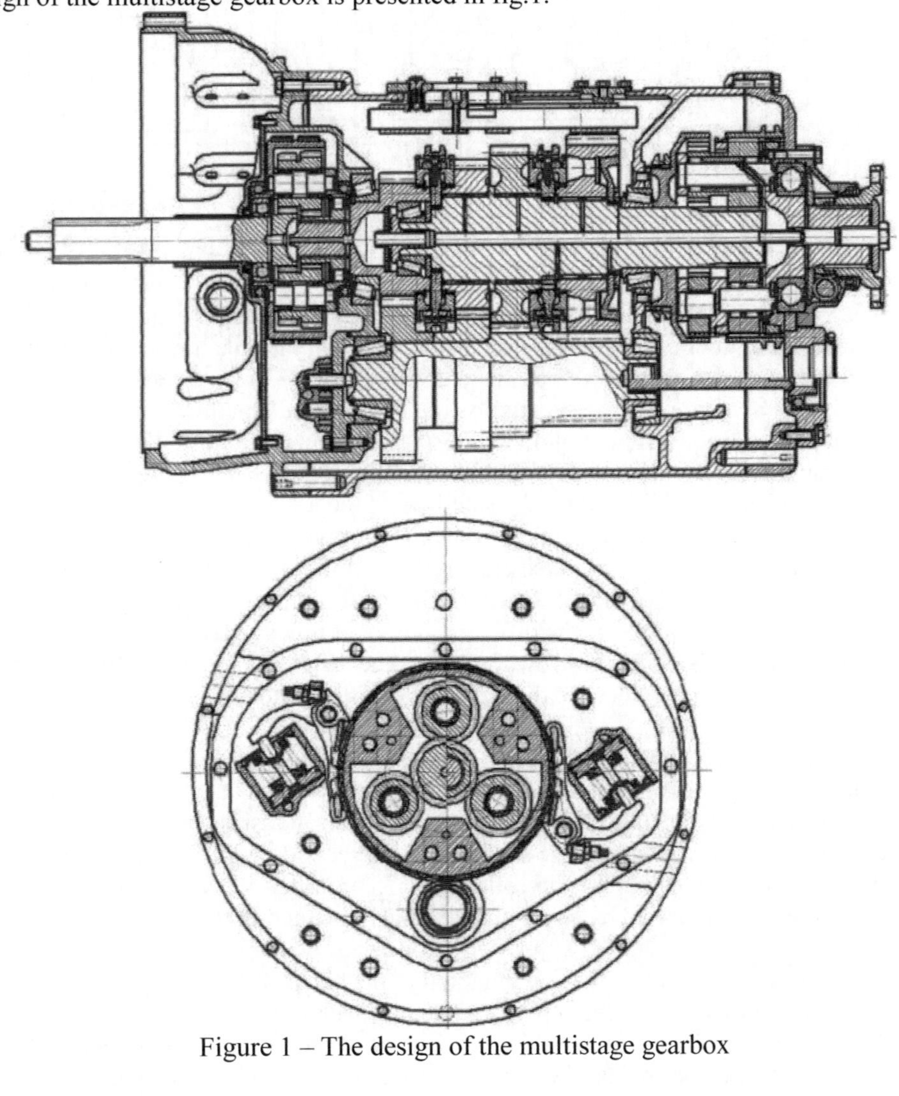

Figure 1 – The design of the multistage gearbox

IOP Conf. Series: Materials Science and Engineering **412** (2018) 012052 doi:10.1088/1757-899X/412/1/012052

2.1. The design of the gearbox.

For fig.2 shows the kinematic scheme of the proposed transmission. The gearbox consists of a differential divider, a main four-speed gearbox and a differential demultiplikator with integrated rear transmission.

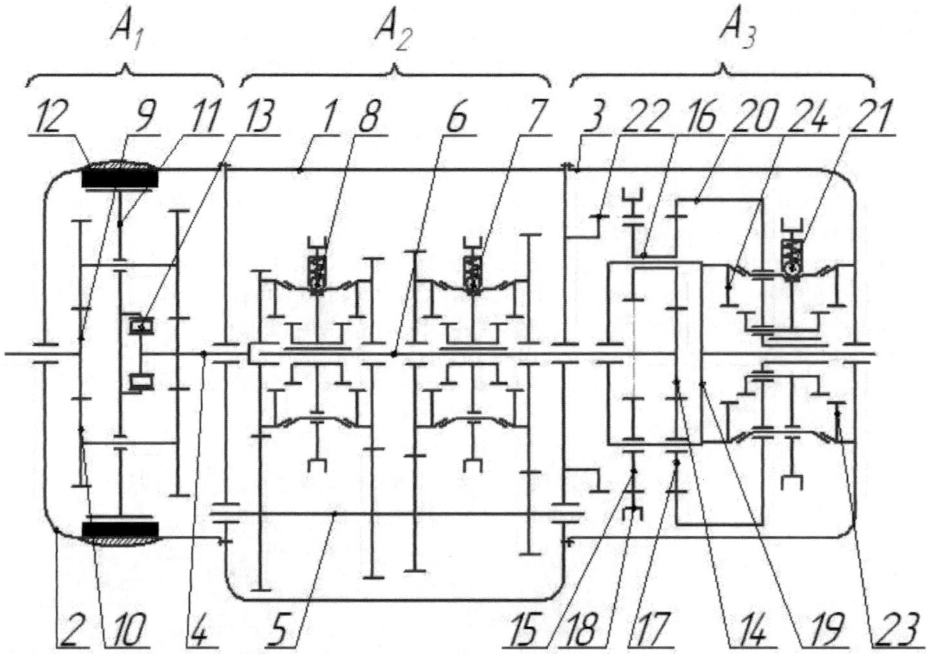

Figure 2 – The kinematic scheme of the multistage transmission: A_1 – differential divider; A_2 – main gearbox; A_3 – differential demultiplikator; 1 – the body of the main gearbox; 2 – clutch housing (crankcase differential divider); 3 – crankcase differential demultiplikator; 4 – the main shaft of the main gearbox; 5 – intermediate shaft of the main gearbox; 6 – the secondary shaft of the main gearbox; 7 and 8 – synchronizers, respectively, the first-second and third-fourth gear of the main gearbox; 9 – input shaft-gear differential divider; 10 – two-crown satellites of the differential divider; 11 – carrier of the differential divider; 12 – band brake; 13 – free-wheeling clutch; 14 – a leading gear ring of the differential demultiplikator; 15 – satellites of the differential demultiplikator; 16 – two-crown satellites of the differential demultiplikator; 17 – revers idler satellites of the differential demultiplikator; 18 – crown wheel (reverse clutch) of the differential demultiplikator; 19 – carrier of the differential demultiplikator (the output shaft of the gearbox); 20 – crown wheel of the differential demultiplikator; 21 – synchronizer of the differential demultiplikator; 22 – ring gear lock the reverse clutch; 23 – toothed crown locking the crown wheel of the differential demultiplikator; 24 – ring gear locking differential demultiplikator

The gearbox contains the body of the main gearbox 1 with holes for fastening the clutch housing 2, which is simultaneously the crankcase of the differential divider, and the crankcase of the differential demultiplikator 3. The body of the main gearbox with clutch housing forms a closed cavity, which contains the main shaft 4, the intermediate shaft 5 and the secondary shaft 6 of the main gearbox. The input shaft is the shaft-gear resting on the roller bearing, installed in the hole of the clutch housing. The intermediate shaft is supported by two roller bearings mounted in the holes of the housing and the clutch housing. The secondary shaft is supported at one end by a roller bearing mounted in the main shaft and at the other end by a roller bearing mounted in the body of the main gearbox.

The differential divider contains the input shaft-gear 9, which is in engagement with the two-crown satellites 10, mounted on the axes pressed into the carrier of the differential divider 11. The carrier of the differential divider rests on the ball bearings installed in the clutch housing and in the cover of the differential divider, and is blocked by the band brake 12. Between the carrier 11 and the main shaft of the main gearbox 4 installed free-wheeling clutch 13.

Differential demultiplikator contains a leading gear ring 14, cut on the secondary shaft of the main gearbox, satellites 15 and two-crown satellites 16, revers idler satellites 17, crown wheel 18, which is simultaneously the reverse clutch, the carrier of the demultiplikator 19, which is simultaneously the output shaft of the gearbox, and the crown wheel 20. On the splines of the hub of the crown wheel mounted synchronizer of the differential demultiplikator 21. The carrier of the demultiplikator rests on a ball bearing mounted in the crankcase of the differential demultiplikator 3. The body of the main gear fixed gear ring 22, designed to lock the crown wheel 18, in the crankcase of the differential demultiplikator fixed toothed crown 23, designed for the locking crown wheel 20, and the carrier fixed ring gear 24 designed to lock the differential demultiplikator.

The 3-D model of the transmission was also developed using the application software of the computer-aided design system. 3-D model of the gearbox is shown in Fig.3.

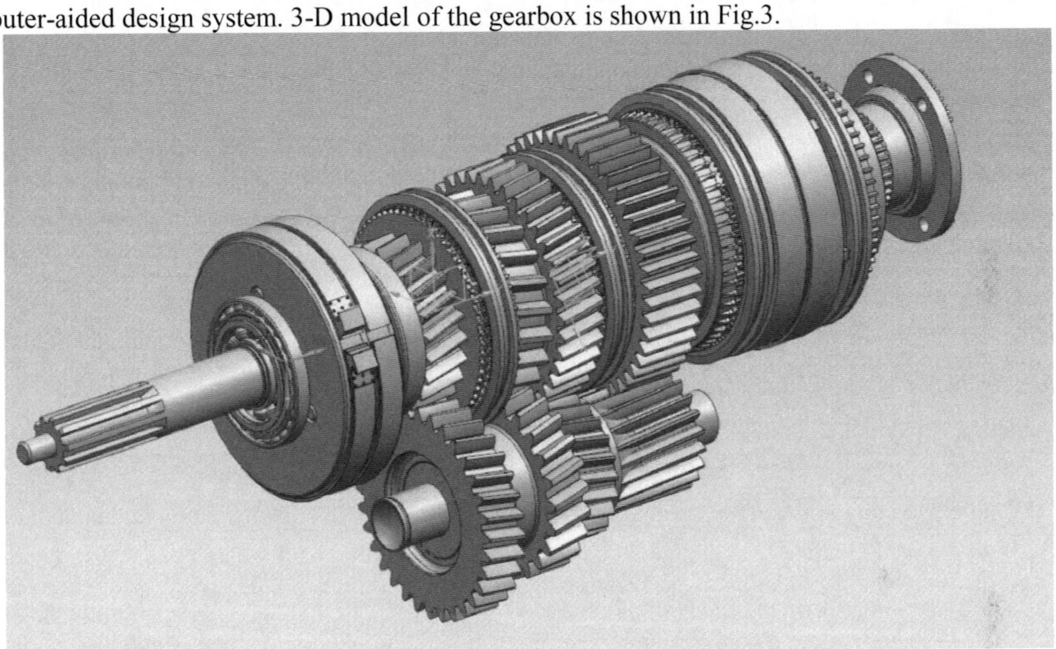

Figure 3 – 3-D model of the gearbox

The offered automobile multistage transmission is competitive, possesses technical characteristics corresponding to requirements of modern automotive industry and completely meets requirements of the state Program of the Russian Federation "Development of the industry and increase of its competitiveness" (the order of the government No. 328 of 15.04.14).

References

[1] Faskhiev H A , Voloshko V V, Mavleev I R , Salakhov I I 2015 *New transmission for vehicles* Handbook. Engineering journal with Appendix No 7 pp 47–51

[2] Voloshko V V , Mavleev I R , Salakhov I I *Automotive multi-stage transmission* Patent. 2508486 RU Application No 2012148057/11

[3] Salaкhov I I, Voloshko V V, Galimyanov I D, Mavleev I R 2015 *Kinematic scheme and design of automatic planetary gear boxes based on a new module* Contemporary Engineering Sciences Vol 8 No 1 pp 1-6

[4] Salaкhov I I, Mavleev I R , Tsybunov E N, Basyrov R R and Salakhov N I 2015 *Car gearbox on the basis of the differential mechanism* Biosciences Biotechnology Research Asia Vol 12 pp 41-44

IOP Conf. Series: Materials Science and Engineering **412** (2018) 012052 doi:10.1088/1757-899X/412/1/012052

[5] Salaкhov I I, Mavleev I R , Shamsutdinov I R, Nuretdinov D I and Salakhov N I 2016 *Development of a gear box of the truck* Biosciences Biotechnology Research Asia Vol 13(2) pp 859-64

[6] Salaкhov I I, Mavleev I R , Voloshko V V, Galimyanov I D, Takhaviev R K 2016 Analysis workflows gear hydraulic machines Biosciences Biotechnology Research Asia Vol 13(2) pp 779-84

[7] Mavleev I R 2007 *Development of efficient schemes and designs high-torque hydromechanical CVTs for vehicles*: Author. dis. Cand. tehn. Sciences. - Naberezhnye Chelny p 19

IOP Conf. Series: Materials Science and Engineering **412** (2018) 012053 doi:10.1088/1757-899X/412/1/012053

Using of functional capability of RTDS hardware and software complex in the educational process of electrotechnical cpeciality master's training in ecomony global digitalization conditions

I S Metelev[1], R G Isakov[1], A V Ferenetz[1], I R Gilmanshin[1,2], S I Gilmanshina[1] , A I Galeeva[1,3]

[1] Kazan national research technical university named after A N Tupolev, K. Marx St. 10, Kazan, 420111 Tatarstan, Russia

[2]Kazan Federal University, 18 Kremlyovskaya street, Kazan, 420008, Russian Federation

[3] Kazan State Power Engineering University, Kazan, 420066, Russian Federation

ivan0893@mail.ru

Abstract. This article deals with problem of renewable enery resource studying during laboratory works of Power Engeneering students (bachelors and masters levels). Hardware and sofware RTDS complex is considered as a base of lab works. Solar panel (PV module) is considered as an example of renewable source. Plan of lab work, which is connectd with practical using, is offered.

Introduction

The renewable energy sources implantation is one of the priority tasks in power supply systems. Many countries gradually refuse of conventional power sources in favor of renewable ones. For example, it is planned to completely abandon of coal-fired power plants in France by 2021 [1], while Britain can cover up to 29% of country's needs in hours of strong winds [2]. Attitudes towards renewable energy sources (RES) are also change in USA (new houses in Miami now will be built with solar panels [3[). According to Finnish scientist's calculations, a complete transition to wind power stations by 2050 is possible [4].

In connection with this, the requirements for electrotechnical students in terms of professional competencies in renewable energy and industry 4.0 are growing. In addition to theoretical course studying, a course of practical and laboratory works, which aimed a practical consolidation of knowledge, are needed.

As a material and technical base for laboratory works carrying out either real stands that simulates power stations work or software (or software and hardware) complexes can be chose. In this paper using of RTDS (Real Time Digital Simulation, RTDS Technologies, Canada) hardware and software complex is proposed. This complex consists of software part – RSCAD software, in which model buildings are done, and hardware part – the cabinet, in which calculations and modeling are done. Software part contains of huge library, which also includes RES.

The structural diagram of laboratory stand is shown on Fig.1.

Content from this work may be used under the terms of the Creative Commons Attribution 3.0 licence. Any further distribution of this work must maintain attribution to the author(s) and the title of the work, journal citation and DOI.
Published under licence by IOP Publishing Ltd

IOP Conf. Series: Materials Science and Engineering **412** (2018) 012053 doi:10.1088/1757-899X/412/1/012053

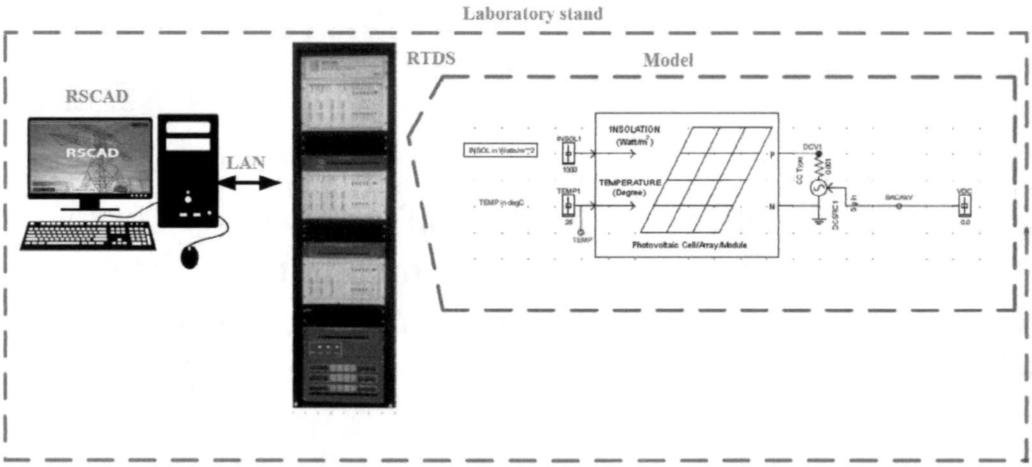

Fig.1. The structural diagram of laboratory stand.

The study of a renewable energy complex based on a solar panel is considered as an example, but it is possible to study wind power stations, energy storage complexes and simulate the work of other power plants in the same way.

Solar panel

The solar panel is a current source with diode connected in parallel and resistors connected in series and in parallel. The model can be constructed in two ways: as a four-parameters model and as a five-parameter model. The difference between them is in a shunt resistance (R_{sh}) (Fig. 2)

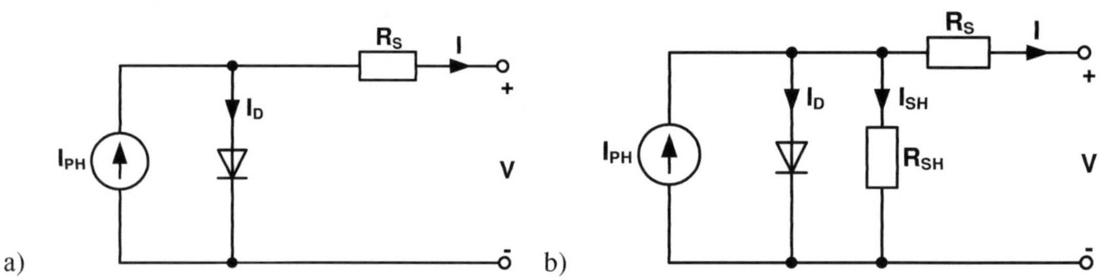

Fig. 2. Four-parameters scheme (a) and five-parameters scheme (b). [5]

The circuit with shunt resistance is more precise, but it requires more complex calculations. To study the basics of solar panel working principles, a simplified four-parameter scheme is quite suitable.

For a such scheme, the equation for current-voltage dependence will be as follows [5]:

$$I = I_{PH} - I_0(\exp(\frac{V + R_S I}{N_C a V_T}) - 1), \quad (1)$$

where I_{PH}- photocurrent generated by a single solar cell, A;

I_0–diode inverse saturation current, A;

N_c- the number of solar cells, connected in series in a solar module,

α – diode ideality factor,

V_t– diode thermal voltage, V.

$$I_{PH} = \frac{G}{G_{ref}}(I_{PHref} + k_i(T - T_{ref})), \quad (2)$$

$$I_0 = I_{0ref} (\frac{T}{T_{ref}})^3 \exp(\frac{E_g}{\alpha V_t}(1-\frac{T_{ref}}{T})), \qquad (3)$$

$$\alpha = \alpha_{ref} (\frac{T}{T_{ref}}), \qquad (4)$$

$$V_t = \frac{kT}{q}, \qquad (5)$$

where
G solar insolation, W/m^2;
k_i– temperature coefficient at short-circuit current, %/^{0};
T - temperature, K;
E_g– photovoltaic element energy gap, eV;
α_{ref}, I_{PHref} – calculated value;
G_{ref} and T_{ref} - insulation and temperature parameters at which solar module parameters were measured (typically T_{ref}= 25^{0}C, G_{ref} = 1000 W/m^3);
q –electron charge, 1,602·10^{-19} C;
k – Boltzmann's constant, 1,38·10^{-23} J/K.
The passport parameters of solar panels are:
V_{oc}– open-circuit voltage, V;
V_m- voltage at maximum power, V;
I_{sc}– short-circuit current, A;
I_m - current at maximum power, A;
k_i– temperature coefficient at short-circuit current, %/^{0}C;
k_v– temperature coefficient at open-circuit voltage, %/^{0}C.
Based on this data, it is possible to calculate the following parameters:

$$I_{PHref} = I_{sc}, \quad (6)$$

$$\alpha_{ref} = \frac{q(2V_m - V_{oc})}{N_c k T_{ref} [\frac{I_{sc}}{I_{sc} - I_m} + \ln(1 - \frac{I_m}{I_{sc}})]}, \qquad (7)$$

$$I_{0ref} = \frac{I_{sc}}{\exp(\frac{V_{OC}}{N_c \alpha_{ref} V_t}) - 1}, \quad (8)$$

$$R_s = \frac{N_c \alpha_{ref} V_t \ln(1 - \frac{I_m}{I_{sc}}) + (V_{oc} - V_m)}{I_m}, \qquad (9)$$

The model of solar panel is contained in the RSCAD library. The appearance of the solar panel model is shown on Fig. 3.

IOP Conf. Series: Materials Science and Engineering **412** (2018) 012053 doi:10.1088/1757-899X/412/1/012053

Input data for this model are solar insolation value (*Insolation*) and temperature (*Temperature*), and outputs are positive and negative DC voltage contacts (*P* and *N*).

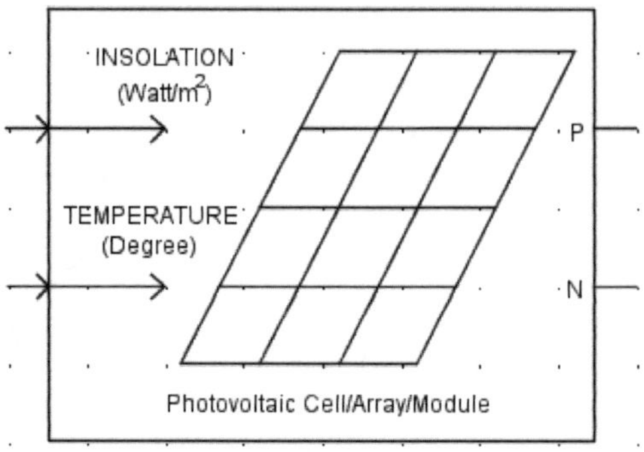

Fig. 3. Solar panel model in RSCAD

Menu of parameters that are needed for modeling is shown on Fig. 4.

a) 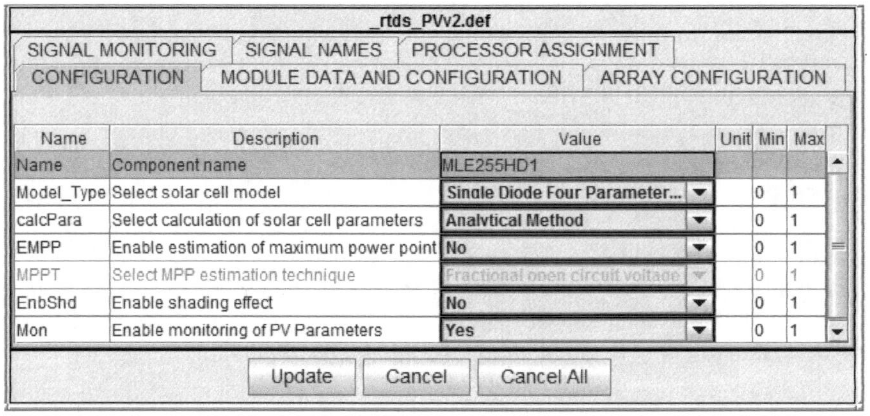

_rtds_PVv2.def

SIGNAL MONITORING | SIGNAL NAMES | PROCESSOR ASSIGNMENT

CONFIGURATION | MODULE DATA AND CONFIGURATION | ARRAY CONFIGURATION

Name	Description	Value	Unit	Min	Max
Name	Component name	MLE255HD1			
Model_Type	Select solar cell model	Single Diode Four Parameter...		0	1
calcPara	Select calculation of solar cell parameters	Analytical Method		0	1
EMPP	Enable estimation of maximum power point	No		0	1
MPPT	Select MPP estimation technique	Fractional open circuit voltage		0	1
EnbShd	Enable shading effect	No		0	1
Mon	Enable monitoring of PV Parameters	Yes		0	1

Update Cancel Cancel All

b) 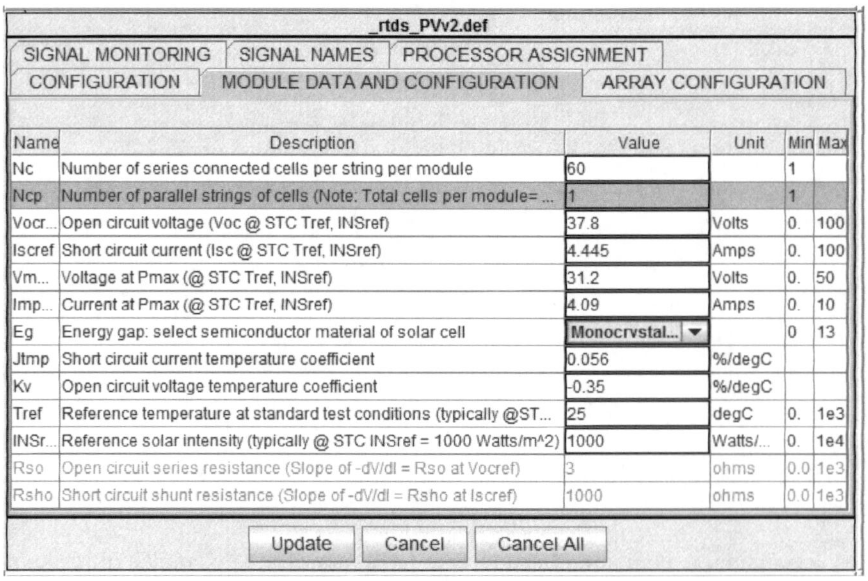

_rtds_PVv2.def

SIGNAL MONITORING | SIGNAL NAMES | PROCESSOR ASSIGNMENT

CONFIGURATION | MODULE DATA AND CONFIGURATION | ARRAY CONFIGURATION

Name	Description	Value	Unit	Min	Max
Nc	Number of series connected cells per string per module	60		1	
Ncp	Number of parallel strings of cells (Note: Total cells per module= ...	1		1	
Vocr...	Open circuit voltage (Voc @ STC Tref, INSref)	37.8	Volts	0.	100
Iscref	Short circuit current (Isc @ STC Tref, INSref)	4.445	Amps	0.	100
Vm...	Voltage at Pmax (@ STC Tref, INSref)	31.2	Volts	0.	50
Imp...	Current at Pmax (@ STC Tref, INSref)	4.09	Amps	0.	10
Eg	Energy gap: select semiconductor material of solar cell	Monocrystal...		0	13
Jtmp	Short circuit current temperature coefficient	0.056	%/degC		
Kv	Open circuit voltage temperature coefficient	-0.35	%/degC		
Tref	Reference temperature at standard test conditions (typically @ST...	25	degC	0.	1e3
INSr...	Reference solar intensity (typically @ STC INSref = 1000 Watts/m^2)	1000	Watts/...	0.	1e4
Rso	Open circuit series resistance (Slope of -dV/dI = Rso at Vocref)	3	ohms	0.0	1e3
Rsho	Short circuit shunt resistance (Slope of -dV/dI = Rsho at Iscref)	1000	ohms	0.0	1e3

Update Cancel Cancel All

Fig.4. Solar panel configuration menu (a) and module parameters menu (b)

IOP Conf. Series: Materials Science and Engineering **412** (2018) 012053 doi:10.1088/1757-899X/412/1/012053

The passport data of solar panel should be used as a menu parameters. For example, parameters of Mitsubishi PV-MLE255HD solar panel are used:

Nc = 60 – number of solar cells connected in series;

N_{cp} = 1 – number of strings with solar cells connected in parallel;

V_{ocref} = 37,8 – open-circuit voltage, V;

I_{scref} = 4,445- short-circuit current, A;

V_{mpref} = 31,2- voltage at maximum power, V;

I_{mpref} = 4,09 – current at maximum power, A;

Material – minocrystalline silicone;

J_{tmp} = 0,056 – temperature coefficient of short-circuit current, %/^{0}C;

k_v = -0,35 – temperature coefficient at open-circuit voltage, %/^{0}C;

T_{ref} = 25 – temperature of parameters measurement, °C;

INS_{ref} = 1000 – insolation, at which parameters were measured, W/m^2.

The main experiments at laboratory works with renewable energy sources are short-circuit and open-circuit experiments, as well as VI characteristics measurements and series and parallel connection of solar panels. Short-circuit and open-circuit experiments can be done more clearly and effectively on a real stands; during the simulation, parameters of open-circuit voltage and short-circuit current are already set.

The experiment of VI-curve drawing of solar panel is easy to implement (Fig. 5). In this case, value of connected resistance can be changing during the simulation.

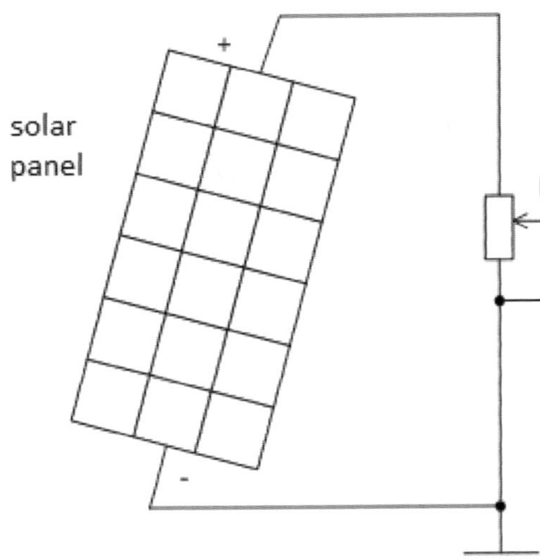

a)

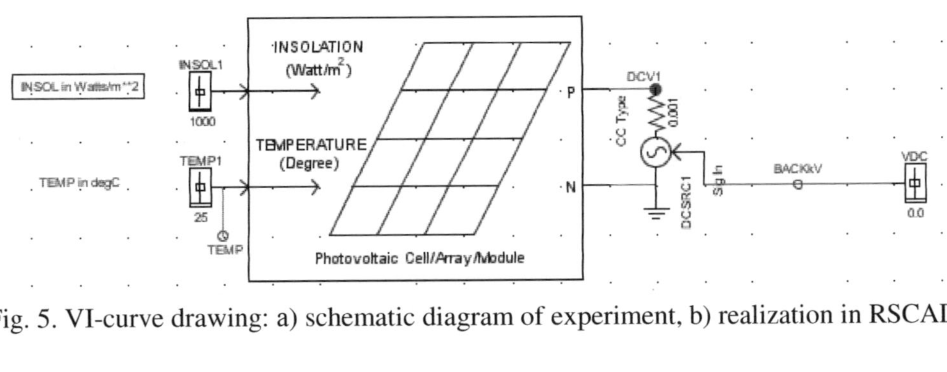

b)

Fig. 5. VI-curve drawing: a) schematic diagram of experiment, b) realization in RSCAD software.

IOP Conf. Series: Materials Science and Engineering **412** (2018) 012053 doi:10.1088/1757-899X/412/1/012053

In the framework of laboratory work on the solar panel studying, students need to calculate VI-characteristic points based on above equations as well as simulate this case on RTDS. Next step is to compare results obtained in the calculations and modeling.

As a result of calculations, the following curve is obtained (Fig. 6):

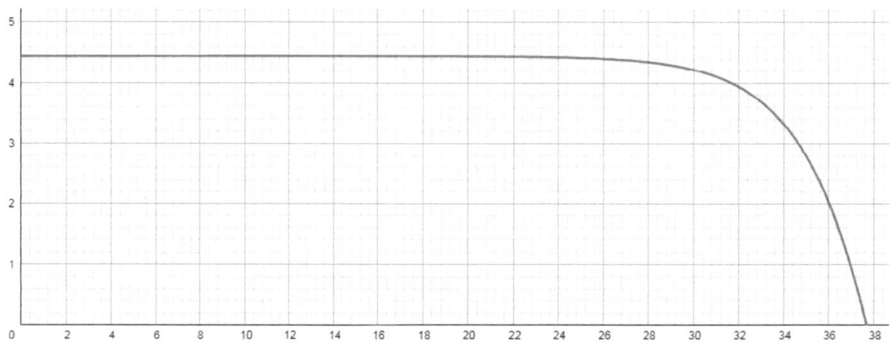

Fig. 6. Solar panel VI-curve, obtained by calculations.

Based on Fig. 5, VI-curve drawing is obtained on RTDS complex. Modeling result is shown on Fig. 7.

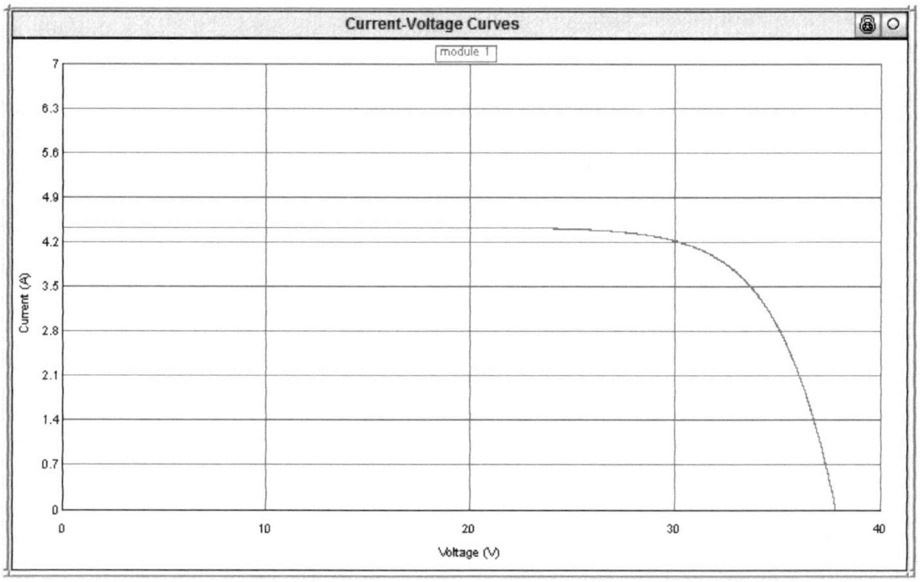

Fig. 7. Solar panel VI-curve, obtained in RSACAD complex.

Un case of solar panel modeling, it is possible to chose parameters of shading conditions of a part of the module, specifying certain parameters (Fig. 8). In the shading parameters menu it is possible to choose will the circuit has a return or/and blocking diode, specify the number of shaded modules and cells, type of shading (constant or varying) and level of isolation on shaded parts.

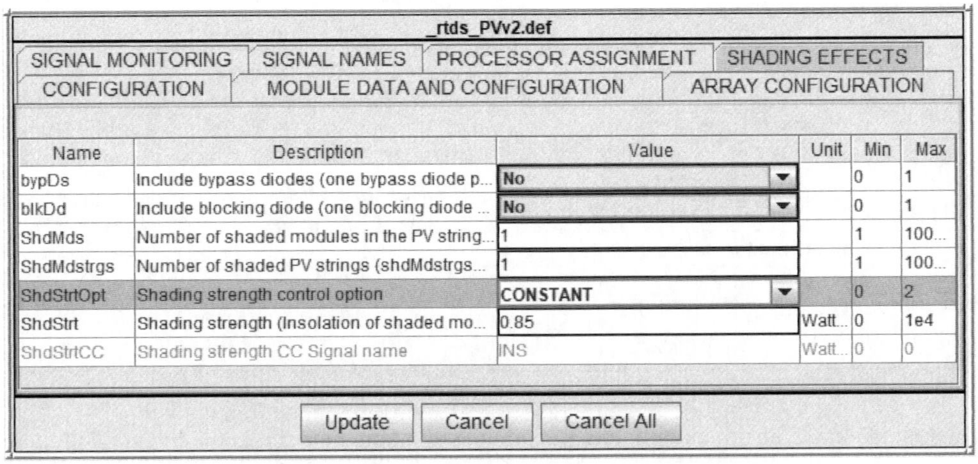

Fig. 8. Shading of the module part parameters.

A typical silicon solar cell generates a voltage about 0.7 V. To achieve the necessary voltage and current levels solar cells are usually connected in parallel and/or in series and creates a solar modules, which generates higher voltage about 16-36 V. For larger voltages (kV) and currents (kA) for practical applications modules can be connected in series and in parallel for create photovoltaic arrays (Fig. 9).

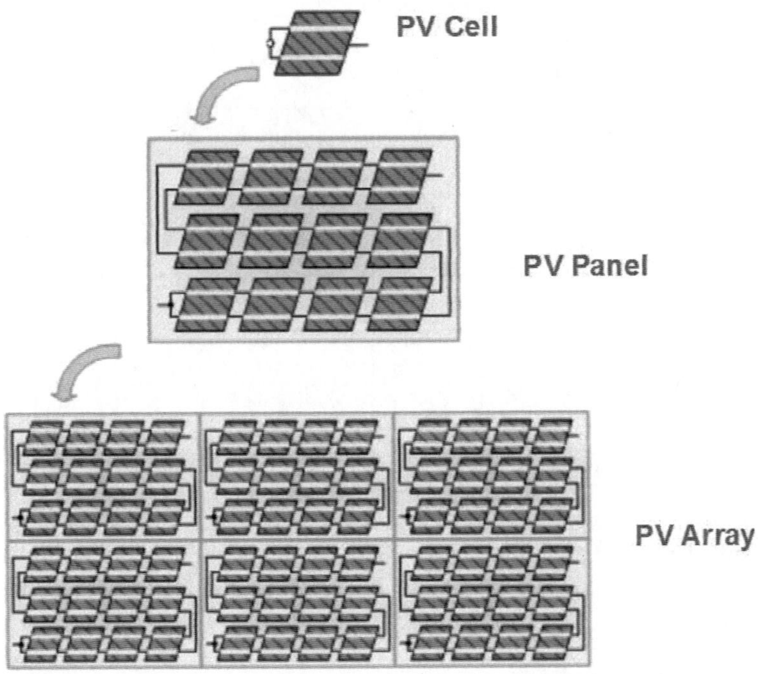

Fig. 9. Photovoltaic arrays.

It is quite easy to implement experiments on series and parallel connection of solar modules: for this case it is enough to set the numbers of connected modules(N_s – number of modules connected in series, N_p – number of modules connected in parallel) in Array Configuration menu (Fig. 10).

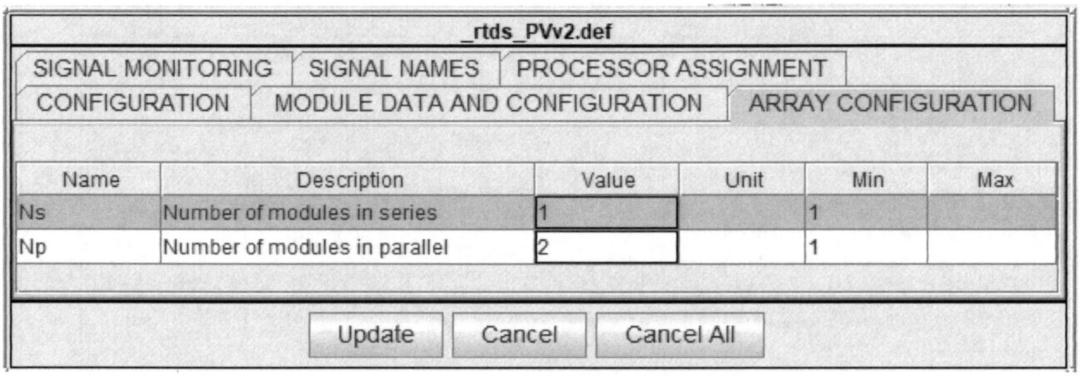

Fig. 10. Array Configuration menu

As a result, it is possible to compare the VI-characteristics of arrays with different number of modules. Four cases are shown in Fig. 11.

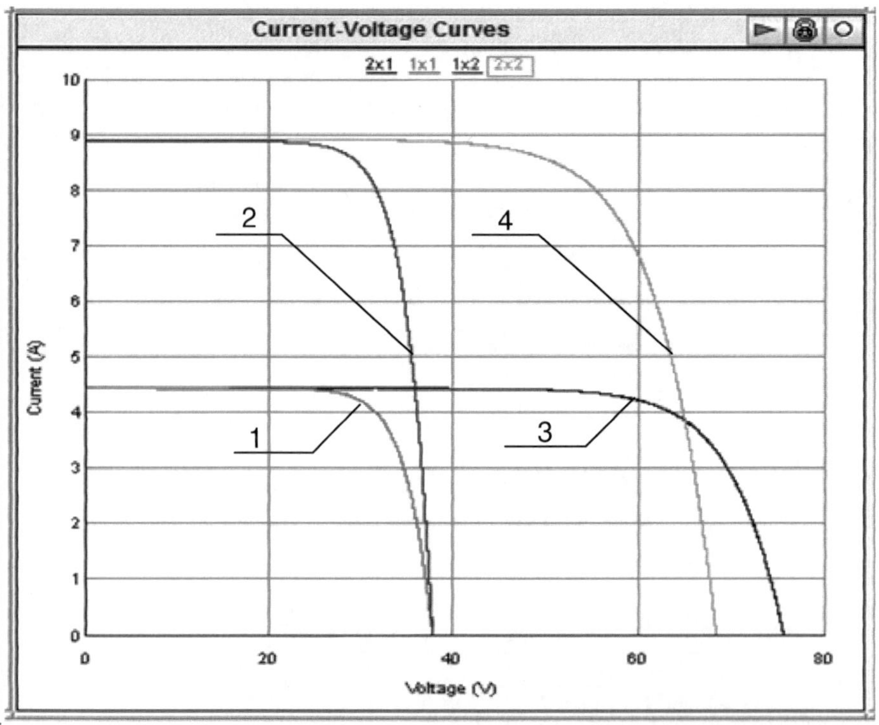

Fig. 11. Series and parallel connection of solar modules: 1- single module; 2 - two modules connected in parallel; 3 – two modules connected in series; 4 - array consisted of two connected in parallel and two connected in series modules.

Such experiments can show work of PV-array modules and will perfectly fit into the studying process of power energy students in bachelor's level. For master level such experiments are too easy.

Proposed laboratory stand

To improve the studying quality of solar panels working principles, the following laboratory experiment's items can be added:

- automatic change of parameters during the day (insolation and temperature) according with statistic data for a specific region;
- modeling of secondary equipment (inverter, battery, battery control system, etc);
-setting the schedule of the load in district, enterprise or household load.

Thus, this laboratory works will allow the following studies:

1. Select or set load schedule in the enterprise, shop or household load;
2. Select a schedule of solar insolation change for a specific area. Such data can be found in free access in the Internet (for example, on weather forecast website);
3. Calculate the parameters of solar array (power, dimensions, number of modules, connection scheme, etc.). In this case geoparameters are also can be teken into account (for example, if it is possible to use only 100 m^2 of area for solar panels);
4. Choose the real model, suitable for calculated parameters;
5. Calculate and choose the secondary equipment (inverters, batteries, etc.);
6. Create model in RSCAD software and test the system's operation. In case if the power generated by renewable energy system is not enough, it is possible to consider the joint operation of the solar system and centralized power source;
7. Perform a feasibility study of the payback period of the project, make conclusions about the economic possibilities of solar panel installation for power supply system.

Additional interest of students can be stimulated by connection of calculations with real objects. Most of students have a job, so they can use parameters from the enterprise, organization, etc. as an input data for calculations.

In the future, it is possible to add one or more other renewable energy sources. Thus, this laboratory work can be a good base for master thesis.

Conclusion

Using of RTDS hardware and software complex is a good basis for a renewable energy sources studying. The advantage of modeling is a combination of theoretical knowledge about RES's main characteristics with practical application.

Despite the fact that solar panel was considered as an example is this article, the principle for studying other renewable sources remains the same: the study of the source under conditions of variable load, variable input parameters, study and connection of additional equipment, such as invertors, batteries, controllers, etc.

This approach will increase the student's interest in the learning process that will lead to calculations and modeling for real power supply facilities and will became a good basis for bachelor's and master's theses.

Also, researches in the field of renewable sources will speed up the process of introducing of this sources in power supply facilities, and will also lead to the possibility of creation electric power systems based on MicroGrid concept.

It is planned to introduce the described laboratory works for *Kazan National Research Technical University n.a. A. N. Tupolev– KAI* Master's student for specialty 13.04.02 on the profiles "Electrical equipment and electric economy of organizations, enterprises and institutions" and the joint program of German-Russian Institute of Advanced Technologies (GRIAT) "Electrical Engineering and Informational Technologies".

References

1. *Independent. France to shut all coal-fired power stations by 2021, Macron declares*. Avaliable at: https://www.independent.co.uk/news/world/europe/france-coal-power-station-emmanuel-macron-davos-shut-2021-a8176796.html (accessed 22 June 2018).
2. *Clean technica. UK Wind Power Output Breaks Record, Tops 13 Gigawatts*. Available at: https://cleantechnica.com/2018/01/18/uk-wind-power-output-breaks-record-tops-13-gw/ (accessed 21 June2018)

3. Digital substation. RES come, Power Supply System without subsidies, IKEA and solar blinds [*Digital Substation. VIE nastupaiyt, SES bez subsidii, IKEA i solnechnye zhaliuzi*].Avaliable at: http://digitalsubstation.com/blog/2017/08/05/vie-nastupayut-ses-bez-subsidij-ikea-i-nbsp-solnechnye-zhalyuzi/. (accessed 21 June2018)
4. Smirnov A 2017 Hiteck. Full transition to renewable energy will occur by 2050. *[Haitek. polnyi perehod na voozobnovlyaemyyu energetiku proizoidet k 2050 godu]* Available at: https://hightech.fm/2017/11/09/renewable-energy. (accessed 22 June 2018) (in Russian)
5. RTDS manuals
6. Nzimako O 2015 *Real-Time Simulation of a Microgrid System with Distributed Energy* Resources Department of Electrical and Computer Engineering Faculty of Engineering University of Manitoba 156 pp
7. Isakov R G 2015 *Using of RTDS hardware and software complex for relay protection and automation devices testing* Energetika Tatrstana [*Tatarstan's energetic*] No 2(38) pp 28-31
8. Mohammed A. Hajahmed 2015 *Protection and Automation of Microgrids for Flexible Distribution of Energy and Storage Resources* Dissertation Presented in Partial Fulfillment of the Requirements for the Degree Doctor of Philosophy in the Graduate School of The Ohio State University. The Ohio State University
9. Kirianova N G 2014 Matematicheskoe modelirovanie istochnikov raspredelyonnoi generatsii v sostave virtualnoi elektrrostancii. VKR bakalavra [*Mathematical modeling of distributed generation sources in virtual power station, Bachelor Thesis*] Novosibirsk, Novosibirsk State Technical University
10. Sharifov B N, Teregulov T R 2015 Solar panel modeling in Matlab Simulink *Vestnik UGATU* [Herald of Ufa State Aviation Technical University] book19 No 4 (70) pp 77-83

IOP Conf. Series: Materials Science and Engineering **412** (2018) 012054 doi:10.1088/1757-899X/412/1/012054

Study of the solubility of die-cast steels in silumin AK12 (AL2)

G F Mukhametzyanova, M S Kolesnikov and I R Mukhametzyanov

Kazan Federal University, Naberezhnye Chelny Institute, 423812, Russia, Naberezhnye Chelny, Prospekt Syuyumbike 10A

gulnara-ineka@mail.ru

Abstract. The results of the research of solubility of die-cast steels in silumin AK12 (AL2), depending on the chemical composition, method of refining, thermal and chemical-thermal treatment. Perspective marks of steels for the manufacturing molds pressure die-casting aluminum alloys are established. The expediency of application for the press- tools of cast alloys, nitriding and methods of arc-plasmous and electroslag remelting on the example of steel DI – 22 is shown.

The processes of die-casting, solid-liquid stamping are characterized by high parameters of the cyclic temperature and force loading of the press-tool and are accompanied by active interaction of press-tools with melts in liquid or solid-liquid states. For the aluminum alloy die-casting press-tool (ALDC) the high physical and chemical activity of aluminum melts with iron-carbon alloys leads to dissolution of the surfaces of the elements of the press-tool, reducing the resistance of the central inserts, splitters, bushes, pressing chamber working surfaces to 3000 – 5000 cycles. The maximum temperature at pressing aluminum alloys can reach 700^0C at a technological temperature of the filled alloy of 640 – 680^0C. The material of ALDC molds also undergoes a significant force action resulting from the pressure of the pressed alloy up to and above 500 MPa [1].

The purpose of this work is to select the die-steels, methods of remelting and hardening for molds of aluminum die-casting under pressure on the solubility index in silumin.

To perform the study we selected the following die-steels: 3H2V8F, 10H4SM4N4TYU, 10H6M5N9YU2, 5HNM, 4H8V2, 35H3V3MF, 28H5M2F, 25H3V3F, 4H2SVMFYU, 4H5M2FS, 5H2MFSD, Kh12M, 3H2V5F, 25H10V7M2K10, 4H3VMF, 4H5V2FS, 5H4SV4MF, 20H12N20F2MT3YUB, 4H4VMFS (DI - 22). Steel 4X4BMФC (DI – 22) was tested in cast and hammered state after electric arc melting (DI – 22EM), and also after electroslag (DI – 22ESR) and plasma arc remelting (DI – 22PAR). The quenching temperature ($T_{harding}$) of the samples was assigned from the conditions of obtaining the maximum hardness while retaining fine grains of austenite (8 – 10 points in accordance with GOST 5639 – 82). The tempering temperature (T_{leave}) was determined experimentally to provide a hardness of 44 – 46HRC. During the tests the temperature of the melt (600 - 700^0C) and the duration of the interaction of the stamping steels with the melt (up to 12 minutes) varied.

Studies of the solubility of steels in silumin, AK12 (AL2) were carried out by immersing specimens of die- steels in the melt and providing hydrodynamic interaction due to cyclic oscillation of samples with a frequency of 100 s^{-1} and an amplitude of 10 mm. For the tests we used cone samples with the following dimensions: length, l = 30 mm; diameter, d = 4mm; taper, Δ 15^0, radius of curvature of cone tops, r = 0,02mm. During the tests the maximum speed of sample movement in silumin was about 2,0 m/s, which corresponds to the contact conditions of the silumin melt on the surface of

Content from this work may be used under the terms of the Creative Commons Attribution 3.0 licence. Any further distribution of this work must maintain attribution to the author(s) and the title of the work, journal citation and DOI.
Published under licence by IOP Publishing Ltd

ALDC molds. As a criterion for estimating the solubility, a change in the sample length, Δl, mm, is accepted.

The results of the experiments are shown in Fig. 1 and 2. Analysis of the data presented indicates a significantly different resistance to dissolution of the investigated die-cast steels in the melt AK12 (AL2). The dissolution rate interval at the melt temperature of 640^0C ranges from 1,0 mm/min (steel 10H4SM4N4TYU, Fig.1, a, curve 2) to 0,09 mm min (steel 25H10V7M2K10, Fig.1, b, curve 14). The increase in the melt temperature of AK12 (AL2) causes the increase in the solubility of all the steels studied. The greatest resistance to solubility is noted in steels with the increased chromium content (Figure 2, b, curves 10, 14) and wolframite (Fig. 2, b, curve 14). Molybdenum and silicon also have a positive effect. The influence of nickel does not play any positive role (Fig. 2, a, curves 2, 4).

The effect of nitriding of die-steels on the solubility index in the melt of AK12 (AL2) has been studied using the example of steels 3H2V8F and 4H3VMF. Nitriding was carried out according to the $T_{nitriding} = 540^0$C, medium - dissociated ammonia (40% dissociation degree), duration – 20 hours, layer depth – 0,18 - 0,22mm, layer hardness – 880 - 900HV. The results of this series are shown in Fig. 1, d and Fig. 2, d. As it follows from the comparison of options 1 and 15, the average rate of dissolution of nitrided steels is characterized by record low values of ~ 0,04 – 0,06 mm/min and is practically independent of the chemical composition of the steels studied.

Influence of methods of smelting, plasma arc and electroslag remelting of die-steels on the solubility in the melt of AK12 (AL2) was investigated for steel 4H4VMFS (DI – 22). The solubility of steels after electroslag remelting (ESR) and after plasma-arc remelting (PAR) is 2-3 times lower than after arc melting.

It has been established that the 4H3VMF, 5H4SV4MF, 4H5V2FS and 4H4VMFS (DI – 22) die-steels are promising for the manufacture of the press- tools of cast alloys for high resistance to dissolution in the melt of silumin; nitriding of die-cast steels increases the resistance to dissolution in the melt of silumin, that is the advantage of die-steels in comparison with the deformed ones.

IOP Conf. Series: Materials Science and Engineering **412** (2018) 012054 doi:10.1088/1757-899X/412/1/012054

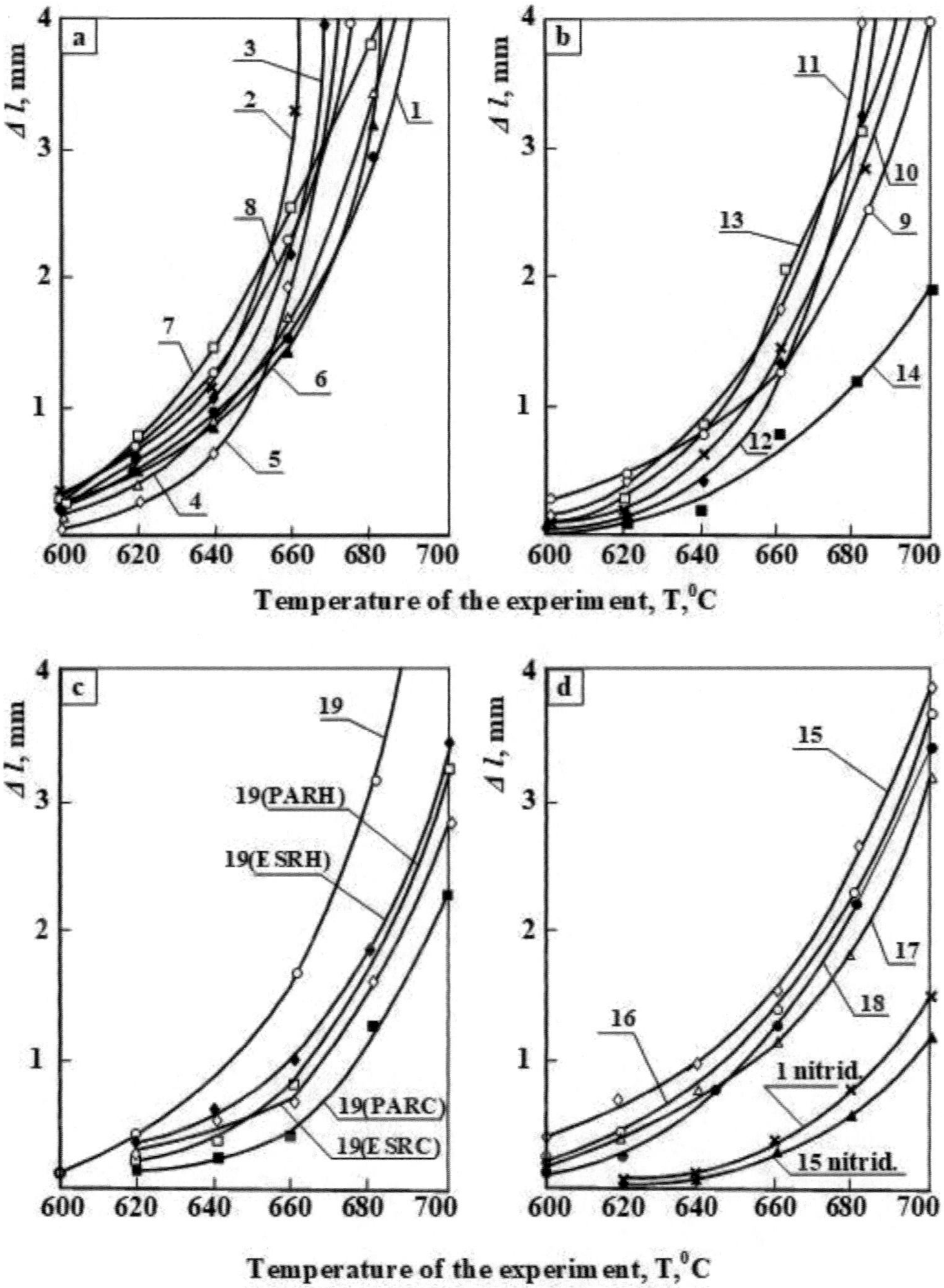

Fig. 1 – Influence of the chemical composition, the method of remelting and nitriding on the solubility ($\Delta\, l$, mm) of steels in the melt of AK12 (AL2) depending on the temperature of the melt. The duration of the interaction, τ = 3 min. 1 – 3H2V8F, 2 – 10H4SM4N4TYu, 3 – 10H6M5N9Yu2, 4 – 5HNM, 5 – 4H8V2, 6 – 35H3V3MF, 7 – 28H5M2F, 8 – 25H3V3F, 9 – 4H2SVMFYu, 10 – 4H5M2FS, 11 – 5H2MFSD, 12 – H12M, 13 – 3H2V5F, 14 – 25H10V7M2K10, 15 – 4H3VMF, 16 – 4H5V2FS, 17 – 5H4SV4MF, 18 – 20H12N20F2MT3YuB, 19 – 4H4VMFS (DI – 22, ESRH – hammered after electroslag remelting, ESRC – cast after electroslag remelting, PARH – hammered after plasma-arc remelting, PARC – cast after plasma-arc remelting).

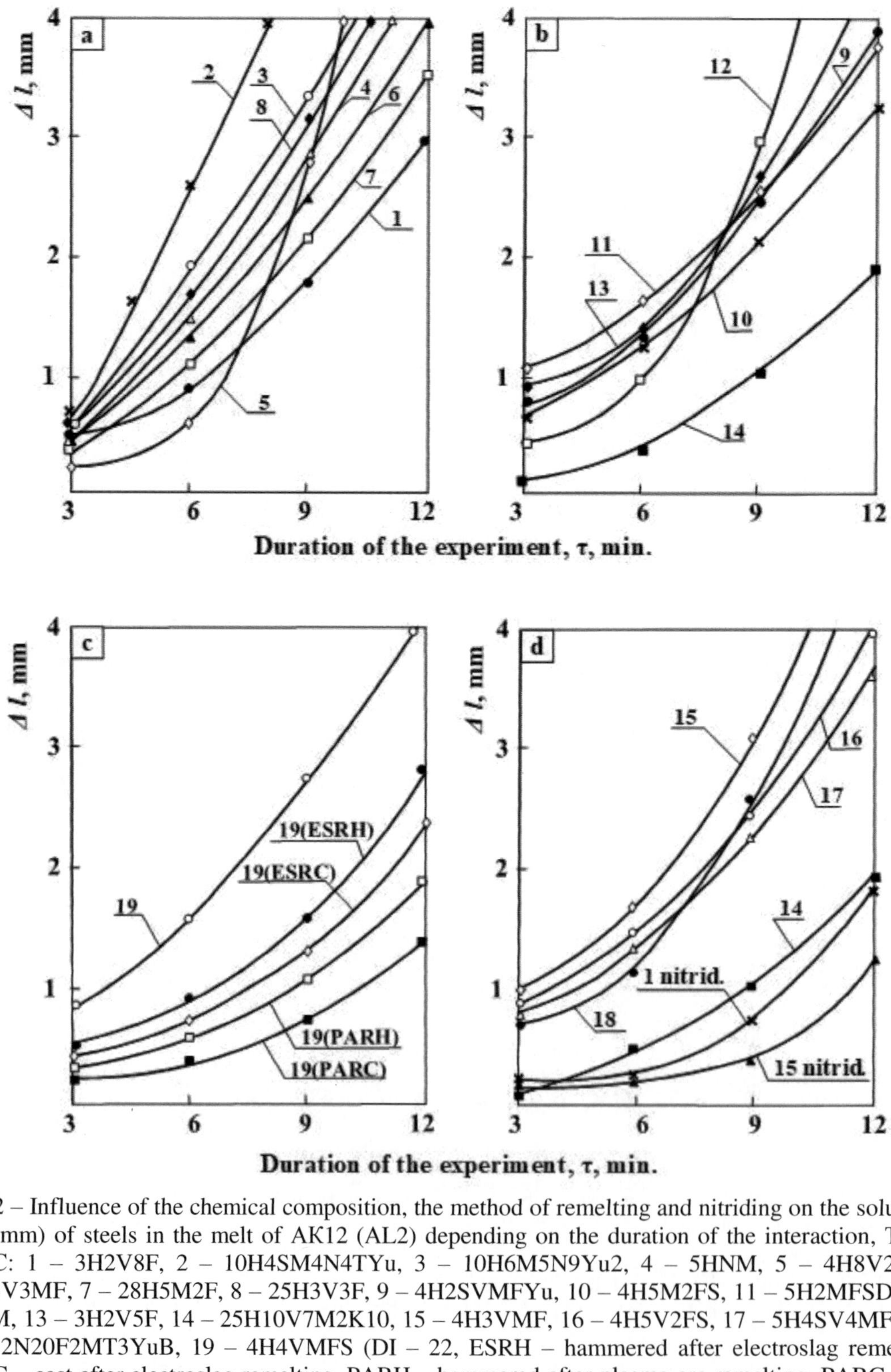

Fig. 2 – Influence of the chemical composition, the method of remelting and nitriding on the solubility (Δl, mm) of steels in the melt of AK12 (AL2) depending on the duration of the interaction, $T_{melt} = 640^0C$: 1 – 3H2V8F, 2 – 10H4SM4N4TYu, 3 – 10H6M5N9Yu2, 4 – 5HNM, 5 – 4H8V2, 6 – 35H3V3MF, 7 – 28H5M2F, 8 – 25H3V3F, 9 – 4H2SVMFYu, 10 – 4H5M2FS, 11 – 5H2MFSD, 12 – H12M, 13 – 3H2V5F, 14 – 25H10V7M2K10, 15 – 4H3VMF, 16 – 4H5V2FS, 17 – 5H4SV4MF, 18 – 20H12N20F2MT3YuB, 19 – 4H4VMFS (DI – 22, ESRH – hammered after electroslag remelting, ESRC – cast after electroslag remelting, PARH – hammered after plasma-arc remelting, PARC – cast after plasma-arc remelting).

IOP Conf. Series: Materials Science and Engineering **412** (2018) 012054 doi:10.1088/1757-899X/412/1/012054

References:

[1] Goryunov I I 1973 Die-casting molds Reference manual L: Mashinostroyeniye 256 p

IOP Conf. Series: Materials Science and Engineering **412** (2018) 012055 doi:10.1088/1757-899X/412/1/012055

Spectral analysis of acoustic car noise

P V Pchenitchnyi[1], V A Strekalov[1], R R Tagirov[1], R R Shaimukhametov [1]

[1]Kazan Federal University, Kazan, 420008, Russian Federation

pavel.pchenitchnyi@kpfu.ru

Abstract. The results of the Fourier spectroscopy of the acoustic noise of an internal combustion engine are presented. The sound was recorded using a voice recorder placed near the engine under varied operating conditions for the four-cylinder petrol engine. The distinctive characteristics of the periodograms where associated with the certain physical processes in the engine.

The sound field is determined by the various physical processes of the car engine. Extraction of useful information from acoustic data allows solving a number of applied problems, in particular, by the nature of acoustic noises to diagnose faults of the car [1], to identify the car by sound. An adequate model of noise makes it possible to improve the ergonomic parameters of the car, at the present time methods of active suppression in the car interior are being developed [2, 3].

The variety and interdependence of sources of acoustic noise causes a complex structure of the sound field. Nevertheless, the frequency of processes in the engine gives hope for the effectiveness of spectral methods of analysis [4].

This paper presents the results of the study of the acoustic noise of the Chevrolet Cruze engine (petrol, 120 hp, in-line, 4 cylinders). Signals were recorded using a voice recorder installed under the hood of the car near the engine cylinders (frequency range 200-7000 Hz, signal-to-noise ratio 37 dB), to eliminate the influence of the AGC of voice recorder on the signal strength determination, a generator was used, at the sound level of which normalization was performed.

The data was recorded for a variety of engine operating modes (city mode, highway, at various engine shaft speeds). Records for spectral analysis had a sampling frequency of 16,000 Hz, periodograms were built on the basis of 2-second realizations (during this time, mechanical details provided by the inertia of the stationarity of the spectral characteristics), energy spectra were calculated with the help of FFT (2048 points, Hamming window).

Content from this work may be used under the terms of the Creative Commons Attribution 3.0 licence. Any further distribution of this work must maintain attribution to the author(s) and the title of the work, journal citation and DOI.
Published under licence by IOP Publishing Ltd

Figure 1 shows a typical periodogram of engine sounds while driving a car along the highway. To the left is the energy spectrum in conventional units (normalization to the maximum), on the right in a logarithmic scale.

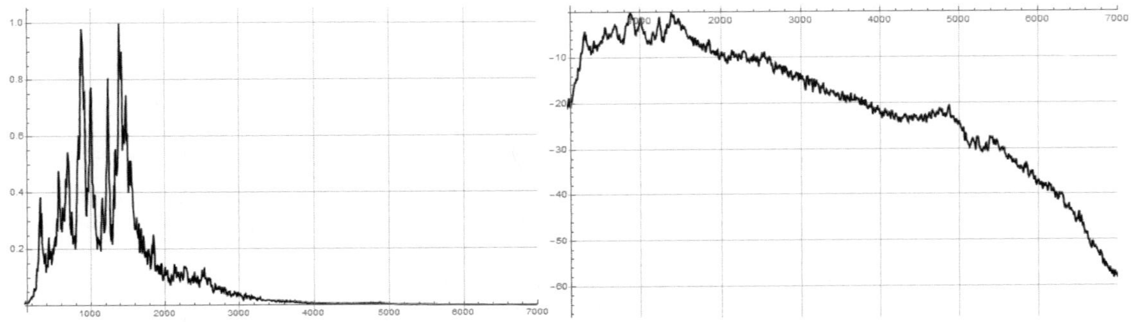

Figure 1. Typical energy spectrum of noise in linear and logarithmic scales.

The width of the spectrum depended on the engine speed of the crankshaft. Figure 2 shows the energy spectra at rotational speeds of 900 and 2300 rpm. On average, the spectrum width at a power level of 0.5 (3 dB) was 1 kHz (band 600-1600 Hz).

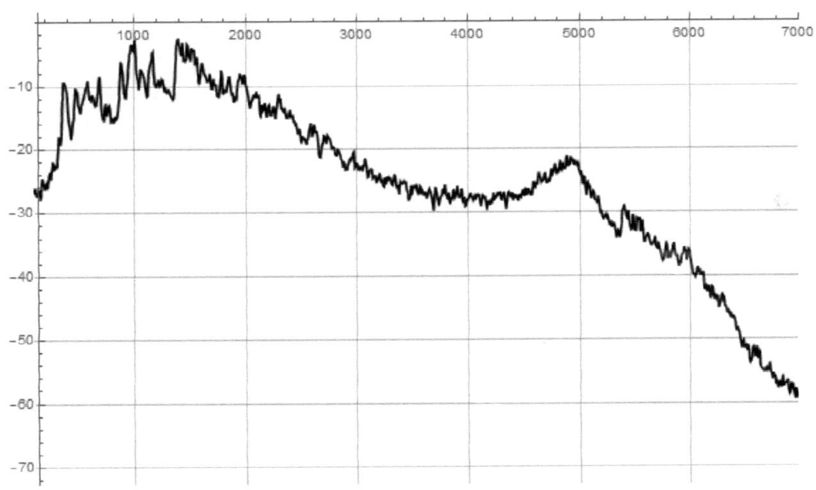

Figure 2. Dependence of the shape of the spectral density on the frequency of rotation of the motor shaft.

A characteristic feature of all spectra is the presence of a local maximum at a frequency of about 5000 Hz, which is clearly visible at a logarithmic scale on the sound power axis (Figure 3).

IOP Conf. Series: Materials Science and Engineering **412** (2018) 012055 doi:10.1088/1757-899X/412/1/012055

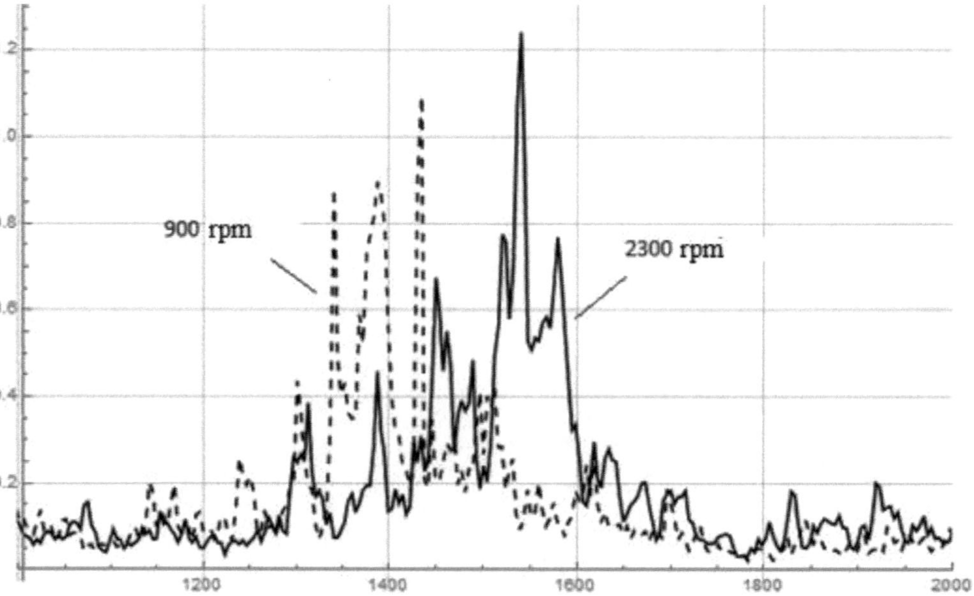

Figure 3. Energy spectrum of acoustic noise in a logarithmic scale.

To compare the sound in different modes of engine operation, a sound generator was installed at the power level under the hood. Figure 4 shows the periodograms at a motor speed of 2300 rpm and 900 rpm, normalized to the power of the sound generator (this procedure eliminated the effect of automatic adjustment of the recorder's gain). It can be seen from the figure that when the engine speed is increased, the sound power is greatly increased.

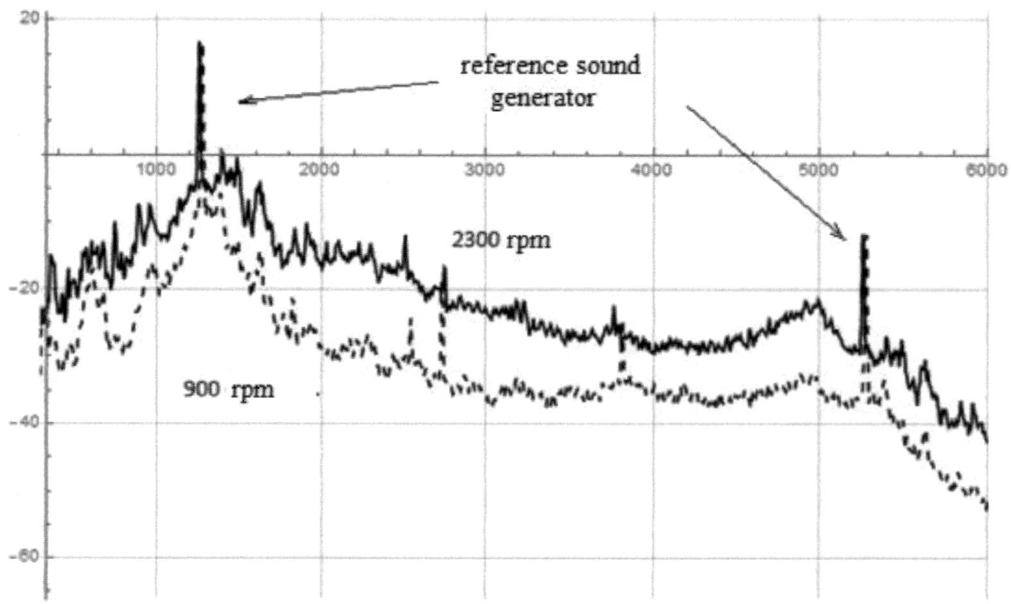

Figure 4. Normalized to the sound level of the reference oscillator periodograms.

IOP Conf. Series: Materials Science and Engineering **412** (2018) 012055 doi:10.1088/1757-899X/412/1/012055

In general, experimental data show that the main energy of the engine sound is concentrated rather in a narrow frequency range. Perhaps the main generation of sound is determined by the combustion processes of the fuel, mechanical oscillations of the pistons have too low frequency to generate the recorded sound, but they can modulate higher-frequency components and be observed as harmonics.

An additional study requires the identification of the nature of the local maximum at a frequency of 5000 Hz. Suitable parameters for generating such a high-frequency sound are detonation waves, hypothetical partial detonation of the fuel-air mixture can cause this component of the spectrum.

References

[1] Dreizin V E, Kasem M M 2009 *Possibilities of diagnostics of automobile engines by analyzing the noise of a running engine* Izvestiya Yugo-Zapadnogo Gosudarstvennogo Universiteta No 2 (27) p 32

[2] Elliott S A 2008 *Elliott, ISVR Technical Memorandum* Technical Report No 981

[3] Hasegawa S, Tabata T, Kinoshita A and Hyodo H 1992 *The development of an active noise control system for automobiles* SAE Technical Report No 922086

[4] Grigoriev E A 2001 *Periodic and random forces acting in the reciprocating engine* (M .: Mechanical Engineering) 272 p

IOP Conf. Series: Materials Science and Engineering **412** (2018) 012056 doi:10.1088/1757-899X/412/1/012056

Surface modification of molds for titanium casting, and its influence on the allocation of chemical elements in boundary layers of molds

E J Petrov, A I Khatmullina, A M Lyahovich

Kazan Federal University, 18 Kremlyovskaya street, Kazan, 420008, Russian Federation

Ezio19@mail.ru

Abstract. Influence of titanium sputtering on the inner surface of the casting mold on the chemical elements' allocating in the boundary layers of the mold is considered. It is established that the sputtering of titanium does not exclude the migration of mold's chemical elements through the titanium to the casting surface.

1. Introduction

One way to reduce the formation of an alpha case during the casting of titanium is the usage of protective technological coatings, for example Al2O3, on the inner surfaces of molds to prevent the formation of an alpha layer. These coatings provide protection against corrosion, decarburization, burnout of alloying elements, surface gas saturation of pieces and stock materials of various configurations and dimensions.

Protective technological coatings allow to save metal up to 8% by eliminating oxide scale and improving the surface quality of parts and semi-finished goods.

Objective: to research boundary layers of the molds used for the casting of titanium, after the deposition of protective coatings on the inner surfaces of the molds.

2. Experimental

We analyzed mold composition: MgO (83.2%), CaO (2.9%), SiO2 (1.9%), Na2O (SiO2) n (12%). The calculated content of chemical elements Mg: O: Si: Ca: Na (at.%) in the form is 41.9 : 50.3 : 2.7 : 1.0 : 4.1. The mold was sintered at a temperature of 1000 ± 50 ° C for 6 hours. Casting mold was saturated with carbon dioxide to grasp its components before sintering. Then it's cooling for 10 hours in the furnace. After excavation from the furnace, a Ti protective layer was applied to the mold by an electric arc method. There were two samples: a form with a deposited Ti (figure 1left) and a form after contact with the casting (figure 1right). Research of the relief, the elemental composition of the mold surface and cross-section were carried out using "EVO 50 XVP" (Carl Zeiss) Scanning Electron Microscope with a probe microanalysis system "INCA Energy-350".

3. Results and discussion

Several surface regions of the form with Ti coating are shown in figure 2 (left) and four cross-section regions are in figure 2 (right). An analysis of the titanium allocating map over the cross-section (figure 2) shows that the thickness of its compact allocation is 200-300 μm. It follows from figure 2 (left) and table 1 that titanium is distributed unequal on the surface. In areas where quantity of titanium is low,

Content from this work may be used under the terms of the Creative Commons Attribution 3.0 licence. Any further distribution of this work must maintain attribution to the author(s) and the title of the work, journal citation and DOI.
Published under licence by IOP Publishing Ltd

the content of the elements is close to original mold without coating [1]. Nevertheless, where titanium is dominant, there are elements from the original mold, whence it follows, the elements of the form Mg, Ca, Si, Na migrate through the titanium layer to the surface.

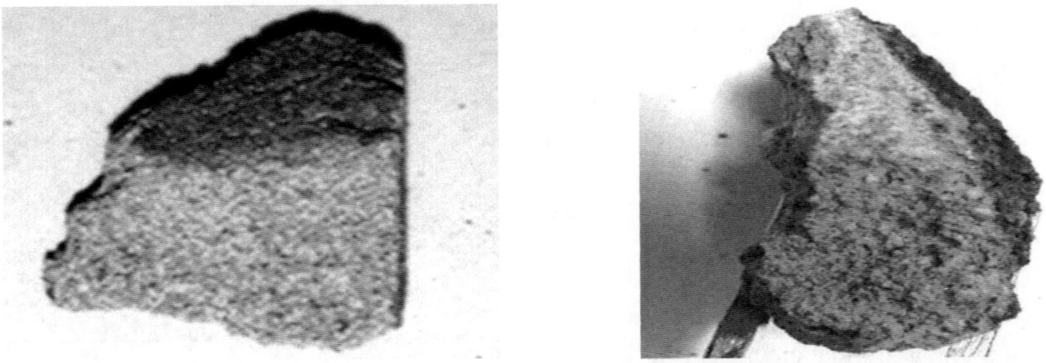

Figure 1. Samples of the mold with a Ti layer (left) and after contact with the casting (right).

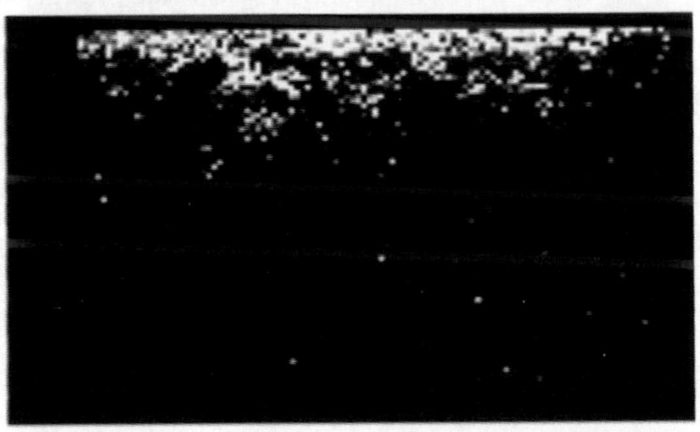

Figure 2. Images of the investigated regions of the surface (left), the cross section (right) and the Ti allocating map of the Ti layer.

According to the data in Table 1, the oxygen content on the surface of the mold exceeds the required amount for the maximum oxidation of the elements. Perhaps it happens due to the formation of carbonic acid salts, which is formed by the reaction of carbon dioxide with water. In addition, we found a large number of pores located throughout the volume in the sample.

Research of the dependence in the chemical elements profile on the depth of the cross section of the sample (figure 3) and the allocating map of Ti (figure 2) shows that the maximum intussusception of titanium into the mold reaches 800 μm. At a distance of 1.5 mm from the surface, the elemental composition of the modified mold approximates the composition of the original mold.

Table 1. Element composition (at.%) of the mold surface with Ti coating.

Elements	Regions of the surface			Original form
	Average	1	2	
O	66,17	61,80	64,76	59,3
Mg	8,66	25,01	4,58	26,6
Si	1,37	5,76	0,77	6,4
Ca	0,57	1,83	0,24	3,2
Ti	21,93	2,06	29,03	-
Na	1,30	2,90	0,61	3,5
O*	55,83	42,28	64,42	42,6

* In case of maximum oxidation of elements.

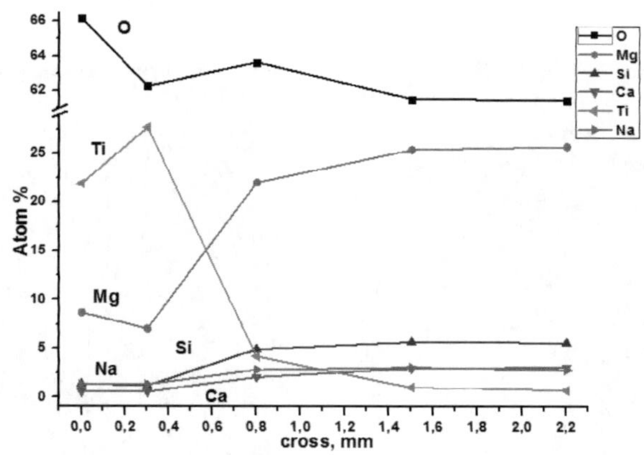

Figure 3. Profile of the elements content in the depth of the cross section of the mold with Ti coating.

After contact of mold with titanium casting, four regions of the separation surface and 8 sections of the cross section were examined. We found several structures on the surface of the mold detachment (figure 4), despite their difference, as follows from Table 2, the first and second structures are close in composition.

Table 2. Element composition (at.%) of the mold surface with Ti coating after peeling off the casting.

Elements	Regions of the surface			
	Surface	1	2	3
O	62,03	64,68	65,24	57,51
Mg	25,89	28,88	28,47	39,44
Si	1,35	1,73	1,32	1,33
Ca	2,61	2,81	3,16	1,40
Ti	2,54	1,91	1,81	0,32
O*	36,28	38,97	37,89	44,14

* In case of maximum oxidation of elements.

IOP Conf. Series: Materials Science and Engineering **412** (2018) 012056 doi:10.1088/1757-899X/412/1/012056

Perhaps this is due to the presence of a similar amount of titanium in both structures. The third structure is close to the composition of the original mold without Ti coating in terms of elemental composition [1]. Study of the cross section showed a pored structure (figure 5 left).

By the content of titanium on the detachment surface (<3%), it can be concluded that separation of the mold from the casting passes along the boundary of sputtered titanium. The homogeneous composition of the form begins at a depth of ~ 800 μm (figure 5 right).

Figure 4. Images of surface structures of the mold with titanium coating after peeling off the casting.

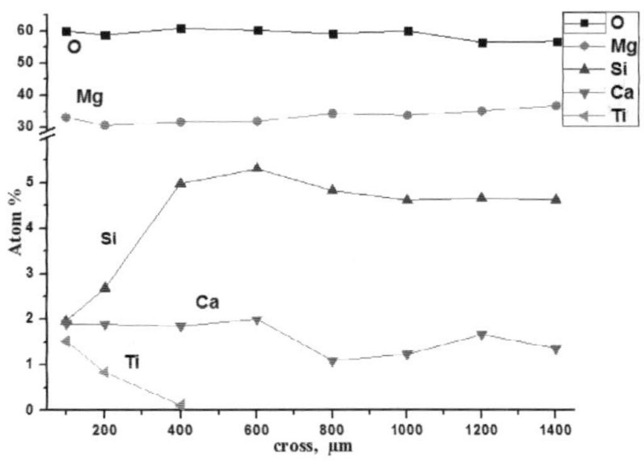

Figure 5. Image of the cross section (left) and profile of elements content by the depth of section (right) after the mold has been detached from the casting.

4. Conclusion

Sputtering of titanium on the surface of the casting mold does not exclude the migration of mold elements through the layer of titanium. After casting, the peeling of the mold from the casting passes along the boundary of sprayed titanium.

5. References

[1] Khatmullina A I, Petrov E J and Lyahovich A M 2017 *8th Int. Science and Technical*

Conf.(Kazan) vol 1 (Kazan: OOO Folianth) pp 357-361

Acknowledgements. This work was funded by the subsidy allocated to Kazan Federal University for the state assignment in the sphere of scientific activities (1.1300.2017/4.6).

IOP Conf. Series: Materials Science and Engineering **412** (2018) 012057 doi:10.1088/1757-899X/412/1/012057

Current state and prospects of development of diagrammatic methods for the calculation of reinforced concrete structures

O V Radaikin[2], L S Sabitov[1,3], N F Kashapov[1], I R Gilmanshin[1,4]
[1]Kazan Federal University, 18 Kremlyovskaya street, Kazan, 420066, Russian Federation
[2] Kazan State University of Architecture and Civil Engineering, Kazan, 420043 Russia
[3] Kazan State Power Engeneering University, Kazan, 420008, Russian Federation
[4] Kazan national research technical university named after A N Tupolev, K. Marx St. 10, Kazan, 420111 Tatarstan, Russia

l.sabitov@bk.ru

Abstract. The article presents the current state and prospects of development of diagramm methods for the calculation of reinforced concrete construction. A brief review obtained by the authors results in this area, where great attention is paid to the improvement of curvilinear diagrams of concrete for different types of static loading of structures. Topical application of the considered methods for the analysis of structures in the zone of joint action of bending moments and shearing force that is not yet reflected in domestic and foreign regulations. In conclusion, the publication identifies ways for further research in the chosen area.

Due to the rapid development of computer technology and information technology, it has become possible in recent years to widely introduce into the engineering practice the design of reinforced concrete structures for nonlinear calculation methods. Among them, the most widely used are diagrammatic methods, using diagrams of deformation of materials in direct form. In 2003, the diagram method called "nonlinear deformation model" entered into the Russian norms - SNiP 52-01-2003 and further in SP 63.13330.2012. Thus, a significant step was taken to harmonize the domestic and European Norms, where the diagrams of the deformation of concrete and steel reinforcement as a calculated basis have been in existence since 1978 [1]. Nevertheless, in Russia in the modern form, the nonlinear deformation model and its physical relationships were first proposed in [2] in the 80s of the last century, and in [3] this model was developed with reference to the calculation of core structures in increments. The development and improvement of the diagram methods and the nonlinear deformation model in particular in our country were dealt with by Baikov V.N., Bondarenko V.M., Bondarenko S.V., Gvozdev A.A., Guscha Yu.P., Dodonov M.I., Dykhovichi_A.A., Il'in O.F., Karpenko N.I., Karpenko S.N., Lazovsky D.N., Lemysh L.L., Mitasov V.M., Murashev V.I., Mukhamediev T.A. ., Rastorguev B.S., Sapozhnikov M.A. and others. Among foreign researchers, Leonhard F., L. Saennz, B. Sinha, P. Desayi, S. Krisnuan, K. Gerstle, L. Tulin, Kabeila and many others were engaged in this issue. etc. As the literature analysis and own studies [4-14] show, the diagrammatic methods allow calculating strength, deformability and fracture toughness on the basis of a unified approach that makes it possible to

Content from this work may be used under the terms of the Creative Commons Attribution 3.0 licence. Any further distribution of this work must maintain attribution to the author(s) and the title of the work, journal citation and DOI.
Published under licence by IOP Publishing Ltd

describe the stress-strain state of structures at all stages of loading - from the beginning of work up to destruction - taking into account the physical nonlinearity of materials and the development of cracks. Moreover, the authors obtained a material where, for the first time using the nonlinear deformation model, the calculation of the width of the opening of normal cracks in bent elements was brought to the final result and compared with the experimental data [15].

In deriving the calculated relations of diagrammatic methods, the equations of equilibrium of internal and external forces are used, the compatibility conditions for deformations in the form of a hypothesis of flat sections (in reinforced concrete elements with cracks in the interpretation of V.I. Murashev and improved by A.A. Gvozdev in relation to the average deformations of reinforcement and concrete in the area between the cracks) and the deformation diagram (in some sources - the state diagram) of the materials. The solution of the system of defining equations thus obtained in most cases is possible only numerically with the use of a computer, for example, by the method of stepwise iterations. For this purpose, the design sections of reinforced concrete elements are divided into elementary areas, and the stresses within each of them are averaged.

Most often, state diagrams relate the stresses and relative deformations arising in the material under simple uniaxial loading-stretching or compression (for example, «σ_{bt}-ε_{bt}» and «σ_b-ε_b» for concrete in Fig. 1).

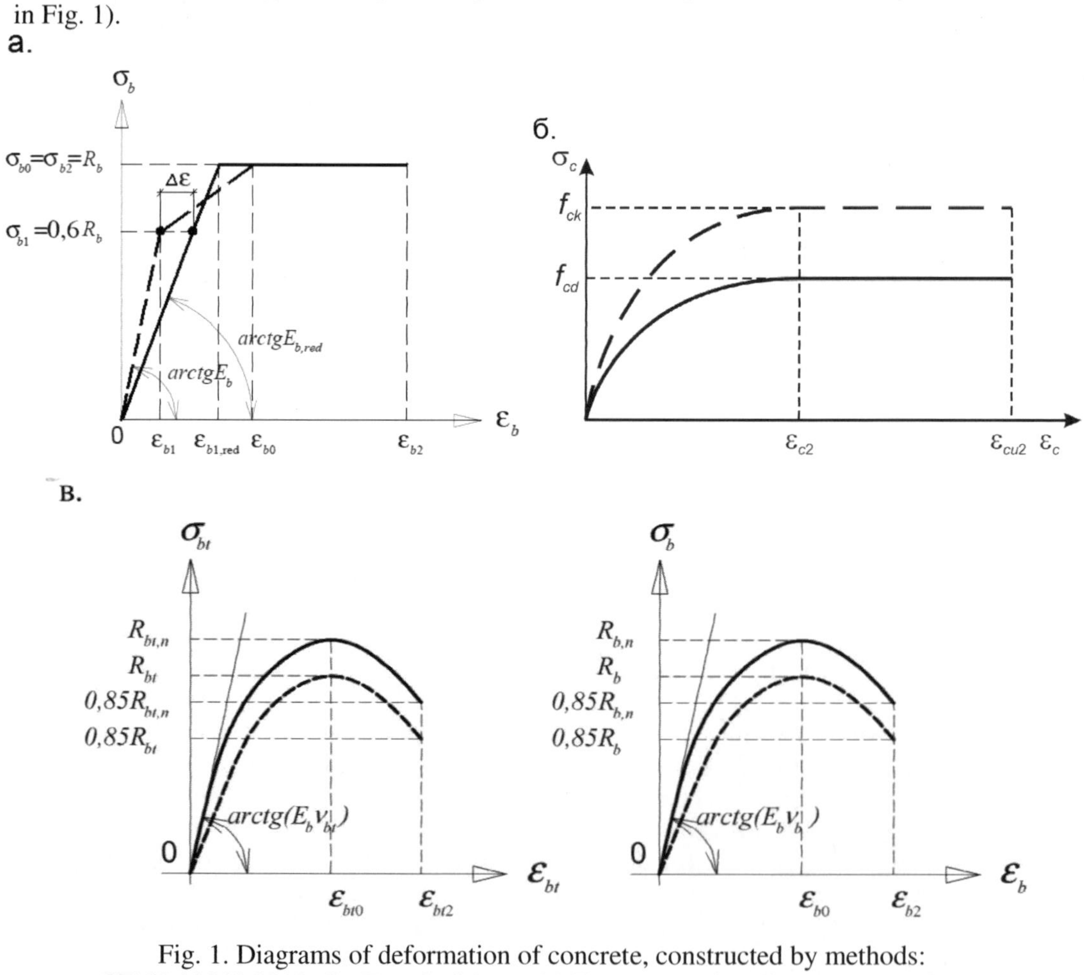

Fig. 1. Diagrams of deformation of concrete, constructed by methods:
а – СП 63.13330.2012; б – EuroCod-2; в – [4,7] (—normative diagrams, --- calculated)

More rarely, the diagrams obtained for two- and three-axis loads [16] are used. For example, in the norms of the Republic of Belarus - SNB 5.03.01-02 "Concrete and Reinforced Concrete Structures" - they are used in the calculation of a compressed-stretched concrete strip between inclined cracks.

In addition, load-balancing and deformation (moving) diagrams are used: for example, «*N*-Δ» for centrally compressed reinforced concrete columns, «*M-f*» or «*M-1/r*» for curved reinforced concrete beams. In this case, the work of the structure as a whole, and the material from which they are made to treat as a homogeneous composite with averaged in volume or cross-section properties, is estimated from them. In "Recommendations for calculating the skeletons of multi-storey buildings, taking into account the compliance of the node interfaces of prefabricated reinforced concrete structures" (TsNIIpromzdaniy, 2008), these diagrams are included.

Below are the data on the topic under consideration, obtained by the author together with prof. Sokolov BS and prof. Karpenko N.I.

At the first stage of the research, the question was solved which diagrams to give preference, since the results of the calculation depend on their type. In [4-6], the diagrams proposed by Academician N.I. Karpenko [16] and included in SP_63.13330.2012, as most fully responding to experiments. Taking into account the new theoretical and experimental data obtained in recent years, improvements have been made to them - in the expressions describing the diagrams, corrective coefficients have been chosen, which made it possible to approximate the calculated values of the cracking moment M_{crc} and the breaking moment M_{ult} in bent ferro-concrete elements, obtained by the diagram technique and by formulas SNIP 2.03.01.84 *. The choice of this SNiP as a "benchmark" for comparison is justified by the fact that this technique has passed more than 30 years of experimental testing and proved to be reliable in practice. After that, in [5], the results of the calculation of deformations of bent ferro-concrete elements are compared using the deformation model and various types of deformation diagrams of concrete and steel reinforcement, including those proposed by the authors. For comparison with the "standard" - with SNIP 2.03.01.84 * - three calculated approaches to the construction of material state diagrams are adopted: SP 52-01-2003; Eurocode-2; own theoretical developments. It is found that the discrepancy between the results in determining the stiffness of the cross-section, the moments M_{crc} and M_{ult} by the compared methods depends on the type of material state diagrams adopted, the concrete strength class, the reinforcement percentage and the loading level. The reasons for the discrepancy are analyzed, which consist in the difference between the analytical prerequisites for constructing diagrams and in the difference between the approaches of domestic and foreign Norms to determining the strength of concrete for tension and compression. When the rigidity of the bent ferro-concrete elements changed with increasing load, better convergence with SNiP 2.01.03-84 * showed the results obtained by the proposed method, which, as established in [6], also has satisfactory convergence with the experimental data of KA. Piradova, S.S. Vatagina and TA. Mukhamediyeva.

For the purpose of differentiated application in the calculation of reinforced concrete elements for the first and second groups of limiting states, the diagrams of concrete deformation are presented in [7] on the basis of unified mathematical expressions in the form of normative and calculated (working) ones. The transition between them was carried out by introducing the reliability coefficients for concrete: $\gamma_{bt}=R_{bt,n}/R_{bt}=1{,}5$ и $\gamma_b=R_{b,n}/R_b=1{,}3$.

In [8], the moment of crack formation in bent ferro-concrete elements was determined using a nonlinear deformation model with an inhomogeneous distribution of deformations along the height of the stretched zone of the element, which was achieved by introducing into the calculated expressions a coefficient that takes into account the gradient of deformations. This allowed obtaining the values of the sought-for moment close to the procedure of SNiP 2.03.01-84 *.

The second stage of scientific research was the consideration of eccentrically compressed concrete elements. In [9], the analysis of the strength, rigidity and fracture toughness of such elements was analyzed using a nonlinear deformation model using various diagrams of concrete deformation. On the basis of a unified approach, all stages of the work of elements are considered - from the beginning of loading up to destruction. To estimate the influence of the longitudinal force on the stress-strain state of eccentrically compressed elements, bending elements having analogous physicomechanical and geometric characteristics were also calculated. It is established that the type of diagrams considered does not practically affect the final result when determining the bending moments of crack formation and fracture, but it affects the value of the calculated stiffness of the cross section. In addition, it was found that the influence of the longitudinal force on the rigidity, especially at the initial stages of loading, as well as on the breaking bending moment, is significant and depends on the percentage of reinforcement and the eccentricity of the longitudinal force. It is concluded that the concrete state diagrams obtained by the authors earlier for bent elements can be used for calculation of eccentrically compressed ones.

The third stage of the research is related to the consideration of the work of reinforced concrete bending elements in the zone of joint action of bending moments (M) and shearing forces (Q). It should be noted that, according to Russian standards, the strength of bent elements in the "shear span" for the action of M and Q is carried out separately, despite the experimentally established mutual influence of these power factors on the VAT of the structures. In addition, in most of the known calculation methods for bent structures, one assumes the assumption of flat sections as one of the prerequisites, which, as experiments show, do not occur with the appearance of inclined cracks in the "shear span". Preliminary work [10] in PC "Ansys" carried out computer simulation of the stress-strain state of bent concrete and reinforced concrete elements in the considered zone, taking into account cracks both normal and inclined to the longitudinal axis. These data served as the basis for the development of a technique for calculating oblique sections using the nonlinear deformation model [11, 12]. For this purpose, using the mathematical analysis software of MathCAD, an algorithm for calculation of concrete and reinforced concrete beams has been developed that takes into account the loading scheme, nonlinear properties of concrete and reinforcement, and the appearance and development of normal and inclined cracks. The algorithm allows us to use the diagrams of the state of concrete obtained by the authors earlier.

Generalizing the obtained data, an attempt was made in [13, 14] to develop, using a nonlinear deformation model, a unified methodology for calculating reinforced concrete bending elements for two groups of limiting states, both for normal and inclined sections. Comparison of the results of the numerical calculation with the experiment showed a satisfactory coincidence of the data - a difference of no more than 10%.

Further, the generalized technique was supplemented by the possibility to calculate the crack opening width a_{crc} from the value of the average deformations in the stretched reinforcement, which was done for the first time using the nonlinear deformation model [15]. Comparison of the calculation results with the experiment showed that the proposed procedure allows to reliably determine the value of a_{crc}.

At the moment, the research will continue. - An attempt is made to consider the influence of different loading schemes of structures, the shapes of their cross sections, the percentage of longitudinal reinforcement, the type of material deformation diagrams, the joint development of normal and inclined cracks, the presence of prestressed reinforcement, the redistribution of forces in statically indeterminate structures, the calculation of pipe-, steel-, fiber-reinforced concrete elements, as well as using diagram methods to study the work of reinforced concrete elements reinforced with composite reinforcement.

Conclusions:

1. The present state of the development of diagrammatic methods for calculating reinforced concrete structures is highlighted. A brief review of the results obtained by the author in this area is given, where great attention is paid to improving curvilinear diagrams of concrete state in relation to various types of static loading of reinforced concrete structures: pure bending, eccentric compression, joint action of bending moments and shear forces. It is shown that the diagrammatic methods make it possible to calculate the strength, deformability and fracture toughness on the basis of a single approach, which makes it possible to describe the stress-strain state of structures at all stages of loading - from the beginning of work up to destruction - taking into account the physical nonlinearity of materials and the development of cracks.
2. The ways of further research on the chosen direction are outlined. The data obtained will serve as the basis for the development of a software package for the automated calculation of reinforced

IOP Conf. Series: Materials Science and Engineering **412** (2018) 012057 doi:10.1088/1757-899X/412/1/012057

concrete structures with various reinforcement schemes and loading applications using the diagram method.

References

[1] Code-sample of ECB-FMP / For the norms on SCR. - T.11. - Euro-International Committee for Concrete. - Moscow: 1984. - 263 p

[2]Karpenko N I, Mukhamediev T A, Sapozhnikov M A 1987 *To the construction of a technique for calculating the rod elements on the basis of material deformation diagrams* Perfection of methods for calculating statically indeterminate reinforced concrete structures pp 4-24

[3]Karpenko S N 2010 *On the construction of connections between the increments of stresses and deformations on the basis of various diagrams* Bulletin of Civil Engineers No 1 (22) pp 60-63

[4]Karpenko N I, Radaykin O V 2012 *To improve the diagrams of concrete deformation for determining the cracking and fracturing moment in bent ferro-concrete elements //* Construction and Reconstruction No 3 (41) pp 10-17

[5]Karpenko N I, Sokolov B S, Radaykin O V 2012 *To definition of deformations of bent ferro-concrete elements with use of diagrams of deformation of concrete and armature //* Construction and Reconstruction No 2 pp 11-20

[6]Radaykin O V 2012 *To the diagram method of calculation of bent ferro-concrete elements // Proceedings of the 1st International (VII All-Russian)* Conference "New in Architecture, Design of Building Structures and Reconstruction" (NASSCR) pp 87-91

[7]Karpenko N I, Sokolov B S, Radaykin O V 2013 *Analysis and improvement of curvilinear diagrams of concrete deformation for the calculation of reinforced concrete structures by the deformation model* Industrial and civil construction No 1 pp 25-27

[8]Karpenko N I, Sokolov B S, Radaykin O V 2013 *Perfection of the calculation procedure for bent ferro-concrete elements without preliminary stress by the formation of normal cracks //* Journal of Building Materials No 6 pp 54-55

[9]Karpenko N I, Sokolov B S, Radaykin O V 2013 *To calculate the strength, rigidity and crack resistance of eccentrically compressed reinforced concrete elements using a nonlinear deformation model* Izvestiya KazGASU No 4 (26) pp 113-120

[10]Sokolov B S, Radaykin O V 2014 *Investigation of the stress-strain state of concrete and reinforced concrete bending elements in the zone of joint action of bending moments and shear forces* Proceedings of the VIII Academic Readings of RAASN - International STC "Mechanics of Destruction of Building Materials and Structures" pp 312-17

[11]Sokolov B S, Radaykin O V 2014 *To calculate the deflections of bent reinforced concrete elements, taking into account the joint action of bending moments and break forces using the nonlinear deformation model* Izvestiya KGASU No 4 p 37

[12]Sokolov B S Radaykin O V 2014 *To the calculation of the stiffness of normal sections of reinforced concrete bending elements under the joint action of bending moments and shear forces* Proceedings of the II International Scientific Conference NASKR-2014 pp 201-205.

[13]Sokolov B S, Radaykin O V 2015 *To the construction of a unified methodology for calculating the strength, rigidity and crack resistance of bent ferro-concrete elements, taking into account the joint action of bending moments and shearing forces with the application of a nonlinear deformation model* Fundamental studies of the RAASN on the scientific support of the development of architecture, urban planning and the construction industry of the Russian Federation in 2014 pp 589-99

[14]Sokolov B S, Radaykin O V 2015 *To the definition of the curvature of concrete and reinforced concrete elements along the span, taking into account the joint action of bending moments and shear forces* Construction and Reconstruction No 2 (58) pp 38-41

[15] *To the evaluation of strength, rigidity, the moment of crack formation and their opening in the zone of pure bending of reinforced concrete beams with the use of a nonlinear deformation model* 2016 Izvestiya Vuzov. Building No 3 (687) pp 5-10
[16] Karpenko N I 1996 *General models of mechanics of reinforced concrete* (Moscow: Stroiizdat) p 416

The way to improvement of the method of calculation of rigidity of bending reinforced concrete elements from conventional ferro-concrete

O V Radaikin[2], L S Sabitov[1,3], N F Kashapov[1], I R Gilmanshin[1,4]

[1] Kazan Federal University, 18 Kremlyovskaya street, Kazan, 420008, Russian Federation
[2] Kazan State University of Architecture and Civil Engineering, Kazan, 420043 Russia
[3] Kazan State Power Engeneering University, Kazan, 420066, Russian Federation
[4] Kazan national research technical university named after A N Tupolev, K. Marx St. 10, Kazan, 420111 Tatarstan, Russia

l.sabitov@bk.ru

Abstract. The aim of the research was to improve the physical model of deformation of a bent ferro-concrete element with cracks on the basis of modern achievements in the theory of damage accumulation and fracture mechanics. The presented model allows to determine the rigidity of the element in the operation stage taking into account the damage, the work of the stretched concrete over the macrocrack, the concentration of stresses at its apex, and the presence of a pre-destruction zone in it. The analysis of the results of calculating the rigidity of elements with different percentages of reinforcement by the proposed method and the diagram method using different diagrams of concrete deformation is performed. It has been established that the work of stretched concrete over the macrocrack has a greater effect on the rigidity at a load exceeding the cracking time by no more than 30%, which is especially pronounced in strongly reinforced elements (up to 48%).

As a rule, when designing bent structures from ordinary reinforced concrete (without prestressing of the armature), the computational requirements of the second group of limiting states, in particular, the requirements for limiting the deflections by limiting values, determine the safety and reliability of their operation. To calculate the deflections of the structures under consideration, it is necessary to determine their stiffness, which in turn is significantly affected by the presence of cracks in the stretched concrete. At the same time, it should be noted that in the operational phase, the presence of cracks in the bend is allowed, only the width of their opening is limited.

The numerical value of the rigidity of the reinforced concrete section, as is known, is included in the calculation of not only deflections, but is also used in determining the bending moments in statically indeterminate structures, taking into account the redistribution of forces. In both cases, allowance for cracks allows to reflect the actual operation of the structure and thus ensure its rational design and efficient material consumption.

In this regard, research aimed at developing and improving the methodology for calculating the rigidity of reinforced concrete bending elements with cracks in operation is relevant.

Content from this work may be used under the terms of the Creative Commons Attribution 3.0 licence. Any further distribution of this work must maintain attribution to the author(s) and the title of the work, journal citation and DOI.
Published under licence by IOP Publishing Ltd

IOP Conf. Series: Materials Science and Engineering **412** (2018) 012058 doi:10.1088/1757-899X/412/1/012058

Normative calculation by the method of limiting states (MPS) considers crack formation, as achievement of limiting value of this or that quantity chosen as criterion of durability [1]. Moreover, the physical processes that take place in the operational stage and precede the final destruction of the element, are either not included here at all, or are indirectly taken into account through empirical coefficients. In particular, the work of the stretched concrete over the macrocrack is neglected, the stress concentration at its apex is not taken into account, as well as the effect of scattered small cracks on the behavior of reinforced concrete under load. Not taking into account these factors can lead to a significant underestimation of the rigidity of the cross sections and, as a consequence, to the overexpenditure of steel and concrete in the design of bent structures and elements.

In the development of the Ministry of Railways in the 1960s. Ya.M. Nemirovsky first considered the problems of calculating the deformations of bent barrels, taking into account the work of stretched concrete over the macrocrack [2]. To this end, an additional linear relation was added to the system of resolving equations describing the model of the reinforced concrete section, which relates the height of the macrocrack l to the bending moment M. This was done on the basis of a small number of experimental data. The latter circumstance, in itself, is the complexity of this approach, as noted in the work of A.S. Zalesova and V.V. Figaro [3], did not allow him at that time to find wide practical application. In addition, the later experiments of K.A. Piradova [4] showed a nonlinear relationship between l and M.

Eliminate these inconsistencies of the Ministry of Railways to the real work of the construction to a certain extent is capable of a diagram method of calculating the SCR [1, 5]. A positive feature of this method is that it makes it possible to calculate the bent RCFs for the I and II groups of limiting states on the basis of a unified approach. Although the diagram method has recently been intensively introduced into the regulatory framework for the design of the concrete plant, its wide distribution in practice has been hampered by the fact that the question of how to approximate the experimental curves for the deformation of concrete and steel reinforcement remains still debatable. In addition, there is another unresolved problem, this is how to take into account the stress concentration at the top of the macrocrack within the framework of this method?

In this connection, both in our country and abroad, research is conducted and methods are developed for a relatively new young science - the mechanics of destruction [6]. In general, the use of its methods makes it possible to supplement the MPS and the diagram method with new practical results in estimating the rigidity of the cross sections. However, they are also unable to take into account the decrease in the resistance of concrete due to accumulation in it of scattered small cracks. The theory of damage accumulation is able to refine the results obtained in this question [7]. It assumes the existence of a continuous isotropic medium in which individual structural defects are distributed-scattered small cracks that obey statistical laws and determine the actual strength and deformability of the material of the structure.

Thus, in work two calculation schemes of the cross-section of the element are taken separately:

- **before the formation of a macrocrack** in the normal section of the element, its stretched zone is represented by a model in the form of a bundle of coupled fibers of H. Daniels [8] (Fig. 1, a). They have different tensile strengths R_i, which is due to the development in the material of initial shrinkage mesotrends of different lengths. In this case, the uneven distribution of the strength of the stretched concrete along the beam section is described by the normal law. The concrete of the compressed zone is considered to be a linearly elastic, undamaged material. The condition of continuity at this stage, according to the assumptions of the theory of accumulation of damages, is provided by the introduction of the Kachanov-Rabotnov damping parameter in the calculation formulas;
- **after the formation of the macrocrack**, the equilibrium of the reinforced concrete section with the mathematical sharp Irvine-Orovan cut modeling the macrocrack (Fig. 1, b) is considered. Thus, the work of stretched concrete over the macrocrack and the concentration of stresses at its apex are taken into account. In addition, microcracking at the top of the cut is taken into account, using the model of the fictitious crack, modified by the author, for this purpose. Sha [6]. This in turn assumes 1) the elongation of the Irvin-Orovan cut (the actual macrocrack) by an amount equal to the size of the prefracture zone d_f, and 2) the

IOP Conf. Series: Materials Science and Engineering **412** (2018) 012058 doi:10.1088/1757-899X/412/1/012058

application to the newly formed cracked surfaces of the cohesion forces p distributed in a triangular manner; and the size d_f is taken as the material constant, which is determined from the experiments.

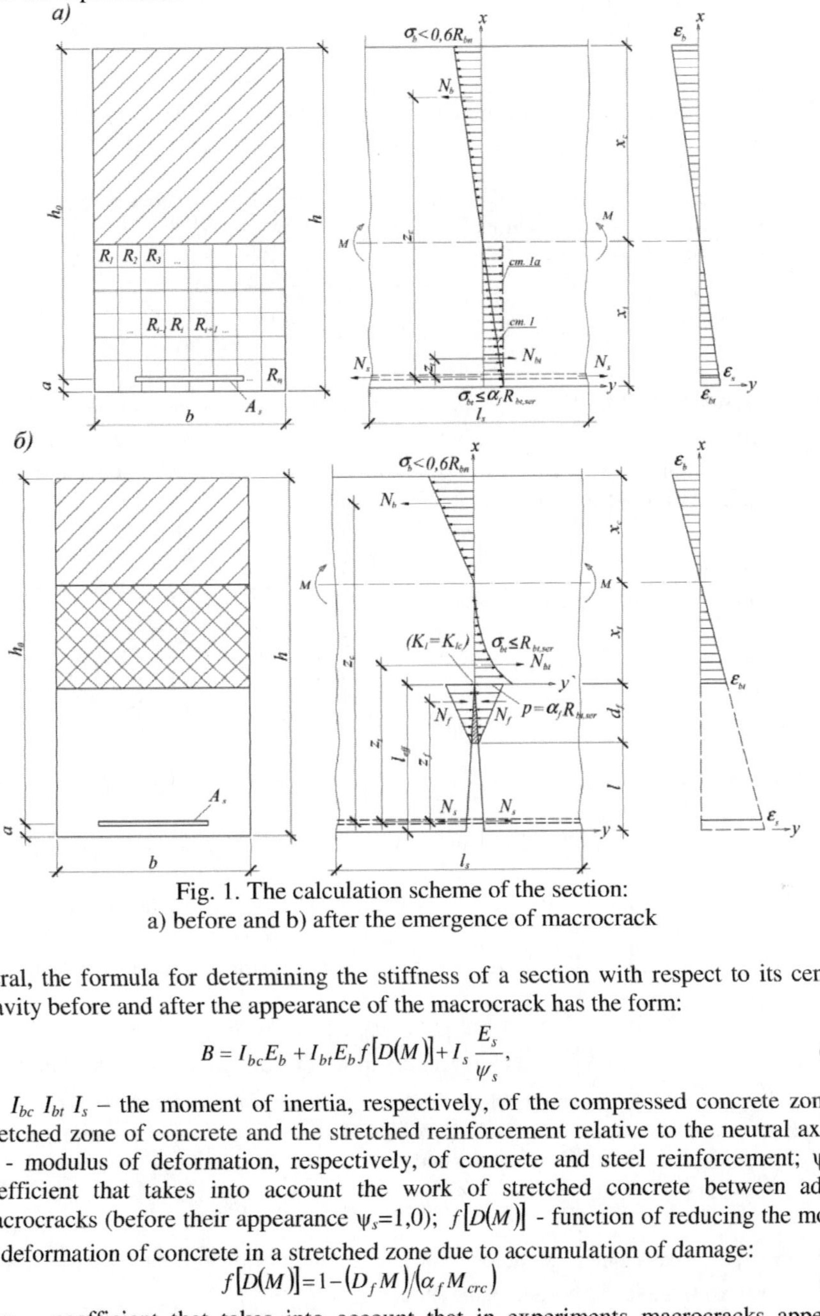

Fig. 1. The calculation scheme of the section:
a) before and b) after the emergence of macrocrack

In general, the formula for determining the stiffness of a section with respect to its center of gravity before and after the appearance of the macrocrack has the form:

$$B = I_{bc}E_b + I_{bt}E_b f[D(M)] + I_s \frac{E_s}{\psi_s}, \tag{1}$$

Where I_{bc} I_{bt} I_s – the moment of inertia, respectively, of the compressed concrete zone, the stretched zone of concrete and the stretched reinforcement relative to the neutral axis; E_b, E_s - modulus of deformation, respectively, of concrete and steel reinforcement; ψ_s is a coefficient that takes into account the work of stretched concrete between adjacent macrocracks (before their appearance ψ_s=1,0); $f[D(M)]$ - function of reducing the modulus of deformation of concrete in a stretched zone due to accumulation of damage:

$$f[D(M)] = 1 - (D_f M)/(\alpha_f M_{crc}) \tag{2}$$

where α_f – coefficient that takes into account that in experiments macrocracks appears at stresses somewhat smaller than $R_{bt,ser}$, D_f – the damaging parameter in this case (α_f=0,88÷0,95, D_f=0,16÷0,32), M_{crc} – normative cracking moment.

In order to find the values of the parameters α_f and D_f, it was required to express LM Kachanov's equation for the effective stresses through statistical parameters and to write it in the following transformed form:

$$1 - sV_m = \frac{\alpha_f}{1 - D_f} \tag{3}$$

where s – reliability indicator; V_m – coefficient of variation in the strength of concrete for tension, installed in accordance with GOST 18105-86. The damping parameter in equation (3) has a probabilistic nature:

$$D(\sigma_{bt}) = \int_{R_{bt,ser}}^{\sigma_{bt}} f(R)dR + q \tag{4}$$

where σ_{bt} – nominal stresses in stretched concrete; $f(R)$ – density distribution of tensile strength by its cross-section, $R_{bt,ser}$ – standard tensile strength of concrete, q – confidence probability.

In order to determine the unknown quantities in (1), the equations of equilibrium and compatibility of deformations for the cross sections in Fig. 1:

- until macrocrack formation:

$$\sum M_{N_s} = 0 \quad => \quad M\text{-}N_b z_c + N_{bt} z_t = 0, \tag{5}$$

$$\sum Y = 0 \quad => \quad N_b\text{-}N_{bt}\text{-}N_s = 0, \tag{6}$$

$$\varepsilon_b / \varepsilon_{bt} = x_c / x_t, \tag{7}$$

where resultant efforts $N_b = \sigma_b b x_c/2$, $N_{bt} = \sigma_{bt} b x_t/2$, $N_s = \sigma_s A_s$ and the corresponding distances between them $z_c = h_0 - x_c/3$, $z_t = x_t/3 - a$ – for stage I, $z_t = x_t/2 - a$ – foe stage Ia; the height of the compressed and stretched zone of concrete is determined by the geometric characteristics of the cross section:

$$x_t = \frac{S_{red}}{A_{red}} = \frac{bh^2/2 + E_s/E_b\, A_s a}{bh + E_s/E_b\, A_s}, \quad x_c = h - x_t \tag{8}$$

- after the formation of the macrocrack:

$$\sum M_{N_s} = 0 \quad => \quad M\text{-}N_b z_c + N_{bt} z_t\text{-}N_f z_f = 0, \tag{9}$$

$$\sum Y = 0 \quad => \quad N_b\text{-}N_{bt} + N_s\text{-}N_f = 0 \tag{10}$$

$$\varepsilon_b / \varepsilon_{bt} = x_c / x_t, \tag{11}$$

where resultant efforts $N_b = \sigma_b x_c b/2$, $N_{bt} = b \int_l^{l+x_t} \sigma_{bt} dx$, $N_f = \alpha_f R_{bt,ser} b d_f/2$, $N_s = \sigma_s A_s$ and corresponding to the figure of the distance between them $z_c = h_0 - x_c/3$, $z_t = l - a + \int_l^{l+x_t} \sigma_{bt} x dx \Big/ \int_l^{l+x_t} \sigma_{bt} dx$, $z_f = l + 2d_f/3$. The height of the compressed and stretched zone of concrete is determined by the geometric characteristics of the reduced section from equation:

$$bx_c^2/2 - bx_t^2/2 - E_s/E_b\, A_s(h_0 - x_c) = 0 \tag{12}$$

where $x_c + x_t + l + d_f = h$, x – the current coordinate of the point along the vertical axis with the origin on the lower extended face of the element.

The tensile stresses over the macrocrack (Fig. 1, b) are distributed according to the curvilinear law and are described by the dependence:

$$\sigma_{bt} = \frac{K_I}{\sqrt{2\pi(x-l)}} + \frac{ME_b}{B}(h - x_c - x) \le R_{bt,ser}, \quad l \le x \le l + x_t, \tag{13}$$

where K_I – stress intensity factor, x – the current coordinate along the vertical axis with the origin on the lower border of the section.

The macrocrack start condition is written in the form:

$$K_I = K_{Ic}, \ \sigma_{bt}\big|_{x=l} = R_{bt,ser}, \tag{14}$$

where K_{Ic} – the critical stress intensity factor, determined experimentally [9].

The joint solution of the equations (5 ÷ 8) allows to estimate the VAT of the reinforced concrete section before the appearance of the macrocrack and, in particular, to determine the height of the compressed and stretched zone of concrete x_c, x_t, the values of which are required for calculating the rigidity by formula (1).

The simultaneous solution of the equations (9 ÷ 14) does not allow to do the same after the emergence of the macrocrack, since the number of unknowns in this system is greater than the number of equations. As an additional equation, the dependence of the macrocrack height on the bending moment is taken as:

$$M(l)=M_0\,\Phi_I(l)/\Phi_I(l_{ult}), \qquad (15)$$

where M_0 is the bending moment when the stretched concrete is completely turned off from work (determined by the well-known formula of A.S. Zalesov [10]); Φ_I - calibration function, constructed based on the results of finite element modeling of cracks with the use of E.M. techniques. Morozova and V.Z. Parton [11]; l_{ult} - the length of the macrocrack is bounded from above by the compressed zone of concrete with complete switching off of the stretched concrete above it from work:

$$l_{ult}=h-\frac{\sqrt{(E_sA_s)^2+4E_bbh_0E_sA_s}-E_sA_s}{2E_bb} \qquad (16)$$

The table presents the results of the numerical determination of the rigidity of a bent reinforced concrete element with a cross section of $b \times h$ = 200 × 400 mm from heavy concrete of class B25 at different percentages of reinforcement: μ = 0,63% (2Ø18 A400), μ = 2,31% (3Ø28 A400) , μ = 3.08% (4Ø28 A400). The table allows you to compare the data obtained by four different methods (see notes to the table). The first three of them are based on the diagrammatic calculation method and the nonlinear deformation model using various diagrams of concrete deformation. In all of the cases considered, in order to evaluate the effect on the rigidity of the stretched concrete only on the macrocrack, the coefficient ψ_s was assumed to be 1.0, that is, its work on the section between macrocracks was not taken into account. The diagram of deformation of compressed concrete and stretched reinforcement in the considered techniques is adopted linear. In Fig. 2 shows the schemes of qualitative distribution of stresses and forces in the cross section of the element (with μ = 2.31%), constructed according to different methods with the same value of the bending moment (M = 23 kN·m).

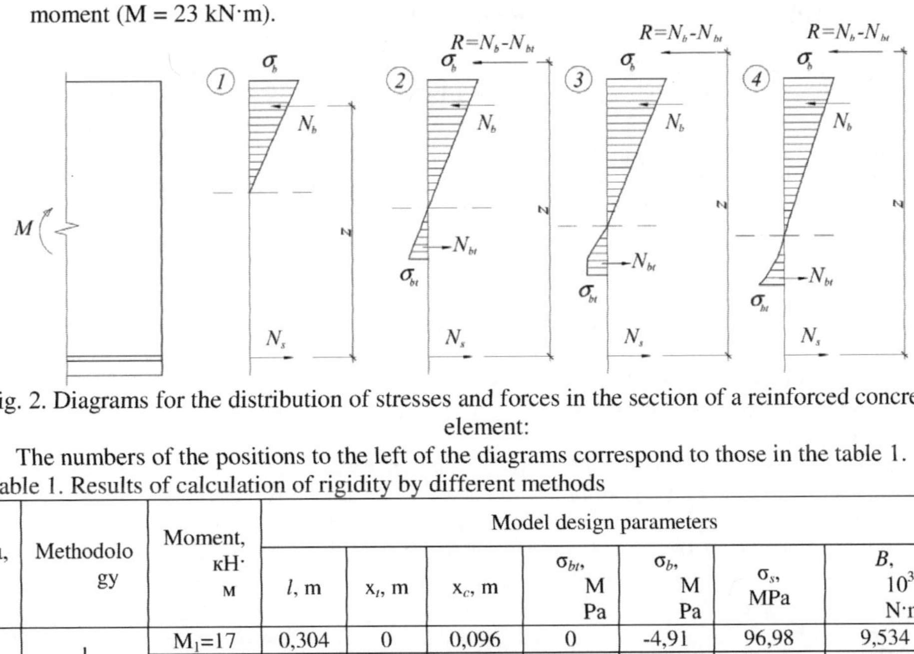

Fig. 2. Diagrams for the distribution of stresses and forces in the section of a reinforced concrete element:

The numbers of the positions to the left of the diagrams correspond to those in the table 1.

Table 1. Results of calculation of rigidity by different methods

μ,	Methodology	Moment, кН·м	Model design parameters						
			l, m	x_l, m	x_c, m	σ_{bt}, MPa	σ_b, MPa	σ_s, MPa	B, 10^3·kN·m^2
0,63	1	M_1=17	0,304	0	0,096	0	-4,91	96,98	9,534
		M_2=50	0,304	0	0,096	0	-14,44	285,24	9,534
	2	M_1=17	0,272	0,032	0,096	1,279	-5,065	95,23	9,647
		M_2=50	0,296	0,008	0,096	0,524	-14,447	284,97	9,594
	3	M_1=17	0,224	0,064	0,112	1,6	-5,33	87,482	10,115
		M_2=50	0,280	0,024	0,096	1,6	-14,663	283,11	9,609
	4	M_1=17	0,008	0,184	0,208	1,6	-2,944	14,841	-
		M_2=50	0,288	0,016	0,096	1,6	-14,617	283,72	9,602
	1	M_1=23	0,24	0	0,16	0	-4,328	38,59	24,961

		M_2=95	0,24	0	0,16	0	-17,878	159,4	24,961
	2	M_1=23	0,168	0,064	0,168	1,585	-4,46	36,115	25,566
		M_2=95	0,224	0,016	0,168	1,245	-17,931	158,97	24,968
	3	M_1=23	0,064	0,144	0,192	1,6	-4,267	26,738	30,509
		M_2=95	0,208	0,032	0,160	1,6	-18,16	158,05	24,999
	4	M_1=23	0,072	0,128	0,2	1,6	-4,232	24,415	31,990
		M_2=95	0,216	0,024	0,16	1,6	-17,974	157,86	24,990
3,08	1	M_1=25	0,224	0	0,176	0	-4,341	32,035	30,013
		M_2=100	0,224	0	0,176	0	-17,362	128,14	30,013
	2	M_1=25	0,144	0,072	0,184	1,588	-4,448	29,727	30,891
		M_2=100	0,208	0,016	0,176	0,992	-17,396	124,87	30,018
	3	M_1=25	0,008	0,176	0,216	1,6	-4,01	19,032	39,982
		M_2=100	0,192	0,032	0,176	1,6	-17,47	127,11	30,052
	4	M_1=25	0,008	0,160	0,232	1,6	-3,831	15,618	44,468
		M_2=100	0,192	0,032	0,176	1,6	-17,509	126,7	30,070

Notes:

1 - calculation of the deformation model without taking into account the work of stretched concrete over the macrocrack;

2 - the same taking into account the work of stretched concrete over the macro crack and the use of the linear diagram " σ_{bt}-ε_{bt} ";

3 - the same with the use of the two-line diagram " σ_{bt}-ε_{bt} " according to [2] ($E_{b1}=E_b$, $\varepsilon_{bt0}=R_{bt,ser}/E_b$, $\varepsilon_{bt2}=2\varepsilon_{bt0}$);

4 - according to the proposed method;

M_1 - the moment exceeding the cracking time by no more than 30%;

M_2 - the moment from the operational load ($\sigma b \leq 0{,}6R_{bn}$).

It was found that the maximum estimate of the effect of stretched concrete over the macrocrack on the stiffness of the element is obtained by methods 3 and 4, which have close results (the divergence in the values of the stiffness of the cross-section is 5 ÷ 12%). In Fig. 2, these methods also correspond to the largest value of the shoulder of the internal pair of forces z. In this case, the use of the diagram method 3 requires calculations by successive approximations, which practically excludes the possibility of calculating the constructions by a "manual" method. The technique proposed by the author allows us to do this without significantly increasing the complexity of computations in comparison with the Normative Approach [1].

It can be seen from the table that the greatest discrepancy in determining the value of B between methods 1 and 4 is observed at a load close to the moment of crack formation (28 ÷ 48%), which is especially pronounced in strongly reinforced elements (up to 48%). Under the action of the operational load, the difference in the stiffness values by all methods is not so significant (less than 5%).

It should be noted that, since the curvature of the longitudinal axis of the reinforced concrete element working with cracks in the stretched zone varies disproportionately to the change in the external torque, in some cases it is necessary to introduce integration of the stiffness over its individual sections in order to accurately determine the deflection of the element. In this case, some sections of the element, in which $M > M_{crc}$, may be in the stage close to the stage of crack formation and account for the work of stretched concrete over the macrocrack in them, as shown above, will lead to an increase in the calculated rigidity of the section in question. This, in turn, will reduce the design deflection of the structure and will give an opportunity to obtain reinforcement savings in the design up to 17%, depending on the percentage of reinforcement and concrete strength class.

Conclusions

1. An improved model of deformation of a bent ferro-concrete element with cracks is proposed, which, unlike the normative approach [1], can be used to determine its rigidity in the exploitation stage, taking into account the damage, the work of stretched concrete over the macrocrack, the stress concentration at its apex, and the presence of a pre-destruction zone in her.

2. Analysis of the results of calculating the rigidity of bent elements with different percentages of reinforcement by the proposed method and the diagram method using different diagrams of concrete deformation. It has been established that the work of stretched concrete over the macrocrack has a greater effect on the rigidity at a load exceeding the cracking time by no more than 30%, which is especially pronounced in strongly reinforced elements (up to 48%).

References

[1] *Manual for the design of concrete and reinforced concrete structures of the heavy concrete without prestressing* (to SP 52-102-2003): date of typing 2004-03-01. – Moscow: TsNIIPromzdany, 2005. – 214 p
[2] Borishansky M S, Nemirovsky J M, Nikolaev J K, Petrova K V et al 1968 *Investigation of the stress-strain state of reinforced concrete elements, with the work of the cracked concrete in tension and revision of this calculation based on the theory of deformation and crack opening* Strength and stiffness of reinforced concrete structures: Coll. of sc. art.– Moscow: NIIZhB p 232
[3] Zalesov A S, Figarovsky V V 1976 *A practical method for calculating the deformations of reinforced concrete structures* (Moscow.: Stroiizdat) p 104
[4] Piradov K A 1998 *Theoretical and experimental foundations of fracture mechanics of concrete and reinforced concrete* (Tbilisi) p 355
[5] Karpenko N I 1996 *General models of the mechanics of reinforced concrete* (Moscow: Stroiizdat) p 416
[6] Zaitsev Y V 1991 *Fracture Mechanics for builders* (Moscow: Vysshaja Shkola) p 288
[7] Lemaitre J A 1992 *Course on Damage Mechanics* (Springer-Verlag) p 210
[8] Daniels H 1945 *The Statistical Theory of the Strength of Bundles of Threads* (I in Proc. R. Soc London A 183) pp 405-35
[9] GOST 29167-91 1992 *Concrete. Methods for determining the characteristics of crack resistance (fracture toughness) under static loading: date of typing 1992-07-01* (Moscow.: State Committee for Construction of the USSR) 38 p
[10] Zalesov A S, Kodysh E N, Lemesh P L, Nikitin I K 1988 *Calculation of reinforced concrete for strength, fracture toughness and deformation* (Moscow: Stroiizdat) p 320
[11] Morozov E M, Nikishkov G P 1980 *The finite element method in fracture mechanics* (Moscow: Nauka) p 256

IOP Conf. Series: Materials Science and Engineering **412** (2018) 012059 doi:10.1088/1757-899X/412/1/012059

Modeling of the complete life cycle of raw materials in the mathematical apparatus of problem-oriented modification of color networks Petri-DN-networks

E R Ruchkina[1], I R Gilmanshin[1,2], N F Kashapov[1], Yu F Kovalchuk[3], Yu I Azimov[1], S I Gilmanshina[1], A I Galeeva[1,2], D R Krainova[1]

[1] Kazan Federal University, 18 Kremlyovskaya street, Kazan, 420008, Russian Federation
[2] Kazan national research technical university named after A N Tupolev, K. Marx St. 10, Kazan, 420111 Tatarstan, Russia
[3] Kastamonu Integrated Wood Industry, 34662, Turkey, Istanbul

is-er@yandex.ru

Abstract. The article considers the issues of increasing the efficiency of waste management through mathematical modeling and description of technological processes of the full life cycle of raw materials. The practical result is the proposed network model of the process of transformation of raw materials in the course of a complete life cycle, using the example of wood in the technological process for the production of MDF panels.

Introduction

At present, the amount of waste is steadily growing in the world, which adversely affects both the environment and human health. It should be noted that recycling of waste is actively used in foreign countries. Dynamically exploring new technologies for their processing. Russia is lagging behind in this area. Unfortunately, in our country, the main way to destroy waste is to dispose of them. It is necessary to improve the waste management system. And this can be achieved on the basis of research stages of the life cycle of waste, as well as the creation of new methods for analyzing the efficiency of the use of raw materials and technologies for recycling production and consumption wastes. Reducing waste through recycling leads to a reduction in energy and raw materials consumption, conservation of natural resources.

The solution of the problem of waste management is to ensure rational nature management, in sustainable development of the territory, both from the point of view of ecology and from the point of view of the economy. A tool is needed to evaluate not individual enterprises and their effectiveness, but a tool for evaluation and comparison across the life cycle of a resource.

For the analysis of the process of production of wood products, the free economic zone "Alabuga" was chosen, in particular, the wood processing plant "Kastamonu Integrated Wood Industries", a subsidiary of the leading Turkish producer of MDF. The enterprise is modern and highly efficient in terms of applied technologies and structure. We set the task to investigate the efficiency of the use of raw materials in the process of transformation from logging to the final disposal of

Content from this work may be used under the terms of the Creative Commons Attribution 3.0 licence. Any further distribution of this work must maintain attribution to the author(s) and the title of the work, journal citation and DOI.
Published under licence by IOP Publishing Ltd

decommissioned finished products and waste at all stages of the technological process. Together with the leading ecologist of the wood processing plant Kastamonu Integrated Wood Industry, Yulia Kovalchuk of the KASTAMONU plant, transformation of raw materials and their parameters.

Figure 1 shows the stages of processing the plant's.

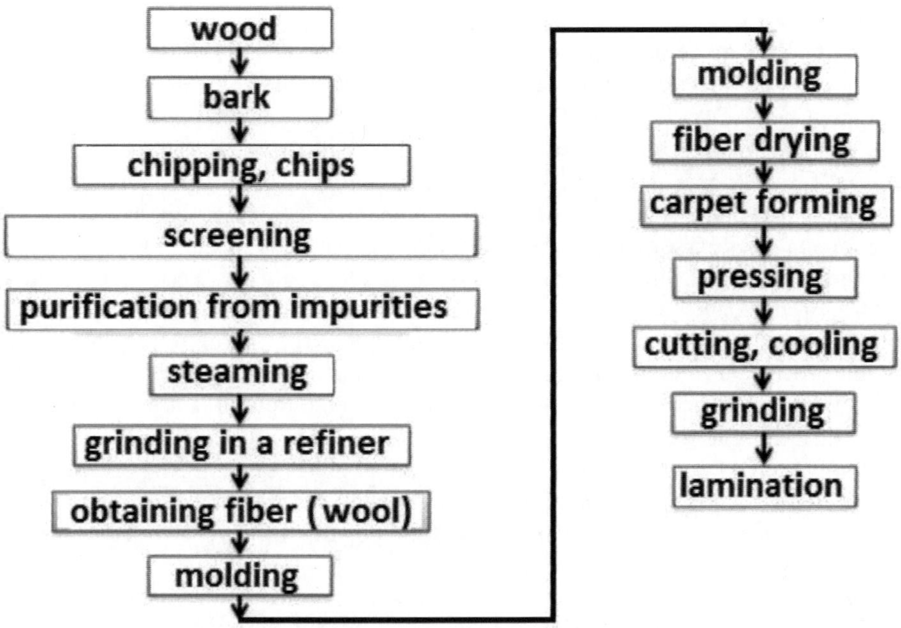

Fig. 1 Stages of processing products factory Kastamonu Integrated Wood Industry

Since it is necessary to develop a model in the full cycle of resource management within the industry, it is necessary to include the logging process in the network model. The scheme of timber processing production is presented in Figure 2.

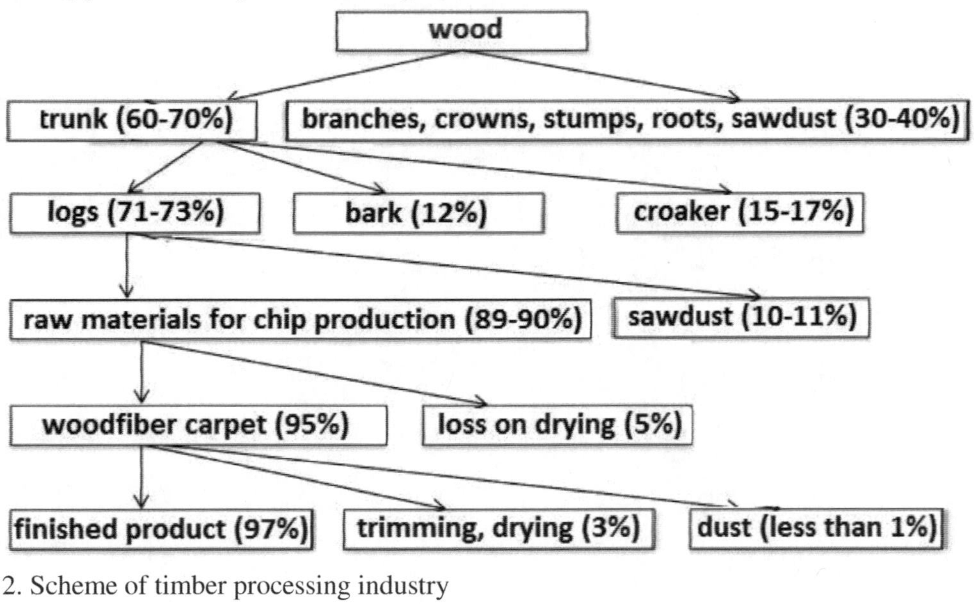

Figure 2. Scheme of timber processing industry

IOP Conf. Series: Materials Science and Engineering **412** (2018) 012059 doi:10.1088/1757-899X/412/1/012059

From the comparative analysis of modeling methods for discrete dynamical systems for visualization of the above scheme, it is expedient to use the mathematical apparatus of the theory of Petri nets. The Petri net is a bipartite oriented multigraph consisting of two types of vertices, positions and transitions, connected by arcs, input and output functions: C = (P, T, I, O), where P = {p1, p2, ..., pn} is a finite set of positions describing the state of the system, $p \geq 0$, T = {t1, t2, ..., tm} is a finite set of transitions characterizing the conditions for changing states, $m \geq 0$, I is the input function, O-output function.

In order to strengthen the modeling capabilities, the modification of the Petri color networks proposed by I.R. Basyrov - DN-networks (discrete-continuous Petri nets) DN network - Petri net of the following type: $C = (P, T, I, O, M, L, F, \tau_1, \tau_2)$, where P = {p1, p2 , ... pn} is the set of positions, T = {t11, t22 ... tml} is the set of transitions. The transition is interpreted as the movement of material flows; M: P - $\{(m_1, m_2, ... m_k) | m_i \in \{0,1,2,...\}, i=1,k\}$ U $\{M(p) \mid p \in P\}$ - DN network marking, given by a constant or marking of some position. When a network is represented as a graphic object, the position denotes a circle, the transition is indicated by a bar. The labels are interpreted as discrete streams (energy, material, information), a set of colors of the Petri net characterizes the nomenclature of the data being processed, the multicolored labels are interpreted as different streams. When a discretely continuous network operates, the color of the labels may change, reflecting the process of data generation. The input arcs are directed from positions to transitions according to the rules specified by the input function I, the output arcs are from the transitions to the positions according to the rules of the output function O. L = {c1, c2, ... ck} is the set of colors for the labels of the DN network. The color of the label $c_j = (e_1, e_2, ... e_l)$, $e_i \in \{0.1\}$, i= 0.l is a vector whose components characterize the presence of one of the characteristics in the intermediate products. The "resulting" color of the label in the final position of the network determines the brand of the finished product.

The main element of modeling is the raw material resource - wood. For the construction of the model, conventional symbols were adopted. The green color indicates the wood, the shade change is its commodity condition, starting with the raw material and ending with the finished product. At the transformation stage, changing the shade, the dynamics of transformation of the state of production (trunk, logs, fiber, sheet, etc.) is reflected. Blue, brown and gray color indicate waste. The model of timber processing production is presented in Fig. 3.

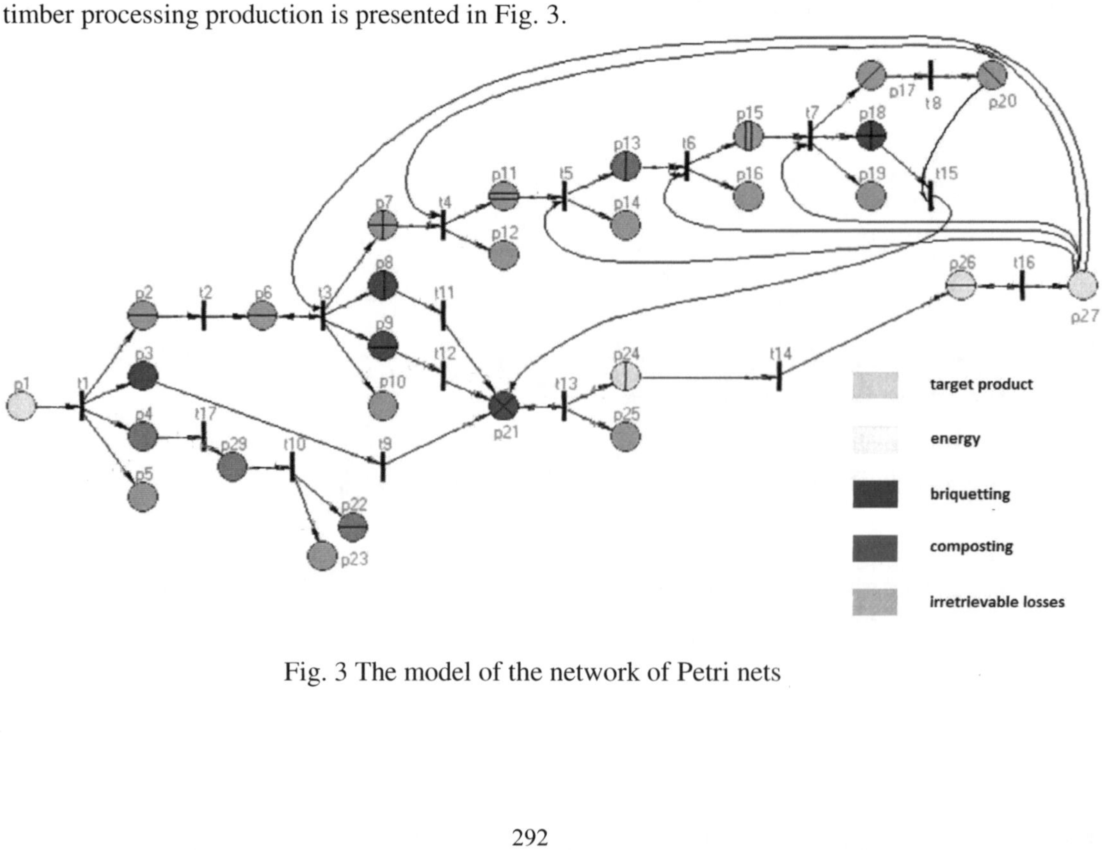

Fig. 3 The model of the network of Petri nets

IOP Conf. Series: Materials Science and Engineering **412** (2018) 012059 doi:10.1088/1757-899X/412/1/012059

The model allows you to visually track conversions, determine the areas of passage of material in one commodity condition, set the parameters of the efficiency of a particular technological process. Thus, summing up all the waste, the overall result is shown for the loss not only of a single production, but of the entire industry, from the raw material to the finished product, down to its further utilization.

Conclusion

The developed model based on the expansion of color networks Petri reflects a multi-stage process of resource transformation throughout the life cycle. On the basis of the model, data on the weight coefficients of energy resources were obtained. Analyzing the work of this model, you can find vulnerabilities, such as the volume of waste generation, taking into account the time. Separately taken branch of the industry will work taking into account the enterprise, manufactured products and processing technology. In one industry, this method allows you to compare the geographical location of different rulers. It is possible to certify this branch of the Republic of Tatarstan in terms of the effectiveness of the whole chain or a single element.

References

[1] Piterson J 1984 *The theory of Petri nets and modeling of systems* Trans. with eng. Moscow: Mir. p 264

[2] Basyrov I R 2001 *Expansion of Petri nets, problem-oriented for modeling multi-assortment production systems* Scientific notes of KFEI. Kazan Is 16 Pp 177-82

[3] Basyrov I R, Akhunov D N 2005 *Petri nets in modeling and production management* Socio-economic and technical systems: Online electronic scientific and technical journal No 8

[4] Basyrova D I, Matrenina O M, Kirpichnikov A P, Shipina T, Kostochko A V 2011 *On one approach to the simulation of the functioning of discretely continuous chemical-technological systems* Bulletin of the Kazan Technological University No 13 pp 146-51

[5] Azimov Yu I, Gilmanshin I R 2016 *Modeling of technological processes of heat supply, as a tool for assessing the state of objects of the property fund and engineering networks of enterprises* Innovative machine-building technologies, equipment and materials - 2016 (MNTK "IMTOM-2016") Materials of the International Scientific and Technical Conference pp 244-48

[6] Gilmanshin I R, Azimov Yu I, Savdur S N, Gilmanshina S I 2017 *Modeling of multilevel structures of power supply in modified networks Petri - DN-networks* Innovative machine-building technologies, equipment and materials - 2017 Materials of the VIII International Scientific and Technical Conference pp 212-16

[7] Gilmanshina S I, Gilmanshin I R, Sagitova R N, Galeeva A I 2016 *The feature of scientific explanation in the teaching of chemistry in the environment of new information of school students' developmental education* International Journal of Environmental and Science Education 11(4) pp 349-58

[8] Gilmanshin I, Kashapov N, Gilmanshina S, Galeeva A 2016 *IOP Conference Series: Materials Science and Engineering* 134(1) **012007**

[9] Azimov Y, Gilmanshin I, Gilmanshina S 2016 *IOP Conference Series: Materials Science and Engineering* 134(1) **012003**

IOP Conf. Series: Materials Science and Engineering **412** (2018) 012060 doi:10.1088/1757-899X/412/1/012060

Direction of chip flow for milling of shaped surfaces by ball end mill

E A Ryabov, R M Khisamutdinov, S Yu Yurasov

Kazan Federal University, NaberezhnyeChelny Institute, 423812, Russia, Naberezhnye Chelny, Prospekt Syuyumbike 10A,

Evgeniy.ryabov@inbox.ru, docfile@yandex.ru.

Abstract. There investigated the direction of chip flow on the front surface by machining with non-cylindrical original instrumental surface end mills.

Currently the design of milling cutters with a non-cylindrical original instrumental surface is based on the convenience of their production. However, there are several ways for grinding shaped cutting units [1]. The use of radial end mills is common nowadays (Figure 1). The choice of radial end mills is determined by the structure, which generates any shape and of a space curve at a point exceeds the radius of sphere.

Figure1. The general form of solidradial end mills

The front surface of such mills comes from a linear displacement of the circle end face at a certain angle, whichdepends on the angle of lead ω and the apparent front angle in standard cross section (Fig.3.). The recommended angle of inclination of screw-type cutters ω lies in a range from 20 ° to 60 °, depending on the material being processed and tool materials. Golubev in his work [2] recommends to use the leading angle for heat-resistant steels, which make up most of the materials of molds.

Content from this work may be used under the terms of the Creative Commons Attribution 3.0 licence. Any further distribution of this work must maintain attribution to the author(s) and the title of the work, journal citation and DOI.
Published under licence by IOP Publishing Ltd

IOP Conf. Series: Materials Science and Engineering **412** (2018) 012060 doi:10.1088/1757-899X/412/1/012060

However, these recommendations refer to cylindrical mills, where standard cross section does not practically change moving along the axis. The changes are caused only by a web taper close to the shank and changes are no more than 1.4-1.7 mm for every 100 mm of the cutting edge [3]. However, it does not lead to a change in geometric parameters of front and relief angles formed by the end surface of the circle, but influences the capacity of chip grooves forming a cone for better chip flow. It can seem rather easy to predict the chip flow [4] in radial cutters with such grinding; it will be directed along the normal line, the same as tool grindingwith a shift of not more than 3 ° due to chip flow. However, in this case the lack of consideration of the fact that the normal angle along the radial part is directed toward the center of original instrumental sphere leads to a shift in chips during the cutting process. As a result (Figure 2.), it turns out that standard front angle varies from a long range of transition to the cylindrical part to smaller ones, up to negative values moving to the end.

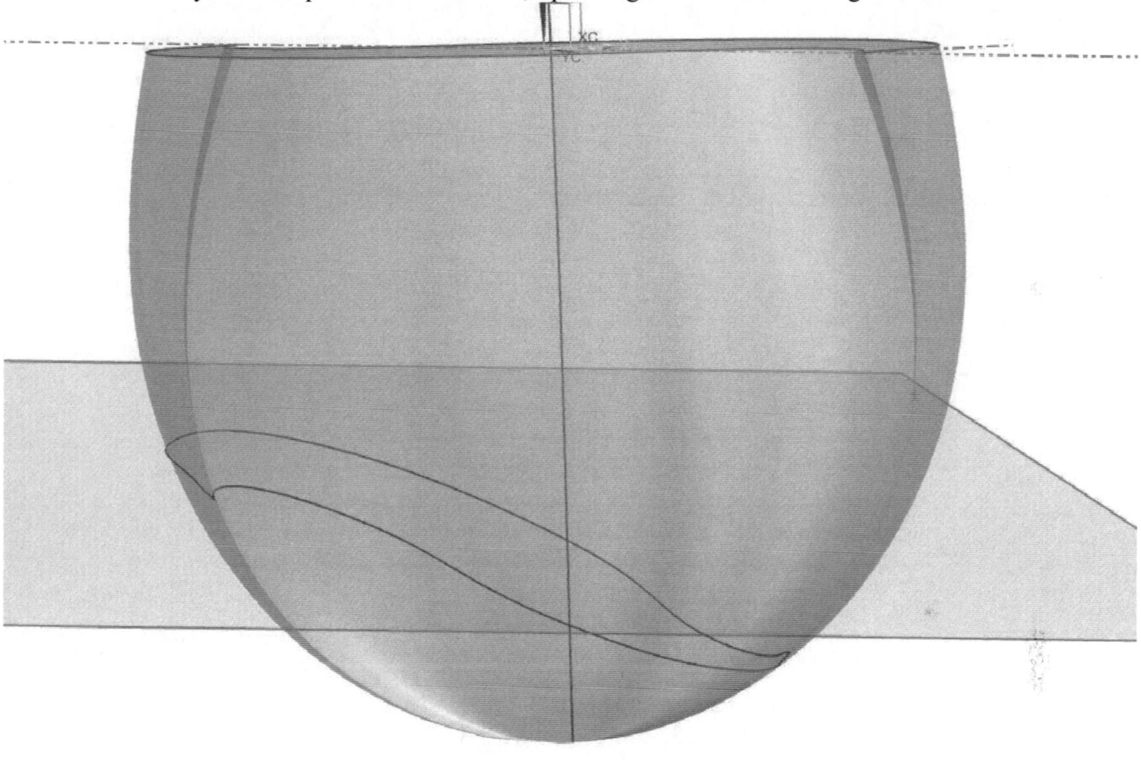

Figure 2.The curve crossing of the cutter with plane, normal front surface

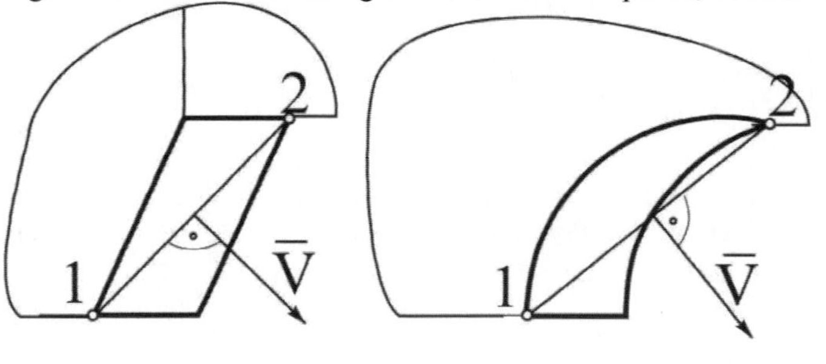

Figure3.The simplifying assumption of chip flow at curved surfaces

The lead angle ω also becomes variable since there should be measured a laxodromy or a great-circle on a spherical surface [5]. In this case, the chip flow has also a variable value which leads to uneven material removal [6], and consequently to a loss of stability: increased vibrations, rapid wear of a cutting edge and, as a result, loss of accuracy of the final profile being formed.The scheme for determining the direction of chip flow was taken from Ushakov's paper [7].

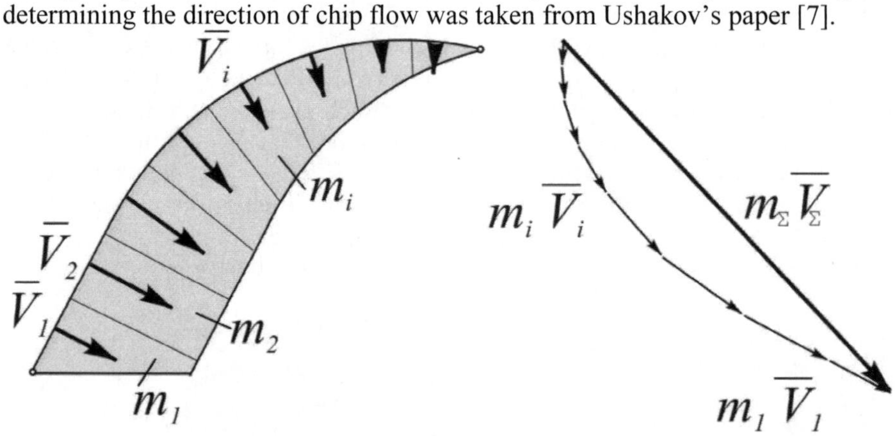

Figure 4.Theforcesactingonthecurvedsurfaceduringthe process of cutting

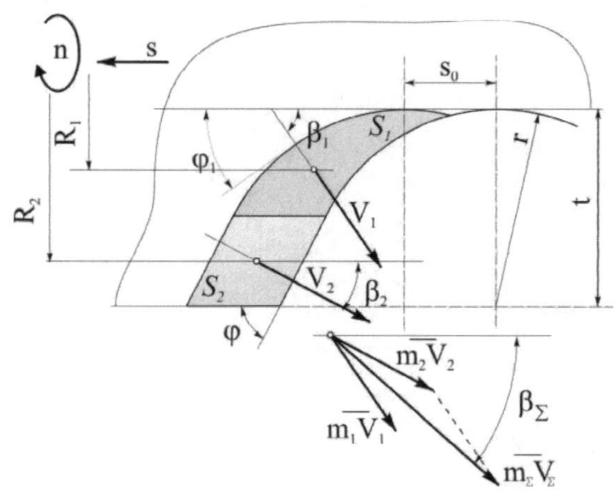

Figure 5.TheschemefordeterminingthedirectionofchipflowbyUshakovM.V.

The sum angle of chip flow is found by the formula:

$$tg\beta_\Sigma = \frac{R_1 S_1 \sin\beta_1 + R_2 S_2 \sin\beta_2}{R_1 S_1 \cos\beta_1 + R_2 S_2 \cos\beta_2}$$

Thus, we can consider modern cutting grinding of radial milling cutters as a technological one, but from the point of operating this tool is not favourable. The angle of chip flow to a greater extent affects the geometric parameters than the kinematic scheme of cutting.

References

IOP Conf. Series: Materials Science and Engineering **412** (2018) 012060 doi:10.1088/1757-899X/412/1/012060

[1] Semenchenko I I, Matyushin V M, Saharov G N 1963 *Proektirovanie metallorezhushchih instrumentov* [The design of metal-cutting tools] (Moskva:Gosudarstvennoe nauchno-tekhnicheskoe izdatel'stvo mashinostroitel'noj literatury) p 951

[2] Golubev S A 1959 *Vliyanie ugla naklona vintovyh zub'ev frez na stojkost'* [The influence of the angle of inclination of the screw teeth of cutters on the resistance] Stankii instrument 10 pp 34-35

[3] Safarov D T, Kondrashov A G 2018 *Stand for monitoring the operational parameters of conjugations "ball support - Body of the tie-rod end" of automotive components* IOP Conference Series: Materials Science and Engineering 289 **012018**

[4] Ryabov E A, Yurasov S Yu 2016 *Creating a rigid model of solid end radius cutter in NX* IOP Conference Series: Materials Science and Engineering 134 **012034**

[5] Khusainov R M, Sabirov A R, Mubarakshin I I 2017 *Study of deformations field in the working zone of vertical milling machine* Procedia Engineering 206 pp 1069-74

[6] Safarov D T, Kondrashov A G, Glinina G F, Safarova L R 2017 *Algorithm of calculation of energy consumption on the basis of differential model of the production task performed on machines with computer numeric control (CNC)* IOP Conference Series: Materials Science and Engineering 240 **012060**

[7] Ushakov M V, Ivanov V V, Pavlova E V 2014 *Opredelenie napravleniya skhoda struzhki po perednej poverhnosti instrumenta* [Determination of the direction of chip flow on the front surface of the tool] Izvestiya Tul'skogo gosudarstvennogo universiteta. Tekhnicheskie nauki 5 pp 118-23

IOP Conf. Series: Materials Science and Engineering **412** (2018) 012061 doi:10.1088/1757-899X/412/1/012061

Material and technical base of the laboratory for the thermal destruction of production and consumption waste for the training of specialists in the field of waste management

R G Sabirzyanov[1], I R Gilmanshin[1,3], N F Kashapov[1], S I Gilmanshina[1], A I Galeeva[1,2], I A Galeev[1], D R Krainova[1]

[1] Kazan Federal University, 18 Kremlyovskaya street, Kazan, 420008, Russian Federation
[2]Kazan State Power Engineering University, Kazan, 420066, Russian Federation
[3] Kazan national research technical university named after A N Tupolev, K. Marx St. 10, Kazan, 420111 Tatarstan, Russia

is-er@yandex.ru

Abstract. In this article, an analysis of the main thermal methods of waste utilization is carried out, a model laboratory for the thermal destruction of production and consumption wastes including medical was developed.

In the process of the life of a person, with increasing demands, a huge amount of waste of various hazard classes appears.

To date, in the Russian Federation, the use, disposal, storage and disposal of waste entails dangerous pollution of the environment, significant economic damage and poses a threat to the life and health of the country's population.

According to Rosprirodnadzor, in 2016, Russia generated 5 billion 441.3 million tons of industrial and solid municipal waste, which is 55% more than in 2006. On the landfills for the neutralization and disposal of industrial and domestic waste and authorized landfills, about 90 billion tons of solid waste have been accumulated.

Until recently, waste was recycled in the Russian Federation in the following ways:

- burning in special furnaces or in open dumps, which leads to air pollution by combustion products;
- burial in cemeteries, which leads to subsidence and pollution of groundwater, as well as intensive formation of methane, which can explode at the slightest spark.
- storage of garbage in specially designated landfills that occupy huge areas, pose a threat to air traffic, attracting flocks of birds.
- processing into secondary raw materials (the problem of this method is the absence of sorting sites and weak involvement of the population in separate collection of garbage)

Wastes containing harmful and toxic substances can not be disposed of by standard methods of utilization and require special plasma technologies.

It is accepted to distinguish three groups of thermal methods of waste destruction: incineration, pyrolysis and gasification.

Content from this work may be used under the terms of the Creative Commons Attribution 3.0 licence. Any further distribution of this work must maintain attribution to the author(s) and the title of the work, journal citation and DOI.
Published under licence by IOP Publishing Ltd

The incineration conditions the possibility of a one-stage utilization of a large volume of waste, which is very convenient for large enterprises and cities. allows you to resort to it as waste is received. The main drawback is that when burning, smoke is formed, saturated with toxic substances, which provokes thinning of the ozone layer and the appearance of ozone holes and, as a result, various diseases in humans. [2]

Table 1. Harmful substances in the combustion of TCR [1].

Substance	mg/m^3
Fly ash, dust	5000
NO_x	210
SO_2	280
CO	50
HCl	-
Cd, Tl	0.5
Hg	0.06
Pb, Co, Cr, Ni, As, Sn, Bi	60
Dioxins	$2*10^{-6}$

By low-temperature pyrolysis is understood a process in which grinded debris undergoes thermal decomposition. This process is mainly intended for the organic part of the waste. The temperature of the process is from 400 to 850 ° C. The advantage of low-temperature pyrolysis compared to traditional combustion is its effectiveness in terms of preventing environmental pollution. The disadvantage of low-temperature pyrolysis is that it does not ensure bacteriological and epidemiological safety of processed products. [2, 5, 6]

Gasification (we will consider high-temperature plasma gasification) is characterized by pyrolysis processes with the formation of pyrolysis gas at high temperatures (from 1300 to 2000 ° C), which is afterburned in the reactor or in a special afterburner. This process differs from the others with a higher temperature, a completely closed technological cycle system and allows to utilize highly toxic hazardous waste including medical. Disadvantages are high energy consumption and the difficulty of cleaning combustion products from heavy metals. [3, 4]

We set the task of scientific and theoretical justification of the composition of the laboratory of thermal destruction of production and consumption wastes. The following wastes were identified as basic: medical, wood.

Medical waste is a dangerous waste that contributes significantly to environmental pollution, second only to industrial emissions, radioactive waste and pesticides, it can be concluded that the most appropriate methods for their disposal are thermal methods: combustion, pyrolysis and plasma gasification, both as in the case of thermal methods of waste processing, most harmful and dangerous compounds, viruses and infections harmful to human life and the environment are rendered harmless. [7]

Wood industry waste accounts for an average of 29% of total logging, of which 11% is used for the production of wood chips and firewood, and the bulk of 18% is exported to landfills. It is more expedient to obtain secondary products from wood industry waste by means of thermal impact than to be disposed to landfills and to be buried. [6]

Table 2. Comparison of the main thermal methods for the utilization of TCR.

	Burning	**Pyrolysis and thermal decomposition**	**Plasma gasification**
Destruction	70% (650 °C)	90% (450 –900 °C)	Full (2000 °C)
The presence of resins and furans	a lot	there are	no
Ash and toxic resins	30% toxic resin	10% ashes	no

IOP Conf. Series: Materials Science and Engineering **412** (2018) 012061 doi:10.1088/1757-899X/412/1/012061

Types of waste	besides certain types of inorganic waste	besides certain types of inorganic waste	any
Waste sorting	require	require a uniform composition of TCR during the year	no
Volume of waste	big	small	big
Emissions of flue gases	high	average	very short
Waste moisture sensitivity	sensitive	sensitive	not sensitive
Output gas quality	Generator gas (technical)	Blasted synthesis gas	High quality of produced synthesis gas
Outcome	Heat, electricity	Synthesis gas, liquid fuels, electricity, heat	Synthesis gas, liquid fuels, electricity, heat

Based on the results of comparison of such process criteria as the percentage of waste destruction, the presence of resins and furans, ash and flue gas emissions, it can be seen that the process of plasma gasification exceeds other thermal methods of waste processing and is suitable for the development of a medical waste treatment plant, treatment of medical waste to control the current serious infectious diseases (and predictable future more aggressive pathogens) due to high temperatures (2000 ° C) Another important advantage is that the process does not depend on the moisture content of the recycled waste.

As a result of the calculations, the parameters of the projected installation were obtained (figure 1): the volume of the plasma gasification furnace V = 5 l (h = 200 mm, d = 180 mm), the lining of the furnace: the 1st layer - chromium-magnesite brick □1 = 230 mm, layer Rockwool □2 = 100 mm, power of the plasma torch P = 10 kW, power source U = 380 V, Area required for the plasma gasification furnace: S = 0.7056 m2, total weight of the furnace: m = 592, 5 kg, height: h = 0.86 m.

For the recycling and recycling of birch bark and birch bark, pyrolysis method is suitable, since with its help it is possible to obtain birch tar. The optimum temperature for obtaining this substance is a temperature of 370 °C (based on the data of [8] on obtaining tar in the laboratory).

As a result of the calculations, the parameters of the projected installation were obtained (Figure 2): furnace volume V = 3 l (h = 200 mm, d = 140 mm), furnace lining: 1st layer - fireclay bricks □1 = 230 mm, chamotte lightweight (ШЛ-0,4) □2 = 115 mm, electric heating element - nichrome wire Х23Ю5Т thickness 1.2 mm, power 3 kW, power source U = 220 V, Area required for pyrolysis furnace: S = 0.6889 m2, total weight of the furnace: m = 448.11 kg, height: h = 0.89 m.

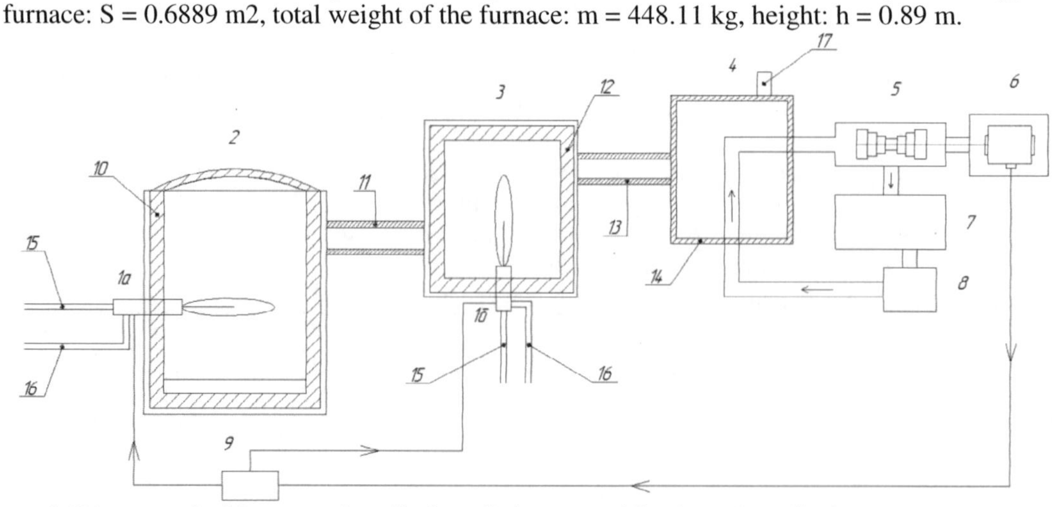

Figure 1. Diagram of a laboratory installation of plasma gasification of medical waste.
1.1a - plasma torch, for plasma gasification of waste;

1b - plasmatron, for afterburning synthesis of gas;
2. Plasma gasification furnace; 3. Afterburning chamber synthesis of gas;
4. Cooling chamber, off-gas; 5. Steam turbine;
6. The generator, 7. Condenser; 8. The pump; 9. Source of power;
10. Two-layer lining furnace;
11. Lining of the pipe for the transition of the synthesis of gas obtained as a result of gasification to the afterburning furnace; 12. Lining of the afterburning furnace;
13. Lining of the transfer pipe of the afterburning products to chamber 4;
14. Lining of the chamber 4; 15. Supply of cooling water to the plasma torch;
16. Supply of working gas (air) into the plasma torch;
17. Pipe, for the exit of residual gases.

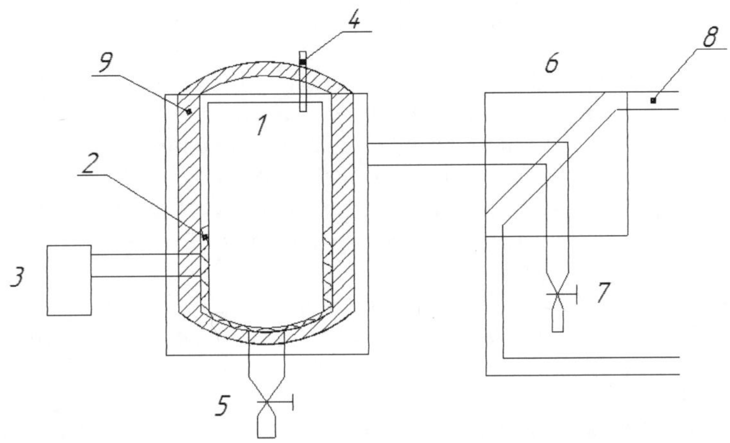

Figure 2. Scheme of a laboratory installation for the processing of woody bark.
1. Camera for pyrolysis of birch bark;
2. Electric heating element - nichrome wire X23Y5T;
3. Power source;
4. Thermometer transmitting data to the heating element;
5. Crane for feeding the obtained tar and residues of the liquid into the reservoir;
6. Pyrolysis gas cooling chamber; 7. Tap for feeding the liquid into the tank; 8. Pipe for passage of running water;
9. Lining the walls of the furnace.

Conclusion

Low-temperature pyrolysis compared with traditional combustion is more effective in terms of preventing environmental pollution, but this method of utilization does not ensure the bacteriological and epidemiological safety of processed products. The pyrolysis process is suitable for the recycling of wood waste in order to reduce the amount of waste and obtain useful secondary products such as tar.

Plasma gasification has a high energy cost, but despite this, it is a highly effective method of waste disposal, as the process differs from the others in a higher temperature, a completely closed technological cycle of the system and allows the utilization of highly toxic hazardous wastes, including medical ones.

As a result of the work, the structure of the laboratory for the "Thermal destruction of production and consumption waste" was selected, the requirements for the component elements of the plants were

IOP Conf. Series: Materials Science and Engineering **412** (2018) 012061 doi:10.1088/1757-899X/412/1/012061

formulated, key parameters of the plant such as capacity, volume, weight and dimensions were calculated.

In terms of perspectives, use for scientific and educational purposes: conducting classes on the discipline "Energy Saving and Energy Audit", scientific research in the field of pyrolysis and plasma waste recycling.

References

[1] Manelis G B 2001 *Energy and Ecology, Institute of Problems of Chemical Physics* Russian Academy of Sciences. Chernogolovka

[2] *Thermal disposal of solid domestic waste* 2006 The concept of NIIStromcomposite. Krasnoyarsk 15 p

[3] Gnedenko V G, Goryachev I V 2011 *Plasmochemical processing of domestic waste in Russia* Solid domestic waste No 10 pp 25-29

[4] Moss A L, Savchin V V, Mossé A L 2015 *Plasma Technologies and Devices for Waste Processing* Minsk: Беларускаянавука p 411

[5] *Processing of solid household waste by pyrolysis method* 1983 Obz. inform. M .: TscNIITEIMS, p 40

[6] Grachov A N, Safin R G, Valeev I A 2006 *Pyrolysis of woodworking enterprises waste* Izv. universities. Chemistry and Chemistry. technology T 49 No 10 pp 104-09

[7] Yanin V S, Yushin V V 2017 Reducing risks when handling medical waste *Education and science in the modern world, Innovations* No 1 pp 376-86

[8] Yakovleva A E, Zabelkin S A, Grachev A N, Bashkirov V N 2014 *Production of birch tar in laboratory conditions and investigation of its solubility* Bulletin of Kazan Technological University T 17 No 1 pp 195-96

[9] Gilmanshina S I, Gilmanshin I R, Sagitova R N, Galeeva A I 2016 *The feature of scientific explanation in the teaching of chemistry in the environment of new information of school students' developmental education* International Journal of Environmental and Science Education 11(4) pp 349-58

[10] Gilmanshin I, Kashapov N, Gilmanshina S, Galeeva A 2016 *IOP Conference Series: Materials Science and Engineering* 134(1) **012007**

[11] Azimov Y, Gilmanshin I, Gilmanshina S 2016 *IOP Conference Series: Materials Science and Engineering* 134(1) **012003**

IOP Conf. Series: Materials Science and Engineering **412** (2018) 012062 doi:10.1088/1757-899X/412/1/012062

The stress-strain state of thin-walled bar of variable cross section with different variants of fastening of the ends that is used in the energy construction

L S Sabitov [2,3], I R Badertdinov[1,4], N F Kashapov [2], I R Gilmanshin[2,5], I L Kuznetsov[1]

[1] Kazan State University of Architecture and Civil Engineering, Kazan, 420043, Russia
[2] Kazan Federal University, 18 Kremlyovskaya street, Kazan, 420008, Russian Federation
[3] Kazan State Power Engineering University, 51 Krasnoselskaya street, Kazan, 420066, Russian Federation
[4] SUE кE "Engineering Center" Industrial Safety ", Kazan, Russia
[5] Kazan National Research Technical University named after A.N.Tupolev, 420111, Kazan, Russia

l.sabitov@bk.ru

Abstract. The authors proposed equations for the calculation of the stability and strength of thin-walled beams with swivel bearing ends can be generalized to the problem with boundary conditions corresponding to the elastic interaction with neighboring elements of the lattice pylons.

1. The basic equations of the problem. Various energy facilities, supports, trestles, various thin-walled rods of variable cross-section, working on compression are widely used. In articles [1, 2, 3], the problem of the stability and strength of hinged supported thin-walled rods of variable cross-section under longitudinal compression was solved. In work [4], a generalization of the calculation technique for rods with fastening of the ends by types "free end + sealing", "hinge + sealing", "sealing + sealing" is given. Here we continue to consider the range of problems associated with the various conditions for fixing the ends of the rod, and we consider a more general case of their elastic fastening.

Variants of elastic fastening of rods in a structure can be various variants, but two main classes can be distinguished: a connection with adjacent thin-walled structural elements and sealing of the ends of a rod or other parts thereof into an elastic array. The second class of problems is more complicated, since it leads to the determination of the elastic interaction coefficients of the elements to the contact problem of the theory of elasticity or to other, also complex, ways of solving the problem. But if these coefficients are defined, the following calculation method is fully applicable to both classes of problems.

Content from this work may be used under the terms of the Creative Commons Attribution 3.0 licence. Any further distribution of this work must maintain attribution to the author(s) and the title of the work, journal citation and DOI.
Published under licence by IOP Publishing Ltd

IOP Conf. Series: Materials Science and Engineering **412** (2018) 012062 doi:10.1088/1757-899X/412/1/012062

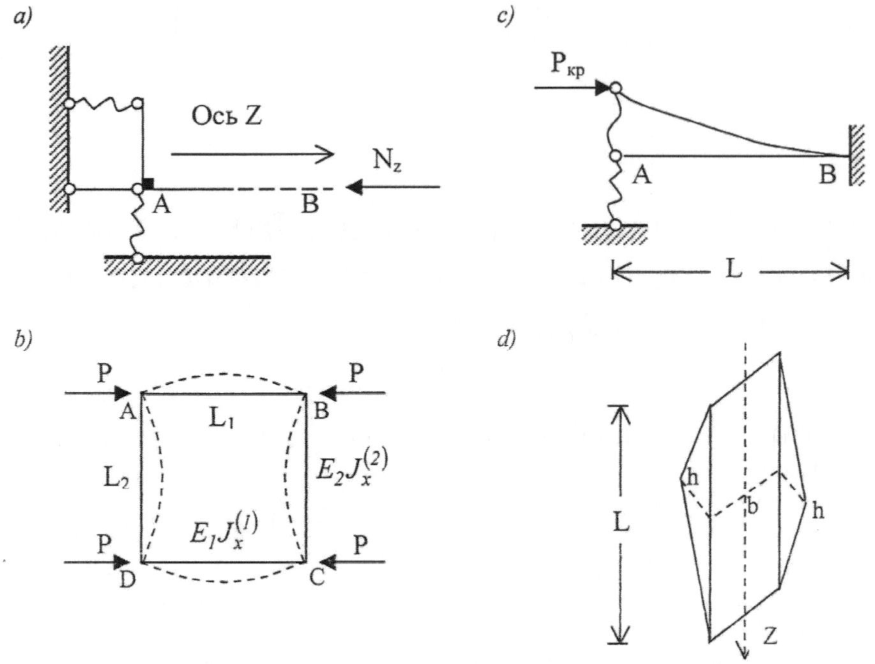

Fig.1 Schemes of elastic fastening of rods and a rod of variable cross-section

As shown in [4], the resolving equations of the problem under different conditions for fixing the rod have the form:

$$EJ_{\omega}\varphi^{IV} + EJ'_{\omega}\varphi''' + k_1\varphi'' + k_2\varphi' + k_3\varphi - p(x_0 - e_x)v'' + p(y_0 - e_y)u'' = 0, (1)$$

$$EJ_y(z)\cdot u''(z) + pu(z) + p[y_o(z) - e_y(z)]\varphi(z) = pe_x(z) + C_1 z + C_2; (2)$$

$$EJ_x(z)\cdot v''(z) + pv(z) + pv(z) + p[-x_0(z) + e_x(z)]\varphi(z) = pe_y(z) + C_3 z + C_2; (3)$$

$$k_1 = -C - 2p\left[e_x(x_0 + R_y) + e_y(y_0 + R_x) - r_0^2/2\right], \quad r_0^2 = x_0^2 + y_0^2 + (J_x + J_y)/F;$$

$$k_2 = -C' + 2p\left[x'_0(x_0 - e_x) + y_0'(y_0 - e_y)\right], k_3 = p\left[x''_0(x_0 - e_x) + y_0''(y_0 - e_y)\right]$$

$$-2p(\alpha')^2[e_x(x_0 - R_y) + e_y(y_0 - R_y) + e_y(y_0 - R_x) + r_0^2/2];$$

$$R_x = \frac{1}{2J_x}\cdot\iint_F y(x^2 + y^2)dF, R_y = \frac{1}{2J_y}\cdot\iint_F x(x^2 + y^2)dF, \quad r_0^2 = x_0^2 + y_0^2 + (J_x + J_y)/F;$$

parameters u,v,φ – the required displacements; x0, y0, – the coordinates of the bending center of the section Z; ex, ey – coordinates of the trace of the loading line in the given section Z; C – torsional stiffness of the rod in section Z; C1, C2, C3 , C4 – are the integration constants; α(z) – is the angle of rotation of the main central axes of the section in naturally twisted rods; for medium- and strongly-structurally twisted rods the flexural stiffnesses EJx(z) and EJy(z) must be calculated in a special way [5]. All the coefficients of the equations, except for the modules E,G and the load P, depend on the longitudinal coordinate Z.

2. The constant of the cross section. As in [4], we first explain the method of solving the problem by the example of a non-thin-walled rod of constant cross section. Let the rod be fixed so that its left

end (Fig. 1a) elastically interacts with the neighboring elements of the structure (Rasco + support belt).

Section A can be vertically intermixed and rotated in a vertical plane with an elastic support resistance. With the central compression of the rod and the conservative direction of the compressive force, the stability equation of the rod can be written in the form [4]:

$$EJ_x \cdot v''(z) + pv(z) = +C_1 z + C_2,$$

where C1 and C2 – constants of integration.

When the point A moves vertically, an external vertical transverse force (reaction of the support) QA(1), appears in it, proportional to the vertical displacement vA :

$$M_A = k_{1A} v'_A.$$

Taking into account the known differential material resistance relations for the internal shear force and the internal bending moment, we obtain two boundary conditions for equation (4) at the end A:

$$EJ_x \cdot v''_A = \alpha_{1A} \cdot k_{1A} \cdot v'_A, (6)$$

$$EJ_x \cdot v'''_A = \alpha_{2A} \cdot k_{2A} \cdot v'_A + P \cdot v'_A \cdot \beta_A. (7)$$

In (6) and (7), for convenience, the coefficients αiA, βA are introduced. The point is that the directions of displacements when the rod is unstable are not known to us in advance, and by changing the signs of the indicated coefficients, we can consider the various forms of stability loss. Secondly, by changing the value of αiA from zero to large values, we can consider various conditions for the elastic fastening of the rod ends - from the case of the free end to the case of rigid embedding, without changing the elasticity coefficients kiA and k2A in the program of the computer. In all cases | βA | =1.. Two other similar boundary conditions must be written for the end face B.

Equation (4) can be solved, for example, by the method of finite differences [4]. To calculate the third derivative in (7) one-sided differences have proved to be well-established:

$$v'''_A \approx \frac{1}{h}(v''_{A+1} - v''_A) = \frac{1}{h^3}(-v_{A-1} + 3v_A - 3v''_{A+1} + v_A)$$

or

$$v'''_B \approx \frac{1}{h}(v''_B - v''_{B-1}) = \frac{1}{h^3}(-v_{B-2} + 3v_{B-1} - 3v''_B + v_{B+1})$$

When the signs of the coefficients αiA, βA are varied in (6) and (7) and the analogous end-to-end coefficients B, the solution of equation (4) gives several different values of the critical load. To choose the true value, you must follow the following rules: First - with increasing rigidity of the support, the critical load must increase; the second - if the first criterion is met by several critical loads, choose the smallest.

Test problem 1. Consider the problem of S.P. Tymoshenko [6]. In the core system - Fig.1b - all the rods are rigidly connected and, consequently, their mechanical action under the load will be elastic. The plane of the least flexural rigidity of each rod coincides with the plane of the bar rectangle. The rods AB and BC are L2, E2Jx(2) and the cross-section area F2. Determine the force P at which the bars AB and CD have a length L1 and bending stiffness E2Jx(1); rods AD and BC - L2, E2Jx(2) and the cross-section area F2. Determine the force P at which the bars AB and CD lose their stability.

The solution of the bending equation for the beam E2Jx(2)*v''(z)=MA for the rod AD under the boundary conditions Z=±L2/2; v'(z)= ± vA' gives MA=2E2Jx(2) vA'/L2.

But since MA=k1A·vA', we obtain the stiffness coefficient k1A =2E2Jx(2)/L2.

If a transverse reaction QA(1)= k1A·vA appears in the cross section A of the rod AB with loss of stability, then the longitudinal deformation of the rod AD is ΔL2=·vA=QA(1)L2/(E2F2). Consequently, the second stiffness coefficient is k2A=E2F2/L2.

It is obvious that in the section B the stiffness coefficients are the same.

Let all the rods of the system be the same, have a rectangular constant section b * t = 4 * 2 (cm), a length L = 100 cm and an elastic modulus E = 2,1 * 105 MPa.

The decision of S.P. Timoshenko: P cr =16,46*EJx/L2=92,18kN.

Solving the finite difference method [4], equation (4) with boundary conditions (6), (7) for sections A and B for 84 segments of the partition, we obtain Pcr = 92.17kN. This result also agrees well with the graphical representation of the solution in [6]. Our solution is obtained for the following values of the additional coefficients in (6) and (7) for the cross sections A and B:

α1A = -α1B = 1; | α2A | = | βA | = | α2B | = | βB | = 1 (the solution does not depend on the sign - transverse forces do not arise or are very small).

Test problem 2. Figure 1c shows a rod of length L, the right end of which is rigidly clamped, and the left one has an elastic transverse support. If the stability is lost, the cross section A can freely rotate and the reactive moment in this section does not arise.

To determine the critical load, we solve equation (4).

At the left end, in the boundary conditions (6) and (7), we set α1A = 0, βA = -1, the value α2A in the calculations is changed to study the influence of rigidity of the support on the critical load. We assume for definiteness, just as in the previous example, k2A =E2F2/L2=1.68*104MPa.

At the right end, we can also use conditions (6) and (7), assuming α1B=10+50, α2B=10+50, β=0 (or any small number). The number of segments of the partition of the rod is accepted for a numerical solution - 84.

The critical loads obtained by us were compared with solutions of an analogous problem, given in [8] - see Table 1 (force values - in kN). The last value of the table Pcr [8] = 13.82 kN corresponds to the free end A and is calculated by the Euler formula.

Table 1. Critical forces (kN)

α2B	Pкр	Pкр[8]
1.0	113.03	113.06
0.1	112.96	112.99
0.01	112.24	112.27
0.001	98.30	98.32
0.0002	39.98	39.98
0.0001	27.19	27.19
0.0	13.82	13.82

The results of solving test problems 1 and 2 confirm the correctness and good accuracy of the proposed solution technique. We have considered these problems in sufficient detail, which makes it possible to dispense with unnecessary explanations in the next, main problem.

3. A thin-walled rod of variable cross-section AB, shown in Fig. 2a and the same - in Fig. 1d, elastically interacts with loss of stability with a rod of a constant circular cross section of radius R; resistance of the rod AC to the force P before the loss of stability of the rod AB is not taken into account. The loading of the rod AB can be either central or eccentric; its end B is rigidly clamped, as is the end C of the rod AC. We determine the critical value of the compressive force P, assuming that the dimensions of the rods and the elastic constants of their materials are given.

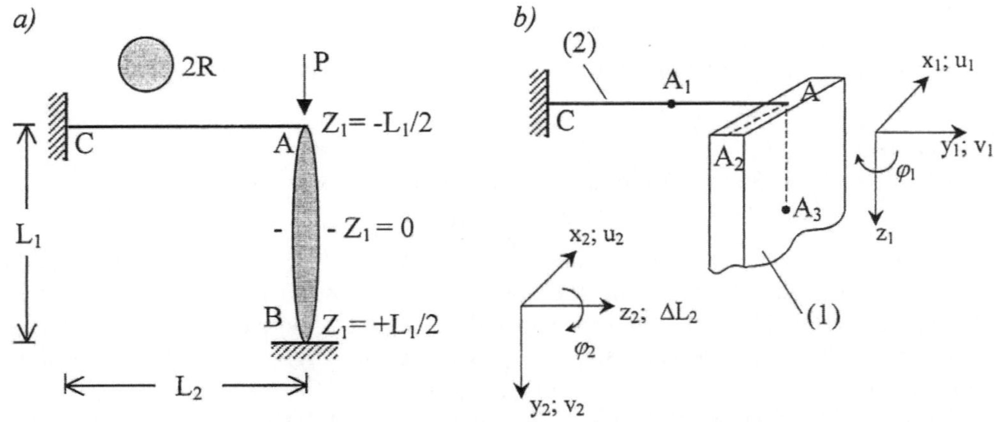

Fig.2 Fastening and loading of a rod of variable cross-section

The deformation of the rod AB under load is described by a system of equations (1) - (3) having the eighth differential order and 4 additional integration constants. For each of the displacements u1, v1, φ1, two boundary conditions at each end (12 in total) must be formulated.

The conditions for pinching the end of B are obvious:

$$v1 = v1' = u1 = u1' = \varphi 1 = \varphi 1' = 0.$$

Formulate for the rod AB six boundary conditions at the end of A, where the rod is expressed in a plate.

1) The rotation of the section A around the axis AA2 (Fig. 2b) at an angle v'A creates in the rod AS a bending moment MA = k1A • v'A. Consequently, the first boundary condition has the form (6):

$$E_1 J_{xA}^{(1)} \cdot v_A''' = \alpha_{1A} \cdot k_{1A} \cdot v_A', (9)$$

Where k1A =E2Jx(2)/L2.

2) The displacement vA of the cross section A of the rod AB in the direction AA1 causes the appearance of a transverse force Qy1= k2A·vA. Then the second condition has the form (7)

$$E_1 J_{xA}^{(1)} \cdot v_A''' = \alpha_{2A} \cdot k_{2A} \cdot v_A + P \cdot v_A' \cdot \beta_{1A}, (10)$$

where k2A =E2F2 /L2.

3) The displacement uA of the rod AB causes the appearance of a transverse force Qx1 = k3A • uA, proportional to the flexural rigidity of the rod AC. Then the third boundary condition is written as

$$E_1 J_{yA}^{(1)} \cdot u_A''' = \alpha_{3A} \cdot k_{3A} \cdot u_A' + P \cdot u_A' \cdot \beta_{2A}, (11)$$

Where k3A =3E2Jy(2)/L2 3.

4) For the rod AB, the quantity uA 'is connected with rotation at the point A around the axis AA1 - that is, with twisting of the rod AC. The torque Mcr = k4A * uA 'is bending for the rod AB in the plane A2AA3. Then the fourth boundary boundary condition is

$$E_1 J_{yA}^{(1)} \cdot u_A'' = \alpha_{4A} \cdot k_{4A} \cdot u_A' (12)$$

where k4A=G2Jp(2)/L2, G2 – the material shear module of the second rod Jp(2) – polar moment of inertia of its cross section.

5) The adopted design scheme allows us to assume that there is no normal voltage from deplanation at the end A of the rod AB. Then the fifth condition for the end A is written as

$$\varphi A'' = 0 \qquad (13)$$

6) When the end A of the rod AB is twisted about the axis AA3, the rod AC bends in the plane A1AA2 by the moment M2=k5AφA(1), where k5A=E2Jy(2)/L2.

IOP Conf. Series: Materials Science and Engineering **412** (2018) 012062 doi:10.1088/1757-899X/412/1/012062

The relative angle of twisting of the rod AB in section A (the end of the plate) is $\varphi'_{1A} = \dfrac{M^{1A}_{кр}}{G_1 \cdot J^{(1)}_t}$,

where for a thin plate $\varphi^{(1)}_t = \dfrac{1}{3} bt^3$. at $M_2 = M^{1A}_{кр}$, we have

G1Jt(1)φ1A'=k5AφA(1) and the sixth boundary condition of the problem:

$$G_1 J^{(1)}_t \cdot \varphi'_A = \alpha_{5A} k_{5A} \cdot \varphi_A (14)$$

As in the previous test problems, under the conditions (9) - (14), the coefficients αiA, βjA; the coefficients βjA can take values ± 1; the coefficients αiA can not only have different signs, but also take different numerical values to change the stiffness of the rod AS in the calculations. Since different critical loads are obtained for different signs of the coefficients αiA, βjA, we use the same two rules for the final choice:

First, with increasing rigidity of the support, the critical load must increase; the second - if the first criterion is met by several critical loads, choose the smallest.

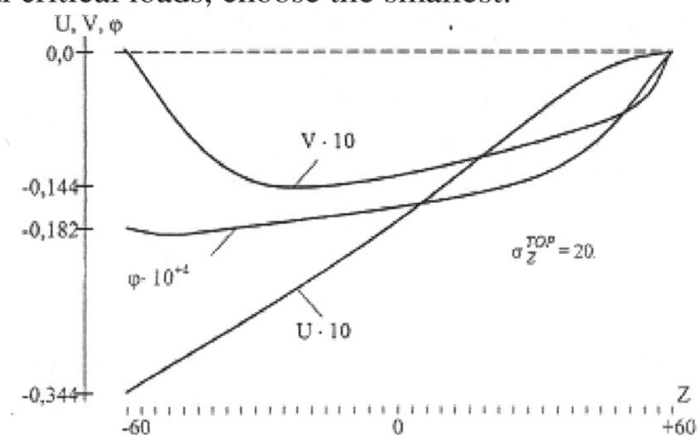

Fig.3 Subcritical displacements in a rod of variable cross-section with an elastic support

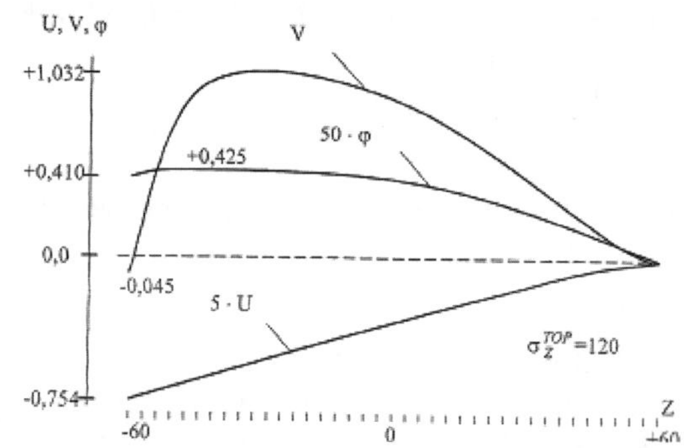

Fig. 4 Subcritical displacements in the rod under a load close to critical

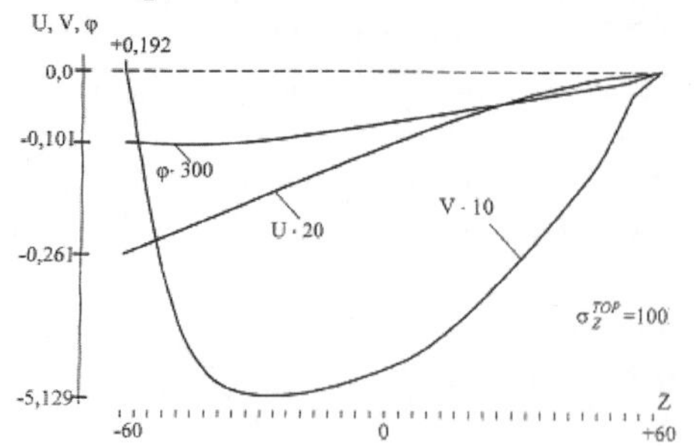

Fig.5 Supercritical displacements in the rod

IOP Conf. Series: Materials Science and Engineering **412** (2018) 012062 doi:10.1088/1757-899X/412/1/012062

A numerical example. Table 2 shows the values of the critical end compressive stresses σor = P / Ftor for the rod AB (Figures 1d, 2a, 2b) with the following numerical data: L1 = L2 = 120cm; b = h = 5 cm; t = 0.5cm (sheet thickness); R = 0.5 cm; E1 = E2 = 2.1 * 105 MPa; G1 = G2 = 8.1 * 104MPa. Equations (1) - (3) with boundary conditions (9) - (14) were solved by the method of finite differences [4] when the rod was divided into N = 50 equal segments (the convergence of the solution with respect to the number N was controlled). The rigidity of the AS support varied in calculations using the coefficients αiA. The problem was solved in a linear formulation, a very small eccentricity of the compressive force at the ends of ≈10-5 cm was introduced into the calculations.

Table 2. Critical stresses (MPa)

αiA	*φ*кр	αiA	*φ*кр	αiA	*φ*кр	αiA	*φ*кр
10+50	212.38	2	116.88	0.05	107.22	0.0005	22.26
100	200.43	1	112.68	0.01	101.30	0.0001	10.42
10	143.13	0.5	110.45	0.005	93.60	0.00005	7.24
5	128.10	0.1	108.13	0.001	43.41	0.0	4.05

Let us follow the change in the displacements as the load increases.

The following results were obtained for the stiffness variant αiA = 1.0 for calculations with one-sided normative eccentricity [2] at the ends en=imin/20+L/750=0.1672cm. The critical value of the compressive stress at the ends in this case in this case is □торкр=112.59MPa.

Figure 3 shows the displacement curves for a small subcritical load **σ**торz=100MPa, which is close to critical. With the growth of the load, the deflection v (z) and φ (z) increase most signically and sharply increase; The displacement u (z) continues to increase monotonically with increasing load, and if the stability of the sign does not change, it does not change.

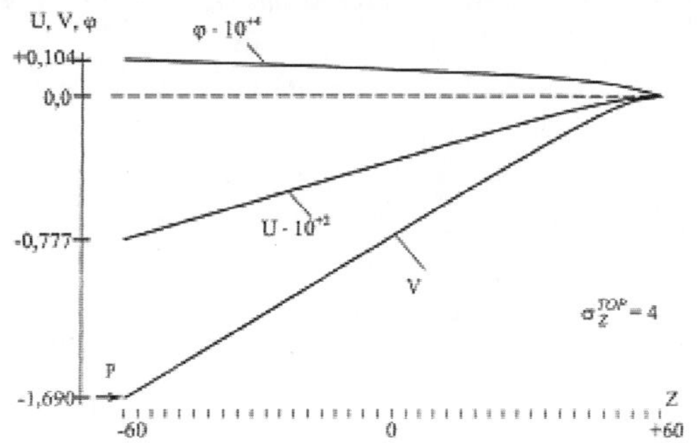

Fig.6 Subcritical displacements in a rod with a free end

With almost zero reference stiffness at the left end (αiA = 2 * 10-5, the other parameters are the same), as follows from Fig. 6 and 7, the rod almost everywhere remains straight both before and after the loss of stability - except for a small zone of strong bending in the vicinity of a blind seal on the right. The critical value of the compressive stress at the end of this version is **σ**торz=5.33MPa.

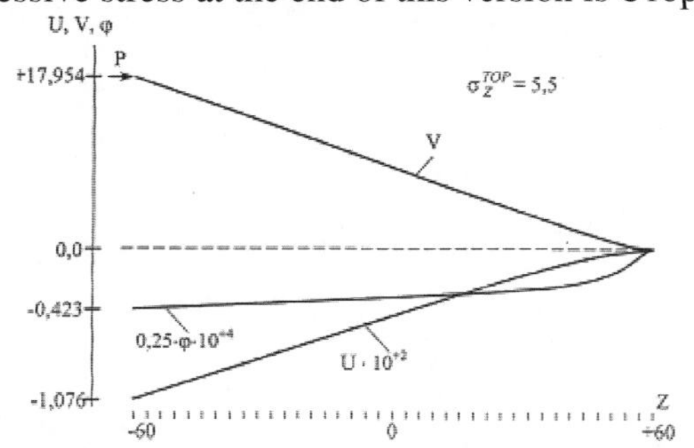

Fig.7 Supercritical displacements in a rod with a free end

IOP Conf. Series: Materials Science and Engineering **412** (2018) 012062 doi:10.1088/1757-899X/412/1/012062

Conclusions. The system of equations (1) - (3) with a reduced differential order of equations (2), (3) is well suited for solving the class of problems considered. The boundary conditions for the elastic sealing of the ends of a thin-walled rod of variable cross-section are proposed to be written in the form (9) - (14). The imposition of elastic bonds on the free end of the rod, even a very small stiffness, sharply increases its stability under longitudinal compression (Table 2). The proposed equations and the method for calculating stability and strength can also be used to solve complex problems related to thin-walled bars of constant cross-section, fixed and loaded at the ends in various ways.

References

[1] Sabitov L S, Kuznetsov I L, Bogdanovich A U, Khusainov D M 2016 *Critical Load Of Thin-Walled Rod With Various Cross Section Shape By* Length Journal of Engineering and Applied Sciences V 11 No 6 pp 1186-90

[2] Khamidullin I N, Sabitov L S, Kuznetsov I L, Ilyin V K 2014 *On the issue of determining the stress-strain state of the power line support* Energy of Tatarstan No 3

[3] Sabitov L S, Kuznetsov I L, Bogdanovich A U 2014 *Stress-strain state of a weakly crooked rod of variable cross-section* Bulletin of IrSTU No 7 (90) pp 71-79

[4] Bogdanovich A U, Kuznetsov I L 2005 *Longitudinal compression of a thin-walled rod of variable cross-section with different variants of fixing the ends* Izvestiya Vuzov. Building, No 10 pp 19-25

[5] Sabitov L S, Khamidulin I N, Kuznetsov I L, Khusainov D M 2015 *Stress-strained state of supports for energy construction* International Journal of Applied Engineering Research (IJAER) Vol 10 No 24 pp 45438-48

[6] Timoshenko S P 1971 *Stability of rods, plates and shells* Moscow: Science p 186

[7] Birger I A and Panovko Ya G 1968 *A reference manual for scientific workers* Strength, stability, fluctuations Vol 3 p 35

[8] Pisarenko G S 1979 *Resistance of materials: Proc. manual for universities* Kiev: Vishcha school p 507

IOP Conf. Series: Materials Science and Engineering **412** (2018) 012063 doi:10.1088/1757-899X/412/1/012063

Development of the method of dynamic tests support of air transmission lines

L S Sabitov[1,2], I Z Gatiyatov[1], N F Kashapov[2], I R Gilmanshin[2], I K Kiyamov[2]

[1]Kazan State Power Engeneering University, Kazan, 420066, Russian Federation
[2] Kazan Federal University, 18 Kremlyovskaya street, Kazan, 420008, Russian Federation

l.sabitov@bk.ru

Abstract. Known methods of testing supports create longitudinal and transverse static load applied to the support. In real conditions the majority of the damage of the supports is connected with the influence on them of dynamic loads, which can exceed the static. The proposed method allows to conduct experimental studies of cascade processes of destruction of anchor site due to a wire breakage, when the potential energy of the tension wires is converted into a dynamic influence on the design of supports.

Metal and reinforced concrete supports have found wide application in the practice of network construction. They have a number of advantages over wooden supports and are built on particularly critical lines, where high operational reliability is required, as well as in large cities and settlements. In addition, metal supports are installed on large crossings through rivers and other engineering structures, in industrial areas and in mountainous areas.

Evaluation of the impact on the strength and deformation of the supports of dynamic loads, for example, connected with the breakage of wires, due to ice accidents or fatigue vibrational damages at the output of the supporting clamp, is currently carried out only with the help of various computer programs.

Known methods of testing the supports create longitudinal and transverse loads applied to the support [1]. The disadvantage of these methods is that the stress-strain state of the support structure is made when static loads are applied to it, while under real operating conditions a significant part of the damages of the supports is due to the impact of dynamic loads on them that can exceed the static ones by almost twice .106 [2].

A close analogue to the proposed one is the method of dynamic testing of the overhead power transmission line supports, which includes fastening the end of the cable rod with a calibrated burst insert and a dynamometer to the top of the support of the end of the cable rod, the other end of which is connected to the winch, the impact on the support by the impulse load created as a result of the rupture inserts when pulling the cable rope to this force and determining the response of the support to the impact of the load impulse [3, 4]. The disadvantage of the known method is that according to it, it is envisaged to carry out dynamic tests of only a single support pillar of overhead power transmission lines without taking into account the vertical loads acting on the support due to the own weight of the wires and ice, as well as the elastic tension of the wires.

Content from this work may be used under the terms of the Creative Commons Attribution 3.0 licence. Any further distribution of this work must maintain attribution to the author(s) and the title of the work, journal citation and DOI.
Published under licence by IOP Publishing Ltd

The technical result of the present invention is to provide dynamic tests of a single support system combined by elastic mechanical links in the form of cable rods with built-in tension springs simulating a real anchor section to determine their dynamic characteristics both under design conditions and in emergency conditions caused by impact of ice loads exceeding the limit values, including the destruction of support structures.

The achievement of this technical result is achieved by the fact that in the known method of dynamical tests of supports comprising, fixing on the upper part of the support of the end of the cable rope with a calibrated burst insert and a dynamometer, the other end of which is connected by a winch, the impact on the support by the impulse load created by the rupture of a calibrated a predetermined insertion force when the winch of the cable rod is tensioned up to this force and determining the response of the support to the impact of the load pulse, according to the invention, On the base of the power rectangular frame of the bench for dynamically testing the supports of at least three intermediate supports simulating the anchor section, in which the side frame legs are the models of the anchor supports, fix the cable rods with insulated calibrated inserts in the insulators installed on the traverses, dynamometers and tension springs in which the relationship between the tensile force and the elongation of the spring corresponds to the dependence of the tension force of the wire in the actual span from its elongation, In addition, the ends of the cable rods through the blocks installed on the upper part of the side frame of the frame are connected to the corresponding winches fixed on the frame base, along with the calibration weight is fixed on each insulator of the support , whose weight is equal to the difference between the weight of the wire in the actual span and the weight of the cable rod, with a calibrated burst insert built into it, a dynamometer and a tension spring fixed in the span between two opt mounted on the stand, tension cable rods with the help of winches before breaking one of the calibrated inserts designed for the minimum cable pulling force, determined by the load due to the weight of the wire in the span, when the calibrated insert is broken, the response of the intermediate supports is determined, the impact of the impulse load , including measurement of the amplitude and natural vibration frequency of the supports by means of displacement sensors mounted on the upper beam of the frame, while strain gauges are determined by the mechanical At the critical points of the support structures during the period of impact on the impulse load supports, the described operations are repeated in the same way after fixing additional cargoes on the insulators, the weight of which corresponds to the weight of the ice with an increasing thickness of its wall, including its maximum value, and taking into account the increase in the tension force wires in the span from the increase in the additional load on it from the weight of the ice, increase the value of the rupture force of the calibrated insert to the value corresponding to the e th additional load and the own weight of the wire, after completing the tests for modeling the impact on the ice wire with the maximum wall thickness, set the value of the rupture force of the calibrated insert equal to the wire breaking force, and by pulling the cable rod up to this effort, the least stable support structure is destroyed, This, due to the reduction of wire traction, destroys the support structures in the adjacent ones with the emergency spans, the main parameters of the destruction process Design supports and determine the coefficients of dynamic supports.

Set of at least three intermediate supports on the basis of power of a rectangular frame, simulating the anchor portion, wherein the side frame strut represent layouts anchor supports, fixing on the insulators installed on the traverse rope linkage with fitted therein calibrated discontinuous inserts, dynamometers and the tension spring, equivalents are elastic wires in the intermediate spans in which the relationship between stretching force and the elongation of the spring force corresponds to the dependence of wire tension in the intermediate span of its elongation, caused by exposure to the load wire from its own weight wire and the weight of ice, as well as the consolidation on insulators supports cargo weight which correspond to the weights applied to the load conductor allowed on a compact stand to examine the work of the supports in the tension section in accounting and emergency modes in conditions as close as possible to real. These circumstances allow to specify method of calculating bearings based on the experimental data, as theoretical methods for describing complex dynamic processes occurring in overhead power lines, with a plurality of influencing factors, are of limited use due to insufficient high precision calculations.

The load of the supports at the end of the tests with the unloading load and the registration of the main parameters of the destruction of the support structures allow us to determine the maximum design load on the support and to identify the least durable components of the support structures.

Figure 1 shows the diagram of the stand for dynamic tests of the support of overhead power transmission lines, in Fig. 2 - the dependence of the wire tension force in the span from its elongation caused by the change in the specific loads from the ice.

The IS-2 stand for dynamic testing of the overhead power transmission supports includes a power rectangular frame 1, on the base of which 2 there are at least three test supports 3 modeling the anchor section, in which the side legs 4 of the frame 1 are mock-ups of the anchor supports. On the insulators 5 installed on the traverses 6, the cable rods 7 with the calibrated discontinuous inserts 8 built in them, the dynamometers 9 (devices 8 and 9 are shown only on one cable rod) and the tension springs 10 representing the elastic equivalents of the wires, including the lightning protection rope , fixed on insulators 5 of the tested supports 3 in real intermediate spans. The ends of the cable rods 7 through the blocks 11 are connected to the respective winches 12 fixed to the base 2. On the upper beam 13 of the frame 1, the displacement sensors 14, the vertices 15 of the supports 3 are mounted.

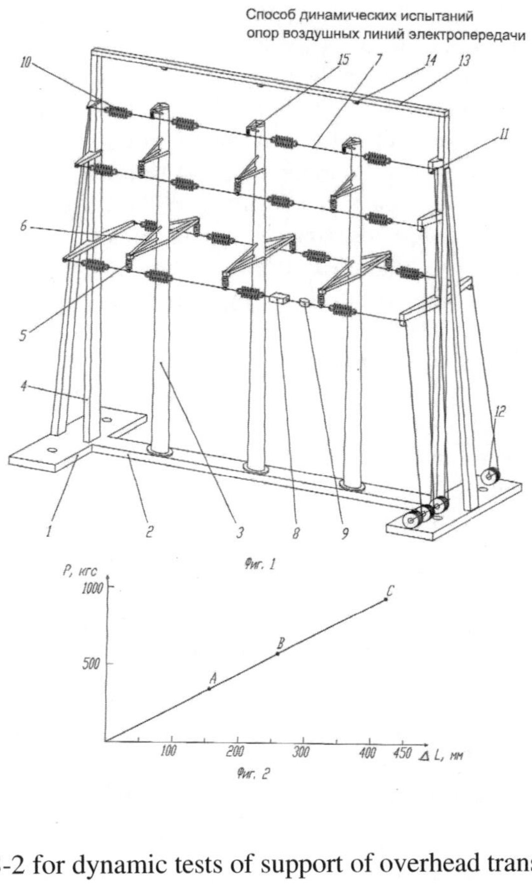

Fig.1 Stand IS-2 for dynamic tests of support of overhead transmission lines

Dynamic tests of supports are carried out as follows. First, to carry out dynamic tests of the supports in full accordance with the real conditions, a calibration weight (not shown) is fixed on each insulator 5 of the test support 3 to be tested, the weight of which is equal to the difference between the weight of the wire in the real span and the weight of the rod 7, with a calibrated discontinuous insert 8, a dynamometer 9 and a tension spring 10 fixed in the span between the two supports 3 mounted on the stand.

After this, the tension of the cable rods 7 is controlled by the dynamometers 9 with the help of the winches 12 until one of the calibrated inserts 8 is broken, calculated for the minimum tension of the tension of the cable rod 7, determined by the load from the weight of the wire in the real span. When the calibrated insert 8 is ruptured, the response of the intermediate supports 3 is determined by the real impulse action corresponding to the operating conditions, namely the wire break in the span. In this case, the amplitude and frequency of natural oscillations of supports 3 connected by cable rods 7 with built-in tension springs 10, which are elastic equivalents of wires, are measured by means of displacement sensors 14.

Also, strain gages (not shown) determine the mechanical stresses at the critical points of the support structures 3 during the period of impact of the impulse load on them. Similarly, the described operations are repeated after fixing additional loads (not shown) on the insulators 5, the weight of which corresponds to the weight of the ice with increasing steps of its wall thickness, including its maximum value. Taking into account the increase in the tension of the wire in the span from the increase in the additional load on it from the weight of the ice, the value of the rupture force of the calibrated insert 8 is increased to the value corresponding to this additional load and the weight of the wire.

After the end of the tests for modeling the impact on the ice wire with the maximum excess wall thickness, the value of the rupture force of the calibrated insert 8 is equal to the breaking force of the wire, and by the tension of the cable rod 7, the less stable structure of the support 3 is destroyed, and, due to the reduction of the wire traction, in the neighboring with emergency spans. Also, the parameters of the process of destruction of the structures of the supports 3 are registered and the coefficients of dynamism of the supports 3 are determined.

On the example of 110 kV overhead transmission line with polygonal supports to which steel-aluminum wires AC-70, II glaze ice area (the maximum thickness of the glaze wall is 10 mm) are suspended with a span length l = 300 m, we determine the dependence of the wire tension force in the span on its elongation caused by the change in specific loads from ice.

Initial data:

- Total cross-section of the wire S = 79.3 mm^2;
- Wire diameter d = 11.4 mm;
- Weight of 1 km of wire 275 kg;
- The maximum permissible stress σmax = 11.6 kgf / mm^2, with a maximum ice wall thickness c = 10 mm and temperature t = -50 C;
- Specific load due to the weight of the wire γ0 = 3,46 • 10-3 kgs / m • mm^2;
- Specific load from the weight of ice with a wall thickness of 10mm

$\gamma_{r\,max} = 7{,}463 \cdot 10^{-3}$ kgs/m·mm^2;

- Specific weight of wire with ice

$$\gamma = \gamma_0 + \gamma_{r\max} = 3.46 \cdot 10^{-3} + 7.63 \cdot 10^{-3} = 11.09 \cdot 10^{-3} \text{kgs/m·mm}^2$$

The mechanical stresses in the wire when the loads from the ice are changed are determined from the equation of state of the wire [5]

$$\sigma - \frac{\gamma^2 \times E \times l^2}{24 \times \sigma^2} = \sigma_0 - \frac{\gamma_0^2 \times E \times l^2}{24 \times \sigma_0^2}, \quad (1)$$

where E = $8{,}25 \cdot 10^3$ kgs/m·mm^2 – modulus of elasticity of wire.

As a result of the solution of equation (1), σ_0 was determined, the voltage in the wire caused by the specific load γ_0 from the intrinsic weight of the wire, $\sigma_0 = 4.21$ kgs / mm^2, and the stress $\sigma_5 = 7.24$ kgs / mm^2 with an ice wall thickness of 5 mm.

From the obtained values of the stresses in the wire for the thickness of the ice wall 0; 5 and 10mm (points A, B, C on the graph) the dependence of the wire tension force in the span P = σ x S on its elongation ΔL caused by the change in the specific loads from the ice was constructed (Fig. 2). The elongation ΔL was determined by the formula

$$\Delta L = l \cdot \frac{\sigma - \sigma_0}{E},$$

its maximum equals

$$\Delta L_{\max} = l \cdot \frac{\sigma_{\max} - \sigma_0}{E} = 300 \cdot \frac{11.6 - 4.21}{8.25 \cdot 10^3} = 268.727 mm.$$

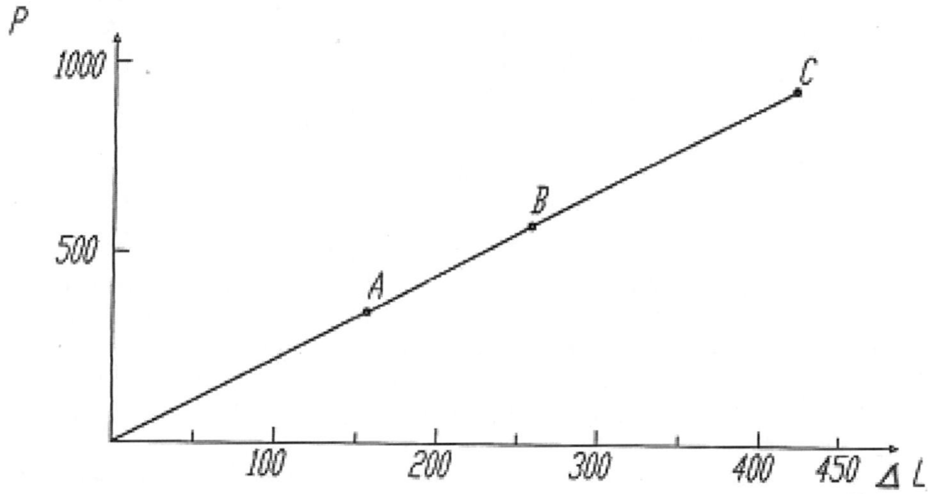

Fig. 2. Dependence of the wire tension force in the span from its elongation, caused by the change in the specific loads from ice.

As follows from the graph of Fig. 2 at its points A, B, C, the wire tension in the span P and the extension of the wire ΔL are equal to $P_A = 333.85$ kgs, $\Delta L_A = 153$ mm, respectively; $P_B = 574{,}1$кгс, $\Delta L_B = 263{,}27$мм; $P_C = 919.88$ kgf, $\Delta L_C = 421.8$ mm. At the same time, in the range of the wire extension equal to 268.73 mm, caused by the influence of ice, its tension force varies from 333.85 to 919.88 kgs practically in linear dependence. Similarly, according to [5], a lightning protection cable is calculated.

Since in the proposed technical solution the wires in the real span are replaced by their elastic equivalents, in which the tension springs are used, the dependence presented in Fig. 2 is the initial one for calculating the load characteristic of the spring.

Calculation of the dimensions and parameters of the spring was carried out according to [6]. The parameters of the calculated spring are shown in the table.

Table.1. Spring parameters.

Format for calculating effort	Calculation of effort in kgs.	
Material	60C2A, 51CHFA	
Diameter of wire (rod), d	14	Mm

Outside diameter, D1	111	Mm
Working number of turns, n	12	piece
Pre-stretching, S1	262	Mm
Working tension, S2	262	Mm
Maximum stretching, S3	268	Mm
Number of springs in a batch	1	piece
Internal diameter of the hook, D2	83	Mm
Length of the spring along the body, L_0	182	Mm
Length of spring between hooks, L	348	Mm
Spring index	6,93	-
Size of the open part of the hook	35	-
Spring Stiffness	3,51	kg/ Mm
Preload, F1	268,33	kg
Work load, F2	919	kg
Maximum load, F3	940,05	kg
Length of spring between hooks under load F1, L1	424,5	Mm
Length of spring between hooks under load F2, L2	610	Mm
Length of spring between hooks under load F3, L3	616	Mm
The maximum tangential stresses	102,84	mPa
Spring weight, m	5,24862	kg

The use of the proposed method of dynamic tests of power transmission line supports will allow to provide dynamic tests of a single system of supports joined by elastic mechanical links in the form of cable rods with built-in tension springs that most fully simulate the real anchor section to determine the dynamic characteristics of the supports in simulating a real impulse influence on them, corresponding to the operating conditions, namely, wire breakage, both in the calculated conditions, and in emergency modes caused by the impact of ice loads exceeding the limit values, including the destruction of support structures.

Also, the proposed method will allow to carry out experimental studies of the processes of cascade destruction of the anchor section due to wire breakage, when the potential energy of wire tension in the span of the anchor section is transformed into a powerful dynamic effect on the structure of the supports, leading to their destruction.

References

[1] The patent of the Russian Federation No 2554285 МПК G01M5 / 00, publ. 05/10/2015.
[2] Birbraer A N 2009 *Extreme effects on structures* (SPb .: Publishing house of Polytechnic University) p 594
[3] The patent of the Russian Federation No 2574419 МПК G01M7 / 02, publ. 10.02.2016.
[4] Kuznetsova V V 1999 *Metal structures* vol 3 (M .: ASV) p 528
[5] Kryukov K P, Novgorodtsev B P 1979 *Constructions and mechanical calculation of power lines* (L .: Energia) p 312
[6] Acherkan N 1968 *Machine parts* vol 2 (M.: Mechanical Engineering) p 408

IOP Conf. Series: Materials Science and Engineering **412** (2018) 012064 doi:10.1088/1757-899X/412/1/012064

Development of the design of a laboratory vibro-grinding machine for preparing samples for metallographic research

D T Safarov[1], A G Kondrashov[1], I I Khafizov[2]

[1] Russia, Republic of Tatarstan, Naberezhnye Chelny institute (branch) of the Kazan (Volga) Federal University, 423812, Naberezhnye Chelny, Mira Ave., 68/19

[2] Kazan Federal University, 18 Kremlyovskaya street, Kazan, 420008, Russian Federation

safarov-dt@mail.ru, khafizov@kpfu.ru

Abstract. The article presents the results of testing a prototype vibro-grinding laboratory machine for making samples for metallographic examination. The effectiveness of the method and its suitability for the preparation of thin sections during laboratory studies in the discipline "Material Science" have been established.

1. Introduction

Any manufacturing enterprise of machine-building products necessarily carries out activities to study the material of manufactured products by groups of quality indicators - the properties of the substance, the presence of internal defects and chemical composition.

One of the most laborious from the point of view of preliminary preparation of samples is a metallographic study of the properties of materials. Preliminary preparation consists of the following technological operations - pieces of the sample, pouring or pressing the sample into the section, several grinding and polishing operations, the number of which depends on the type of material being examined, and also the surface etching operation [1]. The methods of research are standardized, and also the quality requirements of the samples [2]. The main task of abrasive surface preparation is to obtain a surface of a thin section without scratches visible in the metallographic microscope with a minimum deviation from flatness. In the sampling route, the quality of the resulting surface of the thin section depends on each operation, but the key operations are grinding and polishing operations.

2. Theoretical part

In metallographic laboratories, abrasive processing of samples is performed on laboratory grinding machines of metallographic preparation. At present, manufacturers offer a variety of grinding and polishing machines both manual and with automatic control. Offers of modern machines by Russian manufacturers are absent. Chinese analogues, depending on the configuration of auxiliary holding devices are offered at a price of $ 2,600. The equipment of European and Asian manufacturers is offered at a price of up to $ 5,000.

To increase the productivity of image processing, machine tools with manual sample preparation are equipped with additional devices that ensure the retention of thin sections from an automated system by controlling the process of polishing the samples. Laboratory machines of this class are offered at a much higher price. It is, depending on the authority of the manufacturer, from $ 10,000 to

Content from this work may be used under the terms of the Creative Commons Attribution 3.0 licence. Any further distribution of this work must maintain attribution to the author(s) and the title of the work, journal citation and DOI.
Published under licence by IOP Publishing Ltd

IOP Conf. Series: Materials Science and Engineering **412** (2018) 012064 doi:10.1088/1757-899X/412/1/012064

$ 35,000. Successful operation of imported laboratory grinding machines is possible only with branded consumable polishing materials, the cost of which is considerable. This laboratory machine offers a method of vibrating grinding (Figure 1a), in which the site with the abrasive material oscillates at a high frequency, and the samples for changing the cutting abrasive grains and removing the grinding products rotate around the axis of the holder. In this case, the samples can be automatically rotated about their axis because of the unbalance of the resultant forces during the grinding process. The scheme is realized in the form of a constructive solution of a laboratory machine, the three-dimensional model of which is shown in Fig. 1 b.

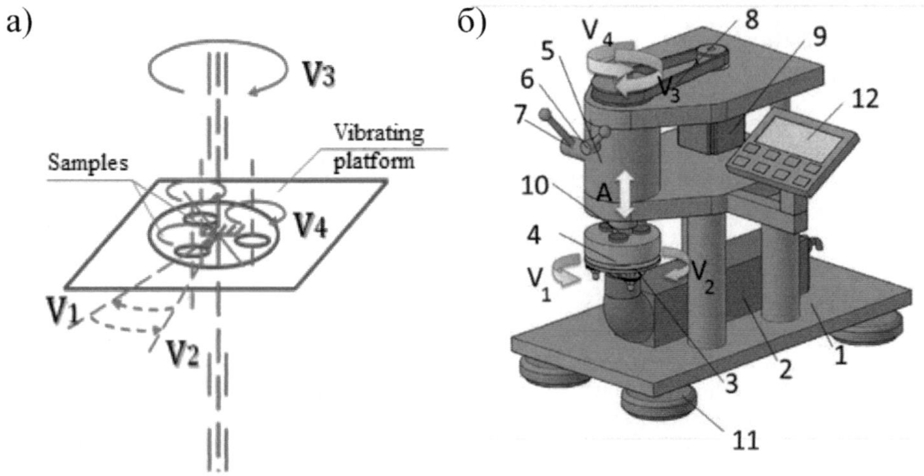

Fig. 1 a - vibration grinding scheme, V1, V2 - the direction of the action of high-frequency oscillations, V3, - forced rotation of the samples, V4 - automatic sample rotation b - a visual model of the working constructive solution of the vibration abrasive laboratory machine (protective screens, coolant supply devices and special polishing solutions are not shown).

The grinding scheme has the following advantages:

1. The absence of a continuously directed vector of the resultant cutting force during grinding. Provides self-alignment of the polished surface of the samples on the abrasive surface of the grinding material.

2. High-frequency vibration ensures that there are no traces of processing from one and the abrasive grains on the same machined areas of the metal sample and provides an effective control of the intensity of material removal.

3. Safe when manually preparing samples.

To pre-test the machine's performance, a prototype would be assembled. Its design is made of affordable and inexpensive materials - MDF, structural plastic and standard fasteners in the form of a base, on which prisms are mounted for attaching a universal grinder equipped with vibrating platforms [3, 4] and also studs in the form of threaded studs, plates in which the sample holder and the spring mechanism are placed, which ensures constant pressing of the samples to the vibrating platform. As a result of testing the prototype of the machine, the surface quality of the samples after grinding and polishing proved to be sufficient for performing metallographic studies in accordance with [1, 2]. In Fig. 2 shows photographs of the surface of a metallic sample after preliminary grinding with a domestic abrasive sheet of the Belgorod combine, followed by polishing with a solution of aluminum oxide. As we see after polishing, the surface of the thin section without scratches visible with a two-hundred-fold increase is suitable for chemical etching of the surface of the working surface. Measurements of deviation from the flatness of polished samples on an altimeter f. Mahr over the entire surface of the section showed values of not more than 0.1 mm. Undoubtedly, this value can be significantly reduced when manufacturing a working copy of the machine. The photographs after etching have identified a clear grain structure of the sample, suitable for further metallographic analysis in a specialized package of graphics programs, both at working magnifications of the study 200 times, and 500 times.

3. Application of the prototype machine

Operation of the prototype of the machine showed the possibility of using abrasive materials of both domestic and foreign production. When using domestic abrasive materials, the cost of grinding and polishing the samples was an order of magnitude smaller, relative to the preparation of the same set on the machine tool with the use of imported abrasive materials.

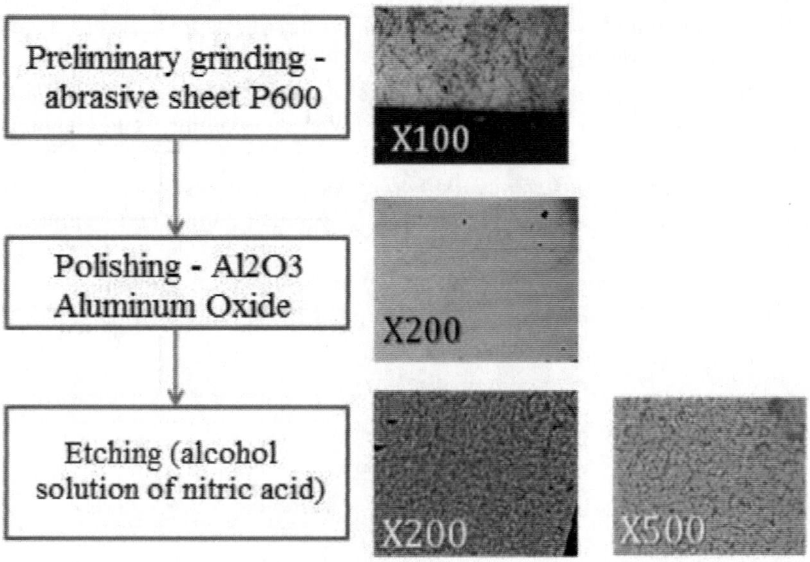

Fig. 2 Results of studies of samples produced on a prototype of a vibration grinder.

4. Conclusion

A high intensity of the removal of the allowance was confirmed and, as a consequence, the duration of grinding of the samples was short. The measurements of the obtained samples showed a flatness deviation of not more than 0.05 mm and a roughness Ra 0.04 in the working area of the section, which corresponds to the established requirements. The prototype of the machine is used for self-preparation of polished sections by bachelor students in the framework of conducting laboratory works on special disciplines in the direction of "Material Science" [5] and for studying the quality of wear-resistant coatings of cutting tools [6,7].

References

[1] Bekkert M, Klemm H 1986 *Handbuch der metallographischen Atzverfahren* (Leipzig: VEB Deutscher Verlag fur Grundstoffindustrie)
[2] GOST 1778-70 Stell. Metallographic methods for the determination of nonmetallic inclusions
[3] Kas'yanov S V, Kondrashov A G, Safarov D T and Faskhutdinov A I 2017 *Durable grinding head for universal machines* Russian Engineering Research V 37 Is 10 pp 912-15
[4] Safarov D T, Kondrashov A G, Faskhutdinov A I and Zairov B F 2017 *Design of grinding plates for a universal grinding machine* Russian Engineering Research V 37 Is 11 pp 1014-16

IOP Conf. Series: Materials Science and Engineering **412** (2018) 012064 doi:10.1088/1757-899X/412/1/012064

[5] Astaschenko V I, Zapadnova E A, Zapadnova N N, Mukhametzyanova G F 2016 *Predicting structure micro-alloyed steel products for different purposes* IOP Conference Series: Materials Science and Engineering V 134 No 1 **012029**
[6] Kasjanov S V, Kondrashov A G, Safarov D T 2014 *Research of characteristics of wearproof coating for cutting tools* INTERFINISH-SERIA 2014: International Conference on Surface Engineering for Research and Industrial Applications pp 124
[7] Kas'yanov S V, Kondrashov A G and Safarov D T 2017 *Rapid Assessment of Wear-Resistant Tool Coatings* Russian Engineering Research V 37 No 11 pp 969–73
[8] Akhmetov I D, Zakirova A R, Z B Sadykov and I I Khafizov 2017 New electrode-tool for the combined kerf of electrically conductive materials *IOP Conference Series: Materials Science and Engineering* V 240 Is1 **012003**
[9] Khafizov I I and Galimov A N 2017 IT-strategy and major aspects of quality management on the market of goods and services *IOP Conference Series: Materials Science and Engineering* vol.240 Is.1 Art. № 012038
[10] Khafizov I I 2017 *Ways of decrease in the material consumption in case of their separation by the combined methods* IOP Conference Series: Materials Science and Engineering V 240 Is 1 **012037**
[11] Khafizov I I 2016 *Economic efficiency and effectiveness of ways of separating materials electro diamond processing* IOP Conference Series: Materials Science and Engineering V 134 No 1 **012014**
[12] Khafizov I I 2015 *Processing methods with imposing of electric field at low- waste division of materials* IOP Conf. Series: Materials Science and Engineering V 86 **012013**

IOP Conf. Series: Materials Science and Engineering **412** (2018) 012065 doi:10.1088/1757-899X/412/1/012065

The method of diagnosing machine systems by measuring the accuracy of manufactured parts

D T Safarov[1], A G Kondrashov[1], I I Khafizov[2]

[1] Russia, Republic of Tatarstan, Naberezhnye Chelny institute (branch) of the Kazan (Volga) Federal University, 423812, Naberezhnye Chelny, Mira Ave., 68/19

[2] Kazan Federal University, 18 Kremlyovskaya street, Kazan, 420008, Russian Federation

safarov-dt@mail.ru, khafizov@kpfu.ru

Abstract. The main provisions of the technique allowing to create a diagnostic complex of the technical state of the machine system, which is informative at the same time of several diagnostic complexes - geometrical accuracy, strain gauge, technological accuracy, the influence of technological heredity - are revealed.

1. Introduction

The Republic of Tatarstan is characterized by a developed automotive industry. Here are concentrated large-scale production of heavy-duty cars. In this case, the issue of the car is provided not by one plant, but by a whole chain of enterprises supplying auto components. Today, suppliers are subjected to stringent requirements for product quality and delivery times. Expensive equipment, often leased. Absence of highly qualified personnel and working capital. An important condition for the competitiveness of an enterprise is the ability to ensure the release of quality products throughout the entire operation cycle of equipment [1].

As the output of products is inevitable wear and tear of the machine components, the displacement of individual modules, and hence the loss of accuracy. This requires a low-cost, simple and effective system for diagnosing the main modules of the technological system.

Requirements for the accuracy of machines and their individual components are set by manufacturers in the passports on the machine. To diagnose various parts of machine systems for compliance with the requirements is carried out by various methods. The diagnostic methods can be internal - built into the CNC machine system and external, for the implementation of which it is necessary to perform measurements with the help of special diagnostic complexes. The most applicable of them are the following: checking the moving nodes for geometrical accuracy, measuring the tensile forces of cutting forces and checking for technological accuracy. For example, the complexes of Renishaw - Ballbar QC20-W, laser system (C-ALS), Kistler 9129AA, etc. These complexes, although accurate, but their cost is comparable to the cost of the machine system.In order to use the complexes it is necessary to readjust the machine system, ie interrupt the technological process. The reason for the use of these systems are the indices of the suitability of technological processes, calculated in accordance with [2].

Content from this work may be used under the terms of the Creative Commons Attribution 3.0 licence. Any further distribution of this work must maintain attribution to the author(s) and the title of the work, journal citation and DOI.
Published under licence by IOP Publishing Ltd

IOP Conf. Series: Materials Science and Engineering **412** (2018) 012065 doi:10.1088/1757-899X/412/1/012065

2. Theoretical part

For a comprehensive assessment of the technical state of the machine system, simultaneous application of several types of diagnostics is necessary, which further increases the cost of this process. The goal of the project is to develop a hardware and software complex for diagnosing the technical accuracy of machine systems, which reveals a set of factors, based on measurements of the accuracy of the machined parts to quickly identify the most significant deviations that affect the quality of manufacturing machine parts.

The diagnostic complex allows:

1. Identify a set of factors recorded by three different methods of diagnostics - geometric accuracy, strain gauges, technological accuracy and calculates the degree of influence on the quality of the detail of the technological inheritance of the error transfer to the part.
2. For measurements, there is no need to reconfigure the equipment,
3. To work with the complex, it is enough to have the skills of personnel who have workplaces.
4. In comparison with foreign with foreign manufacturers, the lower cost of the complex and its operation.

The procedure and results of the experiments are given in [3-6]. The workpiece before processing and the processed part after processing are measured in a single measuring coordinate system from the same measuring bases. An obligatory condition for the calculation is the registration of process conditions. These include:

1. The coordinates of the base of the workpiece when installed in the tool.
2. The values of feed and cutting speed.
3. Status of the cutting edges of the tool.
4. Beating clamping elements rigging.
5. Limit values of key quality indicators of a part.
6. As a result of the complex's work, two 3D models are formed: the workpiece model and the part model in a unified coordinate system related to the coordinates of the position of the workpiece being machined in the machine coordinate system.

According to the 3D models, the actual distribution of the allowance is calculated, which, in the presence of recorded data on the conditions of the cut-off process, calculates the indices for a variety of directions - geometric accuracy, strength, stiffness of the machine system nodes, technological accuracy with calculation of the indices of convergence and reproducibility of the technological process, heredity of indicators of accuracy of preparation. The simultaneous presence of all this set of indicators allows us to formulate what is especially valuable for small enterprises - the content of corrective actions to improve the quality of the workpiece.

Thus, the effectiveness of the work of the complex is much higher, in comparison with the nearest imported analogues. Figure 1 shows a constructive solution of the hardware-software complex for diagnosing technical accuracy.

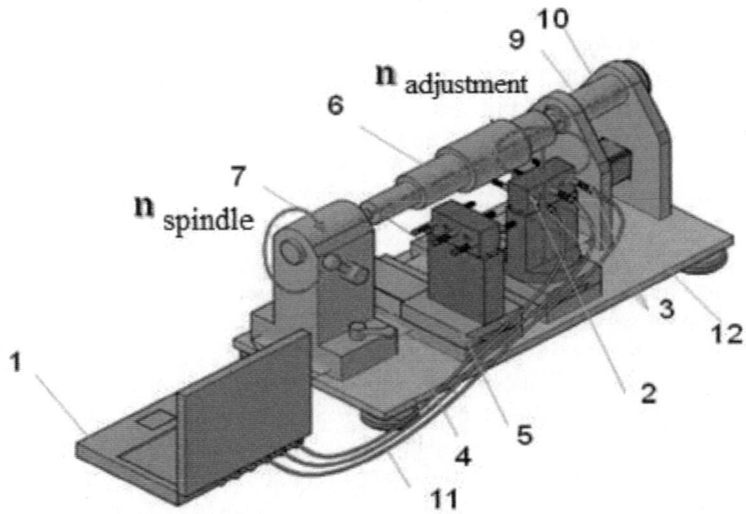

Fig. 1 Constructive solution of the software and hardware complex. nsp. - rotation of the part, nnad. 1 - measuring computer, 2 - capacitive sensor holder, 3 - capacitive sensors, 4 - longitudinal support, 5 - transverse support, 6 - measured part, 7 - front headstock, 8 - belt drive, 9 - stepper motor, 10-spindle drive details, 11 - interface cables, 12 - frame

In this constructive solution, the complex is intended for diagnostics of lathe group machines processing rotating bodies with a diameter of not more than two hundred mm and a length not exceeding four hundred. The machines of this group are most common in the processes of manufacturing parts at machine-building enterprises. For automated measurements, the part is installed in the center, one of which is driven by a stepper motor through a belt drive. The construction of the mathematical models of the workpiece and the workpiece is carried out from measurements taken simultaneously from three capacitive probes. To increase the accuracy of measurements in the design of the complex, it was decided to get rid of moving parts from the sensors.

The sensors have a range of readings that is greater than the depth of cut in the machining transition of the part being measured. The sensor holder is able to rotate about a vertical axis, which allows measurements not only of the outer cylindrical surfaces but also of the end and conical surfaces. Stepper motor control and data processing from the sensors is carried out using a measuring computer.

3. Application of the diagnostic technique

Based on these studies the methodology of diagnostic measurements was defined for accuracy indices. Contrary to the traditional statistical control it is planned to measure in pre-defined points of the machined surfaces. The mathematic treatment allows to define the contribution of each significant factor to the accumulative value of the measured deviation with the reasonable degree of accuracy.

The patents on the methods of measurement, diagnostics and diagnostic systems [7-9].

The experimental comparison of effectiveness and efficiency for a new procedure [2] was performed in the same manufacturing environment as the analysis of statistical methods.

In the nonmachining elements of two work pieces from the experimental sample the coordinate system was placed, the measurements were performed in the specified points before and after machining and the locating coordinates were recorded for each work piece in the manufacturing system

At the first step of measurement data analysis it was established that the critical accuracy indices of products are shaft and bushing circularity deviations and runout. The second step revealed limiting diagnostic constituents of these deviations. The biggest contribution to runout deviation is made by axis misalignment of boring bar and chuck. Therefore, during data analysis the specific corrective action has been already planned.

The technique of diagnosing according to the accuracy indicators of the processed ones was successfully tested in the existing technological processes for processing parts of various automobile units at the KAMAZ enterprise, and small suppliers.

4. Conclusion

For the practical application of established to measurement system. It consists of measuring the plate on which the measuring device linked to a computer and accessories to base parts [18-21]. Calculation of technological components is conducted with the registration in the database, automated issuance of the report on the need for corrective action.

Costs of such a measurement system will cost much cheaper buying any measuring machine , and efficiency will be higher , since the measurements can be performed in the workplace. Such a method of diagnosis is automatically embedded in the process of monitoring the quality of output. Diagnostic measurements performed adjuster with any change process (set-up, change of party preparations and so on.) , As well as planned in relation to the respective modules based cycles .

References

[1] Vasiliev V A, Odinokov S A, Borisova E V and Letuchev G M 2016 *Methods of quality management of innovation process* IEEE Conference on Quality Management, Transport and Information Security, Information Technologies, IT and MQ and IS 2016 pp 233-235

[2] Statistical process control (SPC). Reference Manual. Issued 1992, Second Edition, Issued July 2005. Copyrighht ©1992, ©1995, ©2005. Chrysler Corporation, Ford Motor Company, and General Motors Corporation

[3] Samsonov M A, Kas'yanov S V 2003 *Flexible plant systems for post-examination repairs instead of scheduled preventive maintenance system* Avtomobil'naya Promyshlennost Issue 2 pp 22-26

[4] Safarov D T, Fedorova K A and Ilyasova A I 2016 *Algorithms development of making special techniques in APQP manufacturing process of automotive components* / IOP Conference Series: Materials Science and Engineering V 134 Is 1 **012036**

[5] Safarov D T, Kondrashov A G 2018 *Stand for monitoring the operational parameters of conjugations "ball support - Body of the tie-rod end" of automotive components* IOP Conference Series: Materials Science and Engineering Vol 289 No 1, Is 12 **012018**

[6] Kas'yanov S V, A G Kondrashov, D T Safarov 2017 *Regulation of Geometrical Parameters Deviations of Automotive Components Parts through Diagnostic Measurements Organization* Procedia Engineering Volume 206 pp 1508-14

[7] Kasjanov S V, Kondrashov A G, Safarov D T 2013 Russian patent 2496611

[8] Kasjanov S V, Kondrashov A G, Safarov D T 2013 Russian patent 133040

[9] Kasjanov S V, Kondrashov A G, Safarov D T 2013 Russian patent 133039

[10] Kasjanov S V, Safarov D T 2004 *Diagnosis of technical state of equipment and tools according to indices of technological accuracy* Avtomobil'naya Promyshlennost Issue 5 pp 24-28

[11] Akhmetov I D, Zakirova A R, Z B Sadykov and I I Khafizov 2017 *New electrode-tool for the combined kerf of electrically conductive materials* IOP Conference Series: Materials Science and Engineering vol 240 Is 1 **012003**

[12] Khafizov I I and Galimov A N 2017 *IT-strategy and major aspects of quality management on the market of goods and services* IOP Conference Series: Materials Science and Engineering vol 240 Is 1 **012038**

[13] Khafizov I I 2017 *Ways of decrease in the material consumption in case of their separation by the combined methods* IOP Conference Series: Materials Science and Engineering vol 240 Is 1 **012037**

IOP Conf. Series: Materials Science and Engineering **412** (2018) 012065 doi:10.1088/1757-899X/412/1/012065

[14] Khafizov I I 2016 Economic efficiency and effectiveness of ways of separating materials electro diamond processing IOP Conference Series: Materials Science and Engineering vol 134 No 1 **012014**
[15] I I Khafizov 2015 *Processing methods with imposing of electric field at low- waste division of materials* IOP Conf. Series: Materials Science and Engineering V 86 **012013**

IOP Conf. Series: Materials Science and Engineering **412** (2018) 012066 doi:10.1088/1757-899X/412/1/012066

Feasibility Analysis of Combined Boiler Plant Designs

T R Safin[1], I A Konakhina[1] and G R Khamidullina[1]

[1]Kazan Federal University, 18 Kremlyovskaya street, Kazan, 420008, Russian Federation

rr-088@mail.ru

Abstract. This paper discusses the method of perfecting energy saving actions in boiler plants based on the analysis and optimization of their performance indicators. A boiler-house plant with low-potential secondary energy resource and external combination units is taken as a study object.

Boiler plants are widely used as thermal energy sources to provide loads of heating, ventilation, hot water supply to various groups of consumers. In addition to the specified types of loads, large production boiler houses meet the requirements of industrial undertakings in medium and low pressure steam. In the Russian Federation, the share of boiler plants in heat supply sources exceeds 60%, while the fuel consumption reaches 80% of the cost price of the heat released by them; therefore, the challenge of improving the energy efficiency and performance of the boiler plant equipment is relevant and significant for our economy.

The synthesis of combined boiler plant designs is performed in two options:

1. Reduction of energy losses due to utilization of low-potential renewable energy resources (stack gases, heat radiation, etc.)
2. Addition of external heat components to the boiler plant that make it possible to produce two or more kinds of energy resources (boiler plant turning to mini-CHP mode of operation, installation of heat pumps, hybrid heat collectors, etc.)

Searching for an optimal combination of methods for improving boiler plants and building its effective energy-technology system (ETS) is possible based on the methodology of system analysis and synthesis of complex heat and power plants [1, 2]. As is shown in [3], it is practicable to evaluate the effectiveness of the decisions taken based on the performance indicators.

Similarly to [1] the problem of finding the most advantageous combination of ETS performance parameters of the synthesized combined boiler plant is formulated as follows:

The minimum value of the performance criterion (the reduced costs of the combined boiler plant) is described as

$$Z = f(Y, X)_{\sigma}, \tag{1}$$

given the nonlinear constraint conditions of

$$\varphi_i (Y, X)_{\sigma} = 0; \quad i = \overline{1, n}, \tag{2}$$

and inequality of

Content from this work may be used under the terms of the Creative Commons Attribution 3.0 licence. Any further distribution of this work must maintain attribution to the author(s) and the title of the work, journal citation and DOI.
Published under licence by IOP Publishing Ltd

IOP Conf. Series: Materials Science and Engineering **412** (2018) 012066 doi:10.1088/1757-899X/412/1/012066

$$(f_s)_{min} \leq f(Y,X)_{\boldsymbol{\sigma}} \leq (f_s)_{max}; \quad s=\overline{1,a} \ , \tag{3}$$

and also given the parameter restrictions of

$$\left.\begin{array}{l} Y_{min} \leq Y \leq Y_{max}; \quad \mathrm{Y} \in \boldsymbol{\sigma}_m; \\ \\ X \in L_t. \end{array}\right\} \tag{4}$$

Where Y is a set of ever-changing parameters;
X is a set of discretely changing parameters of the system;
$\boldsymbol{\sigma}$ is a set of characteristics of external accountable factors;
$\boldsymbol{\varphi} = \{\boldsymbol{\varphi}_1, \boldsymbol{\varphi}_2, ..., \boldsymbol{\varphi}_n\}$ is a system of balance equations for all parts of the boiler plant ETS;
$f = \{f_1, f_2, ..., f_a\}$ is a set of ETS characteristics, based on which the limiting conditions are given;
The terms 'min' and 'max' mean minimum and maximum values, respectively;
$\boldsymbol{\sigma}_m$ is the dimension of the concerned space;
L_t is some finite set of discrete elements of dimension t.

The boiler plant ETS consists of systems and subsystems which, when constructing mathematical models, are described in the form of successive chains of connections between the balance equations that define the values of the model parameters. This makes it possible to represent a complete system of n-equations in the form of a strict sequence of subsystems [3]. In this case, it suffices to restrict the defining input and output parameters of the subsystems (the independent parameter vector $\overline{X}$) by means of conditions of the form (4), so that all the dependent internal parameters of the subsystem (vector Y of dimension n) would automatically satisfy this condition.

Hence

$$Y = (\overline{X}, H); \tag{5}$$

provided that

$$(\overline{X}_j)_{min} \leq \overline{X}_j \leq (\overline{X}_j)_{max}; \quad j=\overline{1,k}; \quad k = m\text{-}n. \tag{6}$$

Another feature of these systems is that there can be technical limitations on the nonlinear characteristics $f_p(\overline{X}, H, X)$ the objective function Z of which is clearly independent.

The resulting general sequence of solving the problem is as follows:

1) The parameters X и $\overline{X}$ that ensure the minimum value of the performance criterion will be determined;

2) The calculation of the system of n-equations will determine the vector H from the obtained values X и $\overline{X}$.

In general, the equation for the definition of H is written as

$$H = f(X, \overline{X}) \tag{7}$$

3) Functions Z and f are defined at every step as functions of independent parameters X и $\overline{X}$.

Therefore, the problem is limited to finding the minimum of the reduced cost function

$$Z(X, \overline{X}) \text{ при } (X, \overline{X}) \in R, \tag{8}$$

IOP Conf. Series: Materials Science and Engineering **412** (2018) 012066 doi:10.1088/1757-899X/412/1/012066

where R is the acceptable region given by conditions

$$(f_s)_{\min} \le f_s(X, \overline{X}) \le (f_s)_{\max}, \quad s = \overline{1, a}; \tag{9}$$

$$\overline{X}_{\min} \le \overline{X} \le \overline{X}_{\max}, \quad \overline{X} \in \boldsymbol{\sigma}_s; \tag{10}$$

$$X \in L_t \tag{11}$$

The functions $\Phi(\overline{X})$ and $f_s(\overline{X})$ are assumed differentiable at fixed values X.

Similarly to [1], the solution of the problem is divided into two phases matching the parameters of the object with iterative procedures:

Phase 1. Determination of the aggregate of the most advantageous parameters $\overline{X}$, i.e. parameters that change continuously in the region of definition.

Phase 2. Determination of the aggregate of the most advantageous parameters X, i.e. parameters that change discretely.

The solution of searching for the optimal option lies in the method of coordinate descent through enumeration of possibilities for combining the duty parameters.

The boiler plant of Gazprom dobycha Nadym LLC, equipped with KMB-5.2 MW hot-water boilers was considered as the study object. The synthesis of the combined boiler plant was carried out by improving the thermal conditions of the hot water boilers due to the addition of the external economizers and external combination of the plant by hybrid solar collectors and heat pumps that make it possible to generate auxiliary heat and electric power.

The payback period of the synthesized ETS with optimized parameters was less than 1.5 years suggesting the high efficiency of the proposed solutions.

References

[1] Nazmeyev Yu G and Konakhina I A 2002 *Heat and power systems and energy balances of industrial enterprises* (M.: Publishing house of Moscow Power Engineering Institute)

[2] Konakhina I A, Safin T R and Gilmanshin I R 2013 *Enhanced performance of fuel-using plants due to external waste heat recovery units* (Proceedings of the International Scientific and Technical Conference "Innovative Machine-building Technologies, Equipment and Materials – 2013) Part 2 pp 209-213

[3] Khamidullina G R 2016 *Optimization of energy saving decisions for the large compressor stations of machine-building firms based on the technical and economic indicators* (IOP Conference Series: Materials Science and Engineering) Vol 134 Is 1 **012015**

[4] Khusnutdinova E M, Konakhina I A and Khamidullina G R 2016 *Hybrid Technologies in District Heating Systems* (IOP Conference Series: Materials Science and Engineering) Vol 134 Is 1 **012018**

IOP Conf. Series: Materials Science and Engineering **412** (2018) 012067 doi:10.1088/1757-899X/412/1/012067

Preparation of copper-containing microfertilizers

R N Sagitova, I A Gaisin, S I Gilmanshina

[1]Kazan Federal University, 18 Kremlyovskaya street, Kazan, 420008, Russian Federation

rns19@mail.ru, gilmanshina@yandex.ru

Abstract. The paper considers the preparation of chelate microfertilizers based on copper salts and monoethanolamine. Monoethanolamine increases the solubility of copper compounds in water, which makes it possible to use a wasteless technology for the production of a micronutrient solution with a high mass fraction of a microelement. The resulting microfertilizer solutions have a positive effect on the growth and development of agricultural plants.

At present, in the aspect of solving the problems of effective energy management [1] and renewable energy [2], as well as the utilization of industrial wastes [3], an important task is the development of new materials and compounds, including microfertilizers, using energy and resource-saving technologies.

Microfertilizers are now produced in the form of solid compounds (or a mixture of solid compounds) and in the form of aqueous solutions of the corresponding compounds. Industrial production of microfertilizers in the form of aqueous solutions is more energy- and resource-saving, since this reduces the technological process and there is no need to isolate the final product from the reaction mixture.

An effective form of microfertilizers are not salts, but coordination compounds of microelements (chelates). Consequently, the creation and expansion of the assortment of chelated microfertilizers is an urgent task.

In the presented study, the preparation of a microfertilizer composition based on the compounds of copper (II) and monoethanolamine is considered.

When the compounds of copper and monoethanolamine interact, a chelate is formed. The method for obtaining this microfertilizer is to mix the salts of copper (II) and monoethanolamine in water. The process is one-stage, non-waste.

To prepare the microfertilizers, the calculated amount of water (solvent) is mixed with a copper (II) salt and monoethanolamine. The order of mixing of the components does not affect the quality of the final product, but from the point of view of the energy efficiency of the process, the following order is preferable: 1) obtaining a monoethanolamine solution; 2) dissolving the copper salt in the resulting solution while stirring.

Dissolution of monoethanolamine in water is an exothermic process, the temperature of the reaction mixture rises and the dissolution of copper compounds in a warm solution of monoethanolamine (40-60 ° C) occurs quickly and smoothly. It is not necessary to heat the reaction mixture to speed up the process, which also corresponds to the principle of effective energy saving.

Content from this work may be used under the terms of the Creative Commons Attribution 3.0 licence. Any further distribution of this work must maintain attribution to the author(s) and the title of the work, journal citation and DOI.
Published under licence by IOP Publishing Ltd

The ratio of monoethanolamine and copper salt in the preparation of chelates can vary widely. However, for the formation of soluble coordination compounds, it is necessary to use an excess of monoethanolamine, since in the case of an insufficient amount of this substance, the insoluble basic copper (II) salts precipitate. The minimum ratio (mol) of the copper (II) salt: monoethanolamine, which is necessary for obtaining a stable solution, is 1: 3. The optimum ratio of components (moles) is 1: 4 or 1: 5. In this case, the complexion of copper ions is maximal.

Excess monoethanolamine in comparison with the optimal ratio of components has practically no effect on the stability of microfertilizer, but it positively affects the dissolution rate of copper salts in solution and improves the effectiveness of microfertilizer action. The total content of the components in the microfertilizer solution can also vary widely, but from a practical point of view, it is desirable to obtain formulations with a high concentration of active ingredients. Monoethanolamine increases the solubility of copper compounds in water. As a result, the mass fraction of copper in the final product (microfertilizer solution) becomes comparable with the copper content in solid ethylenediaminetetraacetate cuprate (II) of sodium (Table 1). (This substance is a widespread microfertilizer, where the mass fraction of copper is 16%). Isolation of the final product from the resulting solution is not required.

Table 1

The content of copper in microfertilizer solutions

The copper (II) compound	The weight of the copper (II) compound to produce 1 kg of a microfertilizer solution	Mass fraction of copper in a solution of microfertilizer,%
$CuSO_4 \cdot 5H_2O$	250	6,4
$(CuOH)_2CO_3$	222	12,8
$CuCl_2 \cdot 2H_2O$	252	9,6
$Cu(NO_2)_2 \cdot 3H_2O$	242	6,4
$Cu(CH_3COO)_2 \cdot H_2O$	300	9,6

The obtained solutions of the copper (II) chelate are colored dark blue ($\lambda_{max} = 590$ нм pH = 9.5-10.5) The stability of the complexes obtained depends strongly on the acidity of the medium.The complexes obtained are stable in an alkaline medium, in an acid medium (pH = 5-6) is their destruction.

Synthesized compositions can be used alone or as a basis for the production of mixed-type microfertilizers (copper-boron, copper-molybdenum, copper-boron-molybdenum-containing compounds). In this case, boric acid or molybdic acid (or ammonium molybdenum acid) is introduced into the resulting composition with stirring (Table 2).

Table 2.

The content of microelements in solutions of microfertilizers, if the starting compound $CuSO_4 \cdot 5H_2$

Weight of components in microfertilizers, g / kg				Mass fraction of microelement, %		
$CuSO_4 \cdot 5H_2O$	Mono-Ethanolamine	H_3BO_3	H_2MoO_4	*Cu*	*B*	*Mo*
250	300	62	-	6,40	1,10	-

IOP Conf. Series: Materials Science and Engineering **412** (2018) 012067 doi:10.1088/1757-899X/412/1/012067

250	300	124	-	6,40	2,20	-
125	300	217	-	3,20	3,85	-
250	300	-	82	6,40	-	4,80
250	300	-	41	6,40	-	2,4
125	300	124	41	3,20	2,20	2,4
125	300	62	41	3,20	1,11	2,4

The ratio of trace elements in the formulations can be varied within wide limits. The use of excess monoethanolamine in the formulations makes it possible to increase the solubility of both boric and molybdic acid, as a result of which their content in solutions increases in comparison with their saturated solutions.

When hydroxy-medium (II) carbonate is used as the starting compound, it is possible to obtain compositions with a higher content of trace elements (Table 3).

Table 3.

The content of microelements in solutions of microfertilizers, if the starting compound $(CuOH)_2C($.

Weight of components in microfertilizers, g / kg				Mass fraction of microelement, %		
$(CuOH)_2CO_3$	Mono-Ethanolamine	H_3BO_3	H_2MoO_4	*Cu*	*B*	*Mo*
222	300	62	-	12,8	1,10	-
111	300	124	-	6,40	2,20	-
111	300	217	-	6,40	3,85	-
222	300	-	82	12,80	-	4,80
222	300	-	41	12,80	-	2,4
111	300	124	41	6,40	2,20	2,4
111	300	62	41	6,40	1,11	2,4

All the proposed formulations are ready for use after dilution. The compositions can be used for presowing seed treatment of foliar top dressing of vegetating plants.

The use of microfertilizers obtained positively affects the growth and development of agricultural plants, increases their yield by 18-23% [4, 5].

References

[1] Kashapov N F, Gil'manshin I R, Konahina I A et al. 2014 *System analysis of the energy complex of engineering enterprise as a basic tool of effective energy management* IOP Conference Series: Materials Science and Engineering Vol 69 **012024**

[2] Gilmanshin I R, Kashapov N F, Gilmanshina S I, Galeeva A I 2016 *Landfill energy complex based on the renewable energy installations* IOP Conference Series: Materials Science and Engineering Vol 134 **012007**

[3] Azimov Yu I, Gilmanshin I R, Gilmanshina S I et al. 2016 *Modern technologies of waste utilization from industrial tire production* IOP Conference Series: Materials Science and Engineering, Vol 134 **012003**

[4] Gaisin I A et al 1999 *A method for obtaining compositions for stimulating the growth of agricultural crops* Patent Russian Federation No 2086126

[5] Gaisin I A et al 1995 *A method for preparing a composition for stimulating the growth of crops* Patent Russian Federation No 2086126

Thermal chamber for material testing

T N Savelieva[1], O A Sachenkov[1], I R Gilmanshin[1,2], M M Zakirnichnaya[4], Yu I Azimov[1], N F Kashapov[1], S I Gilmanshina[1], A I Galeeva[1,3], D R Krainova[1]
[1]Kazan Federal University, 18 Kremlyovskaya street, Kazan, 420008, Russian Federation
[2] Kazan national research technical university named after A N Tupolev, K. Marx St. 10, Kazan, 420111 Tatarstan, Russia
[3] Kazan State Power Engineering University, Kazan, 420066, Russian Federation
[4] Ufa State Petroleum Technological University, 1, Kosmonavtov St., Ufa, 450062, Russia

is-er@yandex.ru

Abstract. The problem of ensuring the correct conduct of the experiment with preservation of the required temperature is considered. The construction of the device is proposed that will allow satisfying the conditions for carrying out tensile tests with an increased or lowered sample temperature.

The basis for evaluating the properties of the material is information on how the material reacts to any load. Knowing the magnitude of the strain created by a given load (voltage), it is possible to predict the response of a particular product to its operating conditions [1].

Strength values of polymers are determined on the tensile machines, on which it is possible to perform tensile, bending and compression tests.

To carry out studies on stretching and to determine the physico-mechanical properties of recycled polyethylene waste, samples were prepared. Samples for testing were made on a milling horizontal machine at a speed of 250 revolutions per minute. After manufacturing, the samples were conditioned for at least 16 hours at a temperature of 23 (± 2) ° C and a relative humidity of 50 (± 5)%. To perform a series of tensile tests, the samples were numbered and the samples were held in the freezer at a temperature of plus 4 ° C for 72 hours. Under conditions of testing at minus 40 ° C, the samples were brought to the required temperature immediately before the start of the tests.

The main task of preparing the samples was to hold at certain temperatures in order to obtain the character of the change in the strength characteristics as a result of the tensile tests, depending on the effect of different temperatures.

Tensile tests were carried out in accordance with GOST 11262-80 "Plastics. Method of tensile testing "and GOST 14359-69" Plastics. Methods of mechanical testing. General requirements ", establishing a method for tensile tests of plastic materials for determining the characteristics of mechanical properties: tensile strength, tensile strength, physical tensile yield stress; Young's modulus.

For a series of tensile tests at a temperature of minus 40 ° C, liquid nitrogen was used to cool the samples. When the samples were cooled to minus 40 ° C, they proceeded directly to the tests. To measure the temperature at the time of the tests, a thermal imager DALILT7 was used.

Content from this work may be used under the terms of the Creative Commons Attribution 3.0 licence. Any further distribution of this work must maintain attribution to the author(s) and the title of the work, journal citation and DOI.
Published under licence by IOP Publishing Ltd

IOP Conf. Series: Materials Science and Engineering **412** (2018) 012068 doi:10.1088/1757-899X/412/1/012068

The tensile test was carried out on a two-zonal testing machine model UTS 110M-100 of the floor version [2], certificate of approval of the type of measuring instruments No. 34879. Range of measured loads, kN: 0.001 - 100; load measurement error ± 1% (0.5)% of the reading up to 1/100 of the permissible load of the load cell. The loading rate for the samples was 10 mm / min at -40 ° C and 20 mm / min at -4 ° C and +20 ° C.

According to the test results, loading diagrams in the F-l (strain-strain) coordinates were obtained, over which the tensile strength and yield strength were determined (Fig. 1). The relative uniform elongation and relative narrowing was determined from the measurements of the length of the sample after the test.

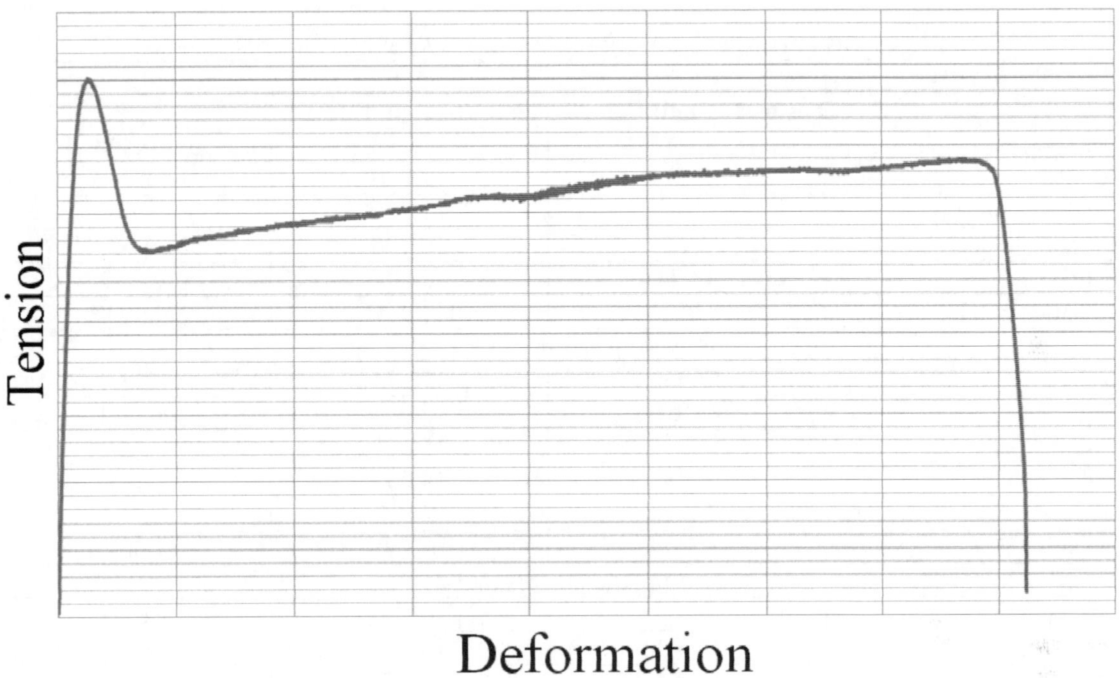

Figure 1. A typical diagram of loading a batch of samples from samples of polyethylene recycled by industrial means

To determine the ultimate strength, yield point, standard formulas were used.

The tensile strength, according to the formula:

$$\sigma_{PM} = \frac{F_{PM}}{A_0} \quad (1)$$

Breaking strength, according to the formula:

$$\sigma_{PP} = \frac{F_{PP}}{A_0} \quad (2)$$

Yield strength in tension, according to the formula:

$$\sigma_{PT} = \frac{F_{PT}}{A_0} \quad (3)$$

where A_0 – area of initial cross section; F_{PM} – maximum tensile load; F_{PP} – load at which the sample collapsed; F_{PT} – tensile load at yield point.

Analyzing the results of the loading diagram of samples from industrial polyethylene recycled at different test temperatures (Figure 1), tests showed that the material has a yield strength.

In all tests, the tensile load at reaching the yield point was greater than the load at which the sample failed, so the tensile strength of the values coincides with the tensile yield strength.

When determining the Young's modulus, the linear section was first determined on the tension diagram, the error of linear approximation was estimated by the least squares (R^2) method, in calculations the linear portion was determined from the condition $R^2 \geq 0.90$.

Table 1 shows the results of tensile tests of polyethylene recycled by industrial processes.

Table 1. Results of tensile tests of polyethylene recycled by industrial process

Parameters	**-40 °C**	**-4 °C**	**NU**
Tensile strength / Tensile yield strength (MPa)			
Test1	28,286	17,222	16,048
Test 2	22,524	18,810	11,310
Test 3	–	21,286	–
Average value	25,405	19,106	13,679
SKO*	4,074	2,048	3,350
Deviation SKO	16,037%	10,719%	24,493%
Breaking strength (MPa)			
Test 1	17,952	15,333	13,571
Test 2	14,810	12,952	8,690
Test 3	–	15,619	–
Average value	16,381	14,635	11,131
SKO	2,222	1,464	3,451
Deviation SKO	13,567%	10,004%	31,007%
Young's modulus (MPa)			
Test 1	320,012	241,06	486,451
Test 2	309,930	186,34	386,22
Test 3	–	344,073	–
Average value	314,971	257,158	436,336
SKO	7,129	80,089	70,874
Deviation SKO	2,263%	31,144%	16,243%

During the research on the fracture, the grippers of the test machine had a large heat capacity, and the test samples had a small thickness, the heat transfer was intensive, the temperature of the stretching region changed rapidly, which directly affects the experimental results. In this connection, the problem arises of ensuring the correct conduct of the experiment while maintaining the necessary temperature.

Based on GOST 14760-69 "Adhesives. The method for determining the strength at break-off (with changes No. 1, 2) "tests at elevated and lower temperatures are prescribed to be carried out in a special chamber intended for these purposes. In this case, the working space of the testing machine should allow the installation of a chamber with extension bars for attaching specimens, which should ensure a reliable centering of the sample in the grippers of the testing machine. The design of the chamber should ensure uniform cooling of the working part of the sample. Cooling the sample to the specified test temperature and the test itself can be carried out in a liquid medium. It is allowed to cool the sample to the specified test temperature and test it in a cooled gaseous medium. As a liquid medium, ethyl rectified alcohol is used in accordance with GOST 18300-87, cooled by carbon dioxide liquid according to GOST 12162-77 or liquid nitrogen in accordance with

GOST 9293-74, or ice, or liquid nitrogen. Periodic monitoring of the temperature of the test sample at elevated and lower temperatures is carried out with a thermocouple with a potentiometer of accuracy class not lower than 0.5 according to GOST 9245-79.

Cryocameras are known to test machines that contain a thermal insulation housing, a device for introducing refrigerant, a screening cylinder with conical bowls [4-9]. These cameras do not exclude the influence of the pressure difference that occurs, for example, when the cavity of the cryocamera is pumped out, to load the sample. However, they have a low efficiency of cooling the sample due to the introduction of heat to the sample and to the non-cooled jaws.

In order to improve the accuracy and reduce the laboriousness of the test by providing the required conditions for carrying out the tensile test with an increased or decreased temperature of the polymer sample, the following device is proposed.

The diagram of the device is shown in Figure 1. It is proposed to produce a glass in the form of a cylinder with a diameter of 150 mm made of steel. At the bottom of this cylinder will be mounted flat bracket, which will clamp the lower jaws of the test machine. Inside the cup is a bracket that will clamp the bottom of the sample. Dimensions are adjusted with screws. The mechanism of fixing on one axis. The upper part of the sample is the clamp in the intermediate sponge, which is fixed through the bracket in the upper jaw of the tensile machine.

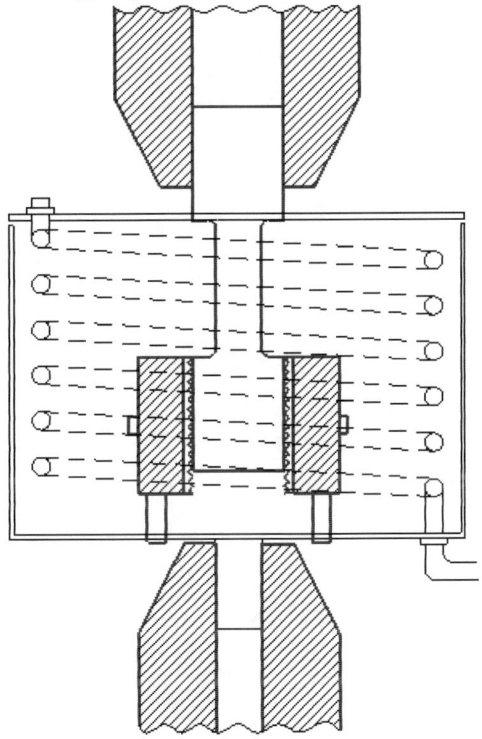

Figure 2 - Diagram of the device

The upper part of the cylinder is closed by a shaped cover with a slot that minimizes heat flows in the inside of the device. The cylinder itself is insulated to reduce losses to the environment. If the lower part is relatively stable, then the upper part is free.

Inside the cylinder is mounted a coil. It is powered by an inverted heat engine. Depending on the requirements, both positive and negative temperatures can be maintained. Additional decrease in temperature is due to liquid nitrogen, whose vapors provide uniform cooling throughout the chamber area. The temperature control is carried out with a thermal sensor mounted in the cylinder.

Thus, the proposed camera allows to increase the cooling efficiency and conditionally stabilize the temperature.

IOP Conf. Series: Materials Science and Engineering **412** (2018) 012068 doi:10.1088/1757-899X/412/1/012068

References

[1] *Methods of testing polymeric materials* [electronic resource] http://komef.ru/metodisppolimer.pdf.
[2] *Universal testing machine UT-110M-100* [electronic resource] https://test-systems.ru/produkciya/universalnye-ispytatelnye-mashiny/110m/uts-110m-100-1u.
[3] Perepechko I I 1977 *Properties of polymers at low temperatures* (M., Chimiya) p 272
[4] Zaitsev N L, Googe S Yu, Waisman L A, Gindin V A 1987 *Method for testing a sample for strength at low temperatures* Patent SU 1538095
[5] Gindin I A, Kravchenko S F, Lebedev V P, Starodubov Ya D 1970 *Testing machine* Patent SU 410286
[6] Deev Yu P, Dyadkin V P, Lebedev Yu P, Semenov L V 1985 *Thermocryometer chamber for statistical testing of materials* Patent SU 1317253
[7] Ermolaev O N, Krapotin V N 1975 *Cryocamera to the test machine* Patent SU 523328
[8] Dunich E A, Dunich O E 1969 *A device for determining the mechanical properties of materials at low temperatures* Patent SU 314107
[9] Pivovarov N K, Zdvizhkov V P 1974 *Cryocamera to test machines* Patent SU 513296

IOP Conf. Series: Materials Science and Engineering **412** (2018) 012069 doi:10.1088/1757-899X/412/1/012069

Aspects of using accelerated weather testing methods for polymeric materials

L N Shafigullin[1], N V Romanova[1], I F Gumerov[1], A R Ibragimov[1], G R Shafigullina[1] and A I Nizamova [1]
[1]Naberezhnye Chelny Institute (branch), Kazan Federal University
Sjujumbike street 10a, Naberezhnye Chelny, 423800 Tatarstan, Russia

misharin_82@mail.ru

Abstract. The paper investigates the aspects of using accelerated weather testing methods for polymeric materials. The need has been identified for consistency of polymer ageing and degradation mechanisms in full-scale and laboratory conditions, as well as for equipment that most closely simulates the actual operating conditions.

1. Introduction

The properties of different polymeric materials change (ageing) as they are used in machine production, construction, electrical and radio engineering, agricultural sector. Ageing of a material during the operation and storage is a result of a series of chemical and physical transformations in complex multicomponent systems [1].

The properties of basically all polymeric materials have low time-dependent stability. Under the influence of heat, oxygen, air, light, mechanical stresses, ionizing radiation and other factors polymeric materials degrade: there are processes running in them, which are accompanied by change in their chemical and physical structure and deterioration of their strength, dielectric and other properties [2].

The factors contributing to ageing can be divided into two main categories: internal (inner stresses, phase transitions, microstructure transformations, changes in chemical composition) and external (change in temperature, humidity, oxygen concentration, visible, ultraviolet (UV) or ionizing radiation, chemical exposure) [3, 9-10].

Ageing of a material can manifest itself in various forms. The form depends not only on the properties and composition of the material but also on the type of influence. The external factors can be divided into six types of ageing which can be combined with each other: thermal ageing; weathering; attack by corrosion and chemical ageing; strain ageing; electrical ageing; UV ageing [3].

2. Body text

There are hundreds of brands of polymeric materials based on synthetic resins, rubbers and high-molecular compounds. They are used in all the manufacturing and construction sectors. Thus far, a lot of research has been done to design new polymer compositions and lower their cost, simplify the production methods, reduce production toxicity, as well as provide a wide operating temperature range. Due to a large number of basic components the properties of resulted polymers can be highly varied. Polymers are widely used in the vehicle industry: noise and thermal insulation, exterior and interior parts, fuel system components [4 - 6].

Content from this work may be used under the terms of the Creative Commons Attribution 3.0 licence. Any further distribution of this work must maintain attribution to the author(s) and the title of the work, journal citation and DOI.
Published under licence by IOP Publishing Ltd

IOP Conf. Series: Materials Science and Engineering **412** (2018) 012069 doi:10.1088/1757-899X/412/1/012069

Today the polymeric material durability and service life prediction remain understudied. The service life of polymer products can vary from 2-3 to 10 years or more depending on their purpose, operating conditions, technical and commercial requirements. However, process designers of polymeric materials have no universal tools to design these materials with the required durability.

In development of new products and scientific research artificial ageing is used. It is a process of simulating the influences on a material to reproduce its ageing process. It may require climatic chambers, environmental simulation equipment and vibration testing systems. It is important to monitor and record any changes in the material properties during development. All the results should be repeatable to help engineers investigate the probability of material degradation [3].

Accelerated weather testing methods are based on transformation of energy values of the operational factors (related to ageing and degradation of materials and products) to the suitable laboratory conditions of accelerated testing. The algorithm of development of accelerated weather testing methods include the following stages: development of an operation chart; transformation of energy values of the operation factors to enhanced simulated laboratory factors; development of laboratory conditions of accelerated weather testing. The main criteria of accelerated weather testing are as follows: maximum reduction of testing time; consistency of the ageing mechanisms in full-scale and accelerated laboratory conditions [2].

We are going to examine the aspects of using accelerated weather testing methods in real life. The claims were found for operation of polypropylene components used in the engine cooling systems in different climatic operation zones between 2013 and 2016 (Figure 1, 2).

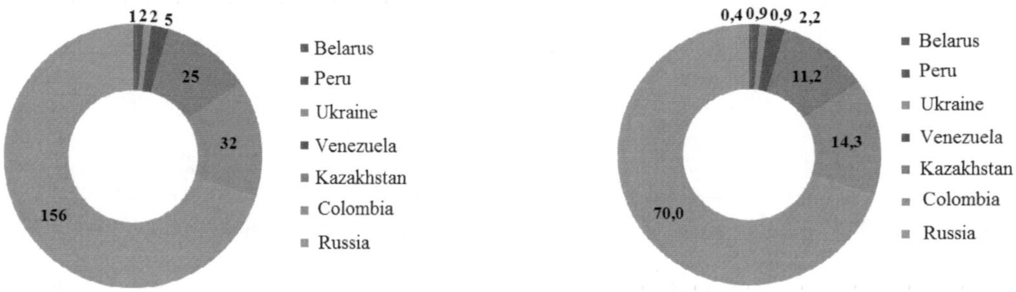

a) b)

Figure 1. Breakdown of the claims for replacement of polypropylene products by climatic zones: a) in units; b) as a percentage

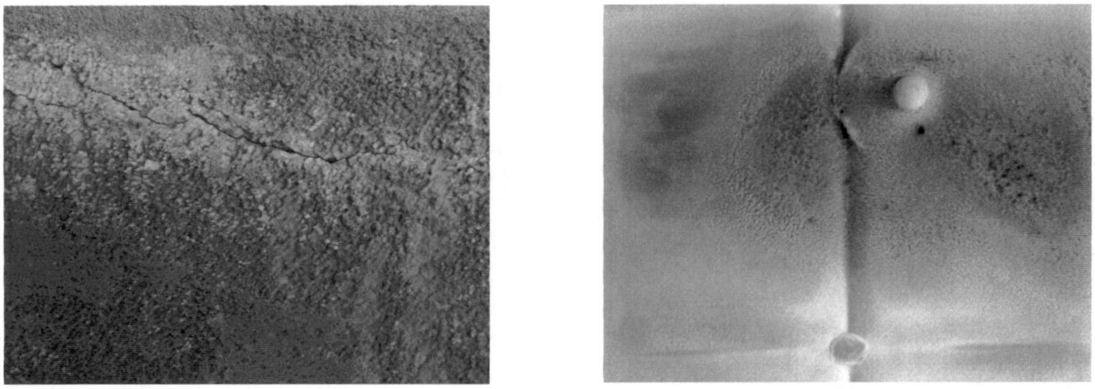

a) b)

Figure 2. Visual appearance of damage in polypropylene components as the result of influence of: a) climatic factors; b) accelerated weather testing

Accelerated weather tests were performed to determine the factors contributing to ageing. For this purpose, the experimental setup with an UV radiation source (1000 W tubular mercury vapour lamp) was used [7]. The operation chart was developed according to GOST 16350-80, Climate of the USSR. Regionalizing and statistical parameters of climatic factors for technical purposes [8]. As the main portion of product claims was found in southern Russia the climatic zone II10 "Warm humid Batumi" was applied. Energy values of the operational factors were calculated into enhanced simulated laboratory factors, the laboratory conditions of accelerated weather testing were developed. 1 year of real-life conditions is equivalent to 29 hours of accelerated weather testing. Figure 2 shows the investigation results.

Figure 2 reveals the significant difference in visual appearance of the defects. When transforming the energy values of the operational factors into enhanced simulated laboratory conditions the polymer ageing and degradation mechanisms under full-scale and laboratory conditions were thought to be different due to different rates of relaxation processes, different rates of competing processes, different synergistic effect. In addition, the activation spectrum initiated by the tubular mercury vapour lamp does not match the real-life spectra. Xenon arc lamps with a Pyrex glass filter should be used to simulate UV radiation. The severe conditions of accelerated weather testing initiated the mechanism of material degradation due to thermal decomposition. In the real-life operating conditions products are simultaneously subjected to different types of ageing: thermal ageing, UV radiation, mechanical stress and attack by corrosion [1, 2].

3. Conclusions

LFI technology helps to increase the production flexibility and quality, reduce the production cost of products, especially large-sized ones, thanks to a high quality of manufacturing and a lower amount of production waste.

This work was supported by the research grant of Kazan Federal University.

References

[1] Pavlov N N 1982 *Ageing of plastic materials in real-life and simulated conditions* (Moscow: Khimiya Publ.) p 220

[2] Sulejmanov A M 2006 Experimental and theoretical basics of prediction and increase in durability of construction soft shell materials. *Abstract doctor of Technical Sciences* (Kazan: Kazan State University of Architecture and Civil Engineering) p 32

[3] Internet resource: Artificial ageing. Predicting material behavior. Available at: https://www.binder-world.com. (accessed 22.10.2017)

[4] Sokolova Yu A, Zharin D E and Shafigullin L N 2009 Development of special-purpose polymeric composite materials *Academia. Architecture and Construction vol 2* (Moscow: Russian Academy of Architecture and Building Sciences) pp 104-7

[5] Jurasov S Yu, Shafigullin L N, Shafigullina A N, Shajahmetova G R and Zharin E D 2017 Sound-Absorbing Polyurethane Foam for the Auto Industry *Russian Engineering Research vol 37 issue 4* pp 38-40.

[6] Kashapov N F, Nafikov M M, Gazetdinov M X, Nafikova M M and Nigmatzyanov A R 2016 *Innovative production technology ethanol from sweet sorghum* IOP Conference Series: Materials Science and Engineering vol 134 issue 1 8 **012012**

[7] Shafigullin L N, Romanova N V and Shafigullina G R 2017 *Investigation of UV resistance in polyurethane foam* Research Journal of Pharmaceutical, Biological and Chemical Sciences vol 8 no 4 pp 373-8

[8] GOST 16350-80 1980 Climate of the USSR. Regionalizing and statistical parameters of climatic factors for technical purposes (Moscow, Izdatelstvo Standartov Publ.).

[9] Fedyaev V L, Galimov E R, Gimranov I R, Takhaviev M S and Morenko I V 2016 *Mathematical modeling of jet flow around of bodies when applying polymer powder coatings* IOP Conference Series: Materials Science and Engineering vol 134 **012005**

[10] Fedyaev V L, Galimov E R, Gimranov I R, Takhaviev M S, Morenko I V and Fazlyev L R 2016 Heat transfer of gas suspension jet flow on the product surface when applying polymer powder coatings Journal of Physics: Conference Series vol 669 **012013**

IOP Conf. Series: Materials Science and Engineering **412** (2018) 012070 doi:10.1088/1757-899X/412/1/012070

Thermal properties of polypropylene and polyethylene blends (PP/LDPE)

L N Shafigullin[1], N V Romanova[1], I F Gumerov[1], A T Gabrakhmanov [1] and D R Sarimov [1]
[1]Naberezhnye Chelny Institute (branch), Kazan Federal University
Sjujumbike street 10a, Naberezhnye Chelny, 423800 Tatarstan, Russia

misharin_82@mail.ru

Abstract. The paper investigates the thermal properties of polymer blends of polypropylene and low-density polyethylene.

1. Introduction

Investigations of polypropylene thermal properties are important from a scientific point of view as they explain the polymer performance at elevated temperatures and define the choice of a suitable material according to the application, as well as predict the structure and methods of creating new materials with required properties [1, 2].

2. Body text

The properties of polypropylene (PP)- and polyethylene (PE)-based blends are known to differ from the properties of primary polymers, but the system data are not compatible. PE crystallizes into more stable orthorhombic lattice and PP – into monocline form, but each component also has a significant influence on the structure and properties of final polymer blends [1, 3-7].

One of the underlying issues regarding blends of mutually soluble polymers is the definition of strength of intermolecular interactions between polymer components. Analysis of melting-point depression is widely used for mutually soluble blends containing partially crystalline polymers. The method includes the comparison between the equilibrium melting points of unblended partially crystalline polymer and the same polymer in different polymer blends [2].

Differential scanning calorimetry (DSC) was used to define the experimental melting points. DSC analysis was recorded with Netzsch differential scanning calorimeter DSC 204 F1 Phoenix. The investigation was performed in dynamic heating/cooling conditions at a rate of 10 °C/min under argon purge with a rate of 50 cm^3/min.

Table 1 shows the DSC data derived from PP and PE melting endotherms and crystallization exotherms. As a result of calculation of crystallinity the heat of fusion for completely crystalline PP is 198 J/g [8, 10-11].

Table 1. Thermophysical properties of PP and PE blends

No of PE/PP sample	Tmaxmelt (PE), °C	Tcr (PE), °C	Tmaxmelt (PP), °C	ΔHfus (PP), J/g	Tcr (PP), °C	Crystallinity (PP), %
0/100	-	-	161.8	72.0	108.1	36.3
10/90	107.3	91.4	162.9	43.5	107.3	22.0

Content from this work may be used under the terms of the Creative Commons Attribution 3.0 licence. Any further distribution of this work must maintain attribution to the author(s) and the title of the work, journal citation and DOI.
Published under licence by IOP Publishing Ltd

20/80	106.9	92.4	161.8	48.5	108.0	24.2
30/70	106.9	92.5	160.8	44.5	107.6	22.4
40/60	107.2	92.6	161.5	35.9	108.6	19.6
50/50	107.1	92.6	160.3	35.5	107.9	17.9

The analysis of table 1 shows that the value of PE melting point (Tmelt) in polymer blends with PP varies within a range of 107.1 ±0.2 °C, and PP melting point (Tmelt) – within a range of 160.3-162.9 °C. The compound PE/PP-10/90 is observed to have higher PP melting point (162.9 °C) in comparison with the melting point of primary PP (161.8 °C). The crystallinity of PP blended with PE decreases as the PE content increases from 36.3 % to 17.9 %: maximum value for the compound PE/PP-20/80 (24.2 %) and minimum value for PE/PP-50/50 (17.9 %). It has to be noted that these polymers in blends have two PP melting peaks, the first peak is small and shifted from the main peak towards the low temperatures by 18.5 °C. It can be assumed that such behavior during melting of PP blend samples with 10 %, 20 % and 40 % is related to recrystallization.

The compound PE/PP-10/90 has the most perfect crystalline structure with higher melting point, which features the lowest number of defects as they have been forced into the amorphous phase during the crystallization.

Therefore, based on the DSC data it can be concluded that the introduction of polyethylene into PP leads to a small shift of PP melting point to the lower-temperature region. It indicates that the structure is changed and low-melting crystals are formed. In addition, as the content of polyethylene increases the propylene enthalpy of fusion as well as its crystallinity decrease, suggesting the partial amorphicity of polypropylene. Of the investigated blends the compound PE/PP-50/50 has the lowest crystallinity.

Degradation processes play a key role for the correct usage of blended PP polymers in the various fields. During the thermal degradation of polymers in the inert atmosphere the molecular mass decreases and highly volatile components are formed [1].

The thermal stability and degradation of PP and PE blends were investigated with Netzsch thermogravimetric analyzer TG 209 F1 Iris. The samples were heated up to 550 °C with a rate of 10 K/min under continuous argon purge.

Table 2 shows the processing results of thermogravimetric analysis. Addition of up to 30 % of PE increases the temperature of loss of 5, 10 и 20 % of PP, and with the PE content of 40 % and 50 % the temperature of maximum degradation rate (Tmax) and final weight loss (Tfinal) increase significantly. These findings indicate the thermostabilizing effect of polyethylene.

Table 2. Processing results of thermogravimetric analysis

PE/PP composition	Tmax	T5	T10	T20	Tfinal
10/90	452	394	419	436	493
20/80	456	414	428	439	493
30/70	449	405	423	438	494
40/60	453	419	431	442	504
50/50	466	374	407	431	504

Therefore, the thermal degradation runs slower in PP and PE blends than in unblended PP due to the influence of the thermally stable PE.

PP is known to be more susceptible to degradation than PE. It is due to the fact that one half of PP carbons is linked to methyl groups which activate a hydrogen atom with a tertiary carbon. As a result, it turns to be more reactive than hydrogen of methylene or methyl group. A higher PE content in the blend leads to higher degradation temperature.

One of the main PP disadvantages is its low resistance to thermo-oxidative degradation, resulting in poor processability, lower stress-strain and electric properties of polymers in products, as well as color

IOP Conf. Series: Materials Science and Engineering **412** (2018) 012070 doi:10.1088/1757-899X/412/1/012070

change. In view of this, the product resistance to thermo-oxidative degradation is the critical property for a material during the operation.

Thermal-oxidative resistance was evaluated by oxidative induction time [8]. It is the time when the oxidation reaction starts, which manifests itself by exothermic effect in the DSC curve (table 3).

Table 3. Oxidative induction time for PP, and PP and PE polymer blends

PE/PP composition	Oxidative induction time, min
0/100	1.00
20/80	1.17
40/60	1.31
50/50	1.63

The analysis of table 3 shows that introduction of PE into PP leads to increase in the oxidative induction time in comparison with PP. Higher PE content (up to 50 %) increases the oxidative induction time from 1 to 1.63 minute.

These findings indicate the thermostabilizing effect of polyethylene, which can result from hindered oxygen diffusion for PP chains due to formation of the interfacial layer.

Thus, thermo-oxidative degradation can be improved by adding 20 %, 40 % of PE.

3. Conclusions

Therefore, the addition of polyethylene to polypropylene gives an insignificant shift of polypropylene melting point to the lower-temperature regions, suggesting the change in structure and the formation of low-melting crystals. In this case, the thermal degradation is slower than in unblended polypropylene due to the influence of more stable polyethylene. Based on the thermal-oxidative resistance polyethylene is found to have thermostabilizing effect which results from hindered oxygen diffusion for polypropylene chains due to formation of the interfacial layer.

This work was supported by the research grant of Kazan Federal University.

References

[1] Kuchmenova L Kh 2014 Thermal properties of polymer-polymer polypropylene-based composites *Cand. eng. sci. diss.* (Nalchik: Kabardino-Balkaria State University. H.M. Berbekova) p.125

[2] Paul D R and Bucknall C B 2009 *Polymer blends: Formulation. Volume 1.* (Saint-Petersburg: Scientific bases and technologies) p 618.

[3] Utracki L A 1998 *Commercial Polymer Blends* (Springer) p 658

[4] Noel O F and Carley J F 1975 Primary spherulite nucleation *Polymer Engineering and Science vol. 15 no. 1* p. 117.

[5] Robertson R E and Paul D R 1973 *Stress-strain behavior of polyolefin blends* Journal of Applied Polymer Science vol 17 no 8 p 2579

[6] Teh J W 1983 *Structure and properties of polyethylene-polypropylene blend* Journal of Applied Polymer Science vol 28 no 2 p 605

[7] Kashapov N F, Nafikov M M, Gazetdinov M X, Nafikova M M and Nigmatzyanov A R 2016 *Innovative production technology ethanol from sweet sorghum* IOP Conference Series: Materials Science and Engineering vol 134 issue 1 **012012**

[8] Kaisersberger E and Mohler H 1991 DSC on polymeric materials *NETZSCH Annual for Science and Industry vol 1* p. 27.

[9] GOST R 56756-2015 2016 Plastics. Differential scanning calorimetry (DSC). Part 6. Determination of oxidation induction time (isothermal OIT) and oxidation induction temperature (dynamic OIT) (Moscow, Standartinform Publ.)

[10] Fedyaev V L, Galimov E R, Gimranov I R, Takhaviev M S, Morenko I V and Siraev A R 2016 *Deposition of polymer powder particles on the product surface at jet spraying* Journal of Physics: Conference Series vol 669 **012012**
[11] Sharafutdinov R F, Snigerev B A, Galimov E R and Galimova N Ya 2016 *Two-layer flow of polymer melts in extruder die channel* IOP Conf. Series: Materials Science and Engineering vol 134 **012024**

Usage of polymeric fuel tanks in the automotive industry

L N Shafigullin[1], G F Muhametzjanova [1], A A Muhamatdinova[1], A M Sotnikov [1] and G R Shafugullina [1]

[1]Naberezhnye Chelny Institute (branch), Kazan Federal University
Sjujumbike street 10a, Naberezhnye Chelny, 423800 Tatarstan, Russia

misharin_82@mail.ru

Abstract. The paper investigates usage of polymeric fuel tanks in the automotive industry.

1. Introduction

A fuel tank in the vehicle is an important component of the fuel system in an internal combustion engine. A tank is used to store motor fuel such as petrol, diesel fuel and gas. The present-day fuel tanks can be made of different advanced materials: ferroalloys, aluminum alloys, plastics.

Steel fuel tanks are manufactured by stamping of sheet metal and welding of upper and lower parts. Steel fuel tanks use steel with high viscosity grade and ultimate strength [1]. When subjected to external influences a fuel tank is deformed due to its viscosity but remains leak-proof. The main disadvantage of steel fuel tanks is high corrodibility. The probability of water ingress into a fuel tank is great as it can find its way inside as a condensate or together with fuel and settle down at the bottom.

Aluminum fuel tanks are manufactured by stamping and welding. They are not corrodible and have lower weight than steel fuel tanks. In comparison with steel tanks aluminum fuel tanks have the following disadvantages: possible fuel leaks due to weld failure during the operation; low deformation capacity (impact resistance); presence of stress risers in a material after short-term strain higher than its strength; high cost.

Today the leading world manufacturers are in process of transition to plastic fuel tanks. Polyethylene and polypropylene fuel tanks are manufactured by rotational molding or extrusion. In comparison with steel and aluminum fuel tanks plastic tanks are not corrodible, resistant to scratches, chipping, have lower weight, require no welding, have lower cost [2-3]. The existing methods of plastic fuel tank production make it possible to manufacture components with complex geometry; use the tank mounting space efficiently; provide the maximum tank volume.

2. Body text

The main reason of fuel tank failures is a surface damage from blasting with abrasive particles flying off the vehicle wheels. These collided particles make the tank walls and brackets thinner. When a vehicle is moving the frame becomes distorted and fuel tank components lose their strength due to abrasive wear, so cracks are formed in fuel tanks resulting in loss of their leakproofness [4]. Damage of a fuel tank can also be caused by the quality of fuel. Diesel fuel is hygroscopic, it can absorb water from air resulting in corrosion of tank walls, which requires the expensive repairs of fuel components (high pressure fuel pump, injectors, fuel pipes) [5].

Content from this work may be used under the terms of the Creative Commons Attribution 3.0 licence. Any further distribution of this work must maintain attribution to the author(s) and the title of the work, journal citation and DOI.
Published under licence by IOP Publishing Ltd

In view of this, the engineers seek to modify the design of fuel tanks and find the suitable materials. The most interesting solution is a fuel tank made of plastic.

The present day plastic fuel tanks for vehicles should meet the following requirements: operating temperature range from minus 60 °C to plus 80 °C; local and short-duration heating up to 110 °C (high temperature near the areas of more intensive sunlight, exhaust system, points of contact with the wire bundles); leakproofness at an excessive pressure up to 0.02 MPa. The design of a fuel tank shall ensure fuel sedimentation and dirt and water removal from the tank (as they are accumulated); ozone resistance; resistance to precipitation, dynamic, impact and vibration loads that occur in a moving vehicle; resistance to short-term exposure to hot detergents, diesel fuel, lubricants outside and inside the cavity; resistance to natural abrasive wear (dust, dirt); high aesthetic characteristics (the surface shall have no cracks, folds, shells, blisters, foreign inclusions and exit points of the reinforcing material nonimpregnated with resin; absence of internal flaws (layer separation, bubbles, porous structure); transportation and storage using transport plugs. The polymeric materials should meet the following requirements: no toxicity and odor; no injury hazard; flammability category G1 (Г1) (self-extinguishing [6]).

The polymeric material was selected to produce tanks. The most suitable material was found to be polyethylene. The polyethylene fuel tank was designed for trucks using the state-of-art CAD and CAE systems. Rotational molding was selected as a production method. It makes it possible to manufacture hollow large components with a volume up to 36 m3, complex components with the wall thickness than can be changed without changing the shape; to integrate metal parts by molding; manufacture several components simultaneously; to ensure no inner stress and polymer orientation for parts [7-10].

The following table shows the results of comparative laboratory tests of plastic and steel vehicle fuel tanks subjected to different types of influences.

Table. Results of comparative tests

Test description	Steel fuel tank (steel 08Yu)	Plastic fuel tank (medium-density polyethylene)
collision of the tank with the steel reservoir in the forklift (impact force and distance are similar)	complete deformation; fuel leakage; scratches on the painted surface and extensive visual damages; the tank shall be replaced immediately due to impossible further operation	deformation recovery: minor temporary deformation; lower risk of fuel leakage; the tank can be used without repair
shot through the plastic and steel fuel tank from the assault rifle with cartridges in 7.62 mm caliber at a distance of 10 m. The test was performed for the tanks filled with water to 70% of the total volume	significant leakages in the two tank walls; impossible repair and further operation; entry hole of 7.98 mm, exit hole of 25 mm.	hole only in the front wall of the tank (3.64 mm); possible temporary repair; shrinkage of the hole within several minutes after the shot and as a result leak reduction; no leakage in the rear wall of the tank
fire resistance of the fuel tank (120-second burning according to ECE34 [9])	not applied	no leak found over the whole surface of the plastic tank

The analysis of the table data revealed that plastic fuel tanks have better characteristics than steel ones. In addition, the advantages of polymeric fuel tanks include:

- easy flushing. It is necessary to ensure that running water flow through the inlet or one of the service openings on the upper surface of the fuel tank and the bottom drain hole is open;
- easy fuel level control: there is a hole for the gage rod in the upper part of the tank;
- efficient recycling of overage parts. The tank material can be reused by granulation to secondary materials for recycling;

IOP Conf. Series: Materials Science and Engineering **412** (2018) 012071 doi:10.1088/1757-899X/412/1/012071

- easy transportation.

3. Conclusions

Due to a variety of forms and configurations and easy installation plastic fuel tanks have established themselves in the truck market. The area of their application is continuously expanding: the manufacturers are attuned to the slightest changes in demand and offer new solutions to make the operation of products easier.

This work was supported by the research grant of Kazan Federal University.

References

Internet resource: Faults of diesel engine fuel systems in KAMAZ vehicles. Available at: http://xn----7sbfkccucpkracijq8iofobm.xn--p1ai./ (accessed 10.10.2017)

[1] Internet resource: Fuel tank: construction, types, operation. Available at: https://autodromo.ru/articles/toplivnyy-bak-ustroystvo-vidy-princip-funkcionirovaniya. (accessed 12.10.2017)

[2] Kashapov N F, Nafikov M M, Gazetdinov M X, Nafikova M M and Nigmatzyanov A R 2016 Innovative production technology ethanol from sweet sorghum *IOP Conference Series: Materials Science and Engineering vol 134 issue 1, 8 number 012012*

[3] Internet resource: Construction of present-day vehicle fuel tanks. Available at: https://auto.today/bok/4303-iz-chego-sostoit-sovremennyy-toplivnyy-bak.html. (accessed date: 11.10.2017).

[4] Repair manual for KAMAZ 5320 and maintenance and operation manual for KAMAZ 5320, 5410, 55102, 55111, 53212, 53211, 53213, 54112, 4310, 43114, 43118, 65111, 53228, 44108, 43115, 65115, 53229, 53215, 54115 with chassis 6x4 and 6x6 (Moscow: Tretij Rim Publ) p. 286

[5] GOST 28157-89 1989 Tests for flammability of plastic materials (Moscow: Izdatelstvo Standartov Publ.)

[6] Internet resource: Rotational molding: new ideas]. Available at: http://polymery.ru/letter.php?n_id=3046&cat_id=2. (accessed date: 25.10.2017)

[7] Internet resource: Rotational molding. Available at: http://www.rototech.it/index.php?evt[page-displayItem]&id=11. (accessed date: 13.10.2017)

[8] European Standard ECE R34 Annex 5, Fire Risks – European Economic Community Regulation – Fire safety of plastic fuel tanks for automobiles (ECE R34 Annex 5, RREG 70/221/EWG, 2000/8/EG).

[9] Fedyaev V L, Galimov E R, Galimova N Ya, Gimranov I R and Siraev A R 2017 *Dynamics of coalescence and spreading of liquid polymeric particles during coating formation* IOP Conf. Series: Journal of Physics: Conf. Series vol 789 **012006**

[10] Fedyaev V L, Galimov E R, Galimova N Ya, Takhaviev M S and Siraev A R 2017 *Mathematical modeling of processes occurring during deposition of sprayed particles of polymeric powder* IOP Conf. Series: Journal of Physics: Conf. Series vol 789 **012007**

IOP Conf. Series: Materials Science and Engineering **412** (2018) 012072 doi:10.1088/1757-899X/412/1/012072

Agile in project management system in mechanical engineering

A Sh Shavaliev, A S Puryaev

Naberezhnye Chelny Institute, Kazan Federal University, 18 Kremlyovskaya street, Kazan, 420008, Russia Federation

aidarp@mail.ru

Abstract. The aim of the work is to highlight the practical application of the Agile concept in the life cycle management of investment projects in the field of engineering. The comparative advantages and disadvantages of the concept of flexible design are revealed. The themes of engineering projects in which the application of the Agile concept can be applied are identified.

1. Introduction

The current level of technology development and progress in almost all sectors imposes special requirements on production management. The pace of technology development worldwide is reaching its peak and is likely to continue to grow.

It is difficult to foresee what will happen to mankind in the next 50-100 years in the field of mechanical engineering and in other sectors [1], but on the example of the picture of the 1900s (see figure 1) it can be stated [2,3] that the pace of technology development will grow at a high rate in the future

Figure 1 - New York, intersection of Broadway streets, 5th Avenue and 23 streets (1905): where is the picture of the car? [1].

The same street about 20 years later (see figure 2).

Content from this work may be used under the terms of the Creative Commons Attribution 3.0 licence. Any further distribution of this work must maintain attribution to the author(s) and the title of the work, journal citation and DOI.
Published under licence by IOP Publishing Ltd

Figure 2 - New York, intersection of Broadway streets, 5th Avenue and 23 streets (circa 1925): where is the horse in the picture? [1].

The pace of development of modern mechanical engineering (machine building) and other industries is directly related to the development of the Internet, computer technology and digital technologies. The progress of the development of society can be observed on the following examples [4]:

- in the 1990s about 1 million people have the mobile phone, in our time in free circulation is 5-6 billion phones, i.e. in 5-6 thousand times more;
- no one has won a chess tournament against a high-tech computer since 2005;
- every 2 minutes we take as many photos as all the people in the 19th century did. Only in 2014, humanity made 880 billion photographs. This is for 123 photos for every man, woman and child on Earth;
- People will create more information in the form of data in the next 2 days than was created in the history before 2003;
- the computing power of the iPhone 5S in 1991 was worth more than $ 3 billion dollars.

2. Discussions

Despite the progress of mechanical engineering technologies (active development of robots, electric vehicles, ADAS systems, etc.) – it can be stated that the next 10 years of mechanical engineering development will be closely related to digital technologies, data flows, artificial intelligence and the increasing speed of changes.

Traditional technologies in the management of production and engineering projects will go to the background, because active development is possible only in conditions of flexible fast-changing systems that have cybernetic possibilities for self-regulation and adjustment.

From a lifecycle perspective, the project management standard highlights [5]:

1. *Predictive life cycles.* Predictive life cycles (also known as fully managed plans) are a type of project life cycle in which project content, as well as the time and cost required to deliver that content, are determined at the earliest possible stage of the life cycle. These projects pass through a series of consecutive or overlapping phases, with each phase, as a rule, focused on a subset of project operations and project management processes.

2. *Iterative and incremental life cycles.* Iterative and incremental life cycles are life cycles in which project phases (also called iterations) intentionally repeat one or more project operations as the project team begins to better understand the product. Iteration determines the development of a product by performing a series of repetitive cycles, while incrementality determines the sequential increase in the product functionality.

3. *Adaptive life cycles.* Adaptive life cycles (also known as change-driven or agile methods) are designed to respond to high levels of change and they require a constant high degree of stakeholder engagement. Adaptive methods can be also iterative and incremental, but differ in that iterations are very rapid (usually 2-4 weeks) and fixed in terms of timing and cost. In adaptive projects, several processes typically run during each iteration, although earlier iterations may focus more on operations scheduling. The overall content of the project is broken down into a set of requirements, and the work that needs to be done is sometimes referred to as a backlog (requirements log). At the beginning of the iteration, the team determines how many high-priority items from the backlog can be retrieved during the next iteration. At the end of each iteration, the product must be ready for customer analysis.

It is obvious that the first model of project management – predictive cycles is a thing of the past. It has one significant drawback: a lack of time for detailed planning of all project characteristics.

The second model is iterative and incremental life cycles, most interesting in terms of speed, compared to predictive life cycle.

Adaptive life cycle, perhaps, has the most important advantage – the product is made for the customer, traditional technical requirements and specifications are replaced by a backlog. The key benefit is time. It will not be spent on the original plan.

No need to say that the most acceptable technology for project management is Agile technology, but there are subtleties of the process itself and a number of circumstances while restraining the process of applying new technologies for the management of engineering projects.

In general, mechanical engineering (machine building) projects can be classified as follows:

1. Food projects (new car, automobile knot, automobile unit).
2. Infrastructure projects (modernization of equipment, replacement of worn-out fixed assets, renovation, construction and modernization of new facilities).
3. Development of trade and service network (construction of buildings, modernization of equipment, construction of warehouses).
4. Research projects.
5. IT projects (CRM systems, PLM systems, MES systems and other software products).

3. Conclusion

Agile application for the projects specified in paragraphs 1-3 raises significant questions at the time of "redevelopment". For these categories of projects, so far, in the management of any technology (including "waterfall", in which the life cycle of the project develops consistently from the concept of research to prepare for the introduction and implementation of the project) there is a point of no return, marked by contracting for fixed assets (equipment, buildings and structures, etc.). Currently, it is difficult to assume what to do with equipment or tooling, which are usually unique, in the case of a sharp change in the course and changes in the market situation. The obvious answer can be the purchase of universal equipment, but it does not remove the questions of efficiency (such equipment is usually more expensive and less productive). Probably, answers to these questions can be found after a certain period of time with the advent and mass applications of 3D printing technologies (in tooling and equipment). With regard to the projects mentioned in paragraphs 4 to 5, there are far fewer questions in the application of Agile and it can be concluded that Agile can and should be used in information technology and research. The key benefit of Agile is time, and as a consequence, the absence of lost profits.

The world is rapidly changing, the speed of change has reached its peak and will continue to increase. Project management technologies from the perspective of life cycle management require more attention. One of the advanced and necessary for the development of engineering management technology, Agile, can be used in engineering projects in the sphere of information technology and research.

References

[1] Gor'kov S N 2017 Open lecture in KFU "Possibilities of Break for the Russian Economy"

Retrieved March 31, 2018, from https://www.youtube.com/watch?v=d6GUlokaePU

[2] Anon 2017 Digitalization of the cities *Cent. Strateg. Res. News* Retrieved March 31, 2018, from http://www.csr.ru/news/tsifrovizatsiya-gorodov/

[3] Danilin A 2017 As "clever" technologies are capable to change life of citizens *ITMO. News* Retrieved March 31, 2018, from http://news.ifmo.ru/ru/education/official/news/7034/

[4] Anon 2014 25 surprising facts about human progress Retrieved March 31, 2018, from https://www.infoniac.ru/news/25-udivitel-nyh-faktov-o-chelovecheskom-progresse.html

[5] Anon 2014 *A Guide to the Project Management Body of Knowledge (PMBOK Guide)* (Moscow: Olimp-Biznes) p 586

Surface quality assurance of ring gear during precision stamping of gearwheels

V G Shibakov[1], D L Pankratov[1], R S Nizamov[1] and R A Hairullin

[1]Kazan Federal University, Naberezhnye Chelny Institute, 423812, Russia, Naberezhnye Chelny, Prospekt Syuyumbike 10A,

vladshib50@gmail.com

Abstract. The article highlights the main factors, defining the surface quality of ring gear during hot bulk precision stamping. It offers enhanced techniques of surface quality of stamped ring gear.

During precision stamping of gearwheels strict requirements are imposed on the surface quality of a stamped ring gear.

In manufactured units of foreign producers the size of overused iron scale amounts to 0,013 mm (picture 1)

Picture 1- overused iron scale on the work piece surface × 1000

While examining the microstructure on the work piece surface, stamped in standard conditions, overused iron scale has been exposed to a depth of up to 0,063mm according to picture 2.

The surface defects of mentioned character will lead to premature wearing-out and cog chipping and drop-out of gear transmission. To provide the performance reliability of a gear pair, it's necessary to find out the reasons of stamping the iron scale and determine its solution strategies. The reasons and methods of problem solution of the quality of nonmachining surface are cited in picture 3.

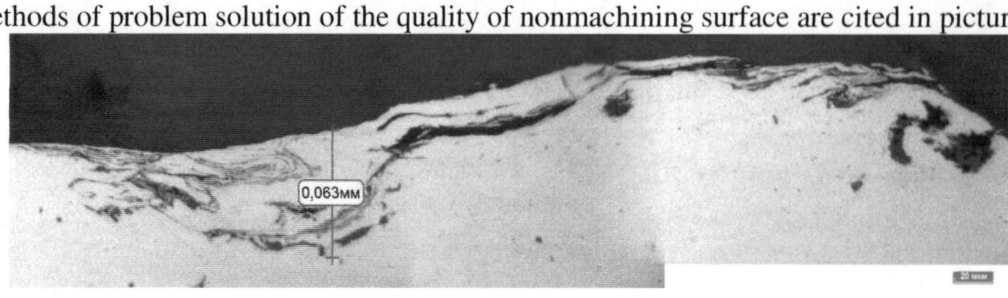

Content from this work may be used under the terms of the Creative Commons Attribution 3.0 licence. Any further distribution of this work must maintain attribution to the author(s) and the title of the work, journal citation and DOI.
Published under licence by IOP Publishing Ltd

Picture 2- overused iron scale on the work piece surface , × 300

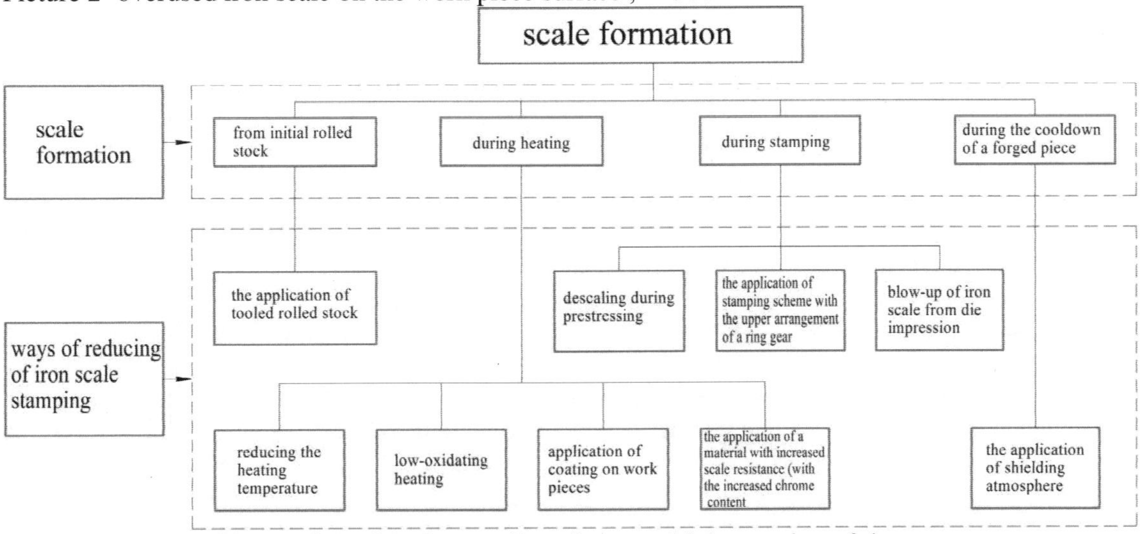

Picture 3 – ways of improving of surface quality of nonmachining cutting of ring gear

According to [1] on the surface of hot-rolled rolled stock into encrusting matter of rolled-in scale to a depth of up to 0,2 mm. Rolled-in scale is difficult to remove by pickling and dressing. Consequently, to exclude the stamping of surface defects in the area of nonmachining part of finished articles the necessary requirement to the surface quality of initial work piece is quality group 1 in surface quality of rolled stock in tooled condition.

The reduction of stamping temperature leads to the reduction in scale formation [2] and the increase of stamping power (pic.4). The increase of stamping power is limited by the capacity of pressure equipment, reduction in resistance of stamping tool and formability of material of initial work piece

Ways of low-oxidating and nonoxidating heating of work pieces before stamping are described in [3].

The application of envelope has a negative impact on the precision of an item as there occurs envelope sticking in the area of a finished surface (pic.5) [4].

The scale resistance of chrome steel is conditioned by chrome content in them (pic.6). If the alloy contains it more than 20% Cr the scale resistance increases most abruptly. The more the chrome content in the alloy is, the higher the steel process temperature. Steel that contains 4—6% Cr and 0,5% Mo has a lot more resistance to creep at 400—600° C, than plain carbon steel, moreover this resistance increases with the increase in alloy 7% Cr and up to 1—1,5% Mo[5].

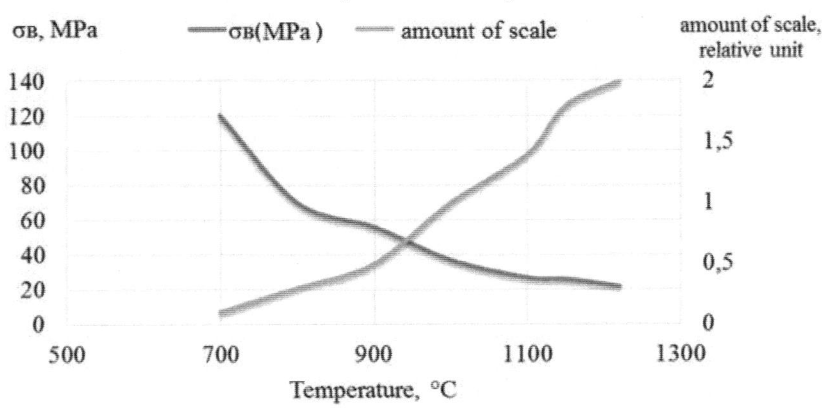

Picture 4 – the example of response of scale formation $\sigma_в$ to metal heating temperature.

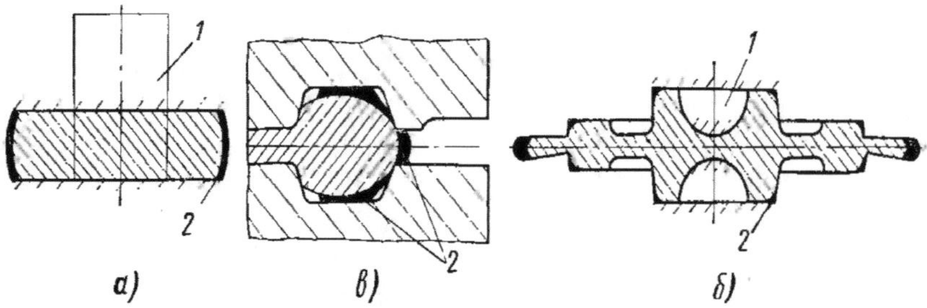

Picture 5 – the scheme of glass distribution in forged pieces: *а*- at the work piece after upsetting ; *б* – in forging of gear wheels; *в* – at the rim during stamping; 1 – the glass film; 2 – glass pileup

Thus, from the line-up of constructional steel applied in standard machinery one should recommend steel resistant to high-temperature oxidation, such as 20H2N4A, X12CrMo5, 20CrNi4 and others.

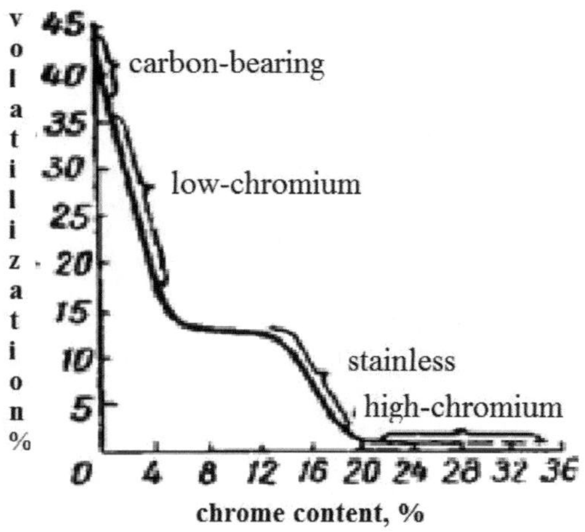

Picture 6 – Scale resistance of steel at 1000° C, 48 h., in Mac-Kwig

Scale formation immediately during hot die forging can be minimized by means of selecting the most effective in terms of scale knocking the upsetting scheme at the first juncture. To prevent the stamping of scale on the ring gear impression it's necessary to foresee the cog formation in the upper forged insert element and forced scale blow-off by compressed air after each forging.

Results:

- The major factors, defining the surface quality of the ring gear during precision hot bulk stamping are: the surface quality of initial rolled stock, scale formation during the work piece heating, during stamping and during forging cooldown.
- The following enhanced techniques of surface quality of stamped ring gear have been offered:

- the application of tooled rolled stock,
- reducing the stamping temperature,
- the application of low-oxidating heating,
- the selection of material with maximum allowable chrome content,
- the selection of rational upsetting technique to knock scale from the work piece surface,
- scale removal from the surface of die impression.

When required, one can evaluate the significance of solutions in terms of resulting precision [6].

References

IOP Conf. Series: Materials Science and Engineering **412** (2018) 012073 doi:10.1088/1757-899X/412/1/012073

[1] State standard (GOST) 4543-2016 Steel products of structural alloy steel.
[2] Gubinsky V I, Minaev A N, Goncharov U V 1981 *Reduction of scale formation during rolled stock production* (K.: Technika) p 135
[3] Petrov P A, Krutina E V, Kalpin U G 2010 *Heating and heaters in forgings: University textbook* (M.: MGTU «MAMI») p 110
[4] Kluev V M, Ostopolovsky A N, Shadek E G 1966 *Work piece stamping, heated in molten glass* Kuzhnechno-shtampovochnoe proizvodstvo No 10
[5] Manjurov S B 2014 *The planning of reduction of surface oxidation and decarburization of steel work pieces during heating with the application of nondurable technological protective layers* dissertation of PhD in Engineering Science
[6] Khairullin R A, Nizamov R S, Vladimirova S G 2017 *The stamping strategy option, providing the acquisition of the functional surface of the tool without machining operation application* IX Kamsky chtenia»: the Russian applied science conference

IOP Conf. Series: Materials Science and Engineering **412** (2018) 012074 doi:10.1088/1757-899X/412/1/012074

Investigation of the joint operation of electric transmission lines with a new type concrete reinforced concrete foundation

Yu M Strelkov[1], L S Sabitov[1,2], N F Kashapov[2], I R Gilmanshin[2, 4]

[1] Engineering Center "ENERGOPROGRESS", Kazan, Russia
[2] Kazan Federal University, 18 Kremlyovskaya street, Kazan, 420008, Russian Federation
[3] Kazan State University of Architecture and Civil Engineering, Kazan, 420043, Russia
[4] Kazan national research technical university named after A N Tupolev, K. Marx St. 10, Kazan, 420111 Tatarstan, Russia

l.sabitov@bk.ru

Abstract. The aim of this work was to study the joint work of structural systems "steel bearing overhead power transmission lines 10 kV – prefabricated concrete Foundation of a new type [1] – ground". For this purpose the technique of computer simulation in a PC "ANSYS" of the system. The method takes into account the spatial work of designs and physical nonlinearity of materials from which they are made. In addition, for steel was used in the theory of Mises, for concrete – Williams-Warnake, for ground – Drucker-Prager. In addition, all the necessary geometric, strength and physical characteristics of the model were obtained on the basis of the existing construction Standards for the design. The analysis of stress shows that its strength, stiffness and stability is provided, and there are reserves: for the metal supports – up to 15% to the base – 30%, ground – 1%.Therefore, further studies should consider the optimization problem for the choice of an effective constructive solution of the support and Foundation (with clarification on how to calculate them), that may be grounds to apply for a patent or utility model. Based on the obtained results it is recommended that a new type of collapsible foundations to apply in the real world.

The economic power of any state is determined by the potential of the power-to-weight ratio of the society. This potential depends on many factors, for example, on the development of energy infrastructure facilities, the material intensity of building structures used in such facilities, etc. In particular, when laying overhead power lines of various capacities and purposes, various steel supports and reinforced concrete foundations are used. From how economical it will be to constructive solutions of supports and foundations depends the cost of transportation of electric power from the source to the consumer, and, ultimately, the cost of all products produced on the basis of electricity. That is, research aimed at finding such constructive solutions has an important economic significance.

In work, the construction system "steel support of the overhead line (overhead line) of 10 kV transmission line - reinforced concrete foundation - ground of the base" is considered as an object of research.

Content from this work may be used under the terms of the Creative Commons Attribution 3.0 licence. Any further distribution of this work must maintain attribution to the author(s) and the title of the work, journal citation and DOI.
Published under licence by IOP Publishing Ltd

IOP Conf. Series: Materials Science and Engineering **412** (2018) 012074 doi:10.1088/1757-899X/412/1/012074

The HVL support has a well-known design solution for the Series 3.407.2-181.09 "Polyhedral steel supports of 6-10 kV overhead line" - anchor-angled support CM10AU, suitable for operation in climatic conditions including Kazan (II-th wind district, II area by glide). The support is intended for suspension of non-insulated steel-aluminum wires in accordance with GOST 839-80, as well as self-supporting insulated wires of type SIP-3 (SAX) according to TU 16.K71-272-98. From the wires presented in the specifications, the wire of the brand AC 185/29 has the greatest load on the pore, which we will take for calculation (the area of the aluminum part is 185 mm2, the steel core is 19 mm2, the wire diameter is 18.8 mm, the running weight is 7.28 kg / m, breaking force - 62,1 kN).

The foundation is collapsible and is performed according to the patent [1]. The general view of the basement is shown in Fig. 2. Its economic efficiency can be ensured not only due to low labor input for installation and dismantling and small transportation costs - this is stated in the patent test - but also in the calculation method when assessing the stress-strain state of the system "steel support - foundation - ground" with taking into account their joint work.

Soil conditions in the place of installation of supports can be different, take the worst option, which allows SP 22.13330.2011 "Bases of buildings and structures": type of soil - clayey soil, non-repellent, non-swelling; coefficient of porosity 0,95; modulus of deformation E = 8 MPa; yield index IL = 0.5; clutch of soil with = 15 kPa; angle of internal friction φ = 170; design resistance R0 = 150 kPa; coefficient of rigidity of the base (coefficient of the bed) k = 10 MPa / m (Designer's Handbook (Calculation and Theoretical), T.2, ed. Umanskogo AA, 1973. P. 307).

The most suitable tool for studying the system under consideration is computer simulation in the ANSYS software package (PC), which allows to take into account the spatial work of the structures and the physical nonlinear materials from which they are made.

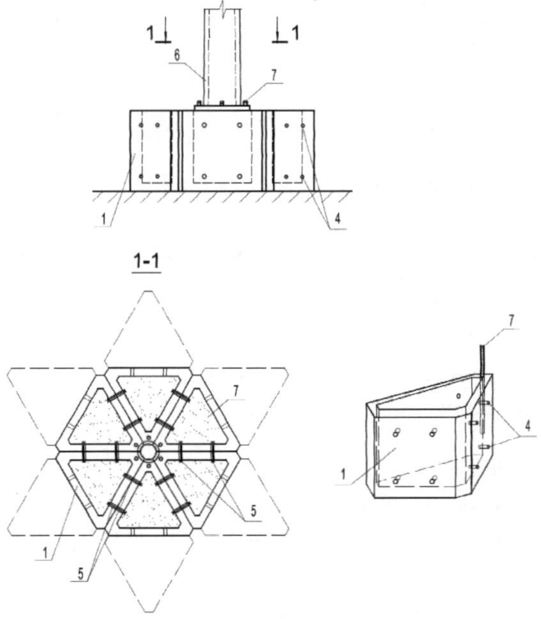

Fig. 1. General view of the collapsible foundation [1]

The essence of the modeling technique is as follows:

1 - definition of structural loads (based on the provisions of SP 20.13330.2011 "Loads and Impacts" and PEU-7);

2 - static calculation of the steel support separately, as a rigidly embedded cantilever column of a hollow section (as a result, the load on the edge of the foundation is determined);

3 - determination of the dimensions of the foundation sole based on the load on it;

IOP Conf. Series: Materials Science and Engineering **412** (2018) 012074 doi:10.1088/1757-899X/412/1/012074

4 - modeling of the unified system "support-foundation-ground", determination of VAT in it;
5 - checking the elements of the system for strength, stability and deformation (respectively, in the 1st and 2nd groups of limit states).

Since we are interested in working with the foundation of only a single support, we can consider it without wires, replacing their action with the appropriate load. Otherwise, it would be necessary to consider the whole so-called "anchor span" of up to 3 km length, within which there would be several intermediate supports; All supports then would be joined by wires having a certain sag (which depends on the type of wire, the length of the intermediate spans and the ambient temperature). Thus, the design support scheme for the first stage of simulation is shown in Fig. 2,

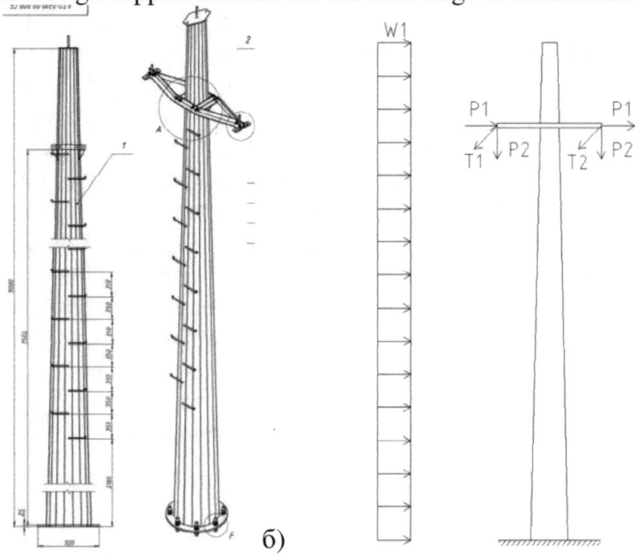

b.a) б)

Fig. 2. Support SM-10AU: a - a general view, b - a design scheme

The calculation takes into account the following types of loads:
1 - depending on the direction of action:
- horizontal (wind on the support, wind on wires and cables, from the tension of wires and cables) and
- vertical (the weight of the support, the weight of the insulator wires (with fittings), the weight of the wires and cables, the ice load, the installation load (the weight of the fitter with the tools));
2 - depending on the duration of the action:
- Constant (own weight of the support, weight of the wires);
- temporary (wind, ice load on the wires and on the traverse).

A separate load is temperature. Since the work of the cables is not considered, therefore, this load is not taken into account in calculations.

According to PEU-7 steel supports are calculated for the following combinations of loads:
1 circuit (normal mode): The wires and the cable are not torn and free from ice. The wind is directed along the axes of the traverse.
2 circuit (normal mode): The wires and the cable are not torn and covered with ice. The wind is directed along the axes of the traverse.
3 scheme (emergency mode): One wire (on the right / left side) giving a torque to the support is broken.
4 circuit (emergency mode): Two wires are disconnected.

The "support-foundation-ground" system includes elements formed from materials with qualitatively and quantitatively differing physicomechanical properties. For their modeling, the corresponding types of finite elements and deformation laws were used.

Preliminary, the calculation of a single support was carried out, as a rigidly embedded cantilever beam, that is, without taking into account the foundation and foundation soil. As a result, the following loadings on the foundation were determined: in normal operation, N = 4078 kg, Q = 599 kg, M = 3727 kg · m; in the emergency - N = 3526 kg, Q = 1923 kg, M = 13657 kg • m. According to the formulas of SP 22.13330.2011, the required dimensions of the foundation are calculated for these loads (see table). After that, a general model of the "support-foundation-soil" system is formed, which is shown in Fig. 3.

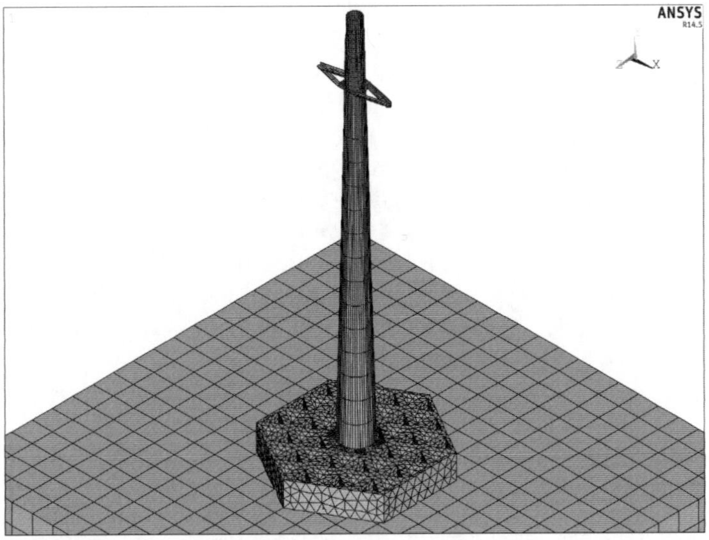

Fig. 3. Finite-element model of the "support-foundation-ground" system

The results of determining the VAT in the model are shown in Fig. 4.

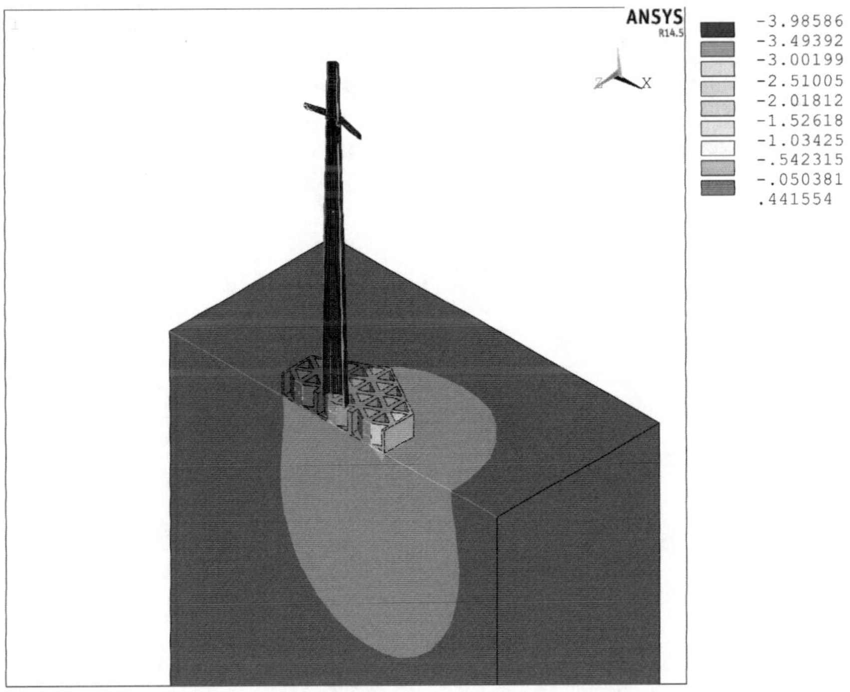

Fig. 4. Assessment of the VAT of the "support-foundation-ground" system

Conclusions.

1. The technique of computer modeling in PC "Ansys" of the constructive system "steel support VLEP - collapsible foundation of a new type [1] - foundation soil" was developed. The technique takes into account the spatial operation of the structures and the physical nonlinearity of the materials from which they are made. In this case, for the steel was used the theory of Mises, for concrete - Williams-Varnak, for ground base - Drucker-Prager. In addition, all the necessary geometric, power and physical characteristics of the model are obtained on the basis of the current building standards for design.
2. The analysis of the VAT system showed that its strength, rigidity and stability are ensured, and there are reserves: for the metal support - up to 15%, for the foundation - 30%, for the foundation soil - 1%. Therefore, in future studies, it is necessary to consider the optimization problem of choosing an effective constructive support and foundation solution (with the specification of the methodology for their calculation), which may be the basis for filing a patent application or utility model.
3. It is established that the joint work of the rack with the foundation is ensured. The suppleness of conjugation of prisms, of which the foundation is made, did not have a significant effect on the VAT of the system as a whole. Proceeding from the above, it is recommended that a new type of collapsible foundations be used in real conditions.

References

1. Sabitov L S, Kuznetsov I L, Khusainov D M, Sabirov S I, Akhunova Z R, Strelkov Yu M 2017 *Patent of the Russian Federation* No 2633604
2. Kryukov K P, Novgorodtsev B P 1979 *Design and mechanical calculation of power lines* (L.: Energia) p 312
3. Karpenko N I, Radaykin O V 2012 *To improve the diagrams of concrete deformation for determining the moment of cracking and breaking torque in bent ferro-concrete elements* Construction and Reconstruction No 2 pp 10-16
4. Sabitov L S, Akhunova Z R, Khusainov D M, Kuznetsov I L, Strelkov Yu M 2016 *A collapsible foundation for support* Materials of the international scientific and technical conference "Innovative machine-building technologies, equipment and materials -2016 " pp 227-30
5. Sabitov L S, Kashapov N F, Gilmanshin I R, Strelkov Yu M, Khusainov D M 2017 *Development and investigation of the stressed-deformed state of the demountable foundation for support* // IOP Conf. Series: Materials Science and Engineering 240 **012005**

IOP Conf. Series: Materials Science and Engineering **412** (2018) 012075 doi:10.1088/1757-899X/412/1/012075

Investigation of plasma metal-ceramic coatings on cutting surfaces of working bodies of machines and mechanisms

N V Titov[1], A V Kolomeichenko[1], Yu A Kuznetsov[1], L V Kalashnikova[1], O O Bagrintsev[1], S N Sharifullin[2,3]

[1] Orel State Agrarian University named after N.V. Parakhin, Russia
[2]Kazan Federal University, Russia
[3]Kazan state agrarian University, Russia

Saidchist@mail.ru

Abstract. The article describes the innovative method of strengthening the working organs of soil-cultivating machinery, which are operated in an abrasive medium.It is performed by carbo-vibro-arc surfacing of working bodies using composite metal ceramic pastes with simultaneous thermodiffusion hardening of basic metal of the working body. Based on the results of the studies, the optimum composition and concentration of pasta components were determined, which ensure an increase in wear resistance of hardened working organs by an average of 1.8 ... 2.0 times.

1. Introduction

Working bodies of national and international tillage machines (shares, chisels, land-sides, shellboards, blades and pointed and chisel-shaped hoes, smooth and toothed disks, etc.) are operated in the conditions of direct impact of abrasive particles. As the result they are strongly worn out, their cutting surfaces grow blunt, considerable changes in form, profile and working dimensions. For example, mean life of plow shares of Russian production is 5…20 ha depending on soil types and their structure, mean life of plow shellboards is 20…100 ha, of land-sides is 20…60 ha, of cultivators pointed hoes and sowing machines is 10…40 ha, of harrow disks and stubble plows is 8…20 ha [1-3]. Utilization of worn working bodies results in reduction of the field works quality; agrotechnical terms violation, increase in machine fault time and growth of costs on soil processing and greases, reduction of amount of the obtained market products.

Modern innovative method carbo-vibro-arc hardening (CVAH) with application of graphite electrode and composite metal ceramic pastes allows the considerable increase of tillage tools wear resistance being operated in abrasive medium. The method subject-matter is in the following. At first, paste is applied on the working element cutting edge, which is dried to hardening. Then, electric arc between the surface with the applied paste under hardening and graphite electrode of the device strikes. When it burns, surfacing of metal ceramic coating obtained from paste components as well as thermodiffusion saturation of working bodies metal with alloy elements which are included into the paste composition and with carbon at the expense of its diffusion that results from graphite electrode sublimation takes place [4-6]. Application of metal ceramic materials in the paste form at carbo-vibro-arc hardening (CVAH) is determined by their high resistance to abrasive and mechano-corrosive wear.

2. Methods.

Content from this work may be used under the terms of the Creative Commons Attribution 3.0 licence. Any further distribution of this work must maintain attribution to the author(s) and the title of the work, journal citation and DOI.
Published under licence by IOP Publishing Ltd

IOP Conf. Series: Materials Science and Engineering **412** (2018) 012075 doi:10.1088/1757-899X/412/1/012075

Paste content includes steel matrix (surfacing powder), aluminum oxide Al_2O_3, silicone dioxide SiO_2, boron carbide B_4C, which are ceramic components, and also the substances containing nitrogen (ammonium nitrate NH_4NO_3, carbamide NH_2CONH_2 etc.), and cryolite Na_3AlF_6, improving stability and quality of arc burning [1, 4]. Adhesive substance is 50% water solution of PVA glue. It was proved that the paste components considerably influence the hardness and wear resistance of the hardened surfaces. To determine paste composition we carried out the corresponding investigations. Pastes with above mentioned components in different proportions were prepared by mechanical mixing. Powder PG-10N-01 was used as matrix material. The content of cryolite Na_3AlF_6 in all analyzed pastes was accepted as 10% [4]. The pastes were applied on the samples produced from steel 65G. This steel type selection was determined by the fact that it is used for the majority of working bodies and other machinery, working in the abrasive wear conditions. Thickness of applied paste layer was 2,5…3,0 mm. Then pastes were dried at temperature 90…95°C to hardening. Usually the setting time does not exceed 8…10 min.

Device VAGH-2 (vibro-arc-graphite hardening) that is developed and is produced in Federal State Budget Research Establishment of State Scientific Technological Institute of All Russian Research Technological University was used for CVAH (carbo-vibro-arc hardening). Sample hardening was done in the following modes: current intensity I=70…80 A, voltage U=60 V, vibration frequency of graphite electrode is 25 Hz. The surfaced coating thickness was on the average 0,8 mm.

3. The results of the research.

The results of studies to determine the microhardness of hardening coatings deposited on samples are presented in Figures 1 ... 4.

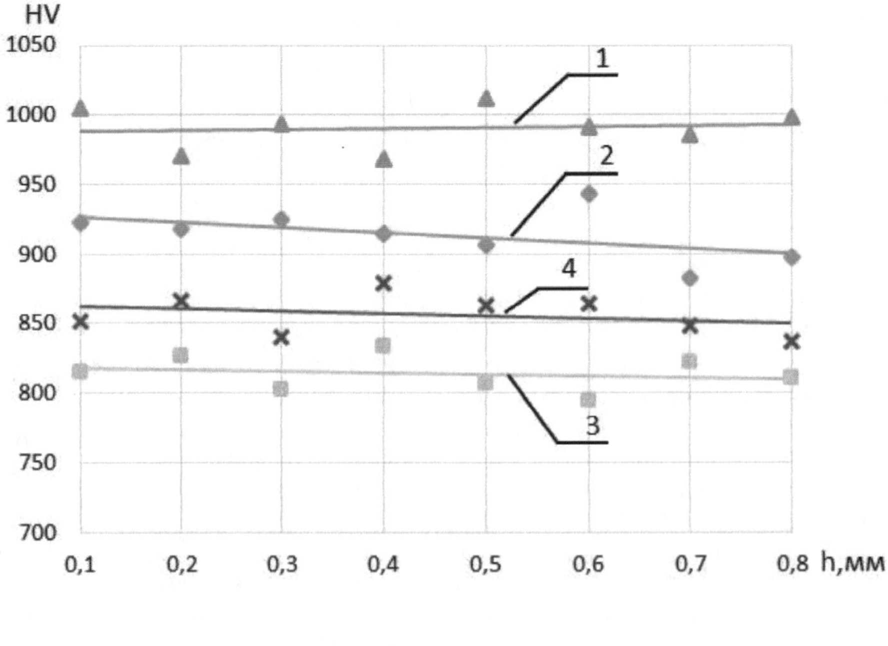

Figure 1 – Changing of HV microhardness of the coating being surfaced in depth h depending on the content of aluminum oxide Al_2O_3: 1 – Al_2O_3 - 30%; 2 – Al_2O_3 - 20%; 3 – Al_2O_3 - 10%; 4 – Al_2O_3 - 35% in paste

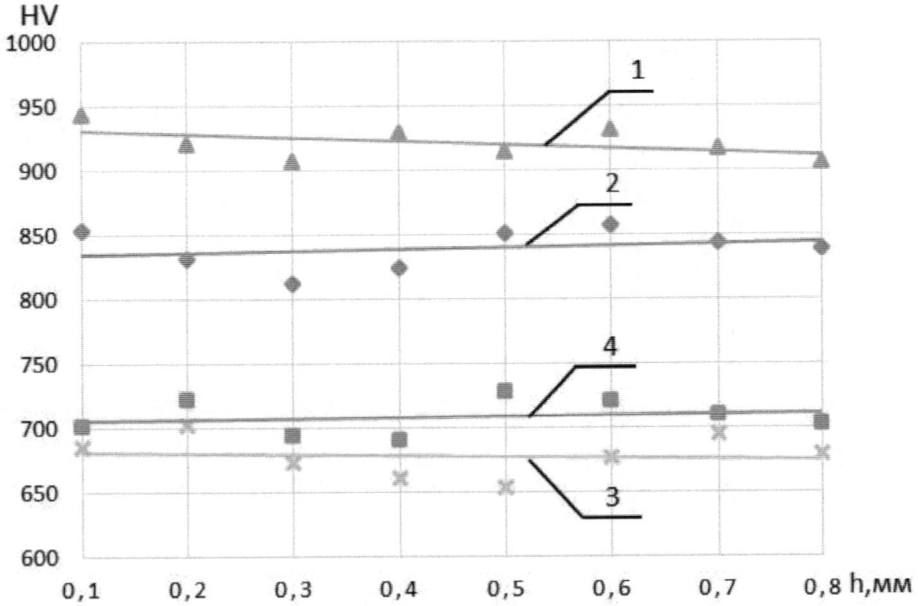

Figure 2 – Changing of HV microhardness of the coating being surfaced in depth h depending on the content of silicone dioxide SiO_2: 1 – SiO_2 - 30%; 2 – SiO_2 - 20%; 3 – SiO_2 - 10%; 4 – SiO_2 - 35% in paste.

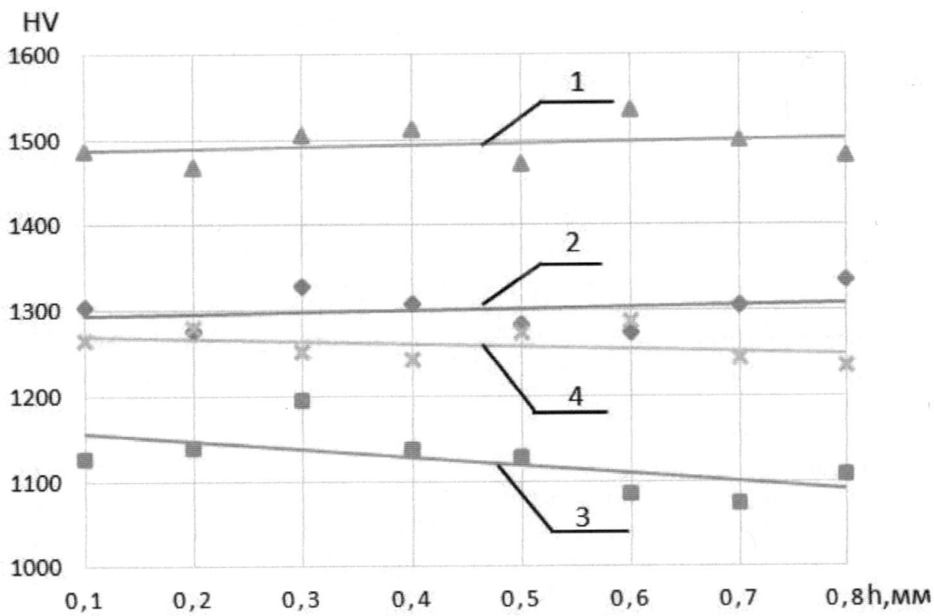

Figure 3 – Changing of HV microhardness of the coating being surfaced in depth h depending on the content of boron carbide B_4C: 1 – B_4C - 30%; 2 – B_4C - 20%; 3 – B_4C - 10%; 4 – B_4C - 35% in paste.

IOP Conf. Series: Materials Science and Engineering **412** (2018) 012075 doi:10.1088/1757-899X/412/1/012075

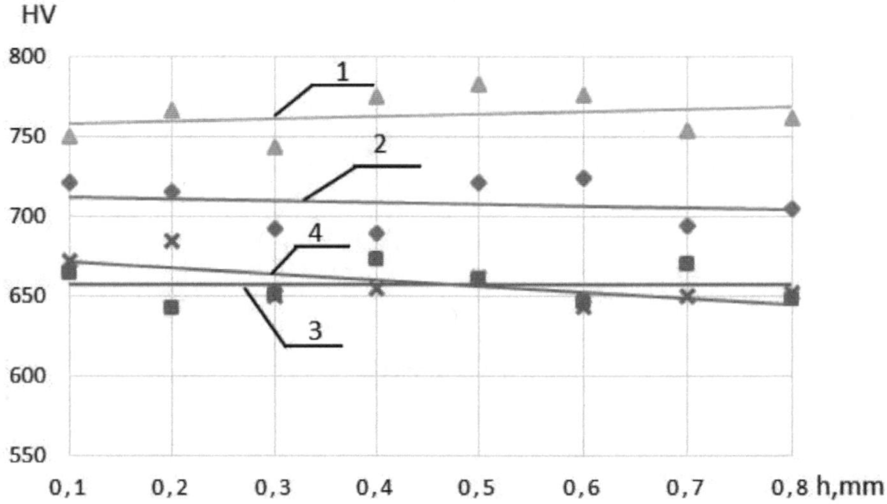

Figure 4 – Changing of HV microhardness of the coating being surfaced in depth h depending on the content of ammonium nitrate NH_4NO_3: 1 – NH_4NO_3 - 30%; 2 – NH_4NO_3 - 20%; 3 – NH_4NO_3 - 10%; 4 – NH_4NO_3 - 35% in paste

Analysis of the data obtained proved that with the increase of percentage content of ceramic components in paste from 10% to 30% microhardness of the surfaced coating also increases regardless the used ceramic component. At the same time the maximum average value of microhardness (1508 HV, that corresponds to microhardness 75 HRC) were showed by the samples being surfaced with the paste containing 30% of boron carbide B_4C (Figure 3, curve 1). The obtained coating hardness on the average by 1,5…1,7 times exceeds hardness of tempered steel 65G, from which the majority of series produced working bodies are made. It is connected with the following that boron carbide is one of the hardest materials and according to its hardness it gives place to diamond and borazon (boron nitride). Pastes containing 30% of aluminum oxide Al_2O_3 and silicone dioxide SiO_2, were characterized by lower values of coatings microhardness – 991 HV (69 HRC) and 922 HV (67 HRC) correspondingly (Figures 1, 2). It can be explained that at surfacing as the result of arc high temperatures influence subsolution of Al_2O_3 and SiO_2 with the corresponding coating hardness reduction takes place.

The samples being hardened with application of pastes containing ammonium nitrate NH_4NO_3, were characterized by the minimum values of coatings microhardness, that does not depend on the above mentioned component content. At the same time the maximum average value of microhardness (763 HV, that corresponds to hardness 62 HRC) was obtained at paste content of 30% NH_4NO_3 (Figure 4, curve 1).

4. Conclusion.

Thus, the described investigations proved that the paste containing 30% of boron carbide, 60% of matrix powder and 10% of cryolite is the most optimal paste for hardening of tillage tools, being operated in abrasive medium. The results of the comparative tests of the samples after hardening by such a paste and unhardened samples from tempered steel 65G showed that wear resistance after hardening increases by 1,8…2,0 times.

IOP Conf. Series: Materials Science and Engineering **412** (2018) 012075 doi:10.1088/1757-899X/412/1/012075

References

[1] Litovchenko N N, Titov N V, Kolomeichenko A V 2013 *Electric short-circuited arc hardening of tillage tools with metal ceramic materials* Tractors and agricultural machines No 2 pp 49-50
[2] Litovchenko N N, Kulikov V N, Titov N V 2013 *Vibro-arc surfacing with a graphite electrode of nanometallic ceramic composites* Welding production No 2 pp 51-53
[3] Titov N V, Kolomeichenko A V, Vinogradov V V, Litovchenko N N 2013 *To the question of application of metal and ceramic materials for hardening cultivator hoes* Works of State Scientific Institution of All Russian Research Technological University V 113 pp 364-67
[4] Kolomeichenko A V, Titov N V, Kondrakhin N A 2015 Investigation of technological capabilities of carbo-vibro-arc method of tillage tools and others *Machinery and equipment for rural area* No 2 – pp 24-26
[5] Titov N V, Kolomeichenko A V, Logachev V N, Kravchenko I N, Litovchenko N N 2015 *Investigation of the hardness and wear resistance of vibro-arc surfacing using cermet materials* Welding International V 29 No 9 pp 737-39
[6] Lyalyakin V P, Titov N V, Litovchenko N N 2014 *Carbo-vibro-arc method of elements hardening of machinery working in the abrasive wear conditions by metal ceramic surfacing (CVASMC)* Works of State Scientific Institution of All Russian Research Technological University V 114 pp 144-49

IOP Conf. Series: Materials Science and Engineering **412** (2018) 012076 doi:10.1088/1757-899X/412/1/012076

Numerical modeling of the powder materials spraying

A L Tukmakov[1,2], **N F Kashapov**[3], **D A Tukmakov**[1,2], **M G Fazlyyakhmatov**[3]

[1] Kazan National Research Technical University named after A.N. Tupolev–KAI, 10 K. Marx St, Kazan, 420111, Russia
[2] Institute of Mechanics and Engineering - Subdivision of the Federal State Budgetary Institution of Science «Kazan Scientific Center of the Russian Academy of Sciences», 2/31 Lobachevsky St, Kazan, 420111, Russia
[3] Kazan (Volga region) Federal University, 18 Kremlyovskaya St., Kazan, 420008, Russia

tukmakovda@imm.knc.ru

Abstract. On the basis of the numerical solution of polydisperse gas suspension dynamics system of equations, the process of powder spraying on a plane surface in an electric field is described. The model includes motion equations of the carrier medium and disperse phase fractions under aerodynamic friction and Coulomb forces action taking into account interfacial exchange of pulse and energy. The system solves by an explicit predictor-corrector method with splitting on the spatial directions and the non-linear correction scheme. The numerical model is applied to gas suspension speed and density fields receiving in interelectrode space and on target electrode surface. The system of equations was written in the generalized curvilinear coordinates: the physical region of a current in variables (x, y) was written in a canonical computational region in variables (ξ,η) and was solved by an explicit second-order Mac-Cormack method. The Poisson equation for the electric field potential was written in the generalized coordinates and was solved by finite difference method with the use of the iterative Seidel scheme on a gas-dynamic computational mesh.

1. Introduction

Movement of the charged gas mixture in the bearing environment under the influence of aerodynamic and electrical forces is the cornerstone of technology of a powder coating [1, 2]. For describing of the deposited layer formation process it is required to solve the problem about the movement of the charged aerosol particles in the space between the spray nozzle, on which the negative potential is supplied, and the painted surface, which potential is the positive or zero. In case of polydisperse composition of particles of pigment powder, their movement in interelectrode space depends on value of a charge, aerodynamic resistance of particles, electrodes layout, electric intensity, speed and concentration of solid fractions.

2. The equations of the movement of the charged gas-suspension and method of their decision

The process model can be based on a system of equations of movement of the multispeed multitemperature medium [2] in which the particle, along with forces of the aerodynamic nature, is influenced by Coulomb force. The system of equations includes motion equations of the carrying medium and disperse fractions accounting of interfacial force interaction and heat exchange and also

Content from this work may be used under the terms of the Creative Commons Attribution 3.0 licence. Any further distribution of this work must maintain attribution to the author(s) and the title of the work, journal citation and DOI.
Published under licence by IOP Publishing Ltd

taking into account interaction of the charged gas suspension with an electric field which is created by both fixed electrodes, and the electrical charge spatially distributed which is transferred by a disperse phase [3, 4]:

$$\frac{\partial \rho}{\partial t}+\frac{\partial(\rho u)}{\partial x}+\frac{\partial(\rho v)}{\partial y}=0,$$

$$\frac{\partial(\rho u)}{\partial t}+\frac{\partial}{\partial x}\left(\rho u^2+p-\tau_{xx}\right)+\frac{\partial}{\partial y}\left(\rho uv-\tau_{xy}\right)=-\sum_{i=2,n}F_{xi}+\sum_{i=2,n}\alpha_i\frac{\partial p}{\partial x},$$

$$\frac{\partial(\rho v)}{\partial t}+\frac{\partial}{\partial x}\left(\rho uv-\tau_{xy}\right)+\frac{\partial}{\partial y}\left(\rho v^2+p-\tau_{yy}\right)=-\sum_{i=2,n}F_{yi}+\sum_{i=2,n}\alpha_i\frac{\partial p}{\partial y},$$

$$\frac{\partial(e)}{\partial t}+\frac{\partial}{\partial x}\left(\left[e+p-\tau_{xx}\right]u-\tau_{xy}v+\lambda\frac{\partial T}{\partial x}\right)+\frac{\partial}{\partial y}\left(\left[e+p-\tau_{yy}\right]v-\tau_{xy}u+\lambda\frac{\partial T}{\partial y}\right)=$$

$$=\sum_{i=2,n}Q_i-\sum_{i=2,n}\left(\left|F_{xi}\right|(u-u_i)-\left|F_{yi}\right|(v-v_i)\right)+\sum_{i=2,n}\alpha_i\left(\frac{\partial(pu)}{\partial x}+\frac{\partial(pv)}{\partial y}\right),$$

$$\tau_{xx}=\mu\left(2\frac{\partial u}{\partial x}-\frac{2}{3}D\right),\ \tau_{yy}=\mu\left(2\frac{\partial v}{\partial y}-\frac{2}{3}D\right),\ \tau_{xy}=\mu\left(\frac{\partial u}{\partial y}+\frac{\partial v}{\partial x}\right),\ D=\frac{\partial u}{\partial x}+\frac{\partial v}{\partial y}.$$

The movement of disperse phase fractions is described by the fraction's average density conservation equation, the impulse components conservation equations and the equation of the energy conservation which are written taking into account heat exchange, impulse exchange with the carrying medium and Coulomb force acting on particles of fraction *i* of the disperse phase:

$$\frac{\partial \rho_i}{\partial t}+\frac{\partial(\rho_i u_i)}{\partial x}+\frac{\partial(\rho_i v_i)}{\partial y}=0,$$

$$\frac{\partial(\rho_i u_i)}{\partial t}+\frac{\partial}{\partial x}\left(\rho_i u_i^2\right)+\frac{\partial}{\partial y}\left(\rho_i u_i v_i\right)=F_{xi}+F_{ex\,i}-\alpha_i\frac{\partial p}{\partial x},$$

$$\frac{\partial(\rho_i v_i)}{\partial t}+\frac{\partial}{\partial x}\left(\rho_i u_i v_i\right)+\frac{\partial}{\partial y}\left(\rho_i v_i^2\right)=F_{yi}+F_{ey\,i}-\alpha_i\frac{\partial p}{\partial y},$$

$$\frac{\partial(e_i)}{\partial t}+\frac{\partial}{\partial x}(e_i u_i)+\frac{\partial}{\partial y}(e_i v_i)=-Q_i,$$

$$\rho_i=\alpha_i\,\rho_{i0},\ e_i=\rho_i C_{vi}T_i,$$

$$\frac{\partial^2\varphi}{\partial x^2}+\frac{\partial^2\varphi}{\partial y^2}=\rho q$$

$$F_{xi}=\frac{3}{4}\frac{\alpha}{(2r)}C_d\,\rho\sqrt{(u-u_i)^2+(v-v_i)^2}\,(u-u_i)+\alpha_i\rho\left(\frac{\partial u}{\partial t}+u\frac{\partial u}{\partial x}+\right.$$

$$\left.+v\frac{\partial u}{\partial y}\right)+0.5\alpha_i\rho\left(\frac{\partial u}{\partial t}+u\frac{\partial u}{\partial x}+v\frac{\partial u}{\partial y}-\frac{\partial u_i}{\partial t}-u_i\frac{\partial u_i}{\partial x}-v_i\frac{\partial u_i}{\partial y}\right),$$

$$F_{yi}=\frac{3}{4}\frac{\alpha_i}{(2r)}C_d\,\rho\sqrt{(u-u_i)^2+(v-v_i)^2}\,(v-v_i)+\alpha_i\rho\left(\frac{\partial v}{\partial t}+u\frac{\partial v}{\partial x}+\right.$$

$$+v\frac{\partial v}{\partial y}\bigg)+0.5\alpha_i\rho\left(\frac{\partial v}{\partial t}+u\frac{\partial v}{\partial x}+v\frac{\partial v}{\partial y}-\frac{\partial v_i}{\partial t}-u_i\frac{\partial v_i}{\partial x}-v_i\frac{\partial v_i}{\partial y}\right),$$

$$C_{di}=\frac{24}{Re_{i1}}+\frac{4}{Re_{i1}^{0.5}}+0.4,\ M_{i0}=\left|\bar{V}-\bar{V}_i\right|/c\ ,\ ,$$

$$Re_{i1}=\rho_1\left|\bar{V}_1-\bar{V}_i\right|2r/\mu,\ Pr=\gamma\, C_p\mu/\lambda$$

$$Q_i=6\alpha_i Nu\lambda\left(T_1-T_i\right)/\left(2r\right)^2\ i{=}2,..n$$

$$Nu_i=2\exp(-M_{i0})+0.459Re_{i0}^{0.55}Pr^{0.33},\ \mathrm{Pr}=\gamma\, C_p\mu/\lambda\ ,0\le M_{i0}\le 2\ ,\ 0\le Re_{i0}<2\cdot 10^5\ .$$

Thermal energy of the solid phase fractions suspended in gas is defined as $e_i = \rho_i C_{pi} T_i$. The system of equations was complemented with the corresponding initial and boundary conditions. On input and output computational region borders, excepting the spray nozzle inlet section, for all gas-dynamic functions, including the carrying medium and disperse phase velocity components Neumann boundary conditions were set. Coulomb force components per unit of gas-suspension volume were defined by its specific charge, solid phase volume density and electric intensity: $F_{ex\,i} = -q_0\rho_i\, {}^{\partial\varphi}/_{\partial x}$, $F_{ey\,i} = -q_0\rho_i\, {}^{\partial\varphi}/_{\partial y}$, where q_0 is the specific charge of solid fraction mass unit, φ – electric field potential.

Electric field potential in computational region was found from the Poisson equation solution with second type boundary conditions. In the right member of Poisson equation the gas-suspension charge density divided by carrying medium absolute dielectric constant [5]:

$$div\,\boldsymbol{E}=\frac{\rho_{el}}{\varepsilon\varepsilon_0},\boldsymbol{E}=-\,\Delta\,\varphi,\rho_{el}=\rho_{sum}\cdot q,\rho_{sum}=\sum_i\rho_i\,,\varepsilon_0=\frac{10^{-9}}{36\pi}\,F/m.$$

System of equations was written in the generalized curvilinear coordinates: physical region of a current in variables (x, y, t) was mapped on canonical computational region in variables (ξ, η, t) [6-9] and was solved by an explicit second order Mac-Cormack method [8] with the subsequent application of the solution non-linear correction scheme. The Poisson equation for the electric field potential was written in the generalized coordinates and was solved by finite difference method by means of the iterative Seidel scheme on a gas-dynamic computational mesh.

Acknowledgments

Works it is execute with financial support of grant of the President of Russia (number of a grant - MK 3497.2018.1)

References

[1] Salyanov F A 1997 *Theory of the low-temperature plasma physics, plasma devices and technologies* (in Russian) (Moscow: Science) 240 pages
[2] Anoshin O A, Belogolovsky A A et al 1999 *High-voltage electrotechnics* ed. I P Vereshchagin (Moscow: MEI)
[3] Kutushev A G 2003 *Mathematical modeling of Wave Processes in Aerodisperse and Powder Media* (Nedra St. Petersburg) 283 pages

[4] Gavrilova V A, Kashapov N F, Tukmakov A L, Tukmakov D A 2012 Numerical modeling of process of evaporation of the charged aerosol on a surface in electrostatic field *Bulletin of the Kazan state technical university named after A N Tupolev* No. 2. pp 177-182

[5] Tukmakov A L 2015 Numerical model of an aerodisperse system's electro-gas dynamics on the basis of double-speed double-temperature gas suspension motion equations *Application-oriented mechanics and technical physics*. Vol. 56 No. 4 pp 112-120

[6] Tukmakov A L, Tukmakov D A 2017 Dynamics of charged gas suspension with non-uniform average density distribution of a disperse phase initial spatially upon transition to an equilibrium state *Thermo physics of high temperatures* Vol. 55 No. 4 pp 509-512

[7] Falkovsky O I 2009 *Technical electrodynamics* (in Russian) (St. Petersburg: "Lan")

[8] Fletcher K 1991 *Computing methods in dynamics Liquids* Vol. 2 (Moscow: World) p 551

[9] Steger J L 1978 Implicit Finite-Difference Simulation of Flow about Arbitrary Two-Dimensional Geometries AIAA J. vol 16 p 679–686

IOP Conf. Series: Materials Science and Engineering **412** (2018) 012077 doi:10.1088/1757-899X/412/1/012077

Numerical simulation of the vapour-droplet systems hydrodynamics taking into account particles' coagulation and fragmentation

N A Tukmakova, V G Tonkonog, A L Tukmakov

Kazan National Research Technical University named after A N Tupolev-KAI 420111, Kazan, K. Marx St., 10

nadejdatukmakova@ya.ru

Abstract. The purpose of the work is description of the processes occurring in the liquefied natural gas heater-gasifier having small duration of transition process upon transition to the stationary mode at the flow volume change on the basis of mathematical modeling of the thermo- and hydrodynamics of multiphase mist flows with swirling of the stream.

1. Introduction

The current of a gas mixture in the Field-tube which is the concentric canal formed by internal and external cylindrical surfaces is researched for the purpose of calculation of parameters of the tube heater of the liquefied natural gas (LNG). The difference of the offered device from the existing analogs consists that on an input in the internal cylinder liquid is sprayed and further moves in the form of a vapour-droplet mixture. Because regasification is carried out as a result of heating and evaporation of a vapour-droplet mixture, process happens without formation of the critical phenomena, to a small inertance regarding the flow volume. The external surface of the device heats up, there is a heating of a moving gas mixture and evaporation of drops of liquid. The device can be used as the heat-absorption element of the LNG tube heater-gasifier intended for heating of natural gas on an input of gas distribution stations. In the course of the gas suspension motion, depending on the mode of a current and value of an external heat flow, there can be processes of coagulation, fragmentation, evaporation of drops influencing flow pattern and flow characteristics of the device. For the heat-absorption element design parameters and its operating regime – dispersibility on an input, input-output pressure differential and external heat flow values – choice it is required to simulate the processes accompanying the vapour-droplet mixture current. In this work the model allowing describing a polydisperse mixture flowing taking into account different fractions particles' coagulation and fragmentation is described.

2. Technological heater

The classical scheme of the re-gasifier heating dripping liquid is quoted in [1]. The heated medium, for example, purified natural gas, from the trunk pipeline comes via the connecting pipe 4 to an internal cavity 6 of the internal heat-exchanging pipe 2, then goes out through the open end of an internal pipe 2, turns and gets into the external cavity 7 formed by a annular gap between external 1 and internal 2 heat pipes. The gas heated in annular gaps of heat-exchange pipes comes to the output connecting pipe 5 and is transported to the reduction block of gas-distribution station (Figure 1).

Content from this work may be used under the terms of the Creative Commons Attribution 3.0 licence. Any further distribution of this work must maintain attribution to the author(s) and the title of the work, journal citation and DOI.
Published under licence by IOP Publishing Ltd

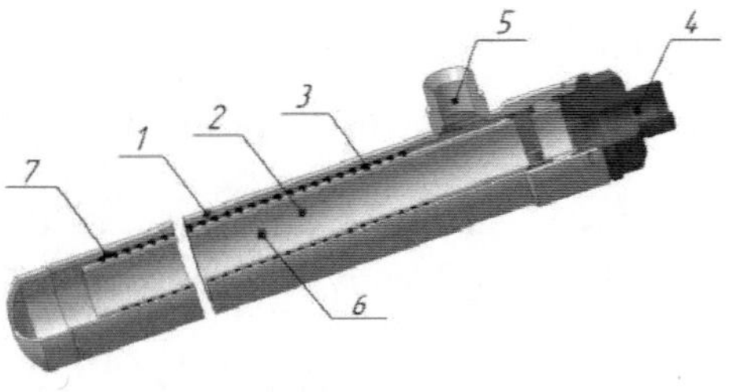

Figure 1. The diagrammatic representation of the technological heater (Field-tube) in section.

3. Mathematical model of the processes occurring in Field-tube

The carrier phase motion equations:

$$\frac{\partial\rho}{\partial t}+\frac{\partial(\rho u)}{\partial x}+\frac{\partial(\rho v)}{\partial y}=-\frac{\rho v}{y},\ ,\ \frac{\partial(\rho u)}{\partial t}+\frac{\partial}{\partial x}\left(\rho u^2+p-\tau_{xx}\right)+\frac{\partial}{\partial y}\left(\rho uv-\tau_{xy}\right)=\frac{\rho uv-\tau_{xy}}{y}-\sum_{i=1,n}F_{xi}+\alpha\frac{\partial p}{\partial x},$$

$$\frac{\partial(\rho v)}{\partial t}+\frac{\partial}{\partial x}\left(\rho uv-\tau_{xy}\right)+\frac{\partial}{\partial y}\left(\rho v^2+p-\tau_{yy}\right)=\frac{\rho v^2-\tau_{yy}}{y}-\sum_{i=1,n}F_{yi}+\alpha\frac{\partial p}{\partial y},\tag{1}$$

$$\frac{\partial(\rho w)}{\partial t}+\frac{\partial(\rho uw)}{\partial x}+\frac{\partial(\rho vw)}{\partial y}=\frac{\rho vw}{y}-\sum_{i=1,n}F_{zi},$$

$$\frac{\partial e}{\partial t}+\frac{\partial}{\partial x}\left(\left[e+p-\tau_{xx}\right]u-\tau_{xy}v+\lambda\frac{\partial T}{\partial x}\right)+\frac{\partial}{\partial y}\left(\left[e+p-\tau_{yy}\right]v-\tau_{xy}u+\lambda\frac{\partial T}{\partial y}\right)=$$

$$=-\frac{\left[e-\tau_{yy}\right]v-\tau_{xy}u+\lambda\dfrac{\partial T}{\partial y}}{y}-\sum_{i=1,\,n}Q_i-\sum_{i=1,\,n}\left(\left|F_{xi}\right|\left(u-u_i\right)-\left|F_{yi}\right|\left(v-v_i\right)\right)+\alpha\left(\frac{\partial(pu)}{\partial x}+\frac{\partial(pv)}{\partial y}\right).$$

Here ρ, u, v, w, u_i, v_i, w_i, E, λ, μ, d_i, σ – density of the carrying medium, the carrying medium and disperse phase fractions velocity components, specific total energy, coefficients of heat conductivity and viscosity of the carrier phase, i-fraction's drop diameter and liquid's surface tension coefficient [2].

The disperse phase consists of a number of fractions formed by same size particles which dynamics is described by the fraction's average density conservation equation, components of an impulse conservation equations and the thermal energy conservation equation:

$$\frac{\partial\rho_i}{\partial t}+\frac{\partial(\rho_i u_i)}{\partial x}+\frac{\partial(\rho_i v_i)}{\partial y}=-\frac{\rho_i v_i}{y},$$

$$\frac{\partial(\rho_i u_i)}{\partial t}+\frac{\partial}{\partial x}\left(\rho_i u_i^2\right)+\frac{\partial}{\partial y}\left(\rho_i u_i v_i\right)=-\frac{\rho_i u_i v_i}{y}+F_{xi}-\alpha\frac{\partial p}{\partial x},$$

$$\frac{\partial(\rho_i v_i)}{\partial t}+\frac{\partial}{\partial x}\left(\rho_i u_i v_i\right)+\frac{\partial}{\partial y}\left(\rho v_i^2\right)=-\frac{\rho v_i^2}{y}+F_{yi}-\alpha\frac{\partial p}{\partial y},\tag{2}$$

$$\frac{\partial(\rho_i w_i)}{\partial t}+\frac{\partial}{\partial x}\left(\rho_i u_i w_i\right)+\frac{\partial}{\partial y}\left(\rho v_i w_i\right)=-\frac{\rho v_i w_i}{y}+F_{zi},$$

$$\frac{\partial e_i}{\partial t}+\frac{\partial}{\partial x}\left(e_i u_i\right)+\frac{\partial}{\partial y}\left(e_i v_i\right)=\mathrm{Nu}_i\frac{6\alpha_i}{\left(2r_i\right)^2}\lambda\left(T-T_i\right),\ \rho_i=\alpha_i\rho_{i0},\ e_i=\rho_i C_{pi}T_i.$$

Here α_i, ρ_i, e_i, T_i – volume content, average density, thermal energy and the disperse phase temperature; C_{pi}, ρ_{i0} – heat capacity and substance density of dispersed fraction i. The system of equations of each dispersed fraction motion contains the equation for an impulse azimuth component, written down in the assumption of azimuth symmetry of a stream at its rotation. The Benedict-Webb-Rubin state equation has been applied as the thermal equation of state for the carrier phase [3]:

$$p = RT\rho + \left(BRT - A - \frac{C}{T} \right)\rho - (bRT - a)\rho + a\alpha\rho + \frac{c\rho}{T}(1+\gamma\rho)\exp(-\gamma\rho), \quad (3)$$

coefficients a, A_0, b, B_0, c, C_0, α, ρ are determined by the known values of density, pressure and temperature of methane in a vaporous state.

The system of equations of two-phase polydisperse gas-suspension motion was written in the generalized movable coordinates [4, 5, 6] and was solved by an explicit second-order Mac-Cormack method with the nonlinear correction scheme [7, 8].

4. Disperse phase fractions' drops coagulation and fragmentation

Drops' coagulation and fragmentation occur depending on the Weber number values for disperse fraction particles. If Weber's number for a particle of *i*-fraction is less than critical value (We_{cr} = 2-10), then at its collisions with smaller fractions' particles, coagulation take place therefore mass, concentration, an impulse and temperature of *i*-fraction particle change. This process is described by Smoluchowski Lagrangian model [6]. If *i*-fraction particle Weber number is more than critical, then the motion of such particle is followed by the gas-dynamic fragmentation leading to reduction of *i*-fraction particle mass. As the drop destruction mechanism the dish-shaped type of fragmentation connected with boundary-layer separation [9, 10], which is implemented at $We_{cr} \leq We < 1.7 \cdot 10^4/Lp^{0.5}$ can be chosen.

Calculations were conducted on the block structured meshes constructed for three subareas (Figure 2). Boundary conditions on blocks docking lines were set taking into account the direction of a current. Pressure on an input to a subarea I and pressure at the output from area III was set. Carrying medium and disperse fractions velocities in the area I was set in initial time point. Medium was assumed motionless in II and III areas at $t = 0$.

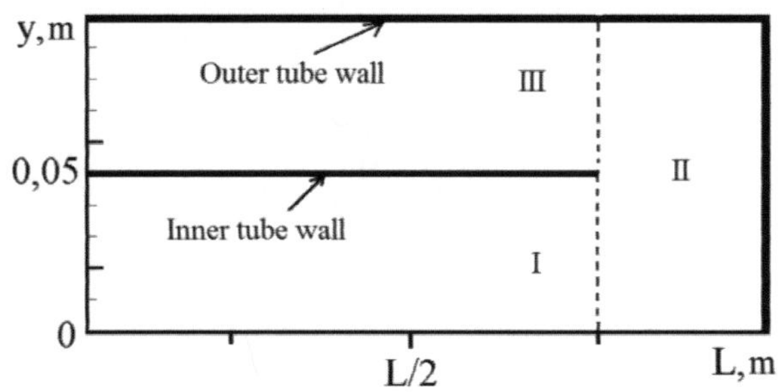

Figure 2. Diagrammatic representation of computational region: I, II, III - block structured mesh's subareas.

As a calculation result gas-dynamic functions fields of the carrying medium and a number of disperse fractions which particles, depending on Weber number participated in coagulation and fragmentation processes have been received.

Acknowledgments

This research is executed with support of the RFBR and Tatarstan Republic government (project No. 18-48-160017) on the state instructions of the Ministry of Education and Science of the Russian Federation to Higher Education Institutions (project No. 9.5542.2017/6.7).

References

[1] Serazetdinov B F, Serazetdinov F Sh, Tonkonog V G 2012 *Technological heater*: stalemate. on an invention No. 2467260. (in Russian)
[2] Nigmatullin R I 1987 *Dynamics of multiphase mediums* (in Russian) P. I. (M.: Science) p 464
[3] Read R, Prausnits J, Sherwood T 1982 *Gases and liquids properties* (in Russian) (Leningrad: Chemistry) 496 pages

[4] Tukmakov A L, Tonkonog V G, Arslanova S N 2016 Wave coagulation of a polydisperse gas-suspension in technology of gasification and cryostatting of liquefied natural gas (in Russian) *Acoustic Journal* V 62 No 1 pp 125-131
[5] Kutushev A G 2003 *Mathematical modeling of wave processes in aero disperse and powdery mediums* (in Russian) (SPb.: Publishing house Nedra) p 283
[6] Alemasov V E et al 1971 *Thermodynamic and thermal properties of combustion products*: the reference book in 5 V. V. 1. Calculation methods (in Russian) (M.: VINITI publishing house) p 267
[7] Fletcher K 1991 *Computational methods in liquid dynamics* (in Russian) V. 2. (M.: World) 551 pages
[8] Steger J L 1978 *Implicit Finite-Difference Simulation of Flow about Arbitrary Two-Dimensional Geometries* AIAA J. V. 16 No. 7 pp 679-686
[9] Arefyev K Yu, Voronetsky A V 2015 *Modeling of process of drops fragmentation and evaporation of nonreacting liquid in the high enthalpy gasdynamic streams* Thermophysics and aeromechanics V 22 No 5 pp 609-620
[10] Samoylovich G S 1990 *Hydraulic gas dynamics* (M.: Mechanical engineering) p 384

IOP Conf. Series: Materials Science and Engineering **412** (2018) 012078 doi:10.1088/1757-899X/412/1/012078

Biomechanical assessment of outputs of surgical treatment of patients with complex fragments of lower limbs

B I Vahitov[1], I O Pankov[2], I Kh Vahitov[1], V R Nagmatullin[2] and L I Vahitov[1]

[1]Kazan Federal University, Russian Federation, Kazan, 18 Kremlin Street, 420008
[2]Kazan State Medical Academy, Russian Federation, Kazan, 36 Butlerova Street, 420012

tggpy-mbofk@mail.ru

Abstract. Studies of static-dynamic disorders, with the use of new, progressive technologies in the post-traumatic period becoming increasingly practical. We conducted studies of the support, dynamic function of the lower limbs, balance in a vertical rack in 20 healthy individuals (comparison group) in order to determine the normal values of the biomechanical research indices. Of all studied women 12, men 8. All subjects were of working age (25-55 years). The median age was 40 years. Of particular interest are data from biomechanical studies of the support, dynamic function of the lower extremities, sensory balance, conducted in 18 patients with fractures of the joint region of the lower leg bones in the dynamics at terms 9-12 and 18-24 months after the operation. In the study of the balance, there were found disturbances in the influence of the somatosensory, visual and vestibular systems on maintaining the balance of the body. Each balance study included a series of open and closed eyelids in a vertical patient stand with lower limbs straight in the knee joints on hard and soft (foamy) surfaces. The application of the hardware-software evaluation of the support-dynamic function of the lower extremities by the type of stabilometry at the rehabilitation stages will allow to identify and carry out the prevention of serious post-traumatic complications at an early stage of the disease.

Introduction

Restoration of patients with intra-articular fractures of the lower limbs is one of the most urgent problems of traumatology and orthopedics at all stages of treatment [1, 2]. Complications and unsatisfactory outcomes of treatment for intraarticular fractures of the bones of lower extremities, according to various authors, reach 30.0% and higher [3, 4]. The most frequent and severe among them are the development of deforming arthrosis and persistent contractures of the joints, disrupting the function of the entire lower limb and leading to a significant decrease in the quality of life [5,6].

Thus, despite the variety of methods and methods of early recovery of intraarticular bone fractures of the lower limbs, a large number of works devoted to improving outcomes, the problem of treating such injuries can not be considered finally solved, which is confirmed by a high percentage of complications and adverse outcomes among patients. In connection with the above, we have attempted to develop a system of diagnostic measures, as well as an integrated approach to the rehabilitation of patients with complex fractures of the lower limbs [7, 8].

Content from this work may be used under the terms of the Creative Commons Attribution 3.0 licence. Any further distribution of this work must maintain attribution to the author(s) and the title of the work, journal citation and DOI.
Published under licence by IOP Publishing Ltd

IOP Conf. Series: Materials Science and Engineering **412** (2018) 012078 doi:10.1088/1757-899X/412/1/012078

Materials and methods of research

Biomechanical studies of the lower extremities in norm, as well as at the stages of restorative treatment of patients with fractures of the knee and ankle joint area were carried out on the NeurocomBalanceMaster® hardware-software version 7.0. The most important components of the BalanceMaster® system are the computer and the platform mounted on the base. When performing this study, the patient stands on the double plate of the platform facing the monitor. The motion sensors under the platform measure vertical movements caused by the pressure of the patient's feet. On the cable, this information is transferred from platform to computer. The computer receives the measurement data from the platform, analyzes the information and generates a display on the screen (Figure 1).

Fig.1. The BalanceMaster® system (NeuroCom® International, Inc.) (general view).

When carrying out biomechanical studies, three types of tests were performed to identify the supporting and dynamic function of the lower extremity, the balance in the vertical rack (holding the body weight, walking with a roll - determining the length of the step, the step width, step speed, and the modified clinical test of sensory interaction of the balance). These studies allow an objective assessment of posttraumatic disorders of the function of the lower extremities, as well as trace the recovery process in the process of medical rehabilitation.

We conducted studies of the support, dynamic function of the lower limbs, balance in a vertical rack in 20 healthy individuals (comparison group) in order to determine the normal values of the biomechanical research indices. Total number of women was 12, men 8. All subjects were of working age (25-55 years). The median age was 40 years.

Statistical processing of the data was carried out using the SPSS software package (v. 13.0). The differences in the quantitative indicators of the two groups were estimated using the Student's test.

Of particular interest are data from biomechanical studies of the support, dynamic function of the lower extremities, sensory balance, conducted in 18 patients with fractures of the joint region of the lower leg bones in the dynamics at terms 9-12 and 18-24 months after the operation.

The data of biomechanical studies of the supporting, dynamic function of the lower extremities, the sensory balance in the vertical rack in patients with complex fractures of the lower extremities at the stages of medical rehabilitation in the dynamics at the periods 9-12 and 18-24 months after the operation are presented in Tables 1, 2 and 3.

Table 1 presents the comparative data of the Weight Bearing / Squat test (Weight retention) in patients after surgical treatment at terms 9-12 and 18-24 months after surgical treatment.

Table 1 - Weight Bearing / Squat (Weight retention) test results in patients at terms 9-12 and 18-24 months after surgical treatment.

Weight retention test	9-12 months	18-24 months
% Difference in position 0^0	10,2±2,0	10,2±3,0
% в Difference in position 30^0	11,4±2,0	11,5±3,0*
% Difference in position 60^0	12,4±3,0	12,8±4,0*
% Difference in position 90^0	13,2±2,0	14,2±2,0*

*Notation: * - statistically significant difference (P <0.001) from the indices to the treatment.*

Table 2 shows the walking with transition test results (WalkAcross (WA)) in patients at terms 9-12 and 18-24 months after surgery.

Table 2 - Data of walking with transition test (WalkAcross (WA)) in patients at terms 9-12 and 18-24 months after surgical treatment.

Test Parameters for walking with transition	9-12 months	18-24 months
Step width (cm.)	26,8±2,5	2,8±2,6*
Step length (cm.)	62,2±4,7	62,1±2,8*
Step speed (cm./sec.)	76,7±6,7	68,4±6,6*
Symmetry of step (%)	5,4±2,0	5,4±3,0*

*Notation: * - statistically significant difference (P <0.001) from the indices to the treatment.*

The results of the sensory balance interaction test (CTSIB (mCTSIB) in patients at terms 9-12 and 18-24 months after surgical treatment are presented in Table 3.

Table 3 - Data obtained from the modified sensory balance interaction clinical test (CTSIB (mCTSIB) in patients at terms 9-12 and 18-24 months after surgical treatment.

Parameters of sensory balance interaction test	9-12 months	18-24 months
Firm-EO (deg/sec)	0,26±0,2	0,26±0,2*
Firm-EC (deg/sec)	0,36±0,3	0,36±0,6*
Foam-EO (deg/sec)	0,64±0,3	0,66±0,3*
Foam-Ec (deg/sec)	1,53±0,2	1,56±0,2*

*Notation: * - statistically significant difference (P <0.001) from the indices to the treatment.*

According to the data of biomechanical studies conducted in patients with complex fractures of the lower limbs, carried out in the dynamics at the intervals of 9-12 and 18-24 months after surgical treatment, presented in Tables 1, 2 and 3, there is a tendency to a certain decrease in the parameters of the reference, dynamic function lower limb, a modified sensory balance in a vertical rack.

Figure 2 presents the comparative data of the study of the support function of the lower extremities of patient B., born in 1975, No. 2467, treated at the clinic of traumatology of the Republican Clinical Hospital with the diagnosis: closed fracture of both ankles, marginal fracture of the posterior edge of the tibia, distal interstice syndesmosis, dislocation of the foot behind and outside, in dynamics at terms 10 and 24 months after surgical treatment.

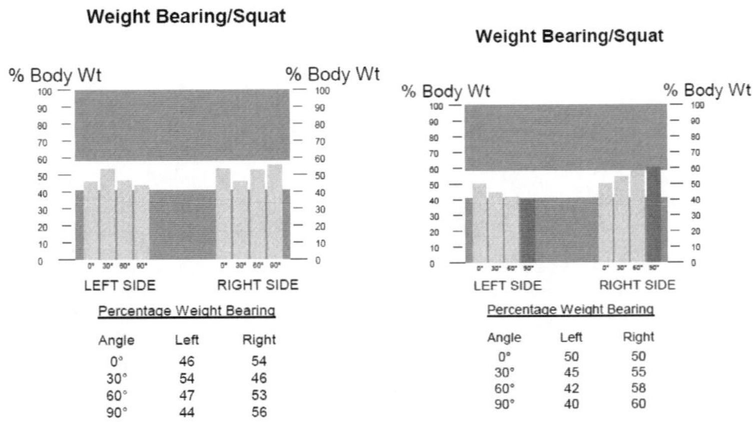

a). b).

Figure 2 - Data of the study of the supporting function of the lower limbs of patient B., born in 1975. through 10 (a) and 24 (b) months after the operation in comparison.

Figure 3 presents the comparative data of the study of the sensory balance in the vertical column of the same patient in dynamics at terms 10 and 24 months after surgical treatment.

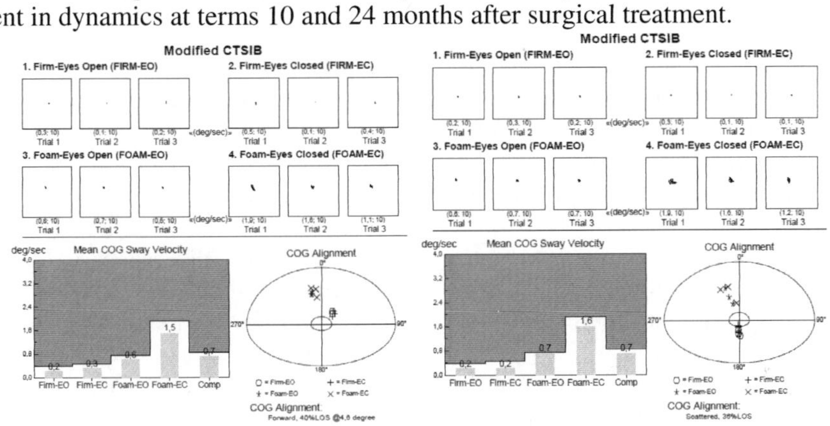

a). b).

Figure 3 - The data of the study of sensory balance in the vertical rack of the patient B., born in 1975 through 10 (a) and 24 (b) months after the operation in comparison.

Figure 4 presents the comparative data of the study of the dynamic function of the lower extremities of patient B., born in 1975, No.2467 in dynamics at terms 10 and 24 months after surgical treatment.

IOP Conf. Series: Materials Science and Engineering **412** (2018) 012078 doi:10.1088/1757-899X/412/1/012078

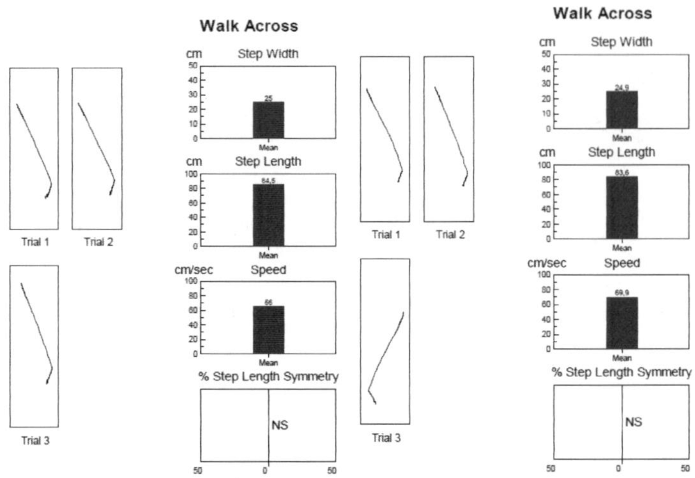

a). b).

Figure 4 - Data of the study of the dynamic function of the lower limbs of patient B., born in 1975. through 10 (a) and 24 (b) months after the operation in comparison.

As follows from the above comparative data of biomechanical studies of the function of the lower extremities in dynamics at terms 10 and 24 months after the operation, the patient experienced a decrease in the support function, with insignificant decreases in sensory balance and dynamic function of the lower limbs in the long-term after trauma. These disorders can be associated with the onset of the development of posttraumatic deforming arthrosis due to the primary severe damage to the capsular-ligament apparatus of the ankle.

Conclusions

The data of biomechanical studies carried out on the NeurocomBalanceMaster® 7.0 system reveal, before the manifest clinical manifestations of varying degrees of severity, various minor disorders of the support (uneven distribution of loads on the left and right lower limbs), dynamic function (decrease in length, increase in the width of the step on the damage side) sensory balance in a vertical rack in the early and distant periods after surgical treatment, which indicates a developing post-traumatic pathology in the form of deformities iruyuschego arthrosis.

The application of the hardware-software evaluation of the support-dynamic function of the lower extremities by the type of stabilometry at the rehabilitation stages will allow to identify and carry out the prevention of serious post-traumatic complications at an early stage of the disease.

References

[1] Davydov O D, Montile A I, Marchuk Y V 2012 *A new way to diagnose the functional state of the musculoskeletal system based on the results of stabilographic studies* Collected theses of the III Congress of Traumatology-Orthopedists of the Ural Federal District "Chaklinsky Readings". – Ekaterinburg pp 45-47

[2] Ryabchikov I V, Pankov I O 2013 *Evaluation of the dynamic function of the lower extremities of patients after surgical treatment of fractures of the distal articular region of the lower leg bones in the process of restorative treatment* Elektr. sci. Journal of Contemporary Problems of Science and Education

[3] Pankov I O, Ryabchikov I R 2013 *Changes in the dynamic function of the lower extremity of patients after surgical treatment of fractures of the distal joint region of the lower leg bones during medical rehabilitation* Elektr. sci. Journal of Contemporary Problems of Science and Education

[4] Pankov I O, Ryabchikov I R 2014 *Our experience in restoring the supporting, dynamic function of the lower extremity, the balance of patients with complex fractures of the distal joint region of the lower leg bones* Elektr. sci. Journal of Contemporary Problems of Science and Education

[5] Ryabchikov I V, Pankov I O 2013 *Evaluation of the impact of medical rehabilitation on the restoration of the dynamic function of the lower limbs of patients after surgical treatment of fractures of the ankle area* Elektr. sci. Journal of Contemporary Problems of Science and Education
[6] Arkhipov L K, Lychagin A V 2000 *Modern aspects of the treatment of posttraumatic deforming arthrosis of the ankle joint* Vestnik travmatol. and orthop. them. N. N. Priorov V 4 pp 64-67
[7] Yu V Antoniadi, Berdyugin, Galyautdinov A F 2006 *Modern technologies in the fracture of the ankles* Traumatology and orthop. Russia No 2 pp 22
[8] Babelo O Yu, Sviridov A I, Bobrova O V 2002 *The device of its own design for the treatment of ankle joint damage* Materials of the VII Congress of Traumatologists-Orthopedists of Russia V 2 pp 387-8

IOP Conf. Series: Materials Science and Engineering **412** (2018) 012079 doi:10.1088/1757-899X/412/1/012079

A Method for Increasing the Efficiency of Assembling Non-detachable Products by Plastic Deformation Using Artificial Intelligence

A M Valiev[1], D L Pankratov[1]

[1]Kazan Federal University, Naberezhnye Chelny Institute, 423812, Russia, Naberezhnye Chelny, Prospekt Syuyumbike 10A

AjMValiev@kpfu.ru

Abstract. The way of achievement of the given parameters of quality of assembling of one-piece products, such as accuracy of geometrical sizes of a product and durability of non-detachable joints at stages of technological preparation and management of technological process with application of an artificial intellect is studied.

Composite metal products obtained by assembly into non-detachable joints by plastic deformation are widely introduced into machine building. The main advantage of their application is the possibility of combining in one single product the technical properties of various steel grades. For example, by using such products, it is possible to solve the problem of saving expensive alloyed steels while maintaining the performance characteristics of the die tool [1].

Among the known ways of assembling non-detachable joints, methods of plastic deformation are the most important, when non-detachable joint is obtained at a temperature below the melting point of the metals to be joined [2, 3]. At the same time, there are practically no zones of structural and chemical heterogeneity which inherent in the methods of obtaining composite parts by means of fusion welding.

Achieving the specified quality indicators for assembly of non-detachable joint, such as the accuracy of the geometric dimensions of the product, the strength of an integral connection, depends on the decisions made, both at the stages of technological preparation and at the stage of management of the technological process itself. The system of technological preparation and management should provide a scientifically justified choice of technological assembling modes, also accurate calculation of the parameters of the connected elements of the product, which allows to take into account their shape change during assembly by stage-by-stage deformation localization in the connected elements of the product in real time.

The solution of the task to improve the efficiency of assembling non-detachable products by plastic deformation is possible only with the use of an automated system for modeling and controlling the technological process of assembly using artificial intelligence elements.

Figure 1 shows a structural model of a multilevel automated system of modeling and control of the technological process of assembly of non-detachable products by plastic deformation, which provides both automated preparation of the technological process and control of the process of assembly of the product in real time.

Content from this work may be used under the terms of the Creative Commons Attribution 3.0 licence. Any further distribution of this work must maintain attribution to the author(s) and the title of the work, journal citation and DOI.
Published under licence by IOP Publishing Ltd

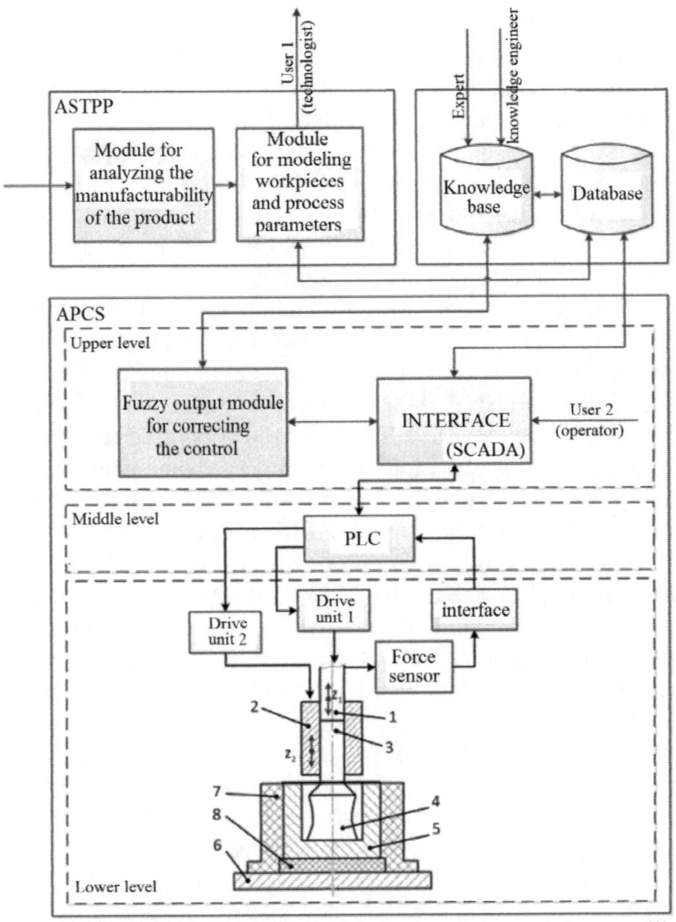

Figure 1 - Structural model of the automated system for modeling and controlling the technological process of assembling non-detachable products by plastic deformation

The initial data of the developed system is the CAD-model of the finished non-detachable product, as well as information about the materials of the connected elements.

According to the initial data, the information system searches for a similar precedent in the database. The database stores ready-made solutions in the form of CAD-models of shaped workpiece and rod, and also in the form of decisions on the choice of equipment and tooling design, formed by the management teams. In the case of absence of a fully coinciding precedent, either a new precedent is created or the closest one is chosen, which is further adapted. When developing a new precedent in the module for analyzing the manufacturability of the product, an analysis is made: on the applicability of the technology according to the size; on the compatibility of strength properties; on the compatibility of metals in heat treatment. Next, the user is offered an alternative for choosing the material of the workpiece. The necessary information for the analysis on rheology and thermal properties of materials, size limitations of the technology, as well as the parameters of hydraulic presses is stored in the database.

In the module for modeling workpieces and process parameters, the geometrical parameters of the shaped workpiece and rod are calculated, process parameters are assigned, recommendations for the selection of equipment and tools are given, information is generated for the tool shop (site) in the form of CAD-models of the shaped workpiece and rod, and input data are generated for the operation of the automated process control system. The module for modeling blanks and process parameters is based on the method for selecting the geometrical parameters of the shaped workpiece.

The automated process of control system (APCS) consists of the upper, middle and lower levels. At the lower level, the sensor signals are coordinated with the inputs of the control device, and the generated commands with the actuators with the help of communication devices with objects. At the middle level, the PLC (Programable Logic Controller) receives information from the sensors about the state of the process and issues control signals to the actuators. The upper level is an intelligent add-in that analyzes and processes information coming from the lower layers and the level of the automated process of control system of technological preparation of production (ASTPP), and also controls the technological process. The intelligent add-on is a fuzzy output module for correcting the control based on the data received from the sensor effort using fuzzy rules.

Input variables characterizing the real state of the control object are subjected to fuzzification and are used directly in the formation of control at the lower level, as well as in the system of fuzzy high-level output. In the system of fuzzy output of the upper level there is the base of fuzzy production rules of the form IF "real state of the object", THEN "correction of the lower level control", which in fuzzy form presents the knowledge of experts to adjust the control parameters of the lower level in order to ensure the gradual localization of deformation in the connected elements of the product in real time.

Approbation of the developed system for the production of a compound tool for punching holes was made, which is an integral product of steels X12M (rod) and U10A (shaped workpiece) with d = 12 mm; D = 31 mm; H = 21 mm; l = 15 mm. When producing a punch made of steel U10A, its resistance is 40-80 thousand strokes, and from X12M - 150-200 thousand strokes.

The rationale for using a compound tool instead of a one-piece of X12M is to maintain the performance of its operational durability with a lower cost of 3-4 times the costly tool steel X12M.

According to the results of mechanical tests of samples, the tensile strength of an integral joint for tearing was 241-265 MPa (MegaPascals) per unit of area of the contact surface of the joining elements of the product. The analysis of geometrical accuracy of the obtained one-piece product is carried out, as a result of which the value of the deviation by weight of the product is equal to 7.57%. Thus, an integral product with a deviation of geometric parameters within the range of accurate hot forging is obtained, which is 5-25% by weight (volume) of the forging.

References

[1] Valiev A M, Shibakov V G, Pankratov D L 2016 *Design and manufacture of a composite punching tool* Russian Engineering Research 36(2) pp 146-148 https://doi.org/10.3103/S1068798X16020246.

[2] Valiev A M, Shibakov V G, Pankratov D L 2016 *Automated control system for the assembly of a composite punching tool* Russian Engineering Research 36(2) pp 142-145 https://doi.org/10.3103/S1068798X16020234.

[3] Shaparev A V, Savin I A 2017 *Calculation of Joint Plastic Deformation to Form Metal Compound in Cold Condition* Solid State Phenomena Vol 265 pp 313-18 DOI: 10.4028/www.scientific.net/SSP.265.313

IOP Conf. Series: Materials Science and Engineering **412** (2018) 012080 doi:10.1088/1757-899X/412/1/012080

Control of electrochemical dimensional processing on the basis of synergy of controllable factors

A R Zakirova and Z B Sadykov

[1]Kazan Federal University, 18 Kremlyovskaya street, Kazan, 420008, Russian Federation

alfiyazakir@gmail.com

Abstract. The article considers the factors that influence the electrochemical dimensional processing, which have synergistic connections. It is revealed that the size of the interelectrode gap is of great importance for ensuring the quality of processing, and methods for adjusting the gap during processing are proposed.

Introduction

Quality indicators of automatic and semi-automatic lines for electrochemical surface treatment are separated into groups of indicators according to:

- their purpose (productivity, specific output of products from the occupied area);
- reliability (trouble-free operating time, life before the first major overhaul, shelf life, service life);
- on the economical use of materials, electricity and labor resources (specific consumption of electricity, water);
- for ergonomics (sound pressure at the workplace, the volume of sucked air);
- on the manufacturability (mass and specific gravity of the line);
- on standardization (coefficient of applicability);
- on patent law (indicators of patent protection);
- safety (protection of electrical equipment).

The performance of automatic and semi-automatic lines for electrochemical surface treatment can be viewed from various aspects: technical, theoretical and operational. Technical productivity is the main technical and economic indicator that characterizes the capabilities of the line, due to the technological process and the design of the equipment. In addition to productivity, the process quality of the electrochemical dimensional processing includes the processing accuracy and roughness of the treated surface.

Analysis of indicators and factors of electrochemical dimensional processing

The carried out investigations on sectional samples of steel 12x13 with dimensions 10 * 10 * 20 mm by electrochemical dimensional processing using a brass cathode from L62 with dimensions of 3*3 mm and electrolyte NaNO3 made it possible to compose the regression equations for the above indices of the quality of the technological process of electrochemical dimensional processing:

У1=142,23-16,69*X1-6,59*X2+20,31*X3+18,21*X4+97,75*X5+110,53*X6 +84,41*X7-56,39X1*X5+72,08*X5*X6+56,73*X5*X7+61,27*X6*X7;

У2=1160,77+271,71*X1+21,21*X2+136,29*X3+39,66*X4+79,52*X5+404,59*X6+426,38*X7+192,49*X1*X5+90,62*X1*X6+151,51*X1*X7-84,27*X2*X3-71,23*X2*X5+67,82*X6*X7;

Content from this work may be used under the terms of the Creative Commons Attribution 3.0 licence. Any further distribution of this work must maintain attribution to the author(s) and the title of the work, journal citation and DOI.
Published under licence by IOP Publishing Ltd

IOP Conf. Series: Materials Science and Engineering **412** (2018) 012080 doi:10.1088/1757-899X/412/1/012080

У3=54,49-9,47*X1-5,73*X2-4,99*X3-7,40*X4-7,13*X5+7,16*X6-4,69*X7+6,32*X1*X2+6,68*X1*X4+6,63*X1*X5+7,18*X3*X4-6,27*X5*X6-6,21*X6*X7.

In these equations У1 - efficiency of process, У2 - accuracy of processing, У3- the surface roughness after processing, X1 - initial interelectrode gap, X2 - electrolyte pH value, X3 - electrolyte concentration, X4 - electrolyte temperature, X5 - electrolyte speed, X6 - treatment time, X7 - operating voltage. [1]

In these equations the variables X1 и X5, X5 и X6, X5 и X7, X6 и X7 (for У1); X1 и X5, X1 и X6, X1 и X7, X6 и X7 (for У2); X1 и X2, X1 и X4, X1 и X5, X3 и X4, X5 и X6, X6 и X7 (for У3) have a synergistic effect. To analyze the synergistic influence of factors, a connection diagram was used. Figure 1 shows the factors by which, according to the regression equation for productivity, there is a synergistic effect.

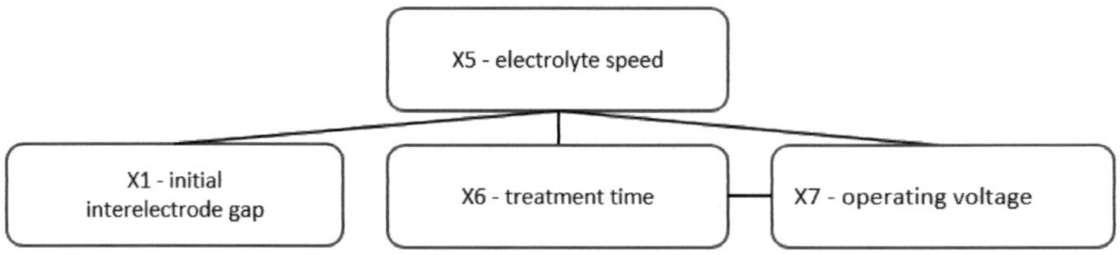

Figure 1 Diagram of the relationships of synergistic factors affecting the performance of electrochemical processing

The increase in productivity is significantly affected by the processing time, but the speed of the electrolyte due to synergistic bonds has a greater impact. Similarly, Figure 2 shows the factors by which, according to the regression equation for accuracy, there is a synergistic effect.

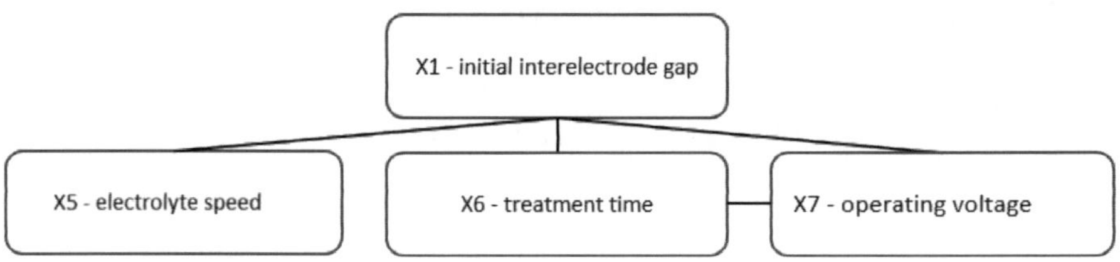

Figure 2 Diagram of the relationships of synergetic factors affecting the accuracy electrochemical processing

The increase in accuracy is significantly affected by the operating voltage, but due to synergistic relationships, the initial interelectrode gap exerts a greater influence.

Figure 3 shows the factors by which, according to the regression equation for roughness, there is a synergistic effect.

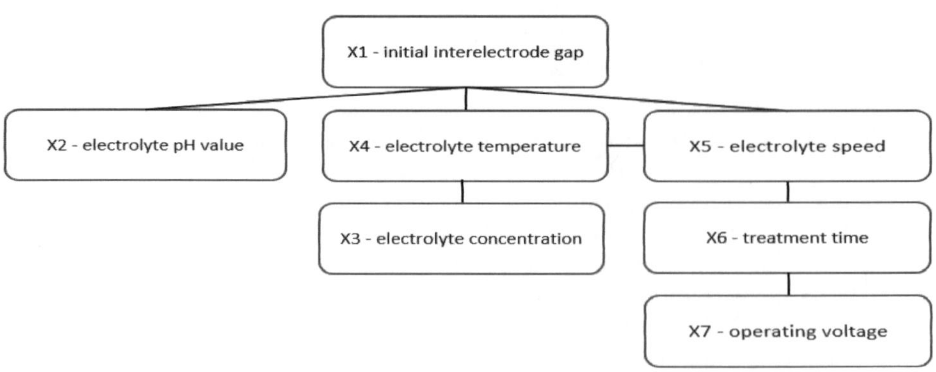

Figure 3 Diagram of the relationships of synergistic factors

IOP Conf. Series: Materials Science and Engineering **412** (2018) 012080 doi:10.1088/1757-899X/412/1/012080

affecting the roughness of electrochemical processing

The decrease in roughness is significantly affected by the temperature of the electrolyte, but due to synergistic bonds, the initial interelectrode gap exerts a greater influence.

Thus, in order to achieve a given level of performance indicators, accuracy and roughness, synergistic factors, different for each indicator, are involved. To ensure the required performance and accuracy, it is necessary to establish factors at the upper levels, with the exception of the hydrogen index of the electrolyte. To reduce the roughness, all factors should also be set at the upper level, except for the processing time.

In modern industry, traditional, for example, mechanical processing, as well as new types, such as ultrasonic, laser, plasma, etc., are used to process natural-deficient, hard metals. At the same time, laser and plasma cutting of materials are the most modern, promising methods and increasingly find practical application in the manufacture of products from hard-to-process materials. These methods provide high productivity of the production process, but they also have a number of disadvantages that do not allow their application in the case of stringent quality processing requirements. [2]

Electrochemical dimensional processing of materials has a number of unique features that allow it to compete with other methods of dimensional shaping. An increase in the level of quality of electrochemical dimensional processing is possible due to the choice of appropriate processing regimes or through the introduction of new technical solutions.

Corresponding treatment regimes should first of all be aimed at achieving the required quality of the treated surface, the performance parameter is, although important, but secondary. Processing modes can be selected in such a way as to ensure that the requirements for both surface quality and productivity are met. Consider the connection diagram for the system of equations under consideration; Figure 4 shows the factors by which the regression equations for performance, accuracy and roughness have a synergistic effect.

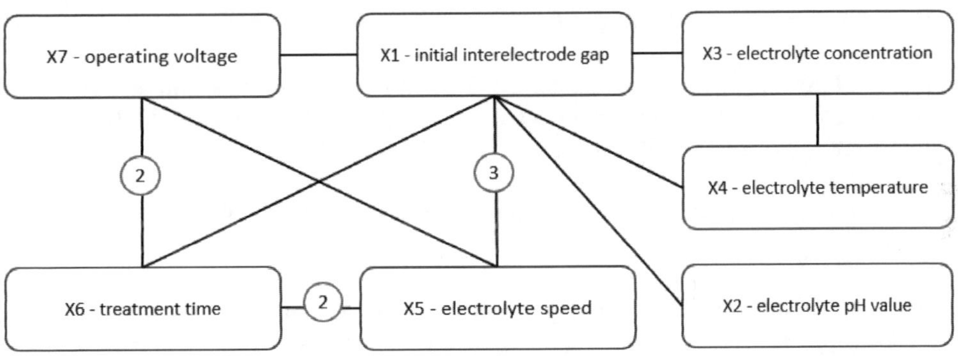

Figure 4 Diagram of the relationships of synergistic factors affecting the productivity, accuracy and roughness of electrochemical processing

The main criteria affecting the process of electrochemical processing with fixed electrodes are the electrolyte velocity, the voltage at the electrodes and due to the synergetic effect, the magnitude of the initial interelectrode gap plays a large role. To improve the quality of the surface, the initial interelectrode gap should be set at the upper level, but in the process of processing, when it increases, the productivity decreases. Maintaining the required value of the interelectrode gap is provided, for example, by the mobility of the tool electrode; as it is realized in the patent RU (11) 2 277 034 (13) C2 Method of electrochemical processing. [3]

In addition, an example of a technical solution is the design of a new electrode tool, which uses nanomaterials and composites. In the proposed embodiment, there is no constant uncontrolled electrical contact between the tool electrode and the workpiece to be machined, so that the side surfaces of the cut channel are plane-parallel. A positive result of the new tool electrode is the

IOP Conf. Series: Materials Science and Engineering **412** (2018) 012080 doi:10.1088/1757-899X/412/1/012080

elimination of additional processing operations to eliminate the non-parallelism of the surfaces of the cutting channel. [4]

Conclusions

The results of the studies allow controlling the electrochemical treatment and providing the required technological parameters. Priority should be given to the accuracy and purity of the treated surface, and then performance. An analysis of the synergistic effect of the regression equations revealed the dependence of the quality indices on the value of the interelectrode gap, which simplifies the control of the electrochemical processing.

References

[1] Akhmetov I D, Zakirova A R and Sadykov Z B 2017 *Management of processes of electrochemical dimensional Processing* http://iopscience.iop.org/article/10.1088/1757-899X/240/1/012002/pdf

[2] Akhmetov I D, Zakirova A R and Sadykov Z B 2017 *The analysis and selection of methods and facilities for cutting of naturally-deficit materials* http://iopscience.iop.org/article/10.1088/1757-899X/134/1/012002/pdf

[3] Gritsjuk Vasilij Grigor'evich (RU), Smolentsev Vladislav Pavlovich (RU), Nekrasov Aleksandr Nikolaevich (RU) and Revin Aleksej Stepanovich (RU) *Electrochemical treatment process* http://www.freepatent.ru/images/patents/191/2277034/patent-2277034.pdf

[4] Akhmetov I D, Zakirova A R, Z B Sadykov and I I Khafizov 2017 New electrode-tool for the combined kerf of electrically conductive materials *IOP Conference Series: Materials Science and Engineering* vol.240 Is1 **012003**

[5] Khafizov I I and Galimov A N 2017 IT-strategy and major aspects of quality management on the market of goods and services *IOP Conference Series: Materials Science and Engineering* vol 240 Is1 **012038**

[6] Khafizov I I, Saveleva T N and Lyubtsov V S 2017 *Estimation of parameters of charge carriers in dielectric materials by CELIV method* IOP Conference Series: Materials Science and Engineering vol 240 Is1 **012040**

[7] Khafizov I I 2017 *Ways of decrease in the material consumption in case of their separation by the combined methods* IOP Conference Series: Materials Science and Engineering vol 240 Is1 **012037**

[8] Khafizov I I 2016 *Economic efficiency and effectiveness of ways of separating materials electro diamond processing* IOP Conference Series: Materials Science and Engineering vol 134 1 **012014**

[9] Khafizov I I and Nurullin I G 2017 *Product quality as the main factor of increase of competitiveness (on the example of JSC «Kazan helicopter plant»)* IOP Conference Series: Materials Science and Engineering vol 240 Is1 **012039**

Automated system of recognition of road signs for ADAS systems

R R Ziyatdinov[1] and R A Biktimirov[1]

[1]Kazan Federal University, Naberezhnye Chelny Institute, 423812, Russia, Naberezhnye Chelny, Prospekt Syuyumbike 10A

rust.kfu@gmail.com

Abstract. Advanced Driver Assistance Systems (ADAS) require an environmental recognition function to inform the driver and for making management decisions. To solve the problems of pattern recognition, it is necessary to use one of the image classification methods. To date, there are quite a few similar algorithms that differ in the quality and speed of recognition. This paper presents a comparative analysis of classification algorithms in the problems of road signs recognition based on the GTSRB dataset. The results of the work showed that the most promising methods of classification in the problems of image recognition are the methods of reference vectors and k – nearest neighbors.

1. Introduction

Pattern recognition is one quite difficult task due to the diversity of both the objects to be recognized and the methods used. The need for such systems is manifested in different areas – from fault detection in diagnostic systems to control of unmanned vehicles [1-4].

One of the problems in the theory of pattern recognition is the problem of classification, that is, the correlation of the image to one of the classes. This paper presents a comparison of classification algorithms in the problem of recognition of road signs.

2. The problem of classification

As a rule, in classification problems there are a lot of objects that are separated in a certain way into classes. For given objects, it is known which classes they belong to. These objects are called training samples. The class affiliation of all other objects is unknown. The task of classification is to determine the class affiliation (classify) an arbitrary object.

Many classification algorithms are known. The main ones are:

1. Support vector method (SVM).
2. EM-algorithm (Expectation-maximization algorithm).
3. The method of k nearest neighbors (k-NN).
4. Decision tree.
5. Naive Bayes classifier.

These algorithms are widely used in computer vision: from searching for features using a trained classifier for tracking, segmenting scenes, to simpler tasks such as object classification and image clustering.

Content from this work may be used under the terms of the Creative Commons Attribution 3.0 licence. Any further distribution of this work must maintain attribution to the author(s) and the title of the work, journal citation and DOI.
Published under licence by IOP Publishing Ltd

There are various open image database road signs. The largest of them are: German (GTSRB, The German Traffic Sign Recognition Benchmark), Swedish (STS), Belgian (BTSD, Belgium Traffic Sign Dataset) and American (LISA, Laboratory for Intelligent and Safe Automobiles).

Table 1 presents the comparative characteristics of these image databases. For integrated training and testing of classifiers, it is necessary that the number of classes of signs and the number of images per class is the maximum.

The Swedish image database (STS) presents a limited number of characters, which makes it more difficult to assess the generalizability of the classification algorithms for more classes.

The Belgian (BTSD) and American (LISA) databases contain a relatively small number of images per class, which complicates the testing of classifiers that require large training samples [5].

Table 1. The characteristics of image databases

	STS	LISA	BTSD	GTSRB
Number of signs	7	47	108	43
Total number of image of signs	3488	7855	13444	51839

Taking into account the above factors, the base of road signs GTSRB [6] was used to test the classification algorithms. Figure 1 shows an example of images of road signs from this database.

Picture 1. Examples of GTSRB database images

Due to the large number of classes and images in this database, only a part of the classification algorithms was used for training and testing: 10590 images divided into 10 classes.

This sample was divided into two parts: for preliminary training, 5,295 images of road signs were used, and further testing for the quality of the algorithm was carried out on a test sample consisting also of 5,295 images.

Preliminary processing was carried out for images:

- reduction to the size of 30 by 30 pixels;
- convert to grayscale;
- adaptive threshold conversion to highlight interesting features of the image [7, 8].

The work of this algorithm in the conversion of images of signs is shown in figure 2.

Picture 2. The result of applying the threshold conversion algorithm

Testing was carried out on the same personal computer.

Table 2 shows the data obtained during the comparison of the training time, the number of correctly recognized figures from the 5295 examples of the test sample, the recognition time of the entire test sample and the size of the resulting classification model (structure).

Table 2. The results of the classification algorithms

Algorithm	Training time, msec	Number of correct recognitions	Time of recognition of the test sample, msec	The size of the classification model (structure) obtained, KByte
SVM	4979	4547	8007	9858
k-NN	2115	4070	33310	5847
Decision Tree	1579	3531	7459	12107
Naïve Bayes classifier	131722	0	290877	35949
EM	39062	511	5813	37545

The SVM algorithm showed a recognition result of 86%, which is the best indicator for the recognition time of 8 seconds. The ability of this algorithm to separate linearly inseparable samples is key advantage with such a high dimension of input vectors of 900 elements.

The K-nearest neighbor algorithm showed 77% accuracy, which is also a very high rate, but the recognition time was 33 seconds. This factor confirms that the search for the nearest neighbors in extra-large samples is difficult and is a bottleneck of the algorithm.

The decision tree algorithm showed an accuracy of only 67% with a recognition time of 7.5 seconds. The low accuracy of the algorithm is due to the lack of depth and optimality of the original tree.

Expectation-maximization algorithm showed an accuracy of 9.7%, with a recognition time of 5.8 seconds. Naive Bayes classifier was not applicable in this problem, as the number of correctly recognized images is zero.

3. Conclusions

IOP Conf. Series: Materials Science and Engineering **412** (2018) 012081 doi:10.1088/1757-899X/412/1/012081

As this comparison shows, the most promising classification methods in image recognition problems are support vector (SVM) and K – nearest neighbor (k-NN) methods.

The method k-nearest neighbors has the fastest learning time and the smallest size of the classification model, but it is inferior to the method of reference vectors, both in time of recognition of the test sample, and in the number of correctly recognized images.

The other three methods, though, are classification methods, but are poorly applicable to this problem, because of the small number of correctly recognized objects of the test sample, the large size of the classification model and the recognition times, especially for the naive Naive Bayes classifier.

References

[1] Ziyatdinov R R, Biktimirov R A and Klochkova K V 2017 *The comparative analysis of classification of algorithms in the problems of pattern recognition* Journal of Fundamental and Applied Sciences v 9 n 1S pp 1790-97

[2] Ziyatdinov R R, Mukhametzyanov V R and Nabiullina G I 2015 *Application of neural networks for the diagnosis of depth sucker rod pumps* International Journal of Applied Engineering Research Vol 10 Is 24 pp 45022-26

[3] Sharafieva L N, Ziyatdinov R R 2016 *Application of artificial neural network in the problems of diagnosis of borehole pumps* Information Technology. Automation. Actualization and solving the problems of training highly qualified personnel: Proceedings of the international conference pp 83-87

[4] Ziyatdinov R, Shigabiev R and Talipov D 2017 *Automated road marking recognition system* IOP Conf. Series: Materials Science and Engineering 240 **012070**

[5] Shahuro V I, Konushin A S 2016 *Russian image database of road signs* Computer Optics T 40 No 2 pp 294–300

[6] GTSRB: German Traffic Sign Recognition Benchmark. Available at: http://benchmark.ini.rub.de/?section=gtsrb&subsection=news (accessed 22 October 2017).

[7] OpenCV: Adaptive Threshholding. Available at: https://docs.opencv.org/trunk/d7/d4d/tutorial_py_thresholding.html (accessed 22 October 2017).

[8] Bradsky G, Kaehler A 2008 Learning OpenCV. Computer Vision with the OpenCV Library – O'Reilly

Study of composite coal–water fuel rheological properties

E R Zvereva[1], L S Sabitov[2], R V Khabibullina[1], G R Akhmetvalieva[1], Y K Mongush[1], D V Ermolaev[3]

[1]Kazan State Power Engineering University, 51, Krasnoselskaya str., Kazan, 420066, Russia

[2] Kazan Federal University, 18 Kremlyovskaya street, Kazan, 420008, Russia

[3]FIC "Kazan Scientific Center of Russian Academy of Sciences", 2/31, Lobachevsky str., Kazan, 420111, Russia

*6elvira6@list.ru

Abstract. The rheological characteristics of composite water-coal fuel, as well as the possibility of reducing the viscosity of composite water-coal fuel due to the use of nanomaterials, are studied. Improved rheological properties of composite fuel containing small additions of these substances: carbon nanomaterial at a concentration of 0.005 wt. % or 0.3 wt. % of the dehydrated carbonate slurry. Possible mechanisms for changing the viscosity properties of fuel are considered.

Introduction

Taking into account the need to increase the share of coal in the fuel and energy balance of Russia and the environmental problems arising from its use, the creation of coal processing and processing technologies that will maximize its benefits and minimize the complexity of its application is a promising direction for the development of energy. In addition, for the needs of small-scale energy, the priority is to use local fuels, which include low-grade coals, peat, shales, bitumen, heavy oil residues. Direct combustion of the listed types of fuels is inefficient and is accompanied by pollution of the environment due to the incompleteness of their combustion. More promising is the preparation on their basis of water-containing liquid composite mixtures, which are much more effective and environmentally friendly in the production of thermal and electrical energy [1,2].

One of the directions for coal power engineering can be the transition from direct coal combustion in various combustion plants to the preparation of coals of various grades, including coal waste, water-coal fuel. The water-coal fuel (CWF) has properties that allow it to replace solid, liquid or gaseous fuels in various fuel-using units without significant reconstruction, and if necessary, the use of coal and other fuels - fuel oil, coal, gas [3- 5].

For water-coal fuel, the most important characteristics that determine its properties as a liquid fuel are rheological parameters. Rheology shows the regularity, to which the fluid is subjected under the action of a deforming force on it. Depending on the values of the rheological parameters, the water-coal suspensions display both the properties of Newtonian and the properties of non-Newtonian liquids. The value of the rheological characteristics of coal suspensions makes it possible to predict

Content from this work may be used under the terms of the Creative Commons Attribution 3.0 licence. Any further distribution of this work must maintain attribution to the author(s) and the title of the work, journal citation and DOI.
Published under licence by IOP Publishing Ltd

and regulate the technological properties of WTT necessary for storage, transportation and its efficient combustion in boiler furnaces. As is known, the temperature has a significant effect on the rheological characteristics of liquids. The behavior of Newtonian fluids during flow depends solely on viscosity, which is only a function of temperature [6,7].

When using VUT, the main problem is the high viscosity of the suspension at the required concentration of crushed coal. Therefore, the development of methods for controlling viscous characteristics due to a change in the composition of the VUT is a very urgent problem [8-15]. One way to change the viscosity is to add nanostructures, which in many cases leads to a radical transformation of the properties of traditional materials, practically without changing their chemical composition, only due to the tendency of the molecules in the presence of nanoparticles to self-organize and self-assemble. These effects lead to the appearance of ordered structures [16, 17] and a change in the functional characteristics of the starting materials. In particular, the use of the mechanisms of spontaneous self-ordering of atoms around nanoparticles in liquid disperse systems can lead to an additional supramolecular structure formation, the consequence of which can be a decrease in viscosity due to the transition to layerwise shear flow.

The purpose of this study was to study the effect on the rheological characteristics of VUT additives of some nanostructured materials and the ability to reduce the viscosity of mixed fuel due to their use.

Materials and methods

As nanoparticles, we chose carbon nanotubes (CNTs), which have already been used with great success in various fields of energy. Despite the fact that carbon nanotubes are biodegradable nanoparticles, they completely burn with fuel, because they consist of pure carbon, and their catalytic properties lead to a more complete combustion of other components of the fuel, which may be important for eliminating the negative consequences of using heavy fuel. At present, the production of carbon nanomaterials, which include different types of carbon nanotubes, has gone beyond laboratory and semi-industrial installations, which makes it possible to assume the economic feasibility of their use. For the experiments, we chose carbon nanotubes of carbon nanomaterial "Taunit" (http://www.nanotc.ru), characterized by their relative cheapness and availability. This choice was also facilitated by the earlier obtained results on reducing the viscosity of fuel oil when CNTs were added, which made it possible to reduce energy consumption and production costs in technological processes [17-20].

Before adding to the fuel to uniformly distribute the volume of the sample and to prevent their re-adherence, multi-walled carbon nanotubes are dispersed in surfactant solutions [21-23]. We used aqueous dispersions of sodium dodecyl sulfate (SDS, anionic surfactant) with a concentration of 100 mM, whose properties in pure form [24-27] and in the presence of CNTs [28-29] have been thoroughly studied. The concentration of the carbon nanomaterial was changed to 0.0125 wt. %.

Another kind of particles used by us that change the rheological properties of fuel and contribute to reducing harmful emissions into the atmosphere [30, 31] are particles of dehydrated carbonate sludge, which is a waste of the process of chemical water purification of water treatment facilities of CHP plants and boiler houses. Previous studies have shown that the use of expensive nanoparticles as additives to hydrocarbon fuels can to some extent be replaced with additives of microparticles of dehydrated carbonate sludge - waste generated during coagulation and liming of natural waters in thermal power plants. As an additive to fuel, a fine fraction of anhydrous carbonate slurry with a particle size of not more than 0.09 mm with a total content of calcium and magnesium carbonates of not less than 85% was used. The choice of carbonate slurry as an additive was dictated by the good results obtained earlier by us on improving the rheological properties of fuel oil using this additive [32-34].

Samples of water-coal fuel were made on the basis of lean coal of the Kuznetsk deposit. For the preparation of samples of water-coal suspension, the crushed coal from the hopper was directed to grinding into a vibrating mill. The obtained coal dust was subjected to screen analysis on the vibration unit and subsequent fractionation. Particles with certain dimensions were weighed and fed into a mixer into which distilled water was dispensed from a measuring container and a water-coal suspension was prepared by mixing the components.

Samples of water-coal fuel, coal-water fuel mixtures with carbonate mud and nano additives were investigated using a Rheomat RM 100 rotational viscometer to determine the values of dynamic viscosity at different shear rates. The determination of the dynamic viscosity consisted in measuring the shear stress arising in the test sample of the VUT. The angular velocity varied from 50 to 300 s-1. Viscosity measurements of VUT were performed at several temperatures in the range of 30-35 ° C and different shear rates. Measurements began with a minimum speed. At all rotational speeds of the rotor, the value of the torque (shear stress) corresponding to the equilibrium value was recorded, i.e. So, when the spinning moment stops changing in time. The work of the viscometer was controlled from a personal computer via the software "VISCO-RM SOFT".

Experimental results and their discussion

The data obtained during the experiments are shown in Fig. 1 and 2. It should be noted that the results observed for VUT are very dependent not only on the concentration of nanoparticles and their dispersion medium, but also on the chemical and fractional composition of the coal dust of the fuel suspension, where a certain role is also played by the features of adsorption interactions on the surface of energy coals [35].

Changes in the dynamic viscosities of the VUT samples with addition of carbonate slurry from the shear rate correspond to the graphs shown in Fig. 1. A fraction of 0.125 mm of coal particles was selected. The carbonate slurry content in all the samples was 0.3 wt. %.

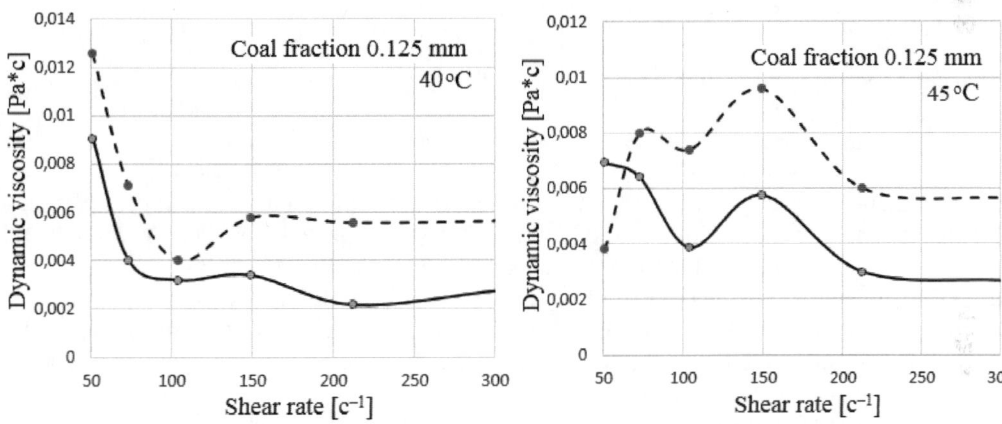

Fig. 1. Dependence of the dynamic viscosities of VUT samples on the basis of coal particles 0.125 mm in size on the shear rate at different temperatures (carbonate slurry concentration is 0.3% by weight)

When analyzing the obtained curves, a rather strong change in the conditional viscosity of composite water-coal fuel was found, depending on the concentration of the additive and the temperature of the sample (Fig. 1). The viscosity of the initial samples of VUT was practically independent of the temperature and was 0.0065-0.009 Pa s. With the addition of 0.1 wt. % of the carbonate slurry, the viscosity of the samples remained almost unchanged (40 ° C, 45 ° C). The increase in the proportion of the additive - carbonate sludge up to 0.3% by weight in water-coal fuel resulted in a marked decrease in the conventional viscosity. However, a further increase in the concentration of the additive again led to an increase.

For VUT with addition of carbonate sludge, the greatest effect is achieved with its concentration of 0.3 wt. %. In the case of the addition of carbonate sludge, one more feature should be noted: an increase in viscosity (with the exception of a temperature of 45 ° C) in the range of low concentrations of this additive. The observed increase in viscosity can be explained by the absorption and binding of water, the most liquid fraction of the dispersion medium, and, correspondingly, to an increase in the viscosity of the VUT [11]. Further increase in the additive concentration to 0.3 wt. % leads to the formation of internal structure formation around the particles of the slurry. This concentration promotes the formation of structures with low-strength coagulation contacts along the planes, which determine the considerable development of plastic deformations leading to layerwise shear flow and a decrease in viscosity.

However, the use of carbonate slurry at this concentration increases the ash content of the fuel. The requirements to reduce ash content and to increase the ecological compatibility of combustion give preference for the concentration of the additive in 0.1 wt. %. In particular, the use of this additive in this concentration to fuel oil allows to reduce the viscosity of fuel oil, as well as to reduce the sulfur content in emissions and improve the structure of deposits, increasing the ash content of fuel oil is negligible. The results of industrial tests conducted for boiler fuel indicate a decrease in the mass fraction of sulfur oxides emitted by 36.5% [30]. It is possible that similar results should be expected when using VUT.

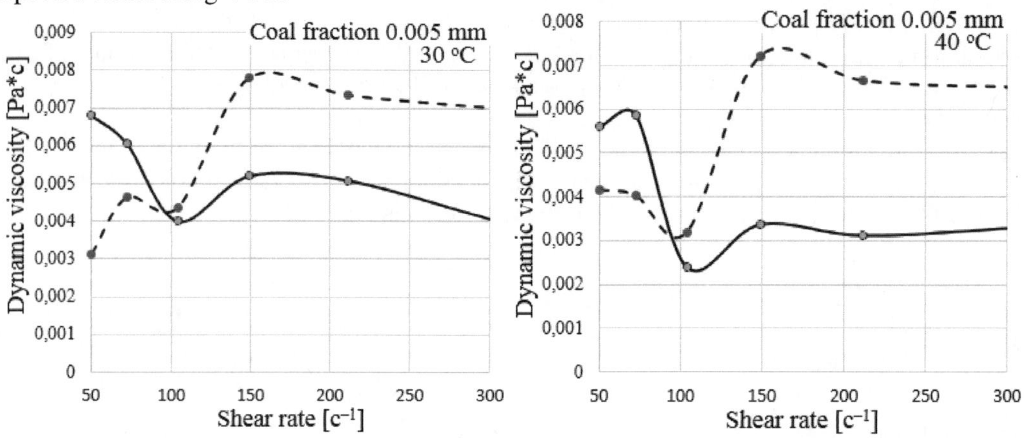

Fig. 2. Dependence of the dynamic viscosities of the VUT samples on the basis of coal particles of 0.005 mm in size with the addition of 0.005 wt. % Of CNT dispersed in a 100 mM sodium dodecyl sulfate solution, on the shear rate at various temperatures

Changes in the dynamic viscosities of the VUT samples with the addition of carbon nanomaterial at a concentration of 0.005 wt. % Of CNT dispersed in a 100 mM sodium dodecyl sulfate solution from the shear rate correspond to Fig. 2. In the presence of carbon nanotubes, the dynamic viscosity of water-coal fuel was higher than the initial dispersion medium (water-coal fuel) at low shear rates and decreased markedly as the flow velocity increased. This behavior indicated the occurrence of additional structure formation in the entire volume of the sample observed when carbon nanotubes were introduced, i.e. the appearance of heterospheres and their adhesion, and then their partial destruction with increasing shear rate and their complete destruction at shear rates greater than 230 s-1.

In addition, it should be assumed that there is a synergistic effect from the combined use of these additives. We hope that the joint use of CNT and carbonate sludge will reduce the amount of harmful emissions into the atmosphere and reduce the content of the pricing additive - the carbon nanomaterial "Taunit" in this sample.

Conclusion

The possibilities of reducing the viscosity of composite water-coal fuel by using nanomaterials: carbon nanotubes and dehydrated carbonate sludge were investigated. Improved rheological properties of composite fuel containing small additions of these substances: carbon nanomaterial at a concentration of 0.005 wt. % or 0.3 wt. % of the dehydrated carbonate slurry. Possible mechanisms for changing the viscosity properties of fuel are considered. When designing industrial technological complexes, one should take into account the obtained viscosity dependence on temperature and, with technical feasibility and economic expediency, it is necessary to use devices for heating the pumped suspension to a temperature of 350 ° C.

References

[1] Glushkov D O, Strizhak P A, Chernetskii M Y 2016 *Organic coal-water fuel: Problems and advances (Review)* Thermal Engineering Vol 63 No 10 pp 707-17
[2] Shamsutdinov E V and Zueva O S 2014 *Nanomaterials and Nanotechnologies in Power-Engineering* (Kazan, KSPEU)
[3] Khodakov G S 2007 *Coal-water suspensions in power* engineering Thermal Engineering. T. 54 No 1 pp 36-47
[4] Trubetskoi K N, Kondrat'ev A S, Murko V I, Zaidenvarg V E, Kassikhin G A, Nekhoroshii I K 2008 *Water-coal fuel: the results of technology development and perspectives of its utilization in Russia* Thermal Engineering T 55 No 5 pp 413-17
[5] Khil'ko S L, Titov E V 2008 *Fuel coal suspensions* Chemistry and Technology of Fuels and Oils Vol 44 No 1 pp 73-79
[6] Khodakov G S 2003 *Rheology of Suspensions. A Theory of Phase Flow and Its Experimental Substantiation* Ross. Khim. Zh V 47(2) pp 33-44
[7] Khodakov G S 2004 *On suspension rheology* Theoretical Foundations of Chemical Engineering T 38 No 4 pp 430-39
[8] Mukherjee A, Pisupati S V 2016 *Effect of additives on interfacial interactions for viscosity reduction of carbonaceous solid–water slurries* Fuel V 180 pp 50-58
[9] Mingaleeva G R, Gaynetdinov A V, Shakirov R R, Akhmetov E A 2015 *Comparative analysis of reduce the viscosity of coal-water fuels* Proceedings of the higher educational institutions. Energy sector problems V 7-8 pp 37-46
[10] Mingaleeva G R, Shamsutdinov E V, Ermolaev D V, Afanas'eva O V, Gil'manov R M, Fedotov A I, Gal'keeva A A 2014 *Mechanism of Influence of Finely Disperse Coal Fraction on Rheological Properties of Water-Suspended Coal Fuels* Chemistry and Technology of Fuels and Oils V 49 (6) pp 459-68
[11] Zvereva E R, Akhmetvalieva G R, Makarova A O, Ermolaev D V, Mongush Y K, Shaikhutdinova A R, Zueva O S 2017 *Alteration of coal-water fuel rheological properties in the presence of nanomaterials* Bulletin of Kazan State Power Engineering University No 3 (35) pp 77-84
[12] Mingaleeva G R, Ermolaev D V, Afanasyeva O V, Timofeeva S S 2012 *An experimental study of the viscosity of coal-water slurry with a bifractional composition of its dispersed phase* Thermal Engineering T 59 No 6 pp 446-48
[13] Zvereva E R, Zueva O S, Khabibullina R V, Yermolaev D V, Ahmetvalieva G R, Salihzyanova D R, Magdeeva A M 2016 *Environmental safety improvement and composition (water-coal) fuel efficiency increase with various additives at fuel and energy complex enterprises* International Journal of Pharmacy and Technology Vol 8 (4) pp 26744-52
[14] Lapin D A et al 2017 *Effect of the fractional composition of the solid components of coal–water fuel on the characteristics of ignition and combustion* Solid fuel chemistry Vol 51 No 2 pp 88-94
[15] Zhao H, Hou Y-B, Liu H-F, Tian X-S, Xu J-L, Li W-F, Liu Y, Wu F-Y, Zhang J, Lin K-F 2014 *Influence of rheological properties on air-blast atomization of coal water slurry* Journal of Non-Newtonian Fluid Mechanics Vol 211 pp 1-15
[16] Zvereva E R, Zueva O S, Khabibullina R V, Makarova A O 2016 *Nanomaterial Effect Study in the Viscosity Characteristics of Fuel Oil and Alternative Fuels Used at Fuel and Energy Complex Enterprises* Journal of Engineering and Applied Sciences Vol 11 pp 2950-54
[17] Zvereva E R, Khabibullina R V, Akhmetvalieva G R, Makarova A O, Zueva O S 2017 *Proc. Int. Conf. "Actual Issues of Mechanical Engineering" (AIME 2017)* Advances in Engineering Research Vol 133 pp 914-920

[18] Zvereva E R, Zueva O S, Khabibullina R V 2016 *Improvement of the Operational Characteristics of Liquid Organic Fuel Oils by Additives* Material Science Forum Vol 870 pp 666–670

[19] Zvereva E R, Zueva O S, Khabibullina R V et al. 2016 *The influence on the rheological quality of liquid fuel oil by nano-additives, including carbon nanotubes* Chem. Technol. Fuels Oils No 5 (597) pp 15–19

[20] Zvereva E R, Khabibullina R V, Zueva O S 2017 *Nano Additives Influence on Fuel Oil Properties* Solid State Phenomena Vol 265 pp 374-78

[21] Vaisman L, Wagner H D, Marom G 2006 *The role of surfactants in dispersion of carbon nanotubes* Advances in Colloid and Interface Science 128-130 pp 37-46

[22] Zueva O S, Osin Y N, Salnikov V V, Zuev Y F 2014 *Research of carbon nanotubes suspensions: the emergence of mesoscopic structures from the self-assembly of surfactant molecules* Fundamental research No 11–5 pp 1021-27

[23] Zueva O S, Makshakova O N, Idiyatullin B Z, Faizullin D A, Benevolenskaya N N, Borovskaya A O, Sharipova E A, Osin Yu N, Salnikov V V and Zuev Yu F 2016 *Structure and properties of aqueous dispersions of sodium dodecyl sulfate with carbon nanotubes* Russ. Chem. Bull Vol 65 No 5 pp 1208–15

[24] Zuev Yu F, Gnezdilov O I, Zueva O S and Us'yarov O G 2011 *Effective self_diffusion coefficients of ions in sodium dodecyl sulfate micellar solutions* Colloid J. 73 pp 59-64

[25] Gnezdilov O I et al 2011 *Self-Diffusion of ionic surfactants and counterions in premicellar and micellar solutions of sodium, lithium and cesium dodecyl sulfates as studied by NMR-diffusometry* Appl. Magn. Reson 40 pp 91-103

[26] Idiyatullin B Z, Potarikina K S et al 2013 *Association of sodium dodecyl sulfate in aqueous solutions according to chemical shifts in 1H NMR spectra* Colloid J. 75 pp 532-37

[27] Gubaidullin A T, Litvinov I A, Samigullina A I, Zueva O S, Rukhlov V S, Idiyatullin B Z and Zuev Yu F 2016 *Structure and dynamics of concentrated micellar solutions of sodium dodecyl sulfate* Russ. Chem. Bull. (Int. Ed.) Vol 65 No 1 pp 158–66

[28] Borovskaya A O, Idiyatullin B Z, Zueva O S 2016 *Carbon nanotubes in the surfactants dispersion: formation of the microenvironment* Journal of Physics: Conference Series Vol 690 **012030**

[29] Zueva O S, Makarova A O, Faizullin D A 2017 *Creating Carbon Nanotubes Microenvironment in Surfactant Water Solutions* Solid State Phenomena Vol 265 pp 342-47

[30] Zvereva E R, Dmitriev A V, Shageev M F and Akhmetvalieva G R 2017 *Results of industrial tests of carbonate additive to fuel oil* Thermal Engineering Vol 64 (8) pp 591-96

[31] Efimov N N, Jatsenko E A, Smoly V A, Kosarev A S, Kopitsa V V 2011 *Ecological aspects and problems of utilization and recycling ashes and the slag waste of thermal power stations* Industrial ecology No 2 pp 40-44

[32] Zvereva E R, Ganina L V, Andryushina I A 2009 *Effects of additives on the working properties of furnace heavy fuel oils* Chem. Technol. Fuels Oils 45 pp 349-53

[33] Zvereva E R, Ganina L V, Andryushina I A 2010 *An experimental study of the effectiveness of an additive for fuel oil* Thermal Engineering 57 pp 530-33

[34] Zvereva E R, Mingaleeva G R, Khabibullina R V, Akhmetvalieva G R 2016 *Improvement in the Viscosity Characteristics of Boiler Oil by Additives* Petroleum Chemistry V 56 (1) pp 65–67

[35] Veselovsckaja E V *Investigation of structure and features of surface adsorption interactions of power-generating coals* University News. North-Caucasian Region. Technical Sciences Series Vol 2 (194) pp 40-45

IOP Conf. Series: Materials Science and Engineering **412** (2018) 012083 doi:10.1088/1757-899X/412/1/012083

Technology of overlay laser welding of durable powdery into blade edge of miller

V V Zvezdin , R M Hisamutdinov, R R Rakhimov, I H Israfilov and R R Saubanov

Kazan Federal University, Naberezhnye Chelny Institute, 423812, Russia, Naberezhnye Chelny, Prospekt Syuyumbike 10A

irmaris@yandex.ru, rafisih88@mail.ru

Abstract. In this paper the laser welding technology features of wear-resistant powders on the hob's cutting edge with laser radiation (LR) focus positioning control are described. It is shown that the quality of the welding process depends on the processing regimes, the energy characteristics of the laser technological complex (LTC), the positioning accuracy of the LR focus and its perpendicularity relative to the plane of a router bit. In this paper we deal with the questions of stabilization of LTC parameters and LR positioning as well.

1. Introduction

The quality of the tool determines the efficiency of production, the competitiveness of its products.

One of the methods of increasing the tool resistance of the, its most important characteristic, is the use of highly concentrated energy sources. The restoration of the tool is possible due to laser surfacing. Automation of the process of laser surfacing of parts in mechanical engineering makes it possible to increase the efficiency of the technological process of tool recovery, with its specified quality indicators [1-3].

Laser surfacing of wear-resistant powders makes it possible to restore the cutting edge of the tool with increased durability. The wear of the tool edge is a consequence of a change in operating conditions, which is characterized by an increase in cutting and cutting forces at high temperatures.

The purpose of the study is to analyze methods and experiments to improve the quality of TP, using the results of the experimental work on the restoration of the tooth milling tool steel R18K5F2 with increased durability.

In this paper, the task is to perform laser surfacing of the tooth milling cutters and microstructural studies.

2. Experimental research

The influence of the instability of LTC parameters, the optical-physical properties of the surface and the positioning of the focus of laser radiation (LR), relative to the edge of the milling cutter, lead to a decrease in the quality of surfacing [4-6].

To stabilize them, ACS LTK with negative feedbacks was developed for the parameters of surfacing (LR energy density, temperature, and positioning accuracy of the focus) [5].

The process LR interaction with metal during surfacing is characterized by the formation of a plasma torch (Figure 1.), which shields the interaction zone, which does not allow us to use optical-

Content from this work may be used under the terms of the Creative Commons Attribution 3.0 licence. Any further distribution of this work must maintain attribution to the author(s) and the title of the work, journal citation and DOI.
Published under licence by IOP Publishing Ltd

physical methods for studying the thermal-effect zone. However, the use of pulsed LI enhances these capabilities.

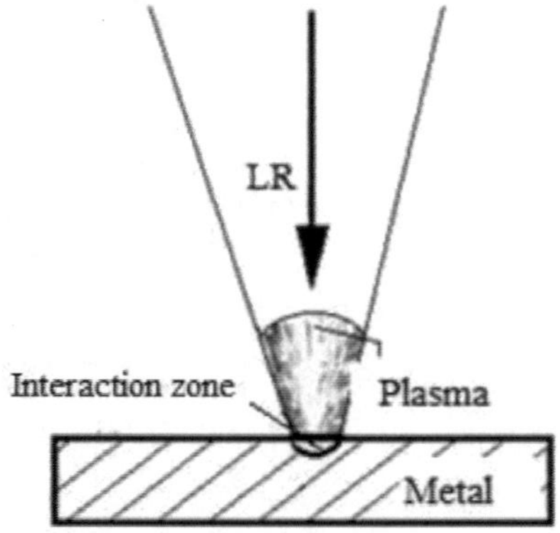

Figure 1. Conditional image of interaction process between LR and metal

The research involved laser surfacing of the BoroTec-Eutalloy® 10009 powder for steel R18K5F2 using flux grade AN-43. A number of changes in the structure of the metal and phase transformations that occurred during heating and subsequent cooling occurred during heat treatment characterized by the type of thermal cycle. The initial heating of a certain volume element to a temperature Tc followed by rapid cooling to a temperature below Tc is necessary for the welding process of steel (according to the Chernov diagram) [7]. This process takes place when the melting point in the zone of processing the temperature of the powder and metal is reached [3, 6]. Restoration of the hob's router bits and the preparation of other structures takes place on condition that the quality of the surface deteriorates. Figure 2 shows a fragment of the tooth of a worm cutter with a defect in the leading edge.

Figure 3 shows the worm hob with a thermal contour of LR. The weld layer has a thickness of less than 0.1 mm, with a depth of thermal influence of about 0.4 mm. The microhardness at the surface is $HV_{0.05}$ 585 and reaches a value of $HV_{0.05}$ 769 at a distance of 0.1 mm from the surface.

The deposited layer consisting of BoroTec-Eutalloy® 10009 powder and flux is clearly distinguishable. Optimum content of alloying elements, silicon and manganese, as well as limiting the content of phosphorus and sulfur in the deposited metal layer is ensured by the use of flux grade AN-43.

Figure 2. Fragment of a hob's router bit with a leading edge defect

Figure 3. Fragment of a hob's router bit with surfacing of Ni-Cr-B-Si-Fe powder, BoroTec - Eutalloy® 10009

Modeling of the hob cutter's surface is necessary for controlling the parameters of the LTC optical system and the size of the flanking [2.6].

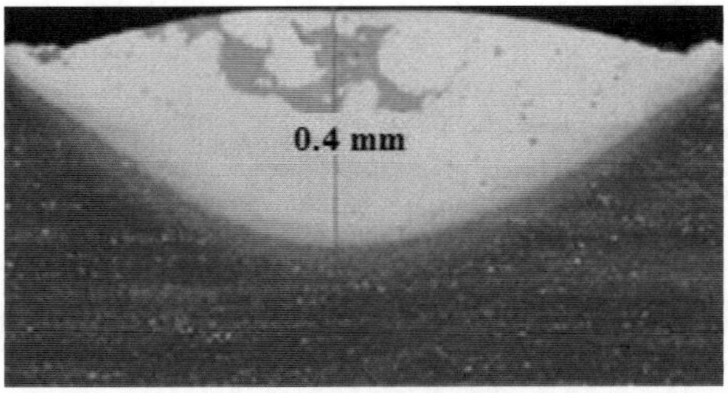

Figure 4. Measurement results of microhardness and microstructure of a worm hob obtained with an increase of x 200

The microstructure of the part (Figure 4) is a hidden-needle martensite and carbides. The carbide heterogeneity in the structure of the milling cutter corresponds to 1A on a scale of 1, which satisfies GOST 19265-73 [8].

In the microstructural investigation, a surface hardened layer with a non-structuring structure and carbides was found on the surface of the working edge of the teeth of the milling cutter. The high hardness of the hardened layer (up to $HV_{0.1}$ 970) has a positive effect on the wear resistance of the mill, which manifests itself at a high temperature of the tool edge.

3. Conclusion.

The physicochemical properties of the materials of a tool, a powder, a flux, and the process medium affect the laser welding process of high-strength coatings having increased wear resistance. Various chemical activity and thermophysical properties leading to a different distribution of the thermal field are main reasons for this phenomena [9]. The specific energy input of the energy source and all types of interaction between the medium and the metal characterize the temperature dependence for the sputtering zone and surfacing. The optimal values and stability of the parameters depending on the physical and chemical properties and temperature of the environment affect the quality of the surfacing process.

References

[1] Chernova M A, Hisamutdinov P M, Zvezdin V V, Simonov L A and Spirin A A 2015 *Metallographic examination of overlay laser welding of durable powdery into* tool Basic researches 11-4 pp 753-758

[2] Khisamutdinov R M, Zvezdin V V, Saubanov Ruz R, Israfilov I H, Rakhimov R R and Spirin A A 2016 *Study of processes of steels surfaces modification with highly concentrated energy flows* IOP Conference Series: Materials Science and Engineering V 669 Is 1 **012024**

[3] Grigoryants A G, Perestoronin A V, Portnov S M, Zvezdin V V and Israfilov I H 2015 *A system for automatic control of precision laser welding in engineering* Welding International T 29 No 10 pp 801-804

[4] Rakhimov R R, Zvezdin V V, Israfilov I H, Nabiullina G I Saubanov R R 2014 *Modeling of the laser heat treatment* News of the Tula State University (Technical science) No 11-2 pp 476-84

[5] Zvezdin V V, Israfilov D I, Portnov S M, Saubanov R R, Rakhimov R R and Zvezdina N M 2015 *Automatic control system of high-precision welding of preparations by the laser radiation at influence of the plasma torch* Proceedings of the higher educational institutions (Physics) No 9-3 V 58 pp 51-54

[6] Zvezdin V V, Samorskiy V V, Pesoshin V A, Aleev R M, Saubanov R R 2015 *Improvement of*

combinated laser-plasma welding Proceedings of the higher educational institutions (Physics) T 58 No 9-2 pp 36-39
[7] Gabdrakhmanov Az T, Israphilov I H, Galiakbarov A T, Samigullin A D, Gabdrakhmanov Al T 2016 *Improving the efficiency of plasma heat treatment of* metals Journal of Physics: Conference Series V 669 **012014**
[8] Rakhimov R R, Saubanov R R and Israfilov I H 2017 *Analysis of the impact of informative heat treatment parameters on the properties of hardening of the surface layers* Journal of Physics: Conference Series V 789 Is 1 **012040**
[9] Saubanov R R, Zvezdin V V, Israfilov I H, Haybullin I I and Rakhimov R R 2014 *Synthesis of oxidic powder in nonequilibrium low-temperature plasma with increase of indicators of quality of process* IOP Conference Series: Materials Science and Engineering V 567 Is 1 **012034**

Technology ability of laser bonding of compacted graphite iron

V V Zvezdin[1], K V Klochkova[1], R R Rakhimov[1], R R Saubanov[1] and V A Pesoshin[2]

[1] Kazan Federal University, Naberezhnye Chelny Institute, 423812, Russia, Naberezhnye Chelny, Prospekt Syuyumbike 10A,
[2] Kazan National Research Technical University named after A. N. Tupolev - KAI, Kazan, Russia

irmaris@yandex.ru, rafisih88@mail.ru

Abstract. Laser welding features of compacted graphite iron are described in this paper. An approach that allows to reduce residual stresses and the probability of occurrence of "hot" cracks in the weld seam as well as to increase the absorption coefficient of laser radiation due to the use of metal powder Ni-Cr-B-Si-Fe, BoroTec-Eutalloy® 10009-chromium-nickel alloy are proposed. Recommendations are given when conducting a welding with compacted graphite iron.

1. Introduction

Compacted graphite iron (CGI) is becoming more and more popular in engineering. They replace gray cast iron, due to their unique physical and mechanical characteristics, economical casting, manufacturability in the producing of parts with high mechanical properties.

Progressive technological processes (PTP), including laser welding (LW), have not been developed at present. Therefore, the problem of laser welding of CVG, providing a welded seam with a given quality, with increased physical and mechanical properties and performance characteristics, is relevant.

With a system approach to the processing of CGI, which is a form of integrating scientific knowledge in the field of laser technologies, it is necessary to consider a set of problems. These are the temperature conditions, the heating and cooling rates of the welded product, the gas medium in which welding takes place and its consumption [1], as well as the positioning of laser radiation (LR) relative to the joint of the welded parts [2]. The solution of the questions as well as the specification of the TP parameters ensure the successful implementation of the required quality of drugs.

2. Welding process

One of the necessary stages of the laser welding process is the preliminary preparation of the welded surfaces [3]. Pollution and moisture create conditions for the formation of porosity, oxide inclusions, and in some cases, cold cracks in the welded iron and the zone of thermal influence due to hydrogen saturation and high cooling rate. After cleaning the surfaces, the parts are degreased.

Figure 1 shows the microstructure of the seam at a high cooling rate, welding was carried out in argon (on top of the seam) and helium (from below the seam).

Content from this work may be used under the terms of the Creative Commons Attribution 3.0 licence. Any further distribution of this work must maintain attribution to the author(s) and the title of the work, journal citation and DOI.
Published under licence by IOP Publishing Ltd

IOP Conf. Series: Materials Science and Engineering **412** (2018) 012084 doi:10.1088/1757-899X/412/1/012084

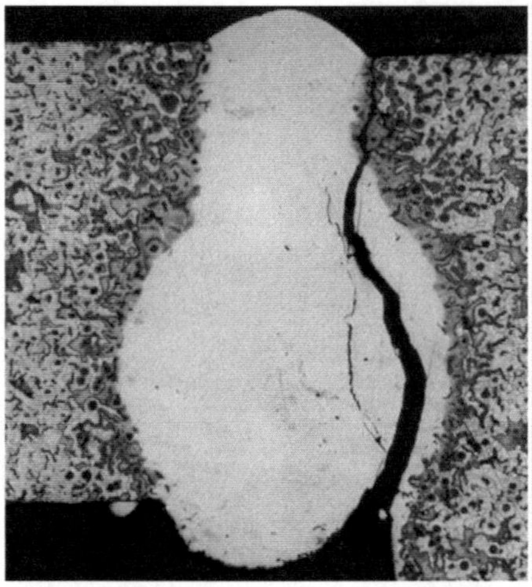

Figure 1. General view of the CGI plates weld 30 (left) and CGI 35, magnification x 50, (grinding is etched)

Investigation of the microstructure was carried out in cross-section of a cross-section microsection relative to the welded seam of the plates. The presence of "hot" cracks is revealed in the welded seam.

The microstructure of cast iron plates was evaluated in accordance with the scales of GOST 3443-87, identical and represents graphite: - the form of inclusions of graphite - spherical regular SHGf5 and meandering VGf2;

The microhardness of the plate amounts:

- ferrit in the main metal plate - 170 HV0.05;
- perlite plate-like in the base metal of the plate - 260 HV0.05;
- martensite in the HAZ - 670 - 702 HV0.05;
- welded seam - 524 - 570 HV0.05.

The effect of the inert gas consumption of argon can be observed from the surface roughness of the weld (Fig. 2.3). It can be seen from the figures that a violation of the nominal TP parameters leads to an increase in the roughness. This is because the increase in argon consumption leads to the formation of turbulent flows and mixing it with air. Hydrogen and oxygen lead to swelling of the welded joint [4-6]. The other technological parameters were the same.

In this case, there is an active interaction of metals with the surrounding gaseous medium present in the joint zone of the two metals because of the surface roughness. Laser welding of the samples was carried out on LTK, based on the ytterbium fiber laser "LS 2". Surface welds from oxidation were protected by a mixture of helium with argon in a ratio of 2: 1, fed through a special nozzle. To increase the absorption coefficient, Ni-Cr-B-Si-Fe metal powder was applied to the interface, BoroTec-Eutalloy® 10009 was a chromium-nickel alloy with an AN-43 flux, with an optimal content of manganese, silicon and other alloying elements, and also limited sulfur content and phosphorus.

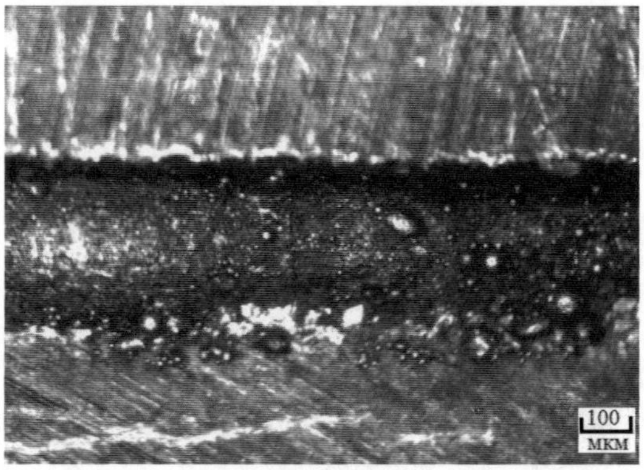

Figure 2. Surface view of CGI plate weld 35 and CGI 30 at a nominal flow rate of argon, magnification x 75

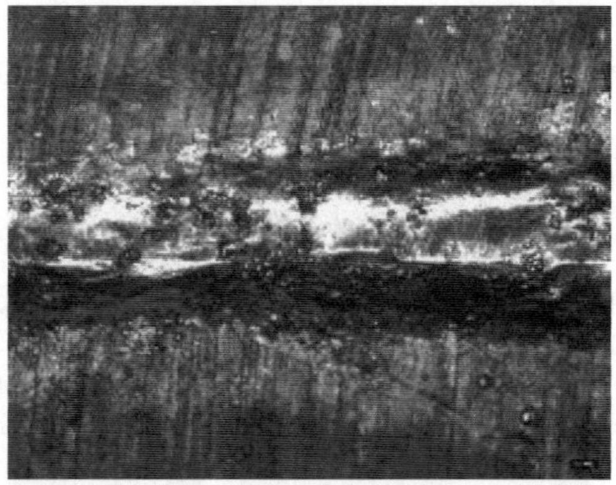

Figure 3. Surface view of CGI plate weld 35 and CGI 30, with an increased flow rate of argon, an increase of x 75

To estimate the depth of penetration, one can use the formula (1):

$$h = \frac{P}{2\pi\lambda_{т}T_{к}} \ln \frac{r_{л} + a/\upsilon_{св}}{r_{л}} , \qquad (1)$$

where λ_T - is the coefficient of thermal conductivity of the material, W / mK; T_k - is the boiling point, K; a - is the thermal diffusivity of the material [1]..

With the initial LR interaction with the metal surface, the efficiency of the useful use of the LI energy depends on the ratio of the reflected and absorbed radiation [7]. The reflected part, as a rule, is irretrievably lost, the other part is almost completely absorbed by electrons in the near-surface layer 10-6 ÷ 10-7 m thick. As a result, the electron temperature Te increases sharply, and the temperature of the crystal lattice Tcr. remains insignificant. With time, the intensity of energy transfer of free electrons of the crystal lattice increases. Beginning with the relaxation time (c):

$$\tau_p = 10^{-9} \div 10^{-11} \text{ c}$$

the temperature difference (Te-Tcrp) becomes minimal and the thermal processes in the metal can be characterized by the total temperature. An integrated characteristic of the thermophysical properties of a metal is the coefficient of thermal diffusivity, which, on the whole, characterizes the rate of temperature equalization for nonstationary thermal conductivity. This coefficient determines the speed with which the energy is transferred to the metal. [8-9].

3. Conclusion

To exclude the "hot" cracks it is necessary to comply with the temperature conditions of heating and cooling of the welded parts, as well as the consumption of inert gas. One should not allow moisture and air to enter the weld seam but should use ultrasonic vibrations during welding [6].

References

[1] Grigoryants A G, Shiganov I N and Misyurov A I 2006 *Technological processes of laser processing (*Moscow: MSTU publishing house of N.E. Bauman*)* p 664

[2] Zvezdin V V, Hamadeev A V, Zagirov R G and Shangaraev I R 2008 *Positioning of lazer radiation towards weld seam as an index od the quality of technological process* Bulletin KSTU N.A. Tupolev number 3 pp 17-21

[3] Khisamutdinov R M, Zvezdin V V, Saubanov Ruz R and Klochkova K V 2016 *Laser technology control of tool hardening and deposition welding* Journal of Advanced Research in Technical Science Issue 3 p 89-95

[4] Zvezdin V V, Israfilov D I, Portnov S M, Saubanov R R, Rakhimov R R and Zvezdina N M 2015 *Automatic control system of high-precision welding of preparations by the laser radiation at influence of the plasma torch* Proceedings of the higher educational institutions

(Physics) number 9-3 Vol 58 pp 51-54

[5] Grigoryants A G, Perestoronin A V, Portnov S M, Zvezdin V V and Israfilov I H 2015 *A system for automatic control of precision laser welding in engineering* Welding International T 29 number 10 pp 801-04

[6] Vagapov I K, Ganiyev M M, Shinkarev A S 2005 *A research of influence of ultrasonic shock processing on value and distribution of residual tension in welded preparation* News of higher education institutions. Aircraft equipmen No 2 p 56-59

[7] Zvezdin V V, Rakhimov R R, Saubanov R R, Israfilov I H, Akhtiamov R F 2017 Management of laser welding based on analysis informative signals *IOP Conference Series: Materials Science and Engineering* Vol 240 Is 1 **012073**

[8] Rakhimov R R, Saubanov R R and Israfilov I H 2017 Analysis of the impact of informative heat treatment parameters on the properties of hardening of the surface layers *Journal of Physics: Conference Series* Vol 789 Is 1 **012040**

[9] Zvezdin V V, Hisamutdinov R M, Rakhimov R R, IsrafilovI H and Akhtiamov R F 2017 Method of control position of laser focus during surfacing teeth of cutters *IOP Conference Series: Materials Science and Engineering* Vol 240 Is 1 **012072**

IOP Conf. Series: Materials Science and Engineering **412** (2018) 012085 doi:10.1088/1757-899X/412/1/012085

Research of bi-nickel coatings obtaining process

B O Daminov[1], N M Lyadov[2], R N Kashapov[1,2], L N Kashapov[1], N F Kashapov[1]and I A Faizrakhmanov[2]

[1] Engineering Institute, Kazan Federal University, Kazan, 420008, Russia

[2] Laboratory of Radiation Physics, Kazan Physical-Technical Institute of the Kazan Scientific Center of the Russian Academy of Sciences, Kazan, 420029, Russia

kashramiln@gmail.com

Abstract. Modern machine building industry can't be without the use of special coatings, which allow increasing the physical and mechanical properties of the product. Nickel coatings are widely used, due to wear resistance and corrosion resistance. Black nickel coatings began to be used in solar power engineering, for efficient conversion of solar energy into thermal energy. Also nickel coatings are used in the manufacture of medical instruments. To improve the physical and mechanical properties of nickel coatings, the task was to study the possibility of creating a bi-nickel coating.

Researches of modern material science are aimed at developing new ways to improve or modify the properties of materials. It is the properties of the coating formed that determine the level of product properties. Economic evaluation shows the attractiveness of coatings in comparison with the volumetric change in properties. The development of new environmentally friendly methods of coating for effective protection and hardening of metal products is an urgent task of modern science and technology [1,2,3]. Chromium-nickel coating substitution for alternative coatings is not inferior in properties, but at the same time environmentally harmlessly obtained, is an important task. Therefore, the goal is to develop a method for obtaining a bi-nickel coating [4,5].

High quality of galvanic coatings is determined by the correct choice of the technological process of preparing the covered surface. The substrates were made of the materials of steel 50, A12, steel 20 and brass LS 59-I. To clean the surface of the parts from the polishing paste, the detergent TMS-3I was chosen. The criterion for choosing the method of preliminary surface preparation was the degree of coating adhesion to the base metal and the appearance of the precipitate. The coating was inspected visually and under a microscope. The procedure for preliminary surface preparation is presented in Tab. 1.

Content from this work may be used under the terms of the Creative Commons Attribution 3.0 licence. Any further distribution of this work must maintain attribution to the author(s) and the title of the work, journal citation and DOI.
Published under licence by IOP Publishing Ltd

IOP Conf. Series: Materials Science and Engineering **412** (2018) 012085 doi:10.1088/1757-899X/412/1/012085

Table 1 Composition of solutions and modes for preliminary surface preparation

Operations	Covered material	Formulations of solutions	Concentration, g / l	Mode of operation		
				Current density, A / dm²	T, °C	τ, min.
Chemical degreasing	steel brass	TMS-3I	60-80	–	70-80	30-60
Electrochemical degreasing	steel brass	NaOH Na_2CO_3 Na_3PO_4	20-40 20-40 20-40	5-10 5-10	60-80	1-3 3-5
Washing in warm water	steel brass	water	–	–	40-50	0,5-1
Washing in cold water	steel brass	water	–	–	15-25	0,5-1
Decaporation	steel	HCl	150-200	–	15-25	0,5-1
	brass	H_2SO_4 $NaNO_3$	600 260-300	–	15-25	0,2-0,5
Washing in cold water	steel brass	water	–	–	15-25	0,5-1

It was experimentally established that for pickling brass parts it is also possible to use a solution of hydrochloric acid (150-200 g/l). Surface preparation according to the above scheme allows to obtain nickel coatings having good adhesion to the base metal..

Electrolytes were selected for the study, providing a decorative, abrasion-resistant and corrosion-resistant two-layer nickel coating. The lower layer of nickel should have good adhesion to the base metal, be non-porous and give the coating corrosion resistance. Electrolytes of nickel plating were prepared on distilled water, systematically subjected to filtration and selective purification. In electrolytes, semi-glossy, shiny and hard nickel plating was used. Table 2 shows the electrolyte compositions for obtaining the lower layer of the bi-nickel coating and the results of research.

Table 2. The composition of electrolytes and the quality of the lower layer of the bi-nickel coating

Electrolyte composition	Concentration of components, g / l	Electrolysis regime			Characteristics of the appearance of the coating	Porosity, number of pores per 1 cm² surface	Cohesion with the substrate
		Current density, A/dm²	T, °C	τ, min			
$NiSO_4{*}7H_2O$ H_3BO_3 NaCl Na_2SO_4	250 30 6 60	1,5 2,0	20 22	20 30	Coverage is uniform, light gray color, semi-shiny	1,0 1,0	
$Ni(SO_3NH_2)_2{*}4H_2O$ $NiCl_2{*}6H_2O$ H_3BO_3	350 5 35	1,2 1,5	33 34	30 30	Coverage is uniform, semi-shiny, light	2 0	

$NiSO_4*7H_2O$ H_3BO_3 NaF NaCl 2,6-2,7 naphthalene-disulphonic acid H_2SO_4	250 25 5 10 3 4	1 1,5	20 20	30 30	Coverage is uniform, light gray, semi-shiny	1	No flaking or peeling
$NiSO_4*7H_2O$ $NaCl_2*6H_2O$ H_3BO_3 Formalin 40% solution	300 50 40 1,0	1,5 3	20 58	30 30	Coverage is uniform, semi-shiny	1,0	
		3	50	20	Coverage is uniform, light, semi-shiny	1,0	
		3,5	45	10	Coverage is uniform light	1,0	
		4	49	10	Coverage light gray, uniform	1,0	
		4	55	20	Coverage is uniform, light gray, semi-shiny	1,0	
		4,5	55	20	Coverage is a dark gray, semi-shiny	1,0	

Based on the results of the data obtained, it can be concluded that the investigated electrolytes of semi-glossy nickel plating make it possible to obtain uniform, bright, practically non-porous coatings in a properly selected regime. However, the half-gloss nickel plating electrolyte contains 0.33 μm / min against 0.1-0.15 μm / min in the other electrolytes tested. Therefore, as a lower layer, a nickel coating was selected from an electrolyte containing $NiSO_4$ * $7H_2O$ 250-280 g / l; NaCl * $6H_2O$ -20-50 g / l; H_3BO_3 -30-40 g / l; formalin 0.5-1.0 g / l; pH = 4.2-5.0. Coating mode: current density 1.5-4 A / dm^2, temperature 35-45 ° C, deposition time when moving - 25-30 min.

To the top layer of nickel more stringent requirements are made. Coatings should be not only decorative and corrosion-resistant, but also have high wear resistance with a small coating thickness. The bi-nickel coating should replace the nickel-chromium coating on the parts that work on wear. For the top layer of the bi-nickel coating, a brilliant nickel plating electrolyte was chosen which is widely used in industry and has high decorative properties. A layer of shiny nickel was worn on a sublayer of semi-glossy nickel, having a thickness of 6 μm. The results of the investigation of the shiny nickel plating electrolyte are shown in Table 3.

As a result of the studies carried out, it can be concluded that a biconical coating with a glossy nickel coating on top has high decorative properties under the next electrolysis regime. a current density of 3 A / dm^2, an electrolyte temperature of 5-60 ° C, a pH of 4.3-4.8, a coating deposition rate of 1 μm for 1.75 minutes. Deposition time 6-7 minutes. To avoid pitting on the surface of the coating, intensive movement is necessary.

Table 3. The modes are shiny nickel plating and the quality of the obtained coating

Electrolyte composition	Concentration of components, g / l	Electrolysis regime			Characteristics of the appearance of the coating	Porosity, number of pores per 1 cm^2 surface	Cohesion with the substrate
		Current density, A/dm^2	T, ^0C	τ, min			

bottom layer $NiSO_4{*}7H_2O$ $NiCl_2{*}6H_2O$ H_3BO_3 Formalin pH=4,0-5,0	250 50 40 1,0	4	45-50	25	Coverage is uniform, semi-shining	1	No flaking or peeling
Upper layer $NiSO_4{*}7H_2O$ $NaCl_2{*}6H_2O$ H_3BO_3 1,4 butyne-diol saccharin pH=4,3-4,8	280 50 40 1 1	2 2 3 3 3,5 3,5 3,5	48-53 60-70 52 52-54 50 55 58	15 7 10 10 10 20 30	On the surface of the coating-pitting Coverage is uniform, shining Coverage is uniform, shining Coverage is uniform, shining On the surface of the coating-pitting Coverage is uniform, shining Coverage is uniform, shining	10 0 3 1 10 0 3	No flaking or peeling No flaking or peeling No flaking or peeling No flaking or peeling No flaking or peeling No flaking or peeling Peeling of the coating

Carrying out corrosion tests showed the resistance of the coating to disinfection and sterilization. Disinfection: wiping the outer surfaces of the bodies with a tampon moistened with a 3% solution of hydrogen peroxide with the addition of 0.5% detergent "Lotus", and then with a pad moistened with 1% chloramine solution. Sterilization: in accordance with TU 64-1-2832-2 three-time wiping with 70% ethyl alcohol. But the tested parts for wear resistance gave negative results. Checking the body for abrasion when carrying out 5000 cycles of fastening and release according to clause 2.7. OST 64-1-167-75. Test conditions: the tip is rigidly fixed in a special device. The fastening and release of the tool in the clamping device was carried out using a roller. Previously, the front body is lubricated with oil "TSIATIM numerous scratches, risks, wear in the form of a groove were formed in the front body as a result of the tests. At the end of the body, burrs formed, and a groove formed in the housing as a result of wear. At the rear housing formed burrs. Therefore, further studies were aimed at finding the electrolyte for the upper wear-resistant layer. As already indicated in the review, nickel coatings with inclusions of phosphorus possess hardness and wear resistance.

Based on this representation, we chose an electrolyte containing phosphorus in the composition. A layer of nickel 3-6 μm thick was then applied to a nickel sublayer made of a semi-shining electrolyte. The results of a study of the mode selection in the electrolyte of solid nickel plating are presented in Table. 4.

Table 4. Modes of solid nickel plating and the quality of the resulting surface

Electrolyte composition	Concentration of components, g / l	Electrolysis regime			Characteristics of the appearance of the coating	Porosity, number of pores per 1 cm^2 surface	Cohesion with the substrate
		Current density, A/dm^2	T, ^OC	τ, min			
bottom layer							
$NiSO_4*7H_2O$	250	3	45-50	25	Coverage is uniform, semi-shining	0	No flaking or peeling
$NiCl_2*6H_2O$	50						
H_3BO_3	40	1,5-3					
Formalin	1,0						
pH=4,0-5,0							
Upper layer							
$NiSO_4*7H_2O$	250	1,5	50-60	15	Coverage is uniform, shining	0	same
$NaCl_2*6H_2O$	50	2	60-65	10		0	
H_3BO_3	40	3	60-70	10		0	”
$NaH_2PO_2*2H_2O$	25	4	60-75	10	Coverage is uniform, shining	0	”
		5	60-65	10	the presence of dendrites	5	Peeling
		6	60-65	10	same	66-7	Same
		10	50-60	10	Coverage isn't uniform	–	Complete exfoliation of the coating

The results show that in the electrolyte of solid nickel plating under the electrolysis regime: the current density is 1.5-3 A / dm^2, the temperature is 60-70 ° C, the time is 10-15 min. it is possible to obtain light, semi-shining coatings 2-3 μm in thickness, which are non-porous, having good adhesion to the bottom layer of nickel and identical in properties to the chromium coating.

This work was funded by the subsidy allocated to Kazan Federal University for the state assignment in the sphere of scientific activities (3.9399.2017/8.9).

References

[1] Kashapov R N, Kashapov L N, Kashapov N F 2015 *The research of anodic microdischarges in plasma-electrolyte processing* IOP Conference Series: Materials Science and Engineering. Vol 86 Is 1 **012019**

[2] Kashapov L N, Kashapov N F Kashapov R N 2013 *Research of the impact acidity of electrolytic cathode on the course of the plasma-electrolytic process* Journal of Physics: Conference Series. Vol 479 Is 1 **012011**

[3] Kashapov L N, Kashapov N F, Kashapov R N 2014 *Influence of plasma-electrolyte discharge to the glass surface* Journal of Physics: Conference Series Vol 567 Is 1 **012024**

[4] Kashapov L N, Kashapov N F, Kashapov R N, Denisov D G 2016 *Plasma electrolytic treatment of products after selective laser melting* Journal of Physics: Conference Series Vol 669 Is 1 **012029**

[5] Kashapov R N, Kashapov L N, Kashapov N F 2017 *Analysis and development of methods for obtaining metallic powders for selective laser melting* IOP Conference Series: Materials Science and Engineering Vol 240 Is 1 **012071**

The microstructure study of CJP products

L N Kashapov [1], R N Kashapov [1,2,3], N F Kashapov [1], R R Farrahov [1]
[1] Engineering Institute, Kazan Federal University, Kazan, 420008, Russia
[2] Laboratory of Radiation Physics, Kazan Physical-Technical Institute of the Kazan Scientific Center of the Russian Academy of Sciences, Kazan, 420029, Russia
[3] Kazan State Medical University, Kazan, 420012, Russia

kashramiln@gmail.com

Abstract. The aim of work is research the microstructure of the products obtained in the ProJet 160 by ColorJet Printing technology. An IR spectrometer and a scanning electron microscope were used to analyze the structure. Analysis of IR spectra showed that when the gluing component was added, redistribution of the intensities of the absorption bands in the 3600-3500 cm^{-1} and 1130-1080 cm^{-1} region and the appearance of new lanes in the 3403 cm^{-1} and 1685 cm^{-1} region were observed. This indicates the phase transformation of the gypsum component under the action of water.

Introdustion.

The technology of additive manufacturing is widely used in the manufacture of unique and single products. One such technology is ColorJet Printing (CJP) [1]. The resulting products consist of gypsum and polymer, while they have a porous structure. Their further impregnation with epoxy resins makes it possible to use these products as models used for casting molds. Epoxy resins provide the necessary strength and have a high durability in comparison with wooden models. This is especially important in the manufacture of magnesite forms, since stronger loads are observed when compacting [2,3,4,5,6,7,8].

Powder material with a roller is applied to the surface of the platform. The print head sprays the binding fluid onto the surface of the applied powder layer. The drop-on-demand method, developed by Hewlett-Packard, uses the same method as the gas bubble method to deliver ink from the powder reservoir to the heating element. However, in the drop-zon-demand method, a special mechanism is additionally applied for the ink supply, while in the gas bubble method this function is assigned exclusively to the heating element. A special mechanism is implemented on the basis of the following physical phenomena. As a rule, the surface tension acts on the particles of the liquid phase to maintain the sphericity of the charged particles of the ink. The surface tension is reduced, which leads to the fission of the particle into smaller parts. This property of the particles to be split is used to produce foggy ink particles that flow to the nozzle outlets controlled by electrical signals. The heating of the ink under this method occurs up to 650 ° C, as a result, all the ink passes into the gaseous state and the color shift occurs at the molecular level. The drop-on-demand technology ensures the fastest ink application, which significantly improves the quality and speed of printing. The

Content from this work may be used under the terms of the Creative Commons Attribution 3.0 licence. Any further distribution of this work must maintain attribution to the author(s) and the title of the work, journal citation and DOI.
Published under licence by IOP Publishing Ltd

IOP Conf. Series: Materials Science and Engineering **412** (2018) 012086 doi:10.1088/1757-899X/412/1/012086

color representation of the image in this case is more contrast. In this technology, the management of ink particles is performed with a constant deflecting field due to the regulation of their electric charge. Therefore, each particle emitted from the nozzle receives its "own" information in the form of a different value of the electric charge, which ensures the printing of high speed and print quality.

After application of the binder, the construction platform is lowered by 150 μm and the printing algorithm is repeated layer by layer. The internal structure of the products is interesting. CJP includes two main components: core and binder. In this case, we can observe a difference in the sprayed doses on the perimeter and core. Therefore, the aim of the work was to study the structure of used powders and obtained products, determination of the binder component influence on the resulting microstructure.

Experimental part:

3D Systems ProJet 160 is a professional 3D printer company 3D Systems, USA. Has become widespread in the field of prototyping for visualization of engineering tasks, design and architecture. The printer's working chamber has a build area of 236 × 185 × 127 mm. The working chamber contains: a cartridge with a reinforcing liquid, a filter, a vacuum pump, a printing head moving system in a horizontal plane, a platform, a box for assembling the powder with a frontal overflow.

As the printhead in the fast axis node, the HP 11 cartridge is used. This cartridge is attached to the carriage as shown in Figure 1.

Figure 1 - The carriage and print head assembly. The image is taken from the source [1].

To start the work of manufacturing the desired product, you need to import a file with a 3D model (possible formats: 3DS, .BLD, .FBX, .PLY, .STL, .SFX, .VRML, .ZBD, .ZCP, .ZPR) into special software 3DPrint. Then the model divides into the thinnest horizontal layers. ProJet 160 prints layers of thickness of 0.15 mm. Data that displays information about these layers is loaded into the printer. Further all the work is done by the ProJet 160 itself.

The principle of work consists in layer-by-layer creation of the model. The minimum topological dimension of one element of the model in the X and Y axes is 0.4 mm. The model creation area is filled with the first layer of powder, then the print head applies a special adhesive to the areas where the layer of the future model is located. After that, the second, third, and all other layers transmitted by the software to the printer are filled in the same way.

In the printing process, the non-glued powder remaining around the product serves as a support function so that it is possible to create complex and structurally fragile models. Also, if you put several models for simultaneous processing, the powder will ensure the safety of each model. After printing, the product needs time to allow the adhesive to completely grasp the powder. Usually it takes 1-2 hours. Sometimes more than 2 hours, depending on the size and number of objects. During the extraction of the part from the powder, it is recommended to use special nozzles in order to not damage the fragile parts of the product.

After that it is necessary to further strengthen the product. For this, a fixer is used. Fixer is a special compound that strengthens the surface of the product. There are a number of fixers, but their strength indicators differ from each other.

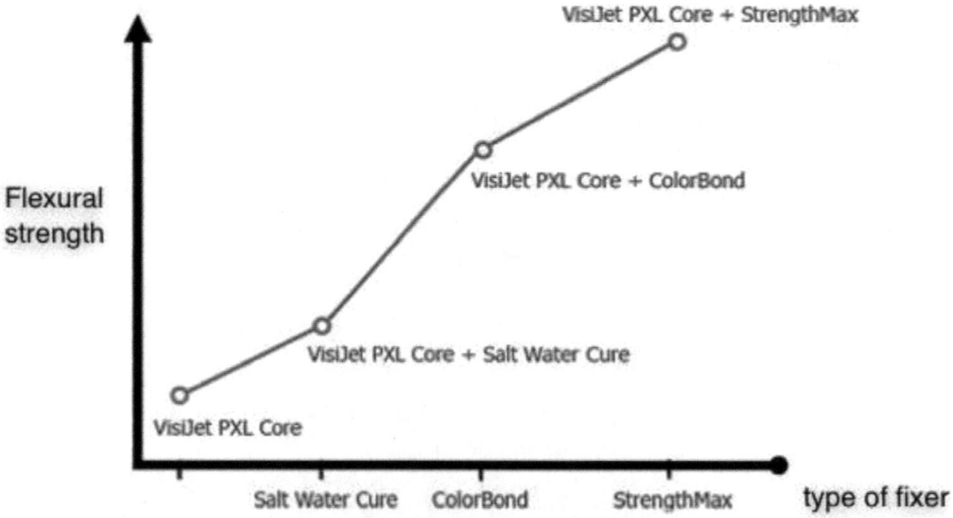

Figure 2 - Graph of flexural strength using different types of fixers. [2]

To verify the print quality of the printer, the structure and composition of the powder and the product are examined after printing. For this, X-ray scanning and IR spectroscopy are performed, respectively. SEM images of the powder were also made using an electron microscope, shown in Fig. 3.

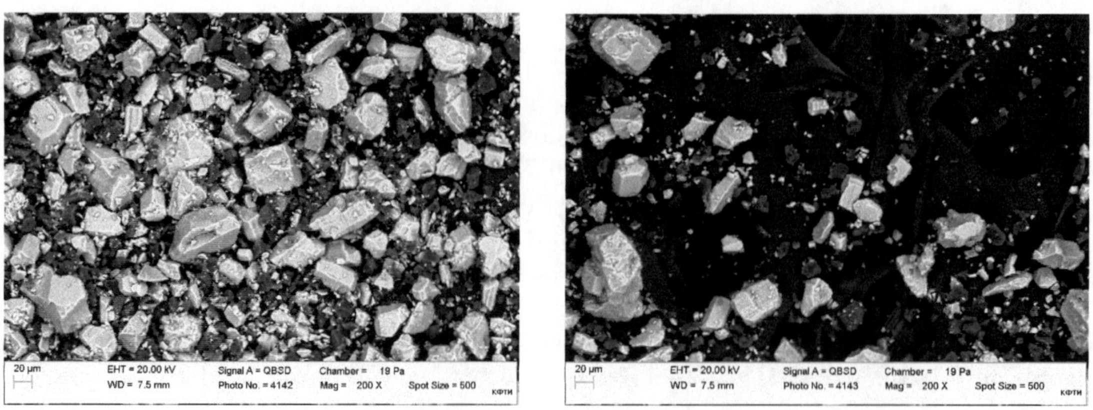

Figure 3 - SEM image of powder.

A detailed analysis of the powder showed the presence of particles of two types, to refine the composition of the powder, the elemental composition was examined using a scanning electron microscope. The results of the studies are shown in Figure 4. It was shown that the powder particles having a dark shade on the SEM image mainly consist of carbon and oxygen, which may indicate their organic origin.

Principle of IR spectroscopy: infrared radiation passes through sample. There is excitation of vibrational motions of molecules or their separate fragments. In this case, the intensity of light passing through the sample is weakened. However, absorption doesn't occur in the entire spectrum of the incident radiation, but only at those wavelengths whose energy corresponds to the excitation energies of vibrations in molecules under study. Consequently, wavelengths (or frequencies) at which the maximum IR radiation absorption is observed can indicate the presence of certain functional groups and other fragments in the sample molecules.

Element	Weight %	Atomic %
C K	4.79	8.17
O K	52.88	67.69
S K	19.66	12.56
Ca K	22.66	11.58
Total	100.00	

Element	Weight %	Atomic %
C K	47.13	60.67
O K	30.97	29.93
S K	9.96	4.80
Ca K	11.93	4.60
Total	100.00	

Figure 4 - Elemental composition of powder particles.

As a IR spectroscopy result of a sample of the VisiJet PXL Core composite material and the details printed from the same material with the addition of a fixing liquid, the following spectra were obtained.

IOP Conf. Series: Materials Science and Engineering **412** (2018) 012086 doi:10.1088/1757-899X/412/1/012086

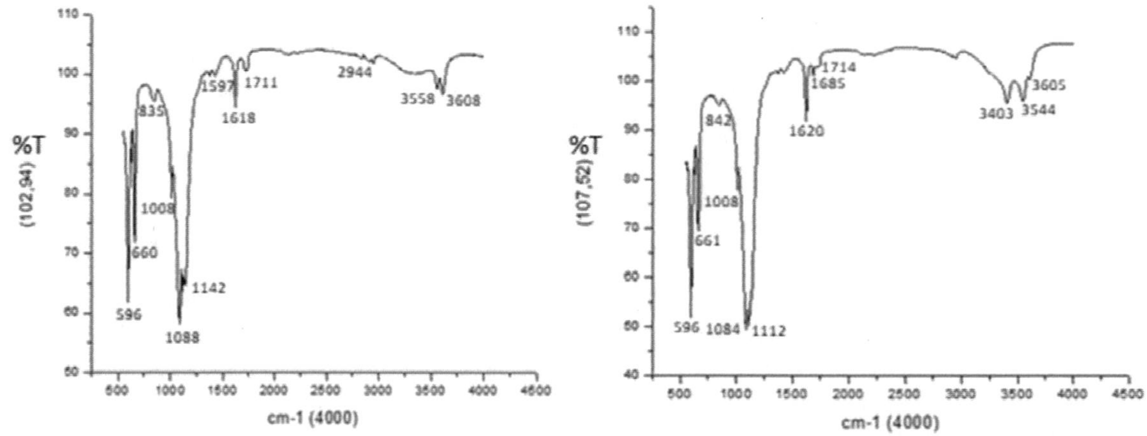

A B

Figure 5 - Graphs of IR spectra a) gypsum and b) details

IR spectroscopy of the reinforcing liquid was also performed, as shown in Figure 6. After analyzing the data, we came to the conclusion that the VisiJet PXL core composite material is a mineral of the sulfate class, in terms of calcium sulfate hydrate ($CaSO_4 \cdot 2H_2O$). The absorption lines in the IR spectrum of the composite can be attributed to the 3600-3500 cm^{-1} and 1618 cm^{-1} region to the crystallization water [4], which is part of the gypsum molecule, the bands in the region of 1130-1080 cm^{-1} and 880-596 cm^{-1} - to SO_4 anions. [3]

Weak lines in the range 3000-2800 cm^{-1} can be attributed to the groups CH_2 and CH_3 of linear polymer chains, in the range of 1600-1500 cm^{-1}, to nitro groups (NH), or to benzene rings. [4]

When analyzing the infrared spectra of a component, a redistribution of the intensities of the absorption lines in the 3500-3600 cm^{-1} and 1130-1080 cm^{-1} regions and the appearance of new lines in the 3403 cm^{-1} and 1685 cm^{-1} region, which may be a consequence of chemical transformations in the composite .

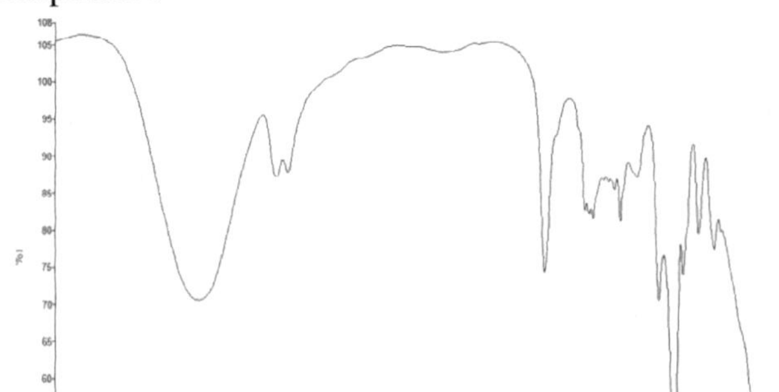

Figure 6 - Graph of the IR spectrum of the strengthening liquid.

An analysis of the composite material using an adhesive liquid was also carried out. After analyzing its structure and porosity, we came to the conclusion that the connection between the VisiJetPXL Core granules is much stronger.

Conclusion

After the tests and the microstructure of the product were examined, minerals belonging to the class of sulfates ($CaSO_4 \cdot 2H_2O$) and anions of SO_4 were found. In addition, polymer chains of the group CH_2 and CH_3, as well as nitro groups (NH), were found in the composition of the part. It was also found that under the influence of a strengthening liquid, chemical transformations occur in the composite. Due to this, the bonds between the molecules of the composite material become more durable.

This work was funded by the subsidy allocated to Kazan Federal University for the state assignment in the sphere of scientific activities (11.1300.2017/4.6).

References

[1] ProJet® x60 Series User Guide [Electronic resource] // 3Dsystems, 2013. URL: http://infocenter.3dsystems.com/product-library/system/files/common/2412_2295043ProJetx60SeriesUserGuide_CET.pdf

[2] Kashapov R N, Korobkina A I, Platonov E V, Saleeva G T 2014 *The method of manufacture of nylon dental partially removable prosthesis using additive technologies* IOP Conference Series: Materials Science and Engineering Vol 69 Is 1 **012026**

[3] Kashapov L N, Kashapov N F, Kashapov R N, Denisov D G 2016 *Plasma electrolytic treatment of products after selective laser melting* Journal of Physics: Conference Series Vol 669 Is 1 **012029**

[4] Globatek.3D [Electronic resource] : Materials for 3D printing: VisiJet PXL Core— URL: http://3d.globatek.ru/3d_printing_materials/composite/visijet_pxl_core/.

[5] Tarasevich B N 2012 *IR spectra of the main classes of organic compounds. Reference materials* Moscow State University named after M.V. Lomonosov

[6] Kuptsov A Kh, Zhizhin G N 2001 *Fourier spectra of Raman scattering and infrared absorption of polymers* Directory. M. Fizmatlit 416 p

[7] Kashapov R N, Kashapov L N, Kashapov N F 2017 *Investigations of the growth of the vapor-air shell of a gas discharge with a liquid electrolytic cathode of sodium hydroxide solution* IOP Conf. Series: Journal of Physics: Conf. Series 927 **012085**

[8] Kashapov L N, Rudyk A N, Kashapov R N 2014 *Applying 3D-printing technology in planning operations of cancer patients* IOP Conference Series: Materials Science and Engineering 69 **012016**

IOP Conf. Series: Materials Science and Engineering **412** (2018) 012087 doi:10.1088/1757-899X/412/1/012087

Study of the possibility three-dimensional printing DLP-technology for casting individual neck telescopic implants

R S Kashapova[1], R N Kashapov[1,2], L N Kashapov[1], N F Kashapov[1], L F Sabirzjanova[1]

[1] Engineering Institute, Kazan Federal University, Kazan, 420008, Russia

[2] Laboratory of Radiation Physics, Kazan Physical-Technical Institute of the Kazan Scientific Center of the Russian Academy of Sciences, Kazan, 420029, Russia

kashramiln@gmail.com

Abstract. The work is devoted to manufacturing possibility of individual telescopic cervical implants by the casting method for burned-out models obtained on a DLP printer. The ProJet 1200 using possibility for the burned model production of a telescopic cervical implant was studied. The results of the model material influence on the molding mass are obtained. Experimental samples of cervical implants on the Inducast installation are cast. The ineffectiveness of this technology in the manufacture of reticular collapsible telescopic implants is established. The possibility of its application for thick-walled structures is established.

Introduction

Individual approach to the treatment of patients is becoming increasingly widespread. Starting from individual selection of medicines based on genetic analysis and ending with individual implants and orthoses. The creation of individual products has always been a more expensive procedure in comparison with mass production. However, with the advent of additive production technologies, the creation of individual parts has become more accessible. Therefore, additive technologies began to be widely used in various fields of medicine. In the additive manufacturing of individual implants, there are two methods: creating models for subsequent casting or direct manufacturing of finished individual implants from metal powders. The most correct is the direct production by selective laser melting of metal implants, thereby reducing the number of manufacturing stages. However, from laboriousness view obtaining a certificate-permission to use this method for medical purposes, it can not be used in the short term. Therefore, you can use only the first option - the symbiosis of additive technologies and traditional - casting. Therefore, the task was to study the combined method possibilities of manufacturing individual cervical telescopic implants [1,2,3]. Namely, how many complex designs can be obtained using DLP-technology of three-dimensional printing, and whether it will be possible to cast them out of metal. This is very important, as there is a tendency to manufacture mesh implants that have a low weight and are not inferior in strength to the standard one. Another advantage of reticular implants is the possibility of filling them with

Content from this work may be used under the terms of the Creative Commons Attribution 3.0 licence. Any further distribution of this work must maintain attribution to the author(s) and the title of the work, journal citation and DOI.
Published under licence by IOP Publishing Ltd

various biomaterials for effective osseointegration. One of the common injuries is spinal injury [4,5]. The greatest danger is a fracture or dislocation of the cervical spine. It is important to replace the damaged vertebra correctly. This is achieved only with the installation of individual telescopic implants, with the possibility of changing the upper platform location in the vertical and horizontal directions [6,7]. Therefore, in work, on the example of this type of implants, the possibility of using a combined casting method and DLP-technology was investigated.

Main part

To create an individual implant, you need to obtain digital data of the fracture site or dislocation. The surgeon decides to install a telescopic cervical implant. Obtaining a three-dimensional model of the fracture site is possible only in installations receiving a object layered structure, such type of equipment include tomographs. Digital data obtained with CT, MRI are needed to create a three-dimensional virtual model. Determine the location of installation and cervical implant attachment. Only after receiving this information will it be possible to begin modeling the geometry and calculating the component dimensions of the telescopic implant. With 3D Slicer software, the CT data was processed and 3D model of the patient cervical spine in STL format was created. In the MeshMagic program, we load the STL file of the fracture site and standard size models-blanks of telescopic cervical implants. We adapt the workpiece model to the fracture site. As a result we have individual three-dimensional virtual implant model in the STL format.

To produce a burned-out model, we used a 3d printer ProJet 1200 equipped by pico projector with a resolution 585 dpi in the X-Y, and Z-step height of 30 μm. This setting applies to the DLP (Digital Light Processing) family of 3d printers. At the heart of the DLP-system is a special device - a DMD-chip (Digital Micromirror Device). This is a complex structure that belongs to the so-called class of microelectronic mechanical systems (MEMS - Micro-Electronic Mechanical System). As a source of light can act as a light bulb (incandescent, luminescent or LED), and lasers. The wavelength extends from the ultraviolet to the infrared range. The device has the form of a micromirror matrix. Each micromirror is 10 microns in size. They are able to reflect both the invisible and visible spectrum of light and reflect light in one of two directions. The direction is determined by the angle of rotation. Direction itself is set by loading the bit "0" or "1" in the memory location. Each cell receives an independent bit stream with a frequency of several kilohertz, as a result of which we have a useful image on one of the outputs, and at the other output there is a light absorber. The GeomagicPrint program splits the printed object into layers with the specified thickness. A photopolymer is poured into the printer's bath with a transparent bottom. At the bottom of the bath the metal substrate is immersed, receding from the bottom to one (first) layer of the product. It turns out in the gap between the substrate and the bathroom is a liquid photopolymer. The projector, located under the bathroom, projects the picture of the first layer onto the bottom of the bath and, thanks to UV radiation, only the plastic that gets the image from the projector freezes. Then the working table rises one more layer and again a new layer is glued, which is attached to the previous one. As a result, a three-dimensional object from the liquid photopolymer grows layer by layer. The material used was the VisiJet FTX Green photopolymer from 3D Systems. It has the following properties: heat resistance - up to 93 ° C, dynamic viscosity (mPa s): 100 - 130 (30 ° C), burn-out - ash residue 0.01%.

Two types of samples were grown in the ProJet 1200 using pre-modeled models: with and without the net filling of the implant body (Figure 1). For casting on burned models, a gating system was modeled, shown in Figure 1b.

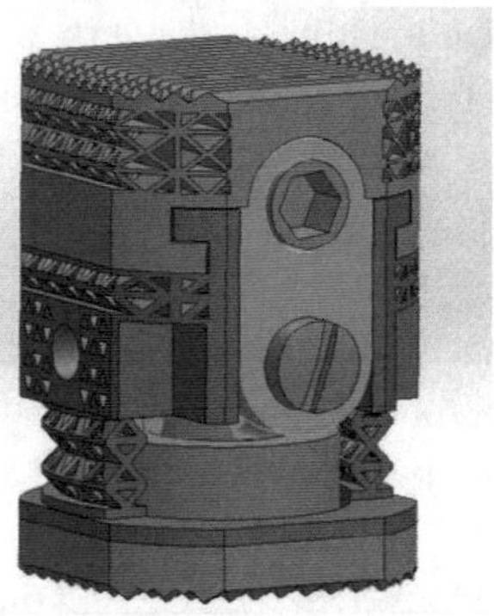
a

b

Figure 1 - a - mesh virtual model of the cervical implant assembly, b – gating system.

The construction of the gating system begins with the establishment of each unit according to the feeder. Elements of the implant are supplied with sprue channels with a diameter of 5 mm. The channels number depends on the number of units cast. After building a gating system, it must be weighed. Based on the known mass of wax, the required amount of metal is determined. The calculation is made by the following formula: $M_{Me} = m_{wax} * 10 + 5$ g. Starbond CoS dental metal was used in the experiments, the following composition of Co - 59%, Cr - 25%, W - 9%, Mo - 3.5%, Si - 1%, other components (C, Fe, Mn, N) - maximum 1.5%. Ultimate tensile strength (Rp0.2) 650 MPa. The tensile strength is 910 MPa. The maximum elongation is 8%. The modulus of elasticity is 200 GPa. Vickers hardness 280 HV 10. Density 8,8 g / cm^3. Solidus-liquidus interval is 1305 - 1400 ° C. The casting temperature is 1500-1550 ° C. The thermal expansion coefficient is 14.0 μm / m. ° C.

In the second stage, the Yeti Expansion refractory molding (90 g) is prepared and kneaded with Yeti Liquid (22/20 ml) + distilled water (0/2 ml). To avoid the formation of air bubbles, a vacuum mixer Fox 88 from OMEC (Italy) was used. The kneading time was 30 seconds, the blade rotation speed was 400 rpm, and the pressure in the bowl was 0.8 Bar. The casting of the gate was made on a vibration table VB 1.1 with 2 vibration frequencies of 3000/6000 min^{-1}. After the molding mass has been filled in, wait 30 minutes before the molding material solidifies completely. The resultant hardening of the molding mass should be heated in the muffle furnace. In the experiments, the method of shock heating of the flask was used. In particular, the Programix 50 Ugin oven was preheated to a temperature of 900 ° C. A mold was used with an X6 size, the minimum warm-up time was 60 minutes. To effectively cast the metal, the crucible with the metal is also heated in a muffle furnace. In the next step, the heated crucible is placed in a casting chamber. The metal is heated by means of high frequency currents. Further, from the muffle furnace, a heated flask is extracted and vertically fixed to the crucible cover and pressed down from above with a clamping mechanism. After that, the vacuum chamber cover is closed, and the evacuation process of air begins with simultaneous high-frequency metal heating. During the heating process, the metal melts, resulting in the formation of a molten metal in the crucible, which is covered with an oxide film on top. At the rupture moment of the oxide film, the caster performs a casting process, namely, a vacuum-casting chamber is rotated, and compressed air is pumped at a rate of 3 bar/s to a pressure of 3.6 bar. As a result of the overturn, the metal from the crucible is poured into the flask and pressurized by excess pressure. The pressure provides the possibility of spilling metal into channels with a thickness more than 0.1 mm. In the experiments, a fore-casting installation was used. The vacuum-casting chamber is in an inverted state for 30 seconds, during which time the metal cools and crystallizes. Molding box must be cooled at room temperature, and mechanically clean the metal frame from the molding mass. A sandblasting unit was used, in which sand particles stream with a grain size of 250-300 μm, bombarded the surface of the sprue system and cleared the molding mass.

Figure 2 shows cast specimens of individual telescopic cervical implants.

Figure 2 - Samples of castings of individual telescopic cervical implants.

As a result of the research, it has been established that implant molding without a mesh structure in the volume is successful. After additional machining, a ready-made disassembly implant can be obtained. At casting wire mesh there are two problems: the difficulty of molding associated with the air extraction from mesh volume and the difficulty of cleaning the molding mass is a mesh space in the implant volume.

Conclusion

The possibility of using the ProJet 1200 system for producing a burned model of a telescopic neck implant with a mesh structure was studied. The results of the model material influence on the molding mass are obtained. Inducast installation cast prototypes of cervical implants. The technology ineffectiveness in manufacture of reticular telescopic implants is established. The possibility of its application for thick-walled structures is determined.

This work was funded by the subsidy allocated to Kazan Federal University for the state assignment in the sphere of scientific activities (11.1300.2017/4.6).

References

[1] Kashapov R N, Kashapov L N, Kashapov N F 2017 *Analysis and development of methods for obtaining metallic powders for selective laser melting* IOP Conference Series: Materials Science and Engineering Vol 240 Is 1 **012071**

[2] Kashapov R N, Kashapov L N, Kashapov N F 2015 *The research of anodic microdischarges in plasma-electrolyte processing* IOP Conference Series: Materials Science and Engineering Vol 86 Is 1 **012019**
[3] Kashapov L N, Kashapov N F, Kashapov R N 2013 *Research of the impact acidity of electrolytic cathode on the course of the plasma-electrolytic process* Journal of Physics: Conference Series. Vol 479 Is 1 **012011**
[4] Kashapov L N, Kashapov N F, Kashapov R N 2014 *Influence of plasma-electrolyte discharge to the glass surface* Journal of Physics: Conference Series. Vol 567 Is 1 **012024**
[5] Kashapov L N, Rudyk A N, Kashapov R N 2014 *Applying 3D-printing technology in planning operations of cancer patients* IOP Conference Series: Materials Science and Engineering 69 **012016**
[6] Kashapov L N, Kashapov N F, Kashapov R N, Denisov D G 2016 *Plasma electrolytic treatment of products after selective laser melting* Journal of Physics: Conference Series Vol 669 Is 1 **012029**
[7] Kashapov R N, Korobkina A I, Platonov E V, Saleeva G T 2014 *The method of manufacture of nylon dental partially removable prosthesis using additive technologies* IOP Conference Series: Materials Science and Engineering Vol 69 Is 1 **012026**

Research methods of hardening the surface of medical instruments

R R Farahov[1], R N Kashapov[1,2], L N Kashapov[1] and N F Kashapov[1]
[1] Engineering Institute, Kazan Federal University, Kazan, 420008, Russia
[2] Laboratory of Radiation Physics, Kazan Physical-Technical Institute of the Kazan Scientific Center of the Russian Academy of Sciences, Kazan, 420029, Russia

kashramiln@gmail.com

Annotation. The use of additive technologies for the manufacture of medical devices and equipment is becoming more widespread. However, manufactured products require additional processing to improve performance. Therefore, the aim was to investigate the processes of borating and electrophysical processing of the working surfaces of articles obtained by selective laser melting.

Introduction

The increased requirements for quality and wear resistance of the medical instrument are the reason for the application of chemical-thermal treatment and electric-spark alloying. Traditional methods of obtaining products of diffuse coatings that increase the resistance and wear resistance of steel include cementation, nitrocarburization, nitriding [1,2,3]. One of the oldest methods of chemical-thermal treatment is cementation in a solid carburizer. On KMIZ (Kazan Medical Instruments Plant) for hardening two-headed mills for cutting teeth on boron, the method of nitrocarburization in corona discharge is used. Also it is reliably known that in 1974 the process of borating ampoule scarifiers was introduced at the Gudermes factory, which made it possible to create a domestic tool that considerably exceeds the resistance of foreign ones [4,5]. Diffusion boride coatings are of great interest primarily as wear-resistant. The rigidity of equipment from various grades of steel is increased by 2-5 times. Also known are the electrophysical methods of hardening, in particular, the use of the process of electric spark alloying has made it possible to create new designs of needle holders with high functional properties.

The use of additive technologies for the manufacture of medical devices and equipment is becoming more widespread. However, manufactured products require additional processing to improve performance. Therefore, the aim was to investigate the processes of borating and electrophysical processing of the working surfaces of articles obtained by selective laser melting.

Main part

The choice of electrodes for the electro-spark hardening of metallic materials is mainly determined by the requirements for the alloyed layer for wear resistance and chemical resistance. The main criteria in our studies are the indications of tests on the hardness in corrosion resistance in sterilization media. High-alloy solid alloys of the brand KBH and SNGN were chosen, which significantly increase the wear resistance and anticorrosive properties of the treated surfaces.

Content from this work may be used under the terms of the Creative Commons Attribution 3.0 licence. Any further distribution of this work must maintain attribution to the author(s) and the title of the work, journal citation and DOI.
Published under licence by IOP Publishing Ltd

IOP Conf. Series: Materials Science and Engineering **412** (2018) 012088 doi:10.1088/1757-899X/412/1/012088

Work has been done on the development of a technical process for the production of electrodes from the alloys KBH and SNGN, manufactured by the powder metallurgy method. As a raw material, KBH powder, TU 48-42-64-71, SNG powder, TU 48-42-60-71, aviation grade B-70, GOST 1012-72, rubber of A grade, GOST 2199-66. For the preparation of furnace charge, powders KBH (SNGN) and plasticizer are used. The plasticizer is a 4% solution of rubber glue in gasoline. For 100 g of powder it is necessary to add 40-50 g of plasticizer. After mixing the furnace charge is unloaded into steel trays. The finished furnace charge is dried in drying cabinets under a hood at a temperature of 40 ± 10°C for 3 hours. Granulation of the charge is carried out by rubbing the charge through a sieve with a No. 0315 mesh. The billets are pressed on a hydraulic press. The specific pressing pressure is 2.5 - 3 t / cm^2. Weights: workpiece on KBH - 8 ± 1 g., workpiece on SNGN - 9 ± 1 g. Cracks, stratifications, and chips are allowed on the surfaces of compacted electrode blanks. The blanks are sintered in a vacuum laboratory electric resistance furnace in a vacuum of $1 * 10^{-4}$ mm. of mercury. The blanks are placed in stands made according to TU 06259-75 and loads the workpieces into a cold furnace, then uniformly raising the temperature (420 °C per hour) to: for blanks from KBH - 1350 ± 50°C; for blanks from SNGN - 1250 ± 50°C. For 3 hours and hold at this temperature for 45 minutes. After soaking, the electrodes are cooled together with the oven under vacuum to room temperature. Electrodes from pure metals Ta, Mo, W, Cr, V, Nb, Cr_3C_2, Cr_3B, $(TiCr)B_2$, graphite, carbides and borides were also investigated.

Table 1. Chemical composition of powdered and cast solid alloys

Alloy	Alloy composition, weight. %											
	Cr	B	C	Ni	Fe	Si	W	Ti	Co	Mo	Al	Mp	P
KBH	44-52	0.7-0.9	4.0-6.0	-	Main	1.4	-	-	-	-	-	-	-
SNGN	14-18	3.5-5.0	0.5-1	Main	Up to 3	3.4-4.5	-	-	-	-	-	-	-
ICH	18-20	-	3.0-3.5	1.5-2	Main	1.8-2	-	-	-	1-1.2	-	2-2.5	0.012
Alloy TZ	30-35	-	1.5-2	1	2	-	10-15	-	Main	≤3.0	-	1-2	-
BKN A-2	4-6	0.01	0.16-0.22	Main	-	-	1.5-2.5	1-1.6	-	0.1	12.2-14	-	-
C-27	23-28	-	3.3-4.5	1.5-2	Main	1-2	0.2-0.4	-	-	0.1-0.2	-	0.8-1.5	-

Table 2.The results of metallographic studies.

Main Material Brand	Brand of electrode material	Alloing regime	Layer height, μm	Hardness of layer, HRC	Layer continuity,%
30Cr13	Electrode from KBH	III	40-50	68	70-80
30Cr13	Electrode from SNGN	II	30-40	66	80-90
20Cr13	TNM-20	IV	30-35	65	90-95
20Cr13	TNM-25	III	25-30	48	50-60
20Cr13	TNM-30	IV	35-40	63	95-98
20Cr13	W	IV	20-25	49	80-85
20Cr13	Ta	IV	70-80	55	100
20Cr13	ICH	IV	25-35	49	60-70

20Cr13	Alloy from "Torez plant of surfacing hard alloys"	III	30-35	60	70-80
20Cr13	Cr3B	III	15-20	65	40-50
20Cr13	VZHKL	III	5-15	80	30-40
20Cr13	VKNA-2	III	15-20	46	60-70
U10A	Ti15C6	II	15-20	45	70-80
U10A	VK-8	II	25-30	62	60-65
U8A	Graphite	II	5-15	63	50

When all the electrodes are hardened, the so-called "white" layer is formed on the surface of the cathode, which is not etched with ordinary etchants, used to reveal the structures of the base materials and the material of the alloying electrode. The thickness and continuity of the doped layer under identical processing conditions are determined by the material of the alloying electrode. Figure 1 shows photos of hardened layers microstructure, which give an idea of the continuity of the "white" layer.

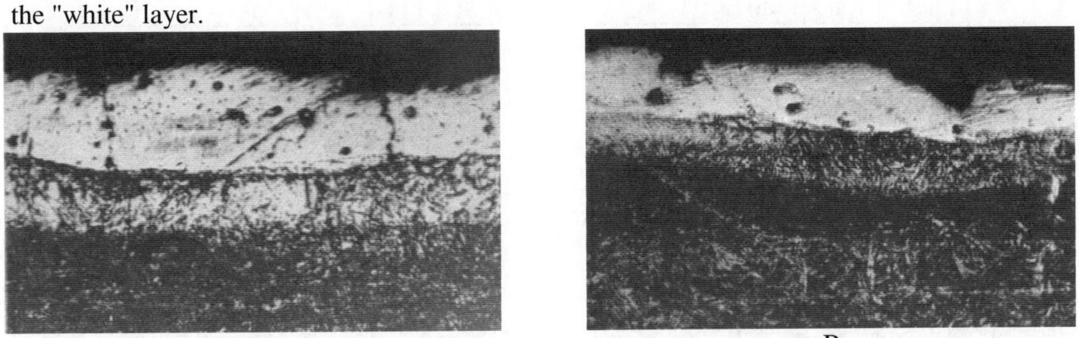

A B

Figure 1 - The microstructure of U8A steel alloyed with graphite (a) and alloy T15K6 (b)

The most qualitative from the point of view of the thickness and continuity of the "white" layer of the coating are obtained when hardening with vanadium. The solidity is close to 100%. At hardening with molybdenum the layer is thin, but its continuity is also quite high (80 - 90%). Satisfactory results were obtained when the anodes were hardened by T15K6, BK8, KBH, and SNG alloys; the continuity of the layers is 70 - 80%, the layer itself does not have shells and voids. When alloying with anodes from non-wolfram solid alloys TMH, the quality of the hardened layer depends on the composition of the alloy. When hardening with THM30 alloy, which contains a large amount of binder, a layer with a large continuity and uniform thickness is obtained. The alloyed layer obtained by processing the above electrodes also has the highest hardness. When the samples are hardened by other electrodes, the layer quality deteriorates noticeably; the doped layer has a low hardness. The highest class of surface frequency was obtained by hardening the metal surfaces with an electrode made of graphite. The value of the layer is 15-20 microns, the hardness reaches 63 units. by Rockwell.

As a result of searching for other electrode materials, positive experimental data were obtained on samples hardened by these alloys VZhKD and VKNA-2. Despite the low hardness of the alloyed layer (HRC 45 ... 46), preliminary technical tests showed an increase in wear resistance by 2-3 times. For hardening tools from carbon steel grades U8A and U10A, it is most rational to use hard alloys of T15K6 and VK8 grades, as well as non-wolfram solid alloys of TNM-25 and TNM-30.

For manufacturing of the basic details of the tooling, the most widely used at medical-instrumental plants are U8A, U10A, 5HNM steels. It was found that the borating must be carried out at a temperature of 930 - 950°C. Table 3 shows the effect of the delay time of the solid borate process

IOP Conf. Series: Materials Science and Engineering **412** (2018) 012088 doi:10.1088/1757-899X/412/1/012088

950OC on the borated layer, its hardness, the grain size and the hardness of the base metal. The value of the borated layer increases smoothly both for steel U8A and for steel 5HNM.

Figure 2 shows photographs of borated layers from steels U8A and 5XNM after different exposure. The borated layers under the chosen modes turn out dense. The metallographic structural components analysis of the layer after etching with reagent (Pecrylic acid - 2-5 g, Caustic soda - 20-25 g, Water - 75-100 ml) showed that the borated layer consists of Fe_2B borides.

The Fe_2B boride has less brittleness than the FeB boride, so obtaining a phase composition with only Fe_2B structure is more preferable for the tooling.

Table 3. Influence of the borating regime and subsequent heat treatment on the basis characteristics of the borated layer and the base metal.

steel grade	The borating regime		Heat treatment mode	Structure	Grain score	Thickness	Hardness, kg / mm^2	
	t, °C	τ, h.					Core	Layer
5XNM	950	4	No	sorbite	4	9-11	310-320	1079-1100
5XNM	950	6	No	sorbite	3	27-30	316.5-320	1320-1420
5XNM	950	8	No	sorbite	3	40-42	310-320	1538-1420
5XNM	950	11	No	sorbite	2-3	42-50	310-320	1420-1670
U8A	950	4	No	sorbite	5-6	27-30	310-320	1320-1420
U8A	950	6	No	sorbite	5	45-46	310-320	1420-1538
U8A	950	8	No	sorbite	5	55-57	316.5-320	1538-1670
U8A	950	11	No	sorbite	4-5	65-70	316.5-320	1538-1670

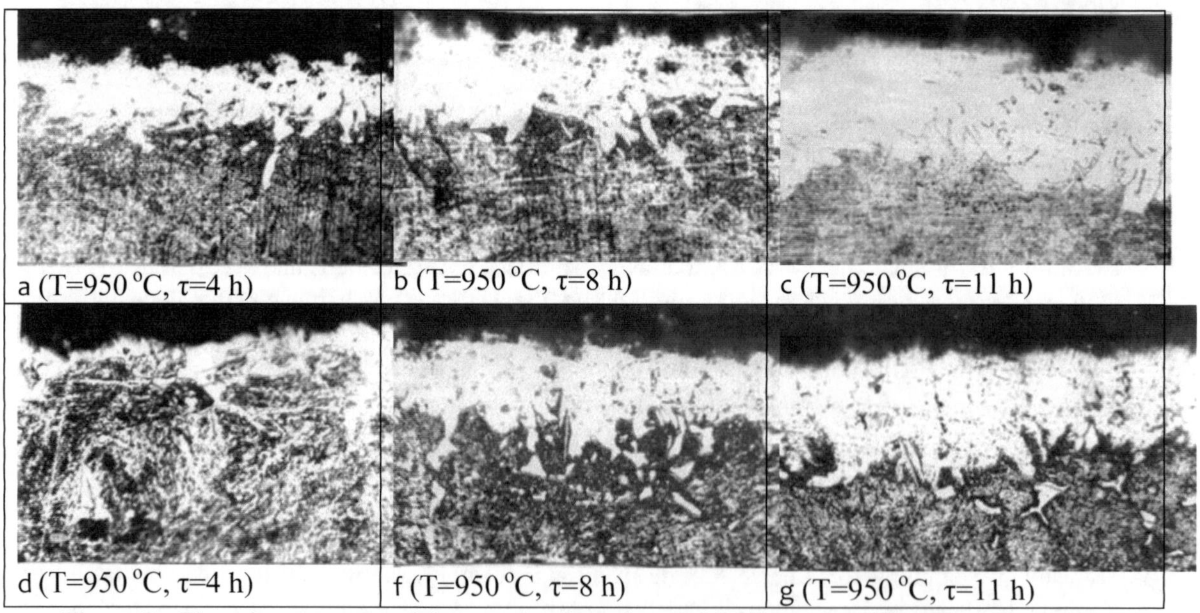

Figure 2 - The microstructure of the samples from the steel YSA (a-c) and SCHNM (d-g) after borating.

The grain growth during borating and the low hardness of the core determine the need for subsequent heat treatment. According to the procedure of the work, heat treatment of the samples was carried out for samples of U8A steel according to the regime: heating temperature = 820 ± 5 °C, holding time 5 min, cooling in oil; low tempering T = 250 °C, holding time 60 min; for samples of steel 5XNM – $T_{heating}$ = 850 ± 5 °C, holding time 5 min, cooling in oil, low tempering - T = 160 °C, τ =

80 min. After heat treatment, the grain score of both steels is 7, the hardness is 560-600 kgf / cm2. The thickness and hardness of the borated layer did not change as a result of heat treatment.

Samples from steel U8A before borating had a slight surface roughness, conditionally corresponding to 8-9 grades, after boriding the roughness increased to 6-7 grades. For steel 5XNM this change is not so significant: before borating 6-7 grades, after borating - 5-7 grades. Subsequent heat treatment does not change the surface roughness.

This work was funded by the subsidy allocated to Kazan Federal University for the state assignment in the sphere of scientific activities (3.9399.2017/8.9).

References

[1] Kashapov R N, Kashapov L N, Kashapov N F 2017 *Analysis and development of methods for obtaining metallic powders for selective laser melting* IOP Conference Series: Materials Science and Engineering Vol 240 Is 1 **012071**

[2] Kashapov R N, Kashapov L N, Kashapov N F 2015 *The research of anodic microdischarges in plasma-electrolyte processing* IOP Conference Series: Materials Science and Engineering Vol 86 Is 1 **012019**

[3] Kashapov L N, Kashapov N F, Kashapov R N 2013 *Research of the impact acidity of electrolytic cathode on the course of the plasma-electrolytic process* Journal of Physics: Conference Series. Vol 479 Is 1 **012011**

[4] Kashapov L N, Kashapov N F, Kashapov R N 2014 *Influence of plasma-electrolyte discharge to the glass surface* Journal of Physics: Conference Series. Vol 567 Is 1 **012024**

[5] Kashapov L N, Kashapov N F, Kashapov R N, Denisov D G 2016 *Plasma electrolytic treatment of products after selective laser melting* Journal of Physics: Conference Series Vol 669 Is 1 **012029**

IOP Conf. Series: Materials Science and Engineering **412** (2018) 012089 doi:10.1088/1757-899X/412/1/012089

Search for the optimal method of making reticular neck telescopic implants

R M Kashapova[1], R N Kashapov[1,2], L N Kashapov[1], N F Kashapov[1], V F Kasimova[1]

[1] Engineering Institute, Kazan Federal University, Kazan, 420008, Russia
[2] Laboratory of Radiation Physics, Kazan Physical-Technical Institute of the Kazan Scientific Center of the Russian Academy of Sciences, Kazan, 420029, Russia

kashramiln@gmail.com

Abstract. The production of individual implants, intended for a specific clinical case, is increasingly used in medicine. The next stage in the development of implants is the creation of a non-monolithic mesh structure of the implant body. This allows to reduce its weight without loss of mechanical characteristics, and the main thing is to fill the free intergrid space with bioresorbable material, which will be replaced by cells in the future. The creation of a reticular three-dimensional structure of the implant is a complex technical task and requires the search for new ways of production. In this paper, a comparative study of the possibilities of selective laser melting and casting on burned-out models of telescopic implants obtained by 3D printing methods is carried out. Various production methods produced prototypes of telescopic cervical implants and carried out their technical and economic analysis. It is established that the best quality of products is observed by Selective Laser Melting technology.

Introduction

The development of medicine is strongly associated with the development and implementation of new technical devices and technologies. It was the development of new robotic instruments of operation that greatly expanded the capabilities of surgeons, and the discovery of superconductivity made it possible to produce affordable for general use MRI systems. At the moment, it is the relationship between medicine and technology that allows solving new problems of human health [1,2,3]. The latest trend of individual approach in the treatment of patients is becoming increasingly widespread. This applies to the individual selection of drugs based on genetic analysis, and in particular the creation of individual implants. In addition to the individuality of the implant site, there is a need to create a retina mesh body. This is necessary to reduce the weight of the structure and fill the bioresorbable material of the intergrid space. In the case of using standard production and molding methods, the creation of this product type is a complex technological task [4,5]. The use of additive technologies solves this problem. In the additive manufacturing of individual implants, there are two methods: creating models for subsequent casting or direct manufacturing of finished individual implants from metal powders. The most correct is the direct production by selective laser melting of metal implants, thereby reducing the number of manufacturing stages. However, from laboriousness view obtaining a certificate-permission to use this method for medical purposes, it can not be used in the short term. But there remains an open question of the

Content from this work may be used under the terms of the Creative Commons Attribution 3.0 licence. Any further distribution of this work must maintain attribution to the author(s) and the title of the work, journal citation and DOI.
Published under licence by IOP Publishing Ltd

possibility of casting a three-dimensional mesh structure of the implant body [6,7]. Therefore, the task was set, to compare and find the optimal method of manufacturing individual cervical telescopic implants. Namely, how many complex designs can be obtained with the help of SLM-technology and the combined technology of casting burn-out SLA- and DLP-models.

One of the common injuries is spinal injury. The greatest danger is a fracture or dislocation of the cervical spine. It is important to replace the damaged vertebra correctly. This is achieved only with the installation of individual telescopic implants, with the possibility of changing the upper platform location in the vertical and horizontal directions. Therefore, the comparison of SLM and the combined method of casting SLA- and DLP-models was carried out in the work.

Main part

To create an individual implant, you need to obtain digital data of the fracture site or dislocation. The surgeon decides to install a telescopic cervical implant. Obtaining a three-dimensional model of the fracture site is possible only in installations receiving a object layered structure, such type of equipment include tomographs. Digital data obtained with CT, MRI are needed to create a three-dimensional virtual model. Determine the location of installation and cervical implant attachment. Only after receiving this information will it be possible to begin modeling the geometry and calculating the component dimensions of the telescopic implant. With 3D Slicer software, the CT data was processed and 3D model of the patient cervical spine in STL format was created. In the MeshMagic program, we load the STL file of the fracture site and standard size models-blanks of telescopic cervical implants. We adapt the workpiece model to the fracture site. As a result we have individual three-dimensional virtual implant model in the STL format, presented in Figure 1.

Fig. 1. A mesh virtual model of the cervical implant assembly.

To produce a burned-out model, we used a 3d printer ProJet 1200 equipped by pico projector with a resolution 585 dpi in the X-Y, and Z-step height of 30 μm. This setting applies to the DLP (Digital Light Processing) family of 3d printers. At the heart of the DLP-system is a special device - a DMD-chip (Digital Micromirror Device). This is a complex structure that belongs to the so-called class of microelectronic mechanical systems (MEMS - Micro-Electronic Mechanical System). As a source of light can act as a light bulb (incandescent, luminescent or LED), and lasers. The wavelength extends from the ultraviolet to the infrared range. The device has the form of a micromirror matrix. Each micromirror is 10 microns in size. They are able to reflect both the invisible and visible spectrum of light and reflect light in one of two directions. The direction is determined by the angle of rotation. Direction itself is set by loading the bit "0" or "1" in the memory location. Each cell receives an independent bit stream with a frequency of several kilohertz, as a result of which we have a useful image on one of the outputs, and at the other output there is a light absorber. The GeomagicPrint program splits the printed object into layers with the specified thickness. A photopolymer is poured into the printer's bath with a transparent bottom. At the bottom of the bath the metal substrate is immersed, receding from the bottom to one (first) layer of the product. It turns out in the gap between the substrate and the bathroom is a liquid photopolymer. The projector, located under the bathroom, projects the picture of the first layer onto the bottom of the bath and, thanks to UV radiation, only the plastic that gets the image from the projector freezes. Then the working table rises one more layer and again a new layer is glued, which is attached to the previous one. As a result, a three-dimensional object from the liquid photopolymer grows layer by layer. The material used was the VisiJet FTX Green photopolymer from 3D Systems. It has the following properties: heat resistance - up to 93 ° C, dynamic viscosity (mPa s): 100 - 130 (30 ° C), burn-out - ash residue 0.01%.

The technology of laser stereolithography is based on photoinitiated laser radiation or the emission of mercury lamp polymerization of the photopolymerizable composition (PPC). In a container with a liquid photopolymer, a mesh platform (elevator) is placed on which the prototype is "grown". Using this technology, a three-dimensional object projected on a computer is synthesized from liquid PPC by successive thin (0.03-0.2 mm) layers formed by laser radiation on a moving platform. The 3D model of the object is preliminarily converted from the STL-file format into a set of stratified sections with the required height step, the array of which is recorded in the executable file with the SLI extension. This file is a set of two-dimensional vector data that provides sequential laser beam orientation control by means of mirrors in the process of object synthesis, commands to turn on the laser and move the platform. A laser is activated that acts on those parts of the polymer that correspond to the target object walls, causing them to solidify. After that, the whole platform is dipped a little lower, by an amount equal to the thickness of the layer. After the printing is completed, the resulting object is immersed in a bath with special compositions to remove excess and cleaning. And, finally, the final irradiation with powerful ultraviolet light for final hardening. Like many other 3D prototyping techniques, SLA requires the construction of support structures that are manually deleted when printing is complete.

On the RSPro 600 (Uniontech) and ProJet 1200 printers, mesh samples of polymeric telescopic cervical implants were grown on simulated models, the photographs of which are shown in Figure 2.

a b

Fig. 2. a) gating system from SLA-models, b)33 gating system from DLP-models.

The construction of the gating system begins with the establishment of each unit according to the feeder. Elements of the implant are supplied with sprue channels with a diameter of 5 mm. The channels number depends on the number of units cast. After building a gating system, it must be weighed. Based on the known mass of wax, the required amount of metal is determined. The calculation is made by the following formula: $M_{Me} = m_{wax} * 10 + 5$ g. Starbond CoS dental metal was used in the experiments, the following composition of Co - 59%, Cr - 25%, W - 9%, Mo - 3.5%, Si - 1%, other components (C, Fe, Mn, N) - maximum 1.5%. Ultimate tensile strength (Rp0.2) 650 MPa. The tensile strength is 910 MPa. The maximum elongation is 8%. The modulus of elasticity is 200 GPa. Vickers hardness 280 HV 10. Density 8,8 g / cm^3. Solidus-liquidus interval is 1305 - 1400 ° C. The casting temperature is 1500-1550 ° C. The thermal expansion coefficient is 14.0 μm / m. ° C.

In the second stage, the Yeti Expansion refractory molding (90 g) is prepared and kneaded with Yeti Liquid (22/20 ml) + distilled water (0/2 ml). To avoid the formation of air bubbles, a vacuum mixer Fox 88 from OMEC (Italy) was used. The kneading time was 30 seconds, the blade rotation speed was 400 rpm, and the pressure in the bowl was 0.8 Bar. The casting of the gate was made on a vibration table VB 1.1 with 2 vibration frequencies of 3000/6000 min^{-1}. After the molding mass has been filled in, wait 30 minutes before the molding material solidifies completely. The resultant hardening of the molding mass should be heated in the muffle furnace. In the experiments, the method of shock heating of the flask was used. In particular, the Programix 50 Ugin oven was preheated to a temperature of 900 ° C. A mold was used with an X6 size, the minimum warm-up time was 60 minutes. To effectively cast the metal, the crucible with the metal is also heated in a muffle furnace. In the next step, the heated crucible is placed in a casting chamber. The metal is heated by means of high frequency currents. Further, from the muffle furnace, a heated flask is extracted and vertically fixed to the crucible cover and pressed down from above with a clamping mechanism. After that, the vacuum chamber cover is closed, and the evacuation process of air begins with simultaneous high-frequency metal heating. During the heating process, the metal melts, resulting in the formation of a molten metal in the crucible, which is covered with an oxide film on top. At the rupture moment of the oxide film, the caster performs a casting process, namely, a vacuum-casting chamber is rotated, and compressed air is pumped at a rate of 3 bar/s to a pressure of 3.6 bar. As a result of the overturn, the metal from the crucible is poured into the flask and pressurized by excess pressure. The pressure provides the possibility of spilling metal into channels with a thickness more than 0.1 mm. In the experiments, a fore-casting installation was used. The vacuum-casting chamber is in an inverted state for 30 seconds, during which time the metal cools and crystallizes. Molding box must be cooled at room temperature, and mechanically clean the metal frame from the molding mass. A sandblasting unit was used, in which sand particles stream with a grain size of 250-300 μm, bombarded the surface of the sprue system and cleared the molding mass.

Figure 3 shows cast specimens of individual telescopic cervical implants.

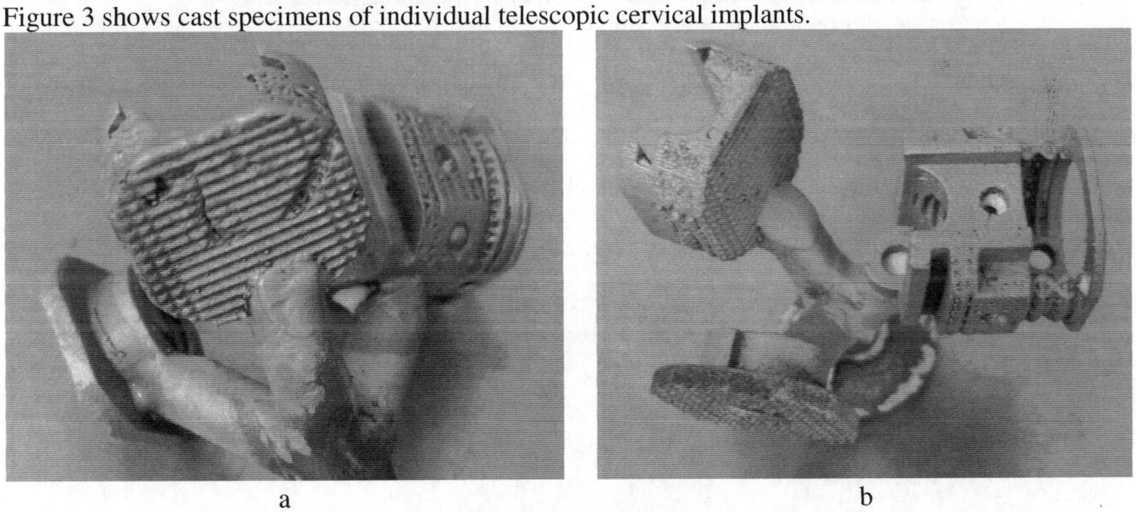

a b

Fig. 3. Castings of individual telescopic cervical implants
a - SLA models, b - DLP models.

At casting wire mesh there are two problems: the difficulty of molding associated with the air extraction from mesh volume and the difficulty of cleaning the molding mass is a mesh space in the implant volume. This shows the relevance of creating new, more efficient methods for pouring the molding mass, which would allow it to effectively penetrate the interplanetary space of the implant. Otherwise, the formation of an air cavity in the intersection space and its subsequent filling with liquid metal is observed. The second demand is to develop an effective method of extracting the molding material after casting the metal. Without solving these two problems, casting implant mesh is difficult to implement.

To create metal prostheses in experiments, a Selective Laser Melting system ProX 300 3D System was used. Figure 4 shows a diagram of the inner chamber, whose length is 1500 mm, height 800 mmThe chamber is hermetic and allows creating an atmosphere of inert gas, on the left is a metal roller for applying powder, it has the ability to move horizontally throughout the chamber. In the middle of the box with powder, inside which the movable piston is located, on the right there is a construction zone consisting of a movable piston with a platform and a laser system installed above it. The cycle of layer construction consists of the following stages: lifting of the central piston by 60 microns, descent of the working platform of the construction by 40 microns, movement of the roller through the central box with the transfer of excess powder to the working platform of construction, laser scanning of the powder. During the scan, combustion products can be generated, which are carried away by the flow of gas into the dust container.

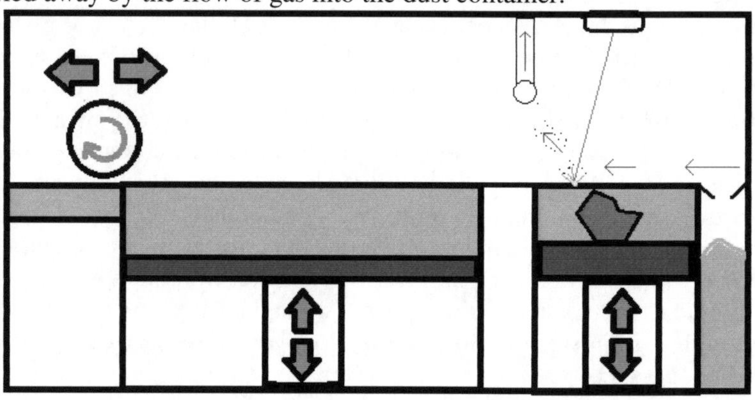

Fig. 4. Diagram of the working chamber ProX 300

The ProX 300 is equipped with a Nd:YAG fiber laser power - 500 W, with a wavelength - 1080 nm, the laser spot area - 20 μm. The size of the construction zone is x - 250 mm, y - 250 mm, z - 300 mm, minimum product size x = 100 μm, y = 100 μm, z = 20 μm. Laser power during the melting supports 100 W, during the melting product - 150 W, the scanning speed is 380 mm/s. Stainless steel powder 17-4PH, dispersed from 0.5 to 40 μm, from 3D Systems was used in the experiments.

Figure 5 shows prototypes of telescopic reticular implants obtained by Selective Laser Melting.

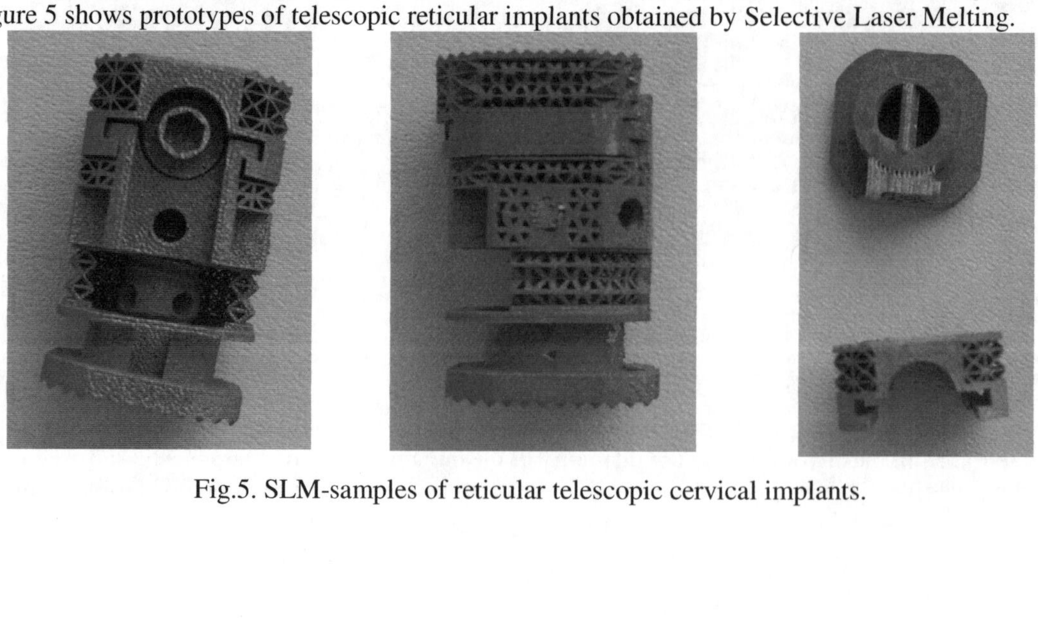

Fig.5. SLM-samples of reticular telescopic cervical implants.

The possibility of growing an article with a layer thickness of 40 μm made it possible to reproduce the geometry of the samples with an accuracy of 15 μm. The grown samples were cut off from the surface of the construction platform. Without additional processing, it was possible to assemble the implant, which is unattainable in casting technology. The removal of powder from the interwoven space is carried out by purging with compressed air. Thus, from the point of view of achieving the required geometric characteristics and constructing a network structure, the most optimal production method is selective laser melting. The limiting factor now is casting, not SLA and DLP-technologies, which are capable of reproducing the network geometry of the implant body.

Conclusion

Comparison selective laser melting technology and the combined casting technology of burned SLA and DLP models has shown the advantages of the first. The possibilities of using the SLA RSPro 600 (Uniontech) and ProJet 1200 for obtaining a burned model of a telescopic neck implant with a mesh structure were studied. The results of the model material influence on the molding mass are obtained. There is a cracking of the flask, which indicates the need to replace the "shock" heating with a "stepped". Experimental samples of cervical implants are cast on an Inducast plant. The ineffectiveness of this technology in the manufacture of mesh collapsible telescopic implants has been established, but it can be used for structures with a thickness of 1 mm. Using SLM technology allows you to get on the platform of construction up to 100 pieces per launch. That is much faster than casting on burned models and printing time of these models. However, the high cost of an SLM installation economically limits the scope of application.

This work was funded by the subsidy allocated to Kazan Federal University for the state assignment in the sphere of scientific activities (11.1300.2017/4.6).

References

[1] Kashapov R N, Kashapov L N, Kashapov N F 2017 *Analysis and development of methods for obtaining metallic powders for selective laser melting* IOP Conference Series: Materials Science and Engineering Vol 240 Is 1 **012071**

[2] Kashapov R N, Kashapov L N, Kashapov N F 2015 *The research of anodic microdischarges in plasma-electrolyte processing* IOP Conference Series: Materials Science and Engineering Vol 86 Is 1 **012019**

[3] Kashapov L N, Kashapov N F, Kashapov R N 2013 *Research of the impact acidity of electrolytic cathode on the course of the plasma-electrolytic process* Journal of Physics: Conference Series. Vol 479 Is 1 **012011**

[4] Kashapov L N, Kashapov N F, Kashapov R N 2014 *Influence of plasma-electrolyte discharge to the glass surface* Journal of Physics: Conference Series. Vol 567 Is 1 **012024**

[5] Kashapov L N, Rudyk A N, Kashapov R N 2014 *Applying 3D-printing technology in planning operations of cancer patients* IOP Conference Series: Materials Science and Engineering 69 **012016**

[6] Kashapov L N, Kashapov N F, Kashapov R N, Denisov D G 2016 *Plasma electrolytic treatment of products after selective laser melting* Journal of Physics: Conference Series Vol 669 Is 1 **012029**

[7] Kashapov R N, Korobkina A I, Platonov E V, Saleeva G T 2014 *The method of manufacture of nylon dental partially removable prosthesis using additive technologies* IOP Conference Series: Materials Science and Engineering Vol 69 Is 1 **012026**

IOP Conf. Series: Materials Science and Engineering **412** (2018) 012090 doi:10.1088/1757-899X/412/1/012090

3D scanning systems for monitoring the size of mesh implants

A A Kaleev[1], L N Kashapov[1], N F Kashapov[1] , R N Kashapov[1,2,3]
[1] Engineering Institute, Kazan Federal University, Kazan, 420008, Russia
[2] Laboratory of Radiation Physics, Kazan Physical-Technical Institute of the Kazan Scientific Center of the Russian Academy of Sciences, Kazan, 420029, Russia
[3] Kazan State Medical University, Kazan,420012,Russia

E-mail: kashramiln@gmail.com

Abstract. The work is devoted to 3D scanning capabilities investigation for controlling the geometric dimensions of cervical spinal implants. Models of cervical implants were modeled in the CAD program "SolidWorks" and were made on rapid prototyping installations RS 6000 and ProJet 1200. Polymer models of implants were used for casting on consumable patterns at the casting unit Inducast. Scanning of polymer models and metal castings was performed using the 3D scanning system Activity 850. The resulting parametric 3D model had a lot of defects and inaccuracies. However, the use of non-contact optical scanning methods greatly simplifies the measurement of the control dimensions of complex profile mesh implants.

Introduction

A rachial implant (Cage) is a special device implanted into the spine of an artificial device of various shapes and sizes that serves as a stabilizer, corrector of the direction and height of the spine, the replacement of individual elements of the spinal column, when they are removed and replaced. These implants should be as close as possible to the spine. At the moment, the greatest number of implants are made using castings [1,2,3,4].

In the process of manufacturing a cage, there are various processes that affect the geometry of the implant. Distortion of the geometry of the device by more than 1 mm leads to rejection of the product. Standard measurement methods are not suitable for obtaining sufficiently accurate implant sizes. The presence of a complex geometric shape makes it difficult to measure the reference dimensions and determine its exact geometry, so the purpose of this work was to explore the possibilities of 3D scanning, the accuracy of the 3D scanner measurement, the comparison of sample geometry after manufacturing on a 3D printer and casting [5,6,7].

Main part

To analyze the scanning capabilities, dental scanner Smart Optics Activity 875 is selected.

Smart Optics Activity 875 is a desktop optical 3D scanner with enhanced performance. The device is designed to produce high-precision digital models used in dental prosthetics. In accordance with its purpose, the scanner has a high resolution, reaching 0.01 mm. The scanning process is fully automated and takes from one to three minutes depending on the size of the object. The received data is exported as STL files.

This scanner does not need to be calibrated when the scan object is changed, and the scanning process itself is automated. Activity 875 is designed for scanning objects with a diameter of up to 85 mm, which is very suitable for us, since the maximum size of the cage does not exceed this value. The

Content from this work may be used under the terms of the Creative Commons Attribution 3.0 licence. Any further distribution of this work must maintain attribution to the author(s) and the title of the work, journal citation and DOI.
Published under licence by IOP Publishing Ltd

main functional elements of the device are the camera and the light source, which structures it in a special way and directs it to the object being scanned. The essence of structured light technology lies in projecting light pattern onto an object and fixing it, analyzing its deformation. The light flux is projected onto a subject by several kinds of light sources: LCD, video projector, diodes, halogen lamps. In this case, Activity 875 uses the LED and camera. [1]

To simulate the cage, the program "SolidWorks" was used (Fig. 1). The walls of the body are a mesh structure, which reduces the weight of the implant. After 3D-cage modeling, the manufacturing process begins.

Two rapid prototyping installations have been chosen for the cage making: RS 600 and ProJet 1200.

Figure 1 - 3D-model of cage.

Figure 2 - Analysis of the geometry of implant number 1 in the program "Geomagic Studio".

UnionTech RS 6000 is an industrial installation for three-dimensional printing using laser stereolithography (SLA). The device is equipped with a working chamber of a large size, allowing you to create models up to 600x600x400mm.

3D ProJet 1200 model printer uses the patented 3D Systems Micro-SLA technology. This technology exactly coincides with the technology of stereolithography (SLA), which is characterized by the highest accuracy of printing. Micro-SLA differs from it only in that the product is created layer by layer from top to bottom, that is, "set", and not in a cell with light-sensitive resin. Resin in this case is applied from the cartridge layer by layer. Since such a method is possible only in conditions of a small working chamber, it was called Micro-SLA.

After the prototyping of the cages, these products were scanned on the Optical 3D Scanner Activity 875. To analyze the geometry of the digital cage model manufactured with the RS 6000 and compare the reference dimensions with the 3D model modeled with the SolidWorks program. Geometry comparison took place in the program "Geomagic Studio". The data of this analysis can be seen in Figure 2.

When scanning implants, difficulties arose with holes and threaded elements. The implant has a shiny surface, so it needs to be processed by the developer, to give a matte surface. In this work, the developer "Sherwin D-100" was used. The influence of the developer on the measurement error is insignificant from 0.01 to 0.1mm. The conditions that were mentioned above could together influence the deviation of the geometry, but it is insignificant, since the maximum value is 0.9 mm, the average is 0.14 mm. The smallest deviation occurred precisely when scanning this implant, from which it can be concluded that the accuracy of manufacturing the RS 6000 is higher than that of the ProJet 1200.

The next geometry comparison takes place between the digital implant model made on the ProJet 1200 rapid prototyping unit and the 3D model modeled with the SolidWorks software. The physical model of the implant has not only a shiny surface, but also a dark green color. All these parameters have a poor effect on the quality of the scan. To improve the scanning process, it is also necessary, as in the first experiment, to process the surface with the developer "Sherwin D-100". Geometry analysis was also carried out in the "Geomagic Studio" program. The results are similar to the first experiment (Figure 3). The maximum deviation is 0.9 mm, the average deviation is 0.14 mm.

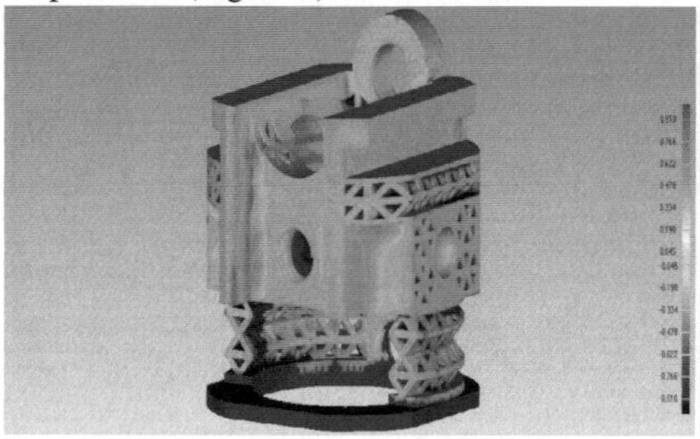

Figure 3 - Analysis of implant geometry number 2 in the program "Geomagic Studio".

Figure 4 – Digital model of cast.

Consumable patterns, created on stereolithographic equipment, replace the traditional wax models. The use of SLA technology makes it possible to quickly produce the most accurate casting products from metal with the best surface quality in comparison with other technologies.

The printed prototypes are coated with a high-temperature suspension in several layers to create a shell shape. Further, the models are burned out at high temperature and molten metal is poured. After curing the metal, the sheath is removed and the casting is polished. Casting cage, in contrast to the implant, manufactured on the RS 6000, has unevenness. Cutouts located on the main part of the implant turned out to be inaccurate. As a result, the surface does not have through cutouts, since the metal completely filled the holes during the casting of the casting.

The metal casting as well as the previous two prototypes was scanned using Activity 850. Then the resulting point clone was processed in the "Geomagic Studio" program for obtaining a polygonal model and comparing the casting with the prototypes, and for quality control. Analysis of the scanned cast showed small deviations, an average of 0.2 mm (Figure 4).

The maximum deviation was 1.3 mm. Such a large deviation is due to the unavailability of the 3D scanner to handle holes and small threaded elements. The minimum deviation was 0.02 mm.

Conclusion

As a result of the work, it was shown the possibility of improving the casting technology, using a consumable pattern, digital scanning with structured light and 3D printing. The equipment and software used influence the quality of the burned-out model. In those cases where it is required to transfer small parts, it is necessary to use equipment and software with the appropriate characteristics.

Shape comparison, surface quality comparison of the consumable pattern and the cast part showed that the surface of the casting has the greatest distortion among all the implants used in this article. Investment casting does not make it possible to obtain mesh structures in implants.

This work was funded by the subsidy allocated to Kazan Federal University for the state assignment in the sphere of scientific activities (11.1300.2017/4.6).

References

[1] Kashapov R N, Kashapov L N, Kashapov N F 2017 *Analysis and development of methods for obtaining metallic powders for selective laser melting* IOP Conference Series: Materials Science and Engineering Vol 240 Is 1 **012071**

[2] Kashapov R N, Kashapov L N, Kashapov N F 2015 *The research of anodic microdischarges in plasma-electrolyte processing* IOP Conference Series: Materials Science and Engineering Vol 86 Is 1 **012019**

[3] Kashapov L N, Kashapov N F, Kashapov R N 2013 *Research of the impact acidity of electrolytic cathode on the course of the plasma-electrolytic process* Journal of Physics: Conference Series. Vol 479 Is 1 **012011**
[4] Kashapov L N, Kashapov N F, Kashapov R N 2014 *Influence of plasma-electrolyte discharge to the glass surface* Journal of Physics: Conference Series. Vol 567 Is 1 **012024**
[5] Kashapov L N, Rudyk A N, Kashapov R N 2014 *Applying 3D-printing technology in planning operations of cancer patients* IOP Conference Series: Materials Science and Engineering 69 **012016**
[6] Kashapov L N, Kashapov N F, Kashapov R N, Denisov D G 2016 *Plasma electrolytic treatment of products after selective laser melting* Journal of Physics: Conference Series Vol 669 Is 1 **012029**
[7] Kashapov R N, Korobkina A I, Platonov E V, Saleeva G T 2014 *The method of manufacture of nylon dental partially removable prosthesis using additive technologies* IOP Conference Series: Materials Science and Engineering Vol 69 Is 1 **012026**

IOP Conf. Series: Materials Science and Engineering **412** (2018) 012091 doi:10.1088/1757-899X/412/1/012091

Plasma electrolyte produce 17-4PH powder for use in 3D MicroPrint technology

R N Kashapov[1,2], **L N Kashapov**[1], **N F Kashapov**[1]

[1] Engineering Institute, Kazan Federal University, Kazan, 420008, Russia
[2] Laboratory of Radiation Physics, Kazan Physical-Technical Institute of the Kazan Scientific Center of the Russian Academy of Sciences, Kazan, 420029, Russia

kashramiln@gmail.com

Abstract. The work is devoted to the investigation of the processes of obtaining fine powders from steel 17-4PH, for the purpose of their further application in 3D MicroPrint technology. This technology is a type of selective high-resolution laser melting of ~ 30 μm. In 3D MicroPrint technology, a powder of less than 5 microns is used, which in turn necessitates the development of new cheap methods for obtaining spherical fine powders. The process of plasma-electrolyte powder production of 17-4PH steel was studied, the discharge burning conditions and conditions for the arrangement of the electrode system for obtaining powders with dimensions less than 5 μm were selected. The proposed method is simple and does not require expensive equipment.

Introduction

The technology of selective laser melting proved itself as capable of manufacturing complex metal products in one production cycle. It is layered formation that makes it possible to manufacture complex and hollow products that can't be done by traditional methods. Despite all the advantages, SLM technology has drawbacks due to high surface roughness, cracking, distortion of geometry and the inability to grow parts with a wall thickness of less than 140 μm. At the moment, commercial machines have a high level, they have eliminated the shortcomings associated with the formation of cracks and distortion of geometry. All this allowed us to start creating factories of additive production. It is the ability to form complex geometric products that makes this technology economically viable and competitive by combining a large number of operations in a single step. The technology of laser melting isn't used only in the manufacture of prototypes, but serial products of small batches.

The need for engineering in the creation of micron products with a microscopic structure only increases. Therefore, part of the research in the field of additive technologies is aimed at creating new devices capable of producing micron-sized parts by selective laser fusion. There are works by the authors Chen J. [1], Exner H. [2], Kathuria Y.P. [3], in which laser fusion of various powders with pulsed and q-switched pulse regimes was investigated. And in the work of Regenfuss P. and the authors of [4,5,6] a new experimental setup was presented, which allows to create details with a resolution of 30 μm and a roughness Ra of 1.5 μm. The installation consisted of a sintering

Content from this work may be used under the terms of the Creative Commons Attribution 3.0 licence. Any further distribution of this work must maintain attribution to the author(s) and the title of the work, journal citation and DOI.
Published under licence by IOP Publishing Ltd

chamber from which a turbomolecular vacuum pump was pumped out, ScanLab scanning system of 25 * 25 mm in size, a Q-switched Nd: YAG laser (λ = 1,064 μm) with an output power of 0.1-1 W in TEM mode and 0.5 - 100 kHz pulsed. In the installation there are three pistons, one - serves as the platform on which the part is grown, and it gradually moves downward, the remaining two are necessary for feeding the powder to the construction zone. These pistons go up and squeeze out the powder, which is carried by a knife to the working platform. As a result of the research, miniature products were obtained, which were tested in practice.

At Formnext 2017 (Frankfurt / Main), 3D MicroPrint GmbH presented the first DMP63 commercial machine capable of producing products with the above listed properties. This unit is equipped with a fiber laser power of 50 W, which works in CW and pulsed mode. Scanning is performed using a high-speed galvanometer, the laser spot diameter is 30 μm, and the device has a built-in laser power measurement system. The size of the platform is 60 * 60 mm, the height of construction is 30 mm. The particle size of the powder used is D90 ≤ 5 μ m, the layer thickness of the applied powder is ≤ 5 μ m, the tolerance of the dimensional deviation is ± 5 μ m by 10 mm. Selective laser melting takes place in an inert argon medium, the gas flow rate is 30 liters per hour, the concentration of oxygen and water vapor is less than 0.1 ppm. Comparison of the claimed characteristics of the obtained products with the capabilities of other machines shows the advantages in the ability to make thin-walled structures with a thickness of 32 microns, a roughness of Rz 13 μm, Ra 2 μm and a bulk density of 99.9%. However, for the operation of this machine, strict requirements are imposed on the powder, namely its dimensions D90 ≤ 5 μ m and the geometry - the spherical shape. The expected widespread use of such DMP63 in the production of precision medical products, aerospace, automation, electronics, energy and chemical industries requires us to develop new alternative methods for obtaining submicron powders. Existing methods of gas, plasma, water atomization make it possible to obtain powders in which the fraction of particles smaller than 5 μm is less than 1%. This shows the ineffectiveness of existing methods. A solution can be found in the joint use of plasma methods and fluid systems. The phenomenon of combustion of a gas discharge with liquid electrodes is known. The use of this phenomenon made it possible to obtain powders with a dispersion up to 100 μm. But it is also known that by changing the combustion parameters of a gas discharge, it is possible to control the size of the particles produced.

The aim of the work is to investigate the possibilities of plasma electrolyte systems for obtaining submicron powders suitable for use in 3D MicroPrint technology.

Main part

The phenomenon of obtaining micro-nanoparticles of metals and their oxides in a discharge with a liquid electrode has become known recently and is actively developing now [7]. However, nanoparticles obtained by this method are partially oxidized, agglomerated and have a wide spread in size. In [8, 9], a method of obtaining nickel oxide particles by Contact Glow Discharge Electrolysis (CGDE) was investigated when a metal electrode is located inside the electrolyte and the discharge burns in a vapor-air shell. In this case, the discharge combustion process takes place under the following conditions: U ~ 30-42 V, j ~ 0.1 A / mm^2. The nanoparticles obtained were 20 to 200 nm in size. As can be seen, there is a large enough spread. A method for obtaining micro- and nanoparticles using Glow Discarge Electrolysis (GDE) was also performed when the metal electrode is located above the surface of the electrolyte and they do not touch. In this way, Ag particles were obtained in [10, 11]. In work [12] it is assumed that it is possible to control the dimensions of nanoparticles by using a rotating electrode.

During the combustion of a discharge with liquid electrodes, various physical phenomena occur, including the formation of shock waves [13], while the electrolyte properties [14] can radically change the combustion process. Surface polishing [15] or the formation of a microrelief can occur. In [16, 17,18], a positive effect was obtained for the production of powder microparticles with sizes up to 100 μm. All this suggests that a gas discharge can be used to produce particles smaller than 5 μm.

IOP Conf. Series: Materials Science and Engineering **412** (2018) 012091 doi:10.1088/1757-899X/412/1/012091

The experimental setup was used in the work, the functional scheme of which is shown in Figure 1. The installation consists of an electric power supply system 1, an electrolytic bath 2, an electrode system 3, an oscilloscope 4, an additional resistance 5, a voltmeter 6, an ammeter 7, thermocouple - 8. With the help of the electrode system, the distance between the anode and the electrolyte solution was monitored. With the help of an oscilloscope 4, the shape of the applied voltage and current was monitored, the voltmeter and ammeter were used to measure the voltage and discharge current. The electric power supply system represents a high-voltage constant current source for creating and maintaining electric discharge burning with smooth regulation of the output voltage in the range from 0 - 3 kV and current 0-10A. Measurement of voltage and discharge current was carried out using two digital universal measuring devices MMH-930 and APPA 109N, the relative error of measurement is 0.8%.

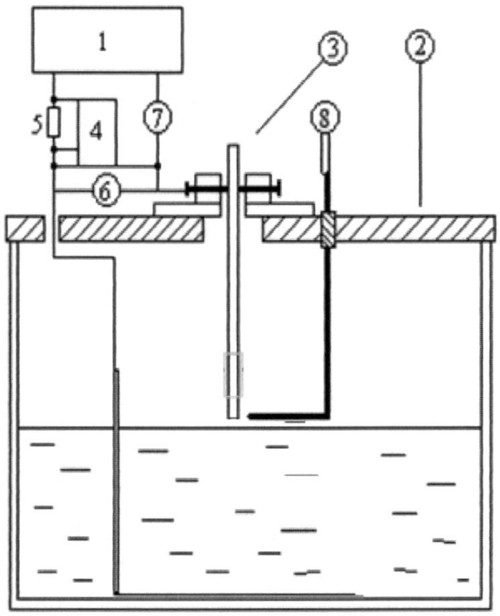

Figure 1 - Diagram of the experimental setup

The combustion of a gas discharge occurs between a metal anode made of 17-4PH steel and an electrolytic cathode. The anode is a metal cylinder with a diameter of 5 mm, located above the surface of the electrolyte at a height of 1 to 5 mm. As the electrolytic cathode, aqueous solutions of NaCl and Na_2CO_3 were used with a concentration of 0.1-1% by weight.

The main parameters of plasma-electrolyte steel powder production are voltage, current, discharge power, used for heat generation, physical and chemical properties of the metal anode. The power put into the discharge is determined by the current-voltage characteristic, on the basis of which it is possible to estimate the energy contribution. All these factors affect the size of the particles obtained [16]. In our work, we added one more factor - the rotation of the electrode. This technique was previously used in [12] to obtain Pt and Au nanoparticles with dimensions of 10 -150 nm. Thus, by increasing the frequency of rotation of the electrode, we must come to reduce the size of the particles obtained.

IOP Conf. Series: Materials Science and Engineering **412** (2018) 012091 doi:10.1088/1757-899X/412/1/012091

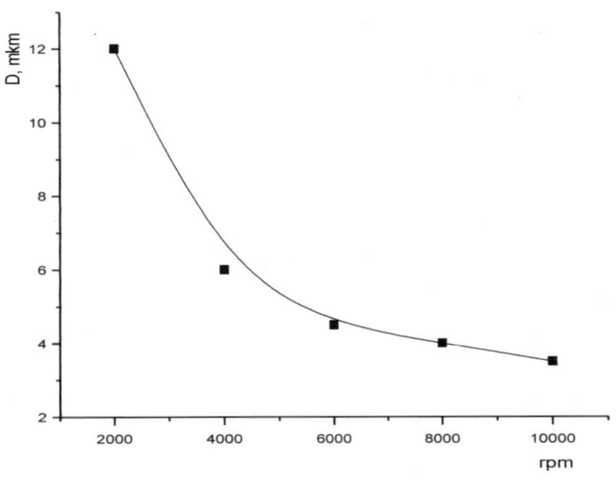

Figure 2 - Dependence of particle size on electrode rotation frequency

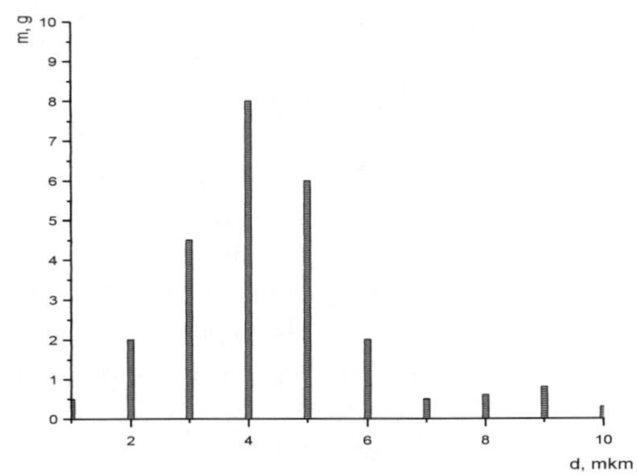

Figure 3- Powder distribution histogram

The significant effect on the discharge combustion process is exerted by the electrolyte concentration, it is established that the electric field strength decreases with increasing concentration. Therefore, for the stability of the results, solutions with a concentration of less than 1% were used.

The formation of the powder occurs when the surface temperature of the anode reaches the appropriate solidus temperature. When the liquidus temperature is reached, the electrode is melted and large drops of metal are formed. Proceeding from this, the powder can be produced in the temperature range of the liquidus and solidus anode. An increase in temperature will lead to an increase in the productivity of the powder, but with an increase in the particle size of the powder. A linear increase in the average particle size with increasing anode temperature is established. With an increase in the rotation frequency of the electrode, a nonlinear decrease in the dimensions of the produced particles is observed. Figure 2 shows the dependence of the particle size on the rotational speed. At a maximum speed of 10,000 rpm, the average particle size was 3.5 μm. Reducing the speed to 2000 rpm leads to a mean particle size of 12 microns.. Figure 3 shows the histogram of the granulometric composition of the powder obtained. The largest amount of powder is obtained by a size of less than 4 μm. The powder has a spherical shape, the smallest particle size is 0.1 μm. The resulting powder should be suitable for use in selective laser melting plants using 3D MicroPrint technology. However, further research is needed to increase the productivity of the process by increasing the anode area and determining the powder dynamic characteristics.

Conclusion

Studies of the combustion of a gas discharge between a metal electrode of grade 17-4PH and liquid electrodes with a rotating electrode have shown the possibility of obtaining a spherical powder with a particle size of 3.5 μm in average. The influence of the gas discharge parameters on the process productivity and the size of obtained particles are determined. In the voltage range from 500 to 800 V, particles of less than 12 μm in size with a productivity of $1 \cdot 10^{-2}$ g/s occur. Above this range, the electrode melts and the process for obtaining the powder ceases. The electrode rotation can be considered as another effective tool for controlling the size of the particles produced.

This work was funded by the subsidy allocated to Kazan Federal University for the state assignment in the sphere of scientific activities (11.1300.2017/4.6).

IOP Conf. Series: Materials Science and Engineering **412** (2018) 012091 doi:10.1088/1757-899X/412/1/012091

References

[1] Chen J, Wang X and Zuo T 2003 *The micro fabrication using selective laser sintering micron metal powder* Proceedings of SPIE Vol 5116 pp 647-51
[2] Exner H, Regenfuss P, Hartwig L, Klotzer S and Ebert R 2003 *Microsintering of miniature and precise components and tools* Proceedings of the Euro-uRapid, Frankfurt/Main
[3] Kathuria Y P 1999 *Microstructuring by selective laser sintering of metallic powders* Surface and coating technology Vol 116-119 pp 643-47
[4] Regenfuss P, Hartwig L, Klotzer S, Ebert R and Exner H 2004 *Microparts by a novel modification of selective laser sintering* SME Technical paper TP04PUB185
[5] Regenfuss P, Hartwig L, Klotzer S, Ebert R, Brabant Th, Petsch T and Exner H 2005 *Industrial freeform generation of microtools by laser micro sintering* Rapid Prototyping Journal Vol 11 No 1, pp 18-25
[6] Regenfuss P, Streek A, Hartwig L, Klotzer S, Brabant Th, Horn M, Ebert R, and Exner H 2007 *Principles of laser micro sintering* Rapid Prototyping Journal Vol 13 Number 4 pp 204-12
[7] Torible Y et al 2007 *Applied Physics Letters* 91(4) **041501**
[8] Allagui A et al 2011 *Electrochemica acta* 58 pp 12-18
[9] Allagui A et al 2014 *Journal of Power Sources* Vol 262 pp 178-82
[10] Richmonds C et al 2008 *Applied Physics Letters* 93 **131501**
[11] Chang F-C et al 2010 *Journal of Vacuum Science and Technology* A28(4)
[12] Lal A et al 2008 *Electrochemistry communications* 10 pp 488-91
[13] Denisov D G, Kashapov N F, Kashapov R N 2015 *The appearance of shock waves in the plasma electrolytic processing* IOP Conference Series: Materials Science and Engineering Vol 86 Is 1 **012005**
[14] Kashapov L N, Kashapov N F, Kashapov R N, Denisov D G 2016 *Plasma electrolytic treatment of products after selective laser melting* Journal of Physics: Conference Series Vol 669 Is 1 **012029**
[15] Kashapov L N, Kashapov N F, Kashapov R N 2013 *Research of the impact acidity of electrolytic cathode on the course of the plasma-electrolytic process* Journal of Physics: Conference Series. Vol 479 Is 1 **012011**
[16] Kashapov R N, Kashapov L N, Kashapov N F 2017 *Research of plasma-electrolyte discharge in the processes of obtaining metallic powders* Journal of Physics: Conference Series Vol 927 Is 1 **012086**
[17] Kashapov R N, Kashapov L N, Kashapov N F 2017 *Analysis and development of methods for obtaining metallic powders for selective laser melting* IOP Conference Series: Materials Science and Engineering Vol 240 Is 1 **012071**
[18] Kashapov R N, Kashapov L N, Kashapov N F 2015 *The research of anodic microdischarges in plasma-electrolyte processing* IOP Conference Series: Materials Science and Engineering Vol 86 Is 1 **012019**

IOP Conf. Series: Materials Science and Engineering **412** (2018) 012092 doi:10.1088/1757-899X/412/1/012092

Development of technology for high-speed branding of titanium alloys

R R Farahov[1], N M Lyadov[2], R N Kashapov[1,2], L N Kashapov[1] and N F Kashapov[1]
[1]Engineering Institute, Kazan Federal University, Kazan, 420008, Russia
[2]Laboratory of Radiation Physics, Kazan Physical-Technical Institute of the Kazan Scientific Center of the Russian Academy of Sciences, Kazan, 420029, Russia

kashramiln@gmail.com

Abstract. Electrochemical stamping produces an oxide film on the titanium surface, which slows down the stamping process. In this paper, the task was set to develop technological methods that accelerate the process of stamping titanium alloys. Studies have shown that markings obtained from an electrolytes given in the work and according to the regime described in the work are stable, clear, contrasting, different in color from the base metal. The resulting marks have withstood repeated sterilization by boiling in water and a moisture resistance test.

Introduction

In industry, the technology of marking has found wide application, almost all types of products are stamped [1,2,3]. This is necessary to identify the manufacturer of the product and protect it from counterfeiting.The technology of laser marking became very popular, because of the low cost of creating an image. However, the development of laser technology did not lead to the disappearance of other methods of stamping, such as electrochemical and impact stamping.Electrochemical stamping technology has high productivity and has been recommended in mass production, where individual marking is required for each part [4]. This technology is well developed for stamping steels, but there is a problem with working with titanium alloys. Oxide film, which slows the process of stamping, is formed on the surface of titanium during electrochemical stamping[5,6,7,8]. The task of developing technological methods that accelerate the process of stamping titanium alloys was posed in this paper.

Main part

As a result of the reactions that occur on the electrodes during the passage of an electric current, the electrons are released at the anode, and are bound at the cathode. Anode processes can proceed in two ways: with dissolution or without dissolution of the anode metalIn the case of a soluble anode, the oxidation reaction, which causes the passage of current at the anode, consists in the release of metal ions into the solution according to the general equation:

$$Me \rightarrow Me^{n+} + ne$$

Anodic dissolution of the metal occurs. The insoluble anode will not transfer any ions to the electrolyte. During the anodic oxidation reaction, the metal does not dissolve. The reaction required for the passage of current from the anode occurs at the boundary of the "metal-solution" phases. The releasing electrodes passing into the metal.Anodic oxidation reactions on insoluble anodes

Content from this work may be used under the terms of the Creative Commons Attribution 3.0 licence. Any further distribution of this work must maintain attribution to the author(s) and the title of the work, journal citation and DOI.
Published under licence by IOP Publishing Ltd

IOP Conf. Series: Materials Science and Engineering **412** (2018) 012092 doi:10.1088/1757-899X/412/1/012092

depend on the composition of the electrolyte and can have very different character. The release of oxygen on insoluble anodes, which passes through the equation in alkaline electrolytes, is of particular importance:

$$4OH^- \rightarrow 2H_2O + O_2 + 4e$$

and acid electrolytes by the equation:

$$4H_2O \rightarrow 4H^+ + O_2 + 4e$$

The behavior of the anodes depends on the anode potential. Many anode reactions can occur simultaneously. Often on soluble anodes with a low current density, there is only one reaction - dissolution of the metal. With increasing current begins to stand out oxygen, resulting in the original anode becomes from soluble to insoluble. As a result of the analysis of literature data and carried out experimental work on the marking of stainless steels, it was revealed that the most productive, providing high quality of the marked marks is the non-stencil method.

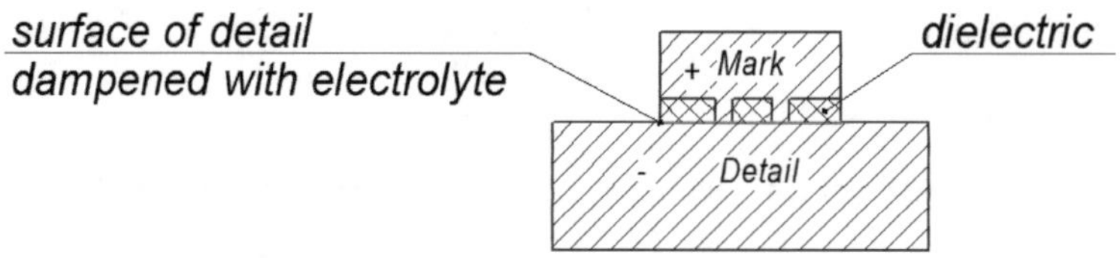

Fig.1. Diagram of the non-stencil method.

The main features of this method is the absence of an artificial integral (gap) between the part and the electrode.

Main factor that determines the process of branding, is a prerequisite of same surface levels of dielectric substrate and electrode. Electrodes with image of brand sign are made of a printing alloy, electrode is poured with an insulating material (one of the fast-hardening plastics: noracryle, styracryl, protakril, etakril, etc.). After solidification, insulating layer from mark of stucco is ground to the same surfaces level of the dielectric substrate and electrode. The process of marking is carried out by imposing head on surface of a DC source connected to various strips of electrolyte pre-wetted with electrolyte.

The process proceeds through following stages: at some initial instant of time an instantaneous maximum current of 3-5 A / cm^2 flows through electrolyte layer, which causes an instantaneous electrochemical reaction; on the parts surface the main metal oxides layer having a large resistance is formed, and a gas shell forms at electrode surface, which sharply reduces the electrical conductivity of interelectrode gap; initial value of the current drops sharply and decreases almost to zero within a short time; the process is terminated.

Formed as a result of an instantaneous electrochemical reaction, the metal oxides strongly bind to the base metal and form the sign itself.

In this case, during the marking process, practically no etching occurs and the signs obtained in this way have the same level with the surface of the parts. Titanium is one of the highly corrosive materials. This is explained by the presence on its surface of oxide films that are resistant in many corrosive environments. The basic electrochemical indices are such that titanium can be considered a metal having a high propensity to passivation than iron, nickel, chromium and stainless steels.

Titanium is stable in atmospheric conditions, waters of different composition, alkalis, solutions of many salts, some organic acids, food environments, oxidizing acids, etc. Corrosion resistance of titanium alloys with multiple boiling or autoclaving is much better than in chromium steels and nickel-base alloys.

IOP Conf. Series: Materials Science and Engineering **412** (2018) 012092 doi:10.1088/1757-899X/412/1/012092

In the experiment, we used: an electrochemical marking device, a sterilizer, an electric tile, a set of stamps, a stamping fixture, a climatic chamber 3001. For the experiments, the PME-2 instrument was modernized. The principle of the PME-2 device is based on the use of the anodic dissolution process. Structurally PME-2 is made as a separate portable device. Autotransformer, board with rectifier, board with relay, rectifier unit, zener diode, located on the chassis of the front of the device. On the front panel of the device there are installed: a voltmeter, changing the output voltage, an ammeter measuring the load current, a signaling device for switching the device on, a signal lamp indicating the end of the voltage supply to the output terminals, a delay time switch, an output voltage regulator, a switch, output terminals, delay time.The PME-2 device is a power rectifier consisting of an autotransformer Tp and a rectifying bridge at D_1-D_4; with the time relay assembled on the lamp L_2 and the control circuit on the relay P_2. The time relay, assembled on the lamp L_2, works as follows. When you press K_{H1}, relay P_1 is activated. At the same time, the load and the delay circuit are switched on via the contacts P_1^1 and P_1^2. The contact P_1^3 is then opened and the capacitor C_3 starts charging. When the lamp L_2 is ignited, the relay P_2 is triggered and for a moment opens its contact P_2

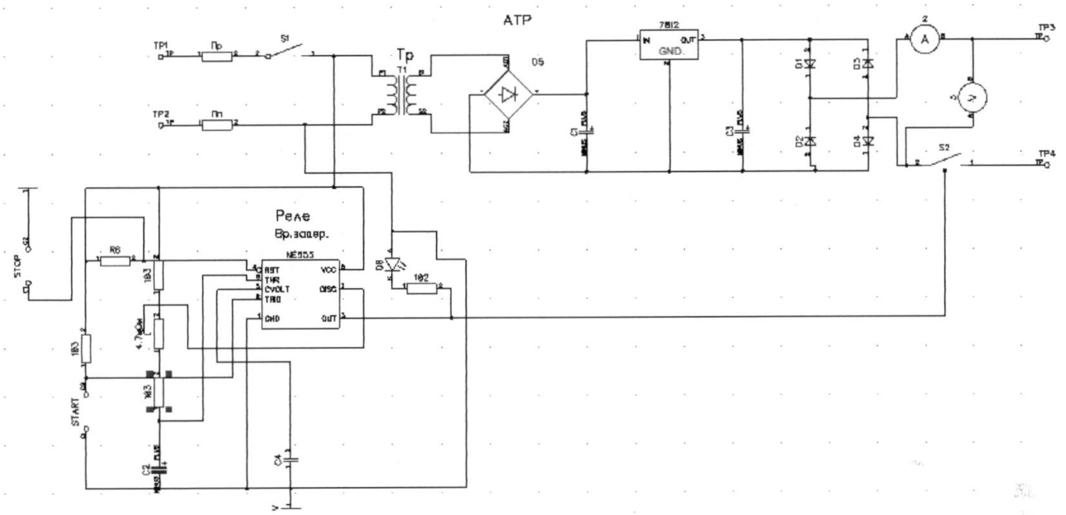

Fig.2. Electrical diagram of the modernized branding device.

Titanium staining by electrochemical method is difficult due to the ability of titanium to permanently passivate the surface. For this reason, the selection of the electrolyte constituted considerable difficulties. In past articles, there was an indication of the use of tap water as an electrolyte for branding titanium. Titanium labeling using tap water as an electrolyte was tested by us first. The received signs were clear, but very light and not sufficiently stable.

In this regard, work was carried out to find new electrolytes, giving a clear and contrast image. The experiments were carried out on plates of OT-4 alloy with preliminary mechanical polishing.

The composition of electrolytes is given in Table 1. Work was carried out on the selection of voltage and time of labeling with various electrolytes. It was found that for marking titanium alloys, an increased voltage is required compared to stainless steels. The voltage is 10-12 volts. This is due primarily to the presence of a passive film on the titanium surface, which has a significant ohmic resistance, as well as high ohmic resistance of the electrolytes themselves. Marking time is 0.1 sec.

Marking modes are shown in Table 1.

Table 1.The composition of electrolytes and the marking modes by electric method on the device.

№	Composition of electrolytes	Marking modes		Characteristics of the sign
		U, V	T, s.	

1	C_2H_5OH 10%, NaCl 0.2%, NH_4Cl 0.2%, distilled water.	10-12	0.1	The sign is dark brown or dark blue, clear
2	C_2H_5OH 10%, NaCl 0.2%, distilled water.	10-12	0.1	The sign is clear brown light
3	NaCl 0.2%, distilled water.	9-11	0.1	The sign is dark brown, clear
4	C_2H_5OH – 15%, NaCl0.2%, CH_4N_2O 0.4%, distilled water.	10-12	0.1	The sign is dark brown, clear
5	C_2H_5OH – 15%, Tap water	10-12	0.1	Sign clear, light brown
6	Tap water	10-12	0.1	Sign light, unstable

The development of additive production technology allows you to create individual things quickly and for short lines. The flagship of this technology is the technology of selective laser melting. To create a stigma-electrode this technology is very attractive. We used a ProX 300 machine with a pressing roller. The model of the three-dimensional stamp is loaded into the installation, into which it is divided into layers with a thickness of 40 μm. Each layer has its own profile. On the platform using a roller, a layer of metal powder 17-4 PH with a thickness of 40 μm is deposited. The laser scans the corresponding profile for each layer.

The grown product is cut off from the surface of the platform, and its working part is poured with the recrystal-65.

A series of stamps with the signs "USSR" and "Stainless" was tested. Tests have shown that sharp, contrasting signs are obtained when setting about 500 characters. When stating more characters, the image turns out to be vague, fuzzy, and the brand itself is rendered useless.

Studies have shown that when marking titanium, the resistance of the stamping electrode was lower than when marking stainless steels (500 characters for titanium and 5,000 marks for stainless steels).

But, considering that the stigma grown with the SLM installation can be re-polished and re-used, as well as the low cost of stamping, we can assume that when replacing the stamping method with electrochemical method, an economical effect will be obtained.

Studies have shown that the markings obtained from the tables given in Table 1. 1. electrolytes, and according to the above regime - stable, clear, contrasting, excellent in color from the base metal. The resulting marks have withstood repeated sterilization by boiling in water and a test for moisture resistance.

This work was funded by the subsidy allocated to Kazan Federal University for the state assignment in the sphere of scientific activities (3.9399.2017/8.9).

References

[1] Knaislová A, Novák P, Pruša F 2016*Preparation of Ti-Al-Si alloys by powder metallurgy* Manufacturing Technology 16(6) pp 1274-78

[2] Ong JL, Lucas LC, Raikar GN 1993*Electrochemical corrosion analyses and characterization of surface-modified titanium*Applied Surface Science

[3] Zhao H,Andrawes B 2017*Mechanical properties of NiTiNb shape memory alloy subjected to a harsh corrosive environment* Journal of Materials in Civil Engineering

[4] Kashapov R N, Kashapov L N, Kashapov N F 2017 *Analysis and development of methods for obtaining metallic powders for selective laser melting* IOP Conference Series: Materials Science and Engineering Vol 240 Is 1 **012071**

[5] Kashapov R N, Kashapov L N, Kashapov N F 2015 *The research of anodic microdischarges in plasma-electrolyte processing* IOP Conference Series: Materials Science and Engineering Vol 86 Is 1 **012019**

IOP Conf. Series: Materials Science and Engineering **412** (2018) 012092 doi:10.1088/1757-899X/412/1/012092

[6] Kashapov L N, Kashapov N F, Kashapov R N 2013 *Research of the impact acidity of electrolytic cathode on the course of the plasma-electrolytic process* Journal of Physics: Conference Series. Vol 479 Is 1 **012011**
[7] Kashapov L N, Kashapov N F, Kashapov R N 2014 *Influence of plasma-electrolyte discharge to the glass surface* Journal of Physics: Conference Series. Vol 567 Is 1 **012024**
[8] Kashapov L N, Kashapov N F, Kashapov R N, Denisov D G 2016 *Plasma electrolytic treatment of products after selective laser melting* Journal of Physics: Conference Series Vol 669 Is 1 **012029**

IOP Conf. Series: Materials Science and Engineering **412** (2018) 012093 doi:10.1088/1757-899X/412/1/012093

Use of polystyrene production wastes in selective laser sintering processes

J N Dementyeva[1], R N Kashapov[1,2], L N Kashapov[1], N F Kashapov[1]
[1] Engineering Institute, Kazan Federal University, Kazan, 420008, Russia
[2] Laboratory of Radiation Physics, Kazan Physical-Technical Institute of the Kazan Scientific Center of the Russian Academy of Sciences, Kazan, 420029, Russia

kashramiln@gmail.com

Abstract. The purpose of this study is to determine the possibility of using polystyrene waste in Selective Laser Sintering (SLS) technology. The powder was subjected to mechanical and chemical influences and was studied in optical and IR microscopes. Test samples were obtained on the sPro 60HD unit (3D Systems, USA).

Introduction

The efficient use problem of natural resources is an urgent task for mankind. Therefore, reduction of production waste solves two problems: saving natural resources and increasing the producer's profit. In the production of polystyrene granules, a large-dispersed powder with a large spread of particle sizes and irregular geometry is obtained as a waste. From waste, it is possible to obtain low-grade products, such as disposable tableware, industrial containers, Petri dishes, etc. Producers are interested in developing new methods of processing for the use of waste in the production of high-quality products.

Additive technologies are one of the modern production methods, where creation of a physical model occurs layer by layer in the CAD model. In technologies such as SLS, SLM, 3DP, powder compositions are used to create models. One of the most popular technologies is selective laser sintering (SLS). In this technology, a polymer or composite powder, sintered by an infrared CO_2 laser with a wavelength of 10.6 μm, is used to create models. In installations of Selective Laser Sintering, a uniform powder with dispersion up to 90 μm with good rheological properties is used. In comparison with other methods of additive production, SLS is highly versatile in terms of the consumables choice. This includes various polymers (polyamide 2200, polyamide 11, polyetheretherketone, polymethyl methacrylate, polycaprolactone, polystyrene [1,2,3], polymeric compositions with metal alloys (steel, titanium, precious metals, cobalt-chromium alloys) and sand mixtures. However, 95% of the powder used in the SLS technology is occupied by polyamide 12. The production uses high-quality and pure raw materials, which leads to a high cost of SLS-powder.

If analogy with standard foundry technologies for the production of polymer products is used, the use of polystyrene is more widespread. It is also known to use polystyrene in the SLS process to create

Content from this work may be used under the terms of the Creative Commons Attribution 3.0 licence. Any further distribution of this work must maintain attribution to the author(s) and the title of the work, journal citation and DOI.
Published under licence by IOP Publishing Ltd

models used in casting for shell molds [5,6,7]. Therefore, a goal of the work is to investigate the possibility of using the polystyrene production waste in Selective Laser Sintering processes.

Main part

Pesticides from Nizhnekamskneftekhim, the Republic of Tatarstan, Russia, were used as feedstock. Using the optical microscope Premiere WF 10X DIN18mm, the geometry of the particles was studied and their size was determined. The dispersion composition was determined by screening through sieves with different mesh sizes at the Vibrogroot GV30 installation. Figure 1 shows photographs of the raw material obtained by an optical microscope. Particles have a large dispersion in size, as well as an irregular geometric shape. Particles measuring from 350 μm to 500 μm were observed. Figure 2 shows the histogram of the particle distribution.

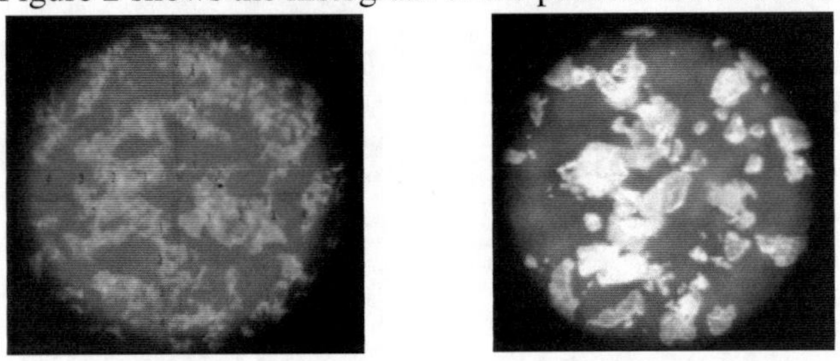

1 division = 100 μm 40x magnification

Figure 1 - Polystyrene starting powder under a microscope.

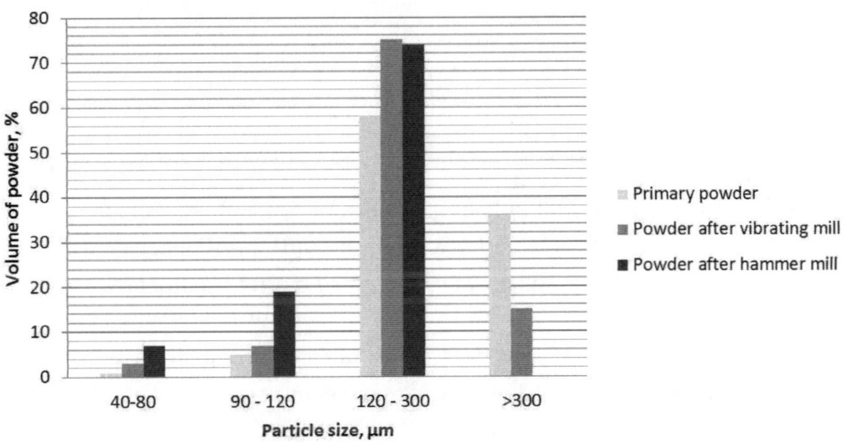

Figure 2 - Disperse composition of the powder.

It was found that the starting powder consist particles with a size from 120 μm to 300 μm. There are also large particles from 300 μm (about 36%).

To determine the fluidity, a glass funnel was used with an glass funnel diameter of 7 mm. A certain amount of material was poured into the funnel, the hole was opened and the time of the material expiration was measured. The original powder did not spill out of the funnel, which indicates its poor flowability, the presence of large particles and their irregular geometry.

To determine the elemental composition of the powder, a PerkinElmer Spotlight 400 infrared microscope was used. An IR imaging system with a fast scanning interferometer, a matrix detector, an ATR probe, and an ATR image system. Simultaneous operation in the IR and visible ranges.

IOP Conf. Series: Materials Science and Engineering **412** (2018) 012093 doi:10.1088/1757-899X/412/1/012093

Modes of operation: transmission, reflection, micro-ATR, ATR image. The spectral range is 7800 - 600cm^{-1} (scanning at a point); 7800 - 720 cm^{-1} (image mode). The geometric accuracy of the resulting image is ± 0.0013%, linearity> 99%. The signal / noise ratio is better than 50000: 1 (point scan), 800: 1 (image mode). The pixel size is 50, 25, 6.25 and 1.56 microns. Scanning time is 1.6 s. The rate of accumulation of spectra is 170 spectra of the full range in min.

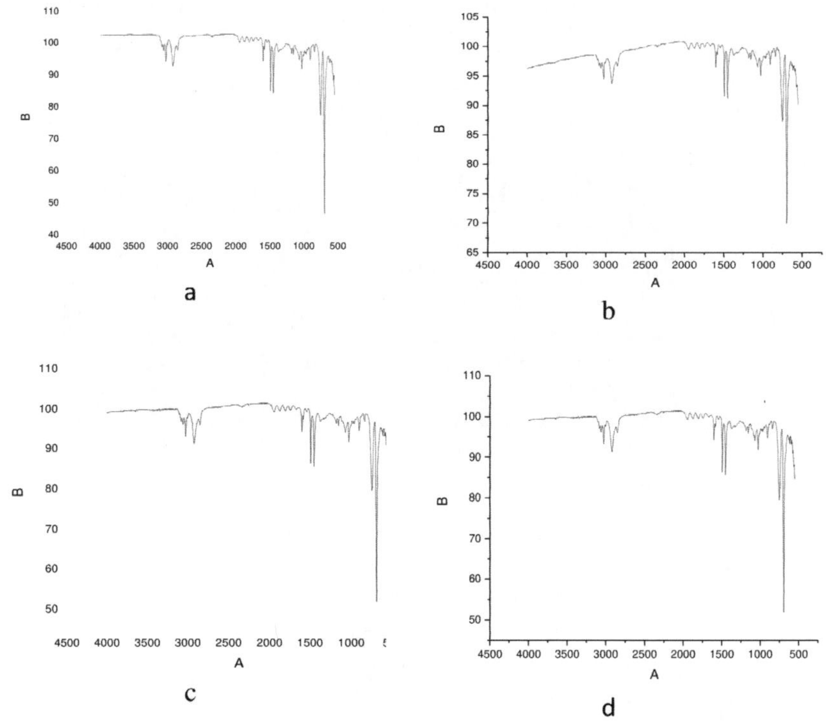

Figure 3 - a) IR spectrum of the initial polystyrene powder of PAO "Nizhnekamskneftekhim", b) IR spectrum of the powder after grinding in a vibratory mill, c) IR spectrum of the powder after grinding on a disintegrator, d) IR spectrum of the powder after treatment with a chemical solution

There are no foreign polymers in the powder, which indicates the qualitative composition of the raw materials, which does not require additional chemical purification. Therefore, it was decided to sinter raw materials, and if necessary, increase the rheological properties. Powder materials used in SLS technology must meet certain characteristics in terms of dispersion, particle shape and surface state, namely: dispersion - 20-90 μm, the shape of the particles must be spherical (for good flow of the powder, the material will not be strongly compacted by the roller), the surface must be hydrophobic.

In experiments on selective laser sintering, we used the sPro 60 HD (3D Systems), USA. sPro 60 HD is an industrial selective laser sintering plant that creates physical models from powder compositions, mostly polymer ones. This installation is actively used in aircraft building, machine building, automotive industry, military-industrial complex, medicine for creating prototype models. Features: the construction area is 381x330x437 mm, the thickness of the layer is 0,08-0,15 mm, the printing speed is 1,0 l / h, the scanning system is ProScan CX (digital), the scanning speed is 6 m / s, the laser power is - 70 W / CO_2 laser. The installation uses commercial materials - polyamide, polystyrene, polypropylene, glass-filled nylon 12.

It is known that the glass-transition temperature of the material is T_g = 89 ° C. Do not exceed this temperature on the surface of the layer to avoid sticking the material to the screed roller. The layer thickness is 150 μm. A laser power of 28 W is required for a layer of 150 μm. Laser scanning speed - 6 m/s.

To improve the rheological properties, several methods of physical action have been applied: grinding in a vibratory mill, grinding in a disintegrator, chemical action to reduce the coagulation process.

Vibrating mills are designed for finely grinding brittle powder materials due to the impact and abrading action of milling bodies, which are located together with the grinding material in a vibrating drum. As grinding bodies, we took ceramic balls of diameters with an average of 1 cm, which fill the volume of the mill by 80%. The vibrating mill drum is driven into oscillatory motion by means of a rotating shaft with unbalance. The vibration frequency was 25 times per second, amplitude 4-5 mm.

The original powder was ground in a vibrating mill for 2 hours. The dispersion composition of the obtained powder is shown in Figure 2. Flowability - 5 grams of powder precipitated from the glass funnel in 5.8 seconds. Analysis on an IR microscope showed no change in the chemical composition. The IR spectrum is shown in Fig. 3b.

In the experiments, the disintegrator DZG-200T-20L was used. The disintegrator consists of two rotating rotors in opposite directions, which are mounted on separate shafts and encased in a casing. On the rotor disks, concentric circles have 2 to 4 rows of circular cylindrical fingers, so that each row of one rotor freely enters between the two rows of the other. The material for crushing is fed to the central part of the rotor and, moving to the periphery, is subjected to repeated impacts of the fingers, which rotate in opposite directions. Technical characteristics of the disintegrator: productivity - 10 ÷ 50 kg / h, the number of rotors - 2 pcs, rotational speed - 2800 rpm.

As can be seen from the diagram in Fig. 2, particles with a size of 400-500 μm were milled and moved to the region of 40-300 μm. Flowability of the powder improved - from the glass funnel 5 grams of powder poured out in 5.5 seconds.

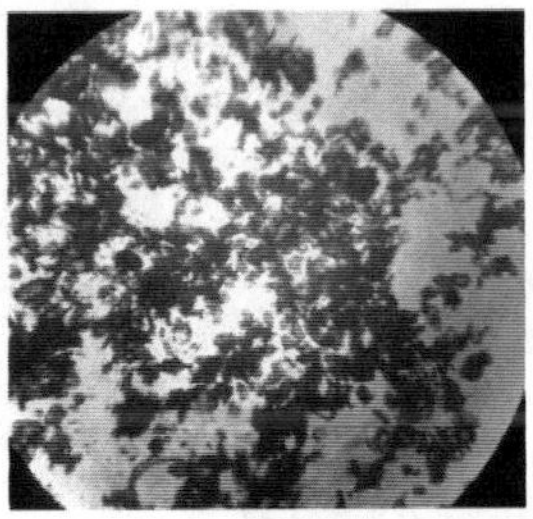

Figure 4 shows a snapshot of the powder after grinding on a disintegrator under a microscope with a 20x magnification. Powder particles larger than 300 μm were not detected.

The treatment with surfactants was carried out in a chemical reactor at room temperature. The surfactant solution was mixed with the powder in a ratio of 1:10 and was periodically mixed for 5 hours. The powder was Figure 4 then filtered off, washed with water and dried at a temperature of 600 ° C. Surfactant treatment was carried out in order to reduce the attraction of particles and increase the fluidity of the powder.

As seen from the IR spectrum, the chemical composition does not remain on the surface of the powder. The precipitation of 5 grams of powder from the glass funnel took place for 12 seconds.

IOP Conf. Series: Materials Science and Engineering **412** (2018) 012093 doi:10.1088/1757-899X/412/1/012093

Thus, a negative result is observed, a decrease in fluidity occurs. The dispersion composition didn't change.

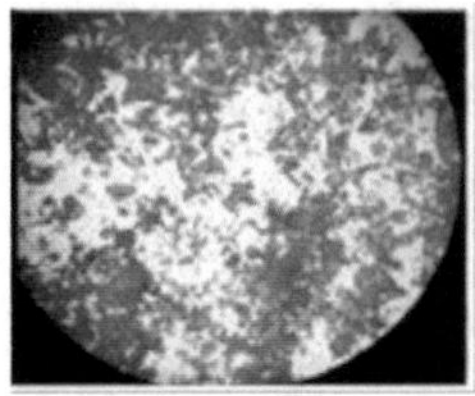

Figure 5 - Sample powder under a microscope (20x magnification)

Carrying out the experiment from the initial polystyrene powder in the Selective Laser Sintering unit sPro 60 HD. Due to the low bulk density and the large particle size, the powder does not cover the entire construction area. This leads to the formation of pits in the central zone of construction (Figure 6a).

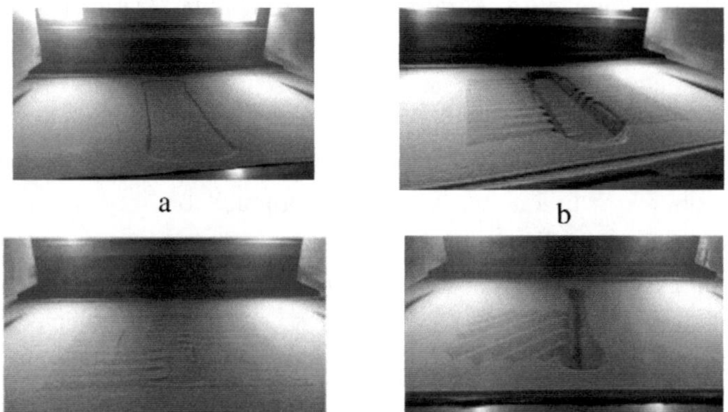

a b

c d

Figure 6 - The sintering process of the powder in the sPro 60HD unit: a) the starting powder; b) powder after the vibrating mill; c) powder after the disintegrator; d) powder treated with surfactant.

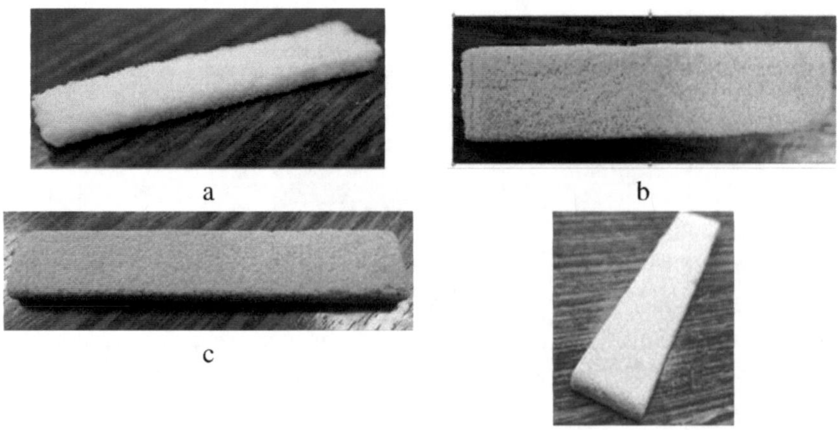

a b

c

d

Figure 7 - The obtained test samples: a) from the original powder; b) from the powder after the vibrating mill; c) from the powder after the disintegrator; d) from a powder treated with a surfactant.

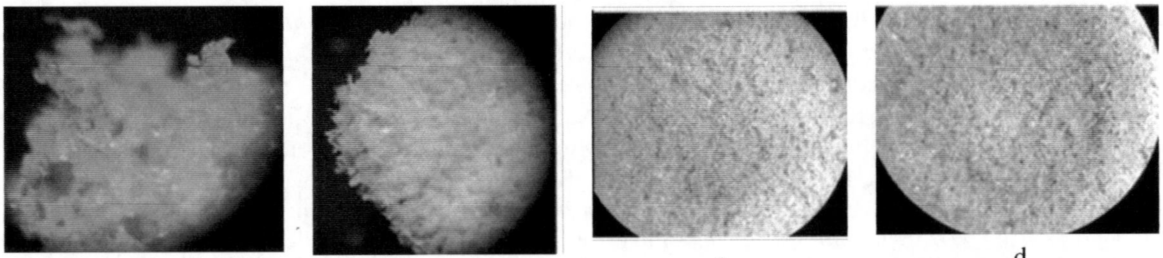

Figure 8 - Obtained test samples under a microscope with a 20x magnification: a) from the original powder; b) after the vibrating mill; c) after the disintegrator; d) after treatment with surfactant.

Figure 7 shows the appearance of the sample. The sample was of low strength and crumbled in the hands. The weight of the test sample was 0.58 g, density - 0.21 g / cm3. It consists of hotel blocks measuring 600-700 microns in size. Between the blocks, voids from 50 μm to 200 μm were formed. This caused high fragility and porosity.

Selective laser sintering of polystyrene powder after grinding on a vibrating mill made it possible to obtain stronger products. When the samples were examined under a microscope, it was found that the samples became more dense, the contours were clear. The weight of the test sample was 0.81 g, density - 0.29 g / cm^3.

Carrying out the experiment from polystyrene powder after grinding on the disintegrator. The weight of the test sample was 0.95 g, the density became 0.34 g / cm3. As can be seen in Figures 7c and 8c, the contours of the sample become more distinct. The voids decreased to 200-400 microns.

Carrying out the experiment from polystyrene powder after surfactant treatment.

The sample weight was 0.94 g. The density of the samples did not change. Treatment of the powder with surfactant solution is effective only when the particle size is less than 30 μm.

Conclusion

It is necessary to find a method for reducing the particle sizes below 100 μm and giving a spherical shape.

As a result of the research it was found out that the best way for mechanical grinding of the powder is the disintegrator. But the dispersion is not enough to use the powder in selective laser sintering. To obtain the necessary dispersion, cryogenic grinding should be used [4].

For good flowability, it is necessary to achieve the sphericity of the powder particles. To obtain spherical particles, an experiment with plasma spheroidization is needed.

This work was funded by the subsidy allocated to Kazan Federal University for the state assignment in the sphere of scientific activities (11.1300.2017/4.6).

References

[1] Way Y, Puwanto H and etc 2014 *Investigation into the differences chemical properties of laser sintering polymer powder materials using fourier transform infrared spectroscopy* Applied Mechanics and Materials pp 548-49

[2] Shcmidt J, Sachs M and etc 2015 *A novel process chain for the production of spherical SLS polymer powder with good flowability* The 7th World Congress on particle technology

[3] Kruth J-P, Levy G *Consolidation of polymer powder by selective laser sintering*

[4] Shcmidt J, Plata M 2012 *Production of polymer particles below 5μm by wet grinding* Powder technology **228**

[5] Kashapov R N, Kashapov L N, Kashapov N F 2017 *Analysis and development of methods for obtaining metallic powders for selective laser melting* IOP Conference Series: Materials Science and Engineering Vol 240 Is 1 **012071**
[6] Kashapov R N, Kashapov L N, Kashapov N F 2015 *The research of anodic microdischarges in plasma-electrolyte processing* IOP Conference Series: Materials Science and Engineering Vol 86 Is 1 **012019**
[7] Kashapov L N, Kashapov N F, Kashapov R N 2013 *Research of the impact acidity of electrolytic cathode on the course of the plasma-electrolytic process* Journal of Physics: Conference Series. Vol 479 Is 1 **012011**

Application of additive technologies in the process of technological preparation of aviation products

N A Lunev[3], A V Shapkin[3], R N Kashapov[1,2], D V Kurylev[3], L N Kashapov[1] and N F Kashapov[1]
[1] Engineering Institute, Kazan Federal University, Kazan, 420008, Russia
[2] Laboratory of Radiation Physics, Kazan Physical-Technical Institute of the Kazan Scientific Center of the Russian Academy of Sciences, Kazan, 420029, Russia
[3] Techproject-Service LLC

kashramiln@gmail.com

Annotation. Preparation of production and development of new science-intensive products is a complex and time-consuming process, including the creation of product models for carrying out a series of experimental studies to refine computational models, and then manufacturing the necessary technological equipment. The use of digital production technologies can reduce the laboriousness of this process. This work researches on decrease in time of development of details of the gas-turbine engine with use of digital manufacture concepts and additive technologies are given.

Technological equipment plays an important role in the technological preparation of new parts production. It largely determines the economic and technical parameters of the part production, including cost and competitiveness of the product on the market.

The main technology of manufacturing tooling is machining from monolithic billets using CNC equipment. And so this version of production is almost completely optimized and does not have a significant resource to reduce labor intensity and cost.

Modern production is characterized by a relatively fast change of products. It is possible to fabricate the forming parts of the tooling in a shorter period by selective laser melting [1, 2, 3, 4].

Additive technologies are a part of modern methods of creating new products, machines, technologies, products that are used at the stage of testing the design and construction of the product or at the stage of preparing the product for production [6].

One of the promising areas of additive technologies application is the production of technological equipment for development parts for casting on investment models (Fig. 1).

Content from this work may be used under the terms of the Creative Commons Attribution 3.0 licence. Any further distribution of this work must maintain attribution to the author(s) and the title of the work, journal citation and DOI.
Published under licence by IOP Publishing Ltd

Fig. 1 Production of parts for by lost wax process

Selective laser melting is the most common method of manufacturing metal parts. The method allows producing parts with greater productivity; however it is limited in the choice of material.

Selective laser melting introduces certain requirements for the design of the part (Fig. 2).

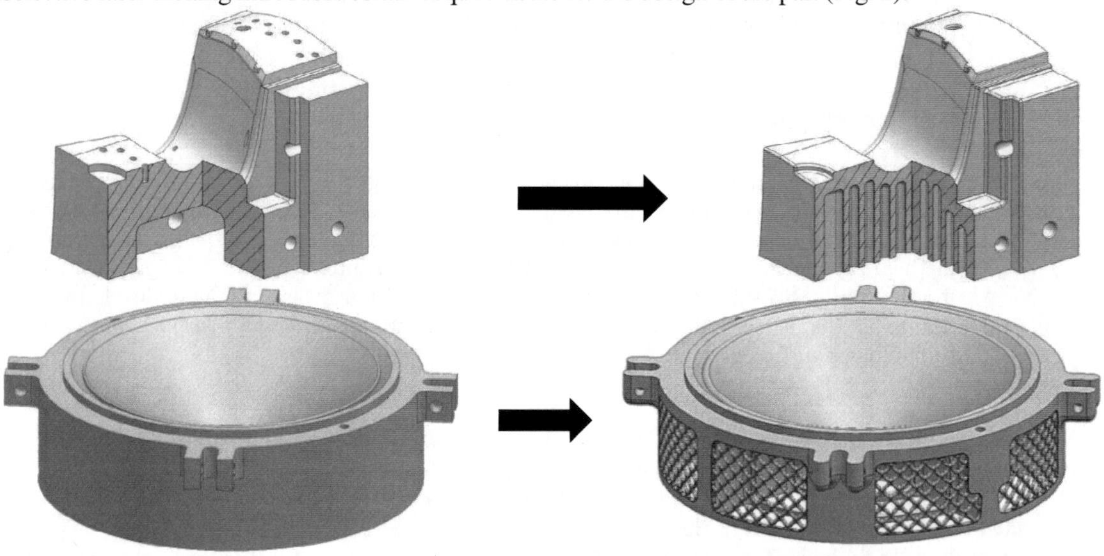

Fig. 2 Changing part design

The technology of laser melting of powder materials consists in a layer-by-layer sintering of a powder (polymer, metal) by a laser beam directed by a scanning system. Details grown by selective laser melting have a high roughness (Fig. 3). Therefore, after manufacturing the part on a 3D printer, it is necessary to mechanically process the parts.

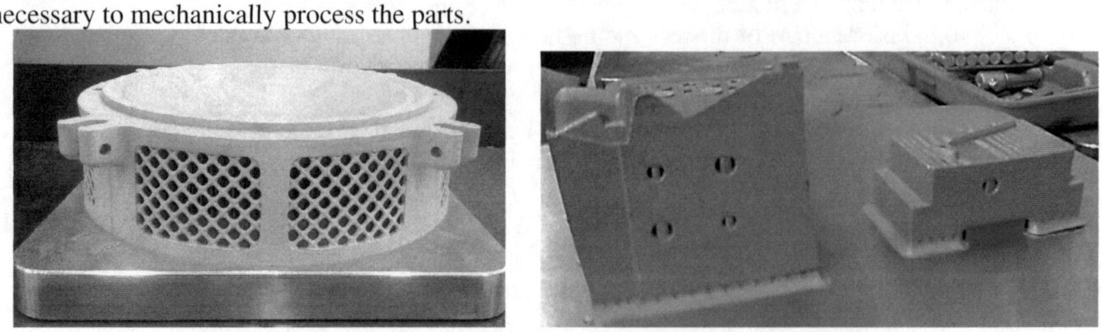

Fig. 3 Details of the mold made by selective laser melting

After machining the parts, measurements were taken on the 3D scanner ATOS III Triple Scan XL and using special templates (Fig. 4)

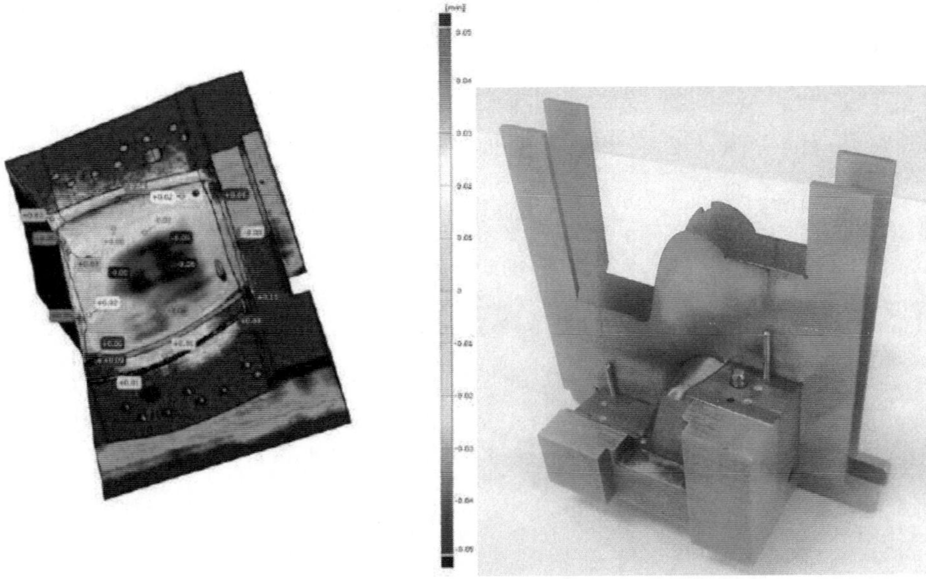

Fig. 4 Measuring the details of grown mold

After the manufacture of all parts, mold was assembled and tested in real production conditions (Fig. 5).

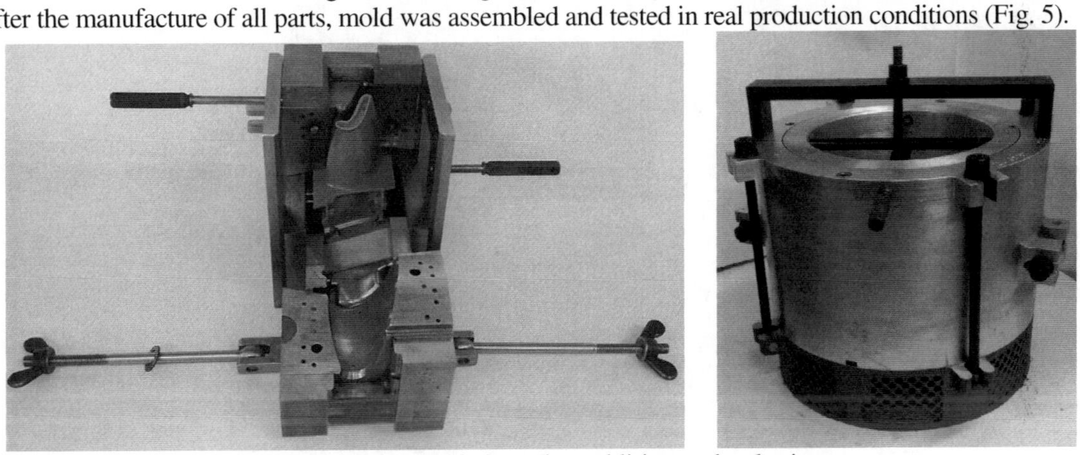

Fig. 5 Molds made by using additive technologies

The following results were achieved during the work:

The duration of work on the manufacture of technological equipment didn't exceed 2- 2.5 months;

The use of additive technologies has reduced the weight of the structure and increased the utilization factor of materials in several times;

Various mesh structures and internal channels and cavities have been used in the mold design, which are impossible or difficult to obtain with conventional machining;

The resulting molds are made from fewer parts and with fewer technological operations, which have reduced the number of mating surfaces and increased accuracy.

The design process was tested and the additive technology of manufacturing reusable tooling was used, which allowed using traditional LWP technology and shortening the cycle of production preparation in real production conditions.

This work was funded by the subsidy allocated to Kazan Federal University for the state assignment in the sphere of scientific activities (11.1300.2017/4.6).

References

[1] Kashapov R N, Kashapov L N, Kashapov N F 2017 *Analysis and development of methods for obtaining metallic powders for selective laser melting* IOP Conference Series: Materials Science and Engineering Vol 240 Is 1 **012071**

[2] Kashapov R N, Kashapov L N, Kashapov N F 2015 *The research of anodic microdischarges in plasma-electrolyte processing* IOP Conference Series: Materials Science and Engineering Vol 86 Is 1 **012019**

[3] Kashapov L N, Kashapov N F, Kashapov R N 2013 *Research of the impact acidity of electrolytic cathode on the course of the plasma-electrolytic process* Journal of Physics: Conference Series. Vol 479 Is 1 **012011**

[4] Zlenko M A, Nagaitsev M V, Dovbysh V M 2015 *Additive technologies in mechanical engineering* M. SSC RF FGUP "NAMI" 220 p

[5] Kablov E N, Evgenov A G, Ospennikova O G, Semenov B I, Semenov A B, Korolev V A 2016 *Metallic powder compositions of the heat-resistant EP648 alloy produced by FSUE VIAM in the technologies of selective laser melting, laser gas-powder overlaying and high-precision casting of polymers filled with metal powders* News of higher educational institutions. Mechanical engineering No 9 (678) pp 62-80

[6] Nyamekye P, Leino M, Piili H, Salminen A 2015 *Overview of Sustainability Studies of CNC Machining and LAM of Stainless Steel* Physics Procedia No 78 pp 367-76

Ultrasound in the manufacture of medical instruments

R M Kashapova[1], R S Kashapova[1], R N Kashapov[1,2], L N Kashapov[1]
[1] Engineering Institute, Kazan Federal University, Kazan, 420008, Russia
[2] Laboratory of Radiation Physics, Kazan Physical-Technical Institute of the Kazan Scientific Center of the Russian Academy of Sciences, Kazan, 420029, Russia

reginasharapova13@mail.ru

Annotation. The article contains information on the properties and possibilities of ultrasonic oscillations of high intensity, the features of the application of ultrasonic vibrations for the intensification of processes occurring in liquid, solid, gaseous media, polymeric materials, as applied to solving problems in the chemical, biotechnological, food industries and related industries. The methods of using ultrasound for cleaning parts on grinding and polishing operations, which are primarily intended for use in the manufacture of medical instruments, are described in more details.

Introduction

Ultrasound is a wave-propagating vibrational motion of particles of the medium and is characterized by a number of distinctive features compared to fluctuations in the audible range. In the ultrasonic frequency range it is relatively easy to obtain directional radiation; ultrasonic vibrations can be easily focused, which increases the intensity of ultrasonic vibrations in certain areas of impact. While expand in gases, liquids and solids, ultrasound generates unique phenomena, many of which have found practical application in various fields of science and technology.

In recent years, industrial use of ultrasonic vibrations has become very important for the implementation of all possible technological processes. The processes consisting in the force action of ultrasonic vibrations on various media and characterized by a relatively high energy intensity: ultrasonic welding and soldering of various products and films, ultrasonic cleaning of complex configuration components from contamination, ultrasonic dimensional processing of various brittle and refractory materials, ultrasonic grinding, polishing and hardening of the surface of metal parts, acceleration of diffusion processes, etc. Studies carried out in recent years have shown that, under certain conditions, excitation of ultrasonic oscillations in cutting tools can improve the quality of the surface and the accuracy of processing, and reduce cutting forces. The aim of the work was to investigate the possibilities of using ultrasound to simplify the manufacture and increase the productivity of medical instruments.

Ultrasonic hardening can be subjected to machine parts and tools from metals of almost any initial hardness. The processing process is easily adjustable. With ultrasonic hardening and finishing of non-hardened steels, the surface roughness is improved by 4 classes; the surface microhardness is increased by 80-180%. When processing hardened and ground steels, the surface roughness is improved by 2 classes, the surface microhardness increases by 30-70% with a hardening depth of 0.05-0.30 mm. When processing cast iron surface is improved by 3-4 classes, surface microhardness is increased by 25-35% at a depth of hard work of 0.3 mm. As a result of ultrasonic machine parts and tools hardening, their longevity increases by 2.2-5.6 times.

Content from this work may be used under the terms of the Creative Commons Attribution 3.0 licence. Any further distribution of this work must maintain attribution to the author(s) and the title of the work, journal citation and DOI.
Published under licence by IOP Publishing Ltd

IOP Conf. Series: Materials Science and Engineering **412** (2018) 012095 doi:10.1088/1757-899X/412/1/012095

Spot and seam welding is widely used in machine building. Ultrasonic welding allows to weld metals with an oxidized and uncleaned surface. Ultrasound reduces own stresses and deformations during welding. Ultrasonic welding can be used in the medical industry, especially with sterilized packaging of medical instruments in film or foil bags.

In the industry, ultrasonic cleaning of complex shape parts in various liquid cleaning media is widely used. This method of purification can be widely used in the medical-instrument industry for mass small medical instruments. Especially promising is the cleaning of medical instruments in methylene chloride (freon-30). Freon-30 is not flammable, explosion-proof, inert to metals, has high degreasing properties.

It is established that the cutting forces during grinding with ultrasonic vibrations are reduced by an average of 3.5 times, and the temperature in the treatment zone is reduced by 30%. It has also been confirmed by experiments that the imposition of ultrasonic vibrations on an abrasive diamond tool when grinding metals and alloys eliminates the salting of the tool and thus provides normal conditions for the cutting process to proceed.

For the medical industry, the most promising is the polishing of small mass products in the chamber with an oil-like suspension, this method can try to remove burrs from injection needles after sharpening. It is also of interest to piece-by-piece surface treatment of flat parts by instant polishing of tool-forming tools. [1]

The field of medicine also widely practices the application of ultrasound. There is a well-known work in which the possibility of using the built-in ultrasound physiotherapy system of the hand orthosis is explored. Using the 3D prototyping method, a customized net orthosis from nylon 12 was fabricated in the selective laser sintering unit SLS SPro 60HD. The applied technology of three-dimensional scanning made it possible to obtain the model of the patient's hand and on the basis of it to construct a virtual model of the mesh framework. In the course of research, the developed ultrasonic exposure system was installed on the orthosis and its tests were carried out. As a result, the healing process acceleration and the reduction in the time of wearing the orthosis were revealed. [2]

At present, as follows from literary sources, a new direction in chemistry is formed: ultrasonic chemistry. Studying the chemical transformations occurring under the action of ultrasound, it is established that ultrasound not only accelerates the oxidation, but in some cases provides a restoring effect. Thus, iron is recovered from oxides and salts.

Good positive results were obtained on the intensification of ultrasound in the following chemical-technological processes:

- electrodeposition, polymerization, depolymerization, oxidation, reduction, dispersion, emulsification, coagulation of aerosols, homogenization, impregnation, dissolution, spraying, drying, burning, tanning, etc.

Electrodeposition - the precipitating metal acquires a fine crystalline structure, reduces porosity. Thus, copper plating, tinning and silvering are carried out. The process goes faster and the quality of the coating is higher than in conventional technologies. The resulting emulsions: water and fat, water and essential oils, water and mercury. The immiscibility barrier is overcome by ultrasound.

Polymerization (the combination of molecules into one) - the degree of polymerization is controlled by the frequency of ultrasound.

Dispersing is the production of ultrathin pigments for the production of dyes.

Drying occurs without heating biologically active substances. In the food, pharmacological industry.

Spraying of liquids and melts. Intensification of processes in spray drying. Preparation of metallic powder from melts. These spraying devices exclude rotating and rubbing parts. Ultrasound enhances the combustion efficiency by 20 times of liquid and solid fuels.

Impregnation. Hundred times faster passes the liquid through the capillaries of the impregnated material. Used in the manufacture of roofing felt, sleepers, cement slabs, textolite, getinaks, impregnation of wood with modified resins. [3]

Data on the method of manufacturing calibrated vesicles on demand in a non-chemically controlled solution using localized microelectrolysis and ultrasound were obtained. The introduction of a feedback loop in the process leads to a point source of stable monodisperse microbubbles. This approach overcomes the inertial constraints encountered in microfluids, with the possibility of obtaining from a single array of calibrated bubbles. In addition, this method avoids the use of an additional surfactant that can modify the composition of the liquid. It affects a wide range of scientific fields of bioengineering, environmental perception. [4]

New data on the effective use of ultrasound is in microbiology. A new method of forming composite nanomaterials based on multilayer and single-walled carbon nanotubes (CNTs) on a silicon substrate has been developed. The formation is accomplished by ultrasonic coating of the silicon substrate by homogeneous dispersion of CNT in the albumin matrix and subsequent irradiation with a continuous laser beam with a wavelength of 810 nm and a power of 5.5 W. High electrical conductivity of CNTs ensures its structuring under the influence of an electric field of laser radiation. The result is a scaffold that provides high mechanical strength of nanocomposite material (250 MPa). To study the biocompatibility of materials in vitro, a method of microscopic analysis of cell growth was developed. Embryonic human fibroblasts (EPP) were used as biological cells. The interaction between nanocomposite material and cells was studied by optical and atomic force microscopy depending on the time of incubation of cells. The study showed that, after 3 hours of incubation, EPP was fixed on the surface of the substrate, avoiding the surface of the composite material. However, after 24 hours of incubation, the EPP is fixed on the surface of the sample, and then begins to grow and divide. After 72 hours of incubation, the cells completely fill the surface of the nanocomposite material sample. Thus, nanocomposite material based on nanotubes in the matrix of albumin does not prevent the growth of cells on its surface and promotes their growth. Nanocomposite material can be used to create implants from soft tissues. [5]

Food industry. In practice, today the following technologies are being implemented: milk processing for homogenization of sterilization, processing for increasing the shelf life and quality of milk in a frozen form, obtaining high-quality powdered milk, obtaining emulsions for bakery, processing yeast by 15% increases their fermentative power, substances, puree, the extraction of fat from the liver, the isolation of tartar, the extraction of plant and animal raw materials. [3] The adoption of new methods of preserving food becomes an alternative strategy of traditional methods of treatment based solely on the use of heat. In this study, the potential application of steam (94-95 ° C) and ultrasound (25-40 kHz), used simultaneously for a short time on mozzarella cheese, was investigated. First, the effect of this technique was evaluated for the restoration of the fluorescence of Pseudomonas and Enterobacteriaceae inoculated on the surface of the cheese in two different concentrations. Then, the processed samples of raw and combined steam ultrasounds (SonoSteam) were packed in plastic trays with brine and stored at 8 ° C. Enterobacteriaceae, Pseudomonas spp., PH and sensory properties were monitored. All data showed that SonoSteam is effective to reduce microbial contamination with a decrease in the range of 0.8 to more than 2 recording cycles, thus preserving the quality of the product. From a sensory point of view, no changes in the processed samples occurred, indicating the potential of SonoSteam for food applications. As a fact, these preliminary results are relevant, because the treatment is very fast (a few seconds) and can cause great interest for future industrial expansion. [6] Researches carried out in biochemistry have also been successful. t was found that ultrasonic treatment can alter the conformation of β-lactoglobulin (β-Lg) and improve the glycation reaction in an aqueous solution in the neutral state. However, the effect of ultrasonic pre-treatment on β-Lg glycation of pentose in the dry state remains ambiguous, and the relationship between glycation and β-Lg allergen pre-treatment with ultrasound is unknown. This study was designed to assess the effect of ultrasound pretreatment on glycation and allergenicity of β-Lg. Apparently, reduced allergenicity of β-Lg was observed after glycosylation with ribose before and after preliminary ultrasound treatment with a minimum found at 400 watts. LC-MS / MS orbital showed that the degree of glycation of some peptides in glycated β-Lg with preliminary and ultrasound pretreatment was different, although the free amino group and molecular weight content were slightly different. Thus, pre-treatment with ultrasound promoted a decrease in allergenicity by improving the degree of glycation of some glycation sites, although it hardly increased the entire degree of glycosylation of β-Lg. [7]

Conclusion

In recent years, ultrasound has begun to play an increasing role in industry and scientific research. Theoretical and experimental studies in the field of ultrasonic cavitation and acoustic currents have been successfully carried out, which made it possible to develop new technological processes that occur under the action of ultrasound in the liquid phase. A new direction of chemistry is developing - ultrasonic chemistry, which allows to accelerate many chemical-technological processes and to obtain new substances. Scientific research contributed to the birth of a new section of acoustics - molecular acoustics, which studies the molecular interaction of sound waves with matter. New application fields of ultrasound appeared: introscopy, holography, quantum acoustics, ultrasonic phasemery, acoustoelectronics. The development and application of ultrasonic technologies opens up new prospects for creating new substances and materials, giving to materials and media new properties (sterility, nanoscience, etc.) and therefore requires an understanding of the phenomena and processes that occur under the ultrasound influence, the possibilities new technologies and prospects for their application.

This work was funded by the subsidy allocated to Kazan Federal University for the state assignment in the sphere of scientific activities (11.1300.2017/4.6).

References

[1] Keilman G 2015 Physics Procedia Vol 87 pp 79-84

[2] Kashapova R M, Kashapov R N, Kashapova R S 2017 *Mesh three-dimensional arm orthosis with built-in ultrasound physiotherapy system* IOP Conference Series: Materials Science and Engineering Vol 240 Is 1

[3] Lopez A, Bacelar R, Pires I, Santos T G, Sousa J P, Quintino L 2018 *Additive Manufacturing* Vol 21 pp 298-306

[4] Achaoui Y , Metwally, K, Fouan D, Hammadi Z, Morin R, Debieu E, Payan C 2017 *Tunable microbubble generator using electrolysis and ultrasound* Vol 7 Is 1

[5] Zhurbina N N, Kurilova U E, Ickitidze L P, Pavlov A A, Gerasimenko A Yu 2016 *Progress in Biomedical Optics and Imaging* Proceedings of SPIE Investigation of cell proliferative activity on the surface of the nanocomposite material produced by laser radiation Vol 9917

[6] Lacivita V, Conte A, Zambrini V A, Nobile D, Musavian M A 2018 *Steam-ultrasound combined treatment*: A promising technology to significantly control mozzarella cheese quality pp 450–55

[7] Liu G-X, Tu Z-C, Yang L, Li X 2018 *Investigation into allergenicity reduction and glycation sites of glycated β-lactoglobulin with ultrasound pretreatment by high-resolution mass spectrometry* Food Chemistry Vol 252 pp 99-107

[8] Kashapov R N, Korobkina A I, Platonov E V, Saleeva G T 2014 *The method of manufacture of nylon dental partially removable prosthesis using additive technologies* IOP Conference Series: Materials Science and Engineering Vol 69 Is 1 **012026**

[9] Kashapov L N, Rudyk A N, Kashapov R N 2014 *Applying 3D-printing technology in planning operations of cancer patients* IOP Conference Series: Materials Science and Engineering 69 **012016**

IOP Conf. Series: Materials Science and Engineering **412** (2018) 012096 doi:10.1088/1757-899X/412/1/012096

Determination of optimal technology for manufacturing dental surgical guides

J N Dementyeva[1], I I Sagitov[3], R N Kashapov[1,2,3], L N Kashapov[1], G T Saleeva[3]
[1]Kazan Federal University, 18 Kremlyovskayastreet, Kazan, 420008, Russia
[2]Kazan E.K. Zavoisky Physical-Technical Institute, 420029, Russia
[3]Kazan State Medical University, Kazan, 420012, Russia

kashramiln@gmail.com

Abstract. This article describes methods of manufacturing surgical guides - their creation in rapid prototyping installations and milling machine grinding. Implantation was carried out using the obtained guides and derating level of installed implants was defined. The work compares existing technologies of surgical guides production and defines a clear advantage of using laser stereolithography technologyforproduct manufacture in quantities greater than one.

Introduction

Dentistryisoneoftheorthogenics,wheretheactive introduction of new innovative technologies takes place.Individualityinherentineachclinicalcase, and it has ledtoactiveimplementationof digitaltechnologies.

One exampleoftheuseofdigitaltechnologiesisthecreationofsurgical guides. The surgical template is a patch with guide cylinders for precise positioning of implants in the oral cavity in accordance with the planned position on the computer for computed tomography. Most often surgical patterns for implantation are required if the patient does not have a large number of teeth. Using a surgical template can improve the quality of planning and the accuracy of the planned outcome [1, 2]. However, the obtained virtual model must be embodied in a real physical phenomenon. There are two ways of creating surgical templates - CNC mill grindingand usage of additive technologies.

The purpose of this work is to compare existing methods and find the optimal one.

Main part

The computer tomography of patient, namely axial sections in the DICOM format, was used for creation of a surgical guide. Scanning of jaw casts was performed on a 3D scanner. It is very important to have a good quality, because the guides are created according to the pick-up impression. If the patient lacks all the teeth, it is necessary to make a radiopaque guide on the CT scan and the scan itself. With its help you can see the thickness of soft tissues on CT.

Computer tomography and impression scans are loaded into the DDS-Pro software program and combined (Figure 1). The guide is created according to pick up impression, CT is used for endosseous topography. We select the implants and arrange them with the help of the software. We must take into account the precise location of the implant and the anatomical features to achieve the optimal aesthetic result and to prevent complications associated with disruption of the maxillary sinus integrity or damage of the neurovascular bundle, and also we must take into account predicted implant angulation to create an axial load on the implant and achieve a biomechanically

Content from this work may be used under the terms of the Creative Commons Attribution 3.0 licence. Any further distribution of this work must maintain attribution to the author(s) and the title of the work, journal citation and DOI.
Published under licence by IOP Publishing Ltd

predicted result [3,4 ,5].

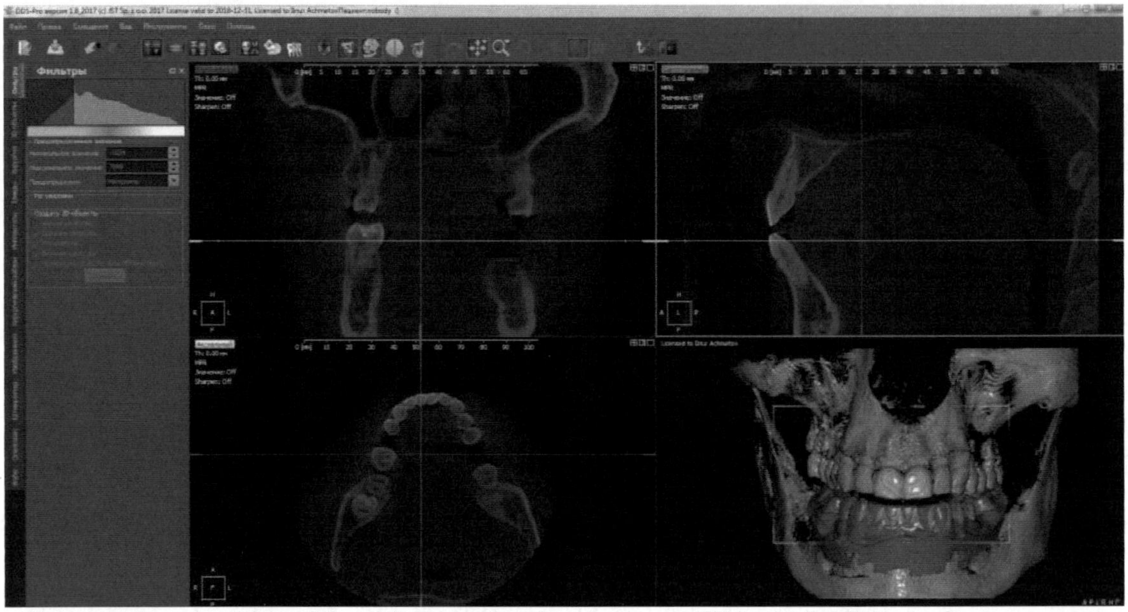

Figure 1 - Superposition of the pick-up impression scans and CT in the DDS-Pro software program

We used a dental 3D-scanner Smart Optics Activity 875 to scan the gypsum models of the jaws. Scanning accuracy is up to 10 microns. The scanning speed of one tooth is approximately 55 seconds, and 90 seconds - of the whole jaw. The test ofmilling possibility was carried out on a 5-axis CNC mill CORiTEC 450i from IMES-ICORE. This miller allows to conduct dry and wet milling of any materials used in dentistry and to manufacture products with high accuracy. The CNC mill CORiTEC 450i works with the following materials: plastics, zirconia, aluminum oxide, glass ceramics, cobalt-chromium alloy, titanium. Performance specifications: five-axis synchronous processing, servo controller, spindle rotation up to 60,000 rpm, cooling system for wet processing (optional), tools with 3 mm collet, automatic change of 12 cutters, built-in high-precision sensor for measuring the length of the instrument, installed holder for blanks with a diameter of 98.5 mm.

Production by stereolithography (SLA). The technology of laser stereolithography is based on the polymerization of liquid photopolymers. A mesh platform is placed in a container with a liquid photopolymer on which the prototype is created. This method of additive manufacturing is one of the most accurate up to date. For the production of the guides we used installation of rapid prototyping RSPro 600 created by company Uniontech. The max.build size is 600 x 600 x 400 mm (full bath), the accuracy is ± 0.1 mm (size <100 mm), ± 0.1% (size> 100 mm.) The accuracy may vary depending on the model parameters, geometry and dimensions of the part, orientation of the part, and subsequent processing.The installation uses a CO2 laser with a power 1000 mW, a wavelength 354.7 nm, the laser spot size has a nominal diameter 0.12 mm, a maximum diameter 0.20 mm, a scanning speed 12 m / s (max) 6-10 m / s (min). The thickness of the layer is 0.05 mm (min) and 0.25 mm (max).

Manufacture bydirect light processing (DLP) method.The method is based on the usage of photocurable resin, whichsolidify when irradiated with ultraviolet light. Unlike laser systems that scan the surface of a material with one or more laser heads, DLP printers project an image of the whole layer until the polymer resin solidifies, after what a new layer of material is applied and a new layer of the digital model is projected. To create a guide, we used the ProJet 1200 (3D Systems), USA. The print quality is 0.03 μm, the layer thickness is 30 μm, the productivity is 14 mm / hour. The build size is 43x27x150 mm. The installation is used in dentistry, jewelry industry, design and engineering.

Manufacture by selective laser sintering (SLS) method. Selective laser sintering is a method of additive manufacturing, where a polymer powder is used to create details, which is sintered by an infrared CO2 laser with a wavelength 10.6 μm.

The surgical guide was made with the help of a selective laser sintering system sPro 60 HD (3DSystems, USA) of polyamide 12. The max.build size is 381x330x437 mm, layer thickness is 0,08 - 0,15T mm, productivity is 1,0 l / h. The laser power is 70 W.

Manufacture by fused deposition modeling (FDM) method. FDM technology means the creation of three-dimensional objects by applying consecutive layers of material that repeat the outline of the digital model. Typically, as materials for printing are thermoplastics, supplied in the form of coils of yarns or rods. For manufacture of a surgical guide of ABS plastic, the Fortus 400mc (Stratasys, USA) was used. The max.build size is 406x355x406 mm, layer thickness is from 127 to 330 μm.

Results and comparison

Grinding of one surgical guide on the milling machine takes approximately 1 hour. Grinding of 20 surgical guides takes about 20 hours. Only one guide can be placed on one blank. Whereas one surgical guide on the rapid prototyping RSPro 600 can be also produced in 1 hour, but the production of 20 guides takes 3,5 hours. This happens due to the fact that the installation has big building platform.

The manufacturing of the guide with the ProJet 1200 takes about 50 min, but it has a small build platform, so only one guide can be placed on the platform. The quality of the obtained guide is as good as the guide obtained on the SLA or CAD / CAM system.

Figure 2 - Process of grinding on the milling machine.

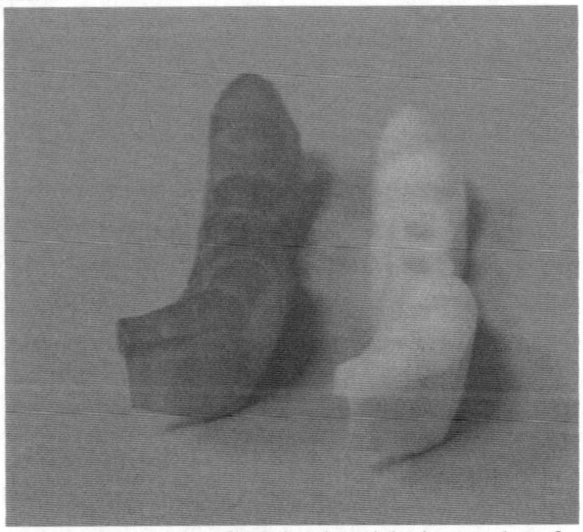

Figure 3 - The finished guide is made of plastic on the CNC mill (left). On the right - a guide, made with RSPro 600 of a photopolymer.

Obtained surgical guide, made of polymer powder, with the help of SLS system is accurate, but flexible, which can be the reason for inaccurate implantation. In addition, the manufacture of a small amount is not rational, since the process chamber must be completely filled with powder. For manufacture of one or 20 surgical guides the same volume of powder is used. The deviation of implants, installed through the guide made with the ABS method, turned out to be higher than deviation of implants installed through the guide, manufactured with the method of SLA, DLP, or on a milling machine.

The research included 16 patients aged from 38 to 60. The bone volume and anatomical features of the patients were studied using CBCT tomography of Planmeca, placed in the dental clinic of the KSMU. All in all, 48 implants of the size from 3,75x10 to 4,2x13 of AB dental were placed. 30 implants were placed through a guide resting on the teeth, 18 implants were placed through a guide resting on the oral mucosa, mainly to patients with adentia. All implants were placed without separation of the sinus membrane. Comparison of materials for the manufacture of surgical guides was carried out in two groups of patients who had implants placed in size from 3.75x10 to 4.2x13. In the first group, there were 3 patients who had 7 implants, with usage of a surgical guide made from photopolymer resin, and in the second there were 3 patients who had 8 implants placed through a guide made of ABS plastic. To compare the results immediately after the operation, a second CBCT was performed. In the planning program, the postoperative image was superimposed on the preoperative image with the implant models placed.

To measure the deviation of the implant from the planned localization, two points were set on each of them. The first point was established in the center of the coronal part (in the collar region) of the implant, the second one - in the region of the root apex center. Also, the angular deflections of the implants were calculated.

When the guide was produced on the SLA printer, the deviation of the implants placed through the guide resting on the teeth was 0.27 ± 0.24 mm in the collar region and 0.37 ± 0.35 mm in the root apex region. The average angular deviation of the implants, placed through the guide with the support on the teeth, was 1.72 ± 1.67, through the guide with the support on the mucosa was 2.71 ± 2.58.

The results of our research testify that surgical guides can improve the functional and aesthetic results of implantation, improve the positioning accuracy, angle and depth of immersion of dental implants. Surgical guide resting on teeth can be more accurate than guide resting on the oral mucosa.

The deviation of the placed implants through a guide made of ABS plastic was 0.45 ± 0.37 mm in the collar region and 0.62 ± 0.56 mm in the root apex region.

When making one guide, it is better to use the CAD / CAM system. When making a large number of guides it is SLA.

Conclusion

The results of our research testify that surgical guides can improve the functional and aesthetic results of implantation, improve the positioning accuracy, angle and depth of immersion of dental implants. Surgical guides resting on teeth can be more accurate than guide resting on the oral mucosa. Surgical guides made of photopolymer plastic are more accurate than surgical guides made of ABS plastic.

It is better to use the rapid prototyping RSPro 600 for making a large number of guides. For single-piece - CAD / CAM system or DLP.

This work was funded by the subsidy allocated to Kazan Federal University for the state assignment in the sphere of scientific activities (3.9399.2017/8.9).

References

[1] Nickenig HJ, Eitner S 2007*Reliability of implant placement after virtual planning of implant positions using cone beam CT data and surgical (guide) templates* J CraniomaxillofacSurg35 pp 207–11

[2] Ozan O, Turkyilmaz I, Ersoy AE, McGlumphy EA, Rosenstiel SF 2009*Clinical Accuracy of 3 Different Types of Computed Tomography-Derived Stereolithographic Surgical Guides in Implant Placement* J Oral MaxillofacSurg67 pp 394–401

[3] Chibisova M A, Saleyeva G T, Yarulina Z I, Mikhalev P N 2007 *Planirovaniyeianalizkhirurgicheskogoetapadental'noyimplantatsii*Nauchno-prakticheskiyzhurnalInstitutStomatologii 36 pp 142-143

[4] Yarulina Z I, Saleyeva G T, Chibisova M A, Mikhalev P N, Sagitov I I 2010 *Opredeleniyerentgenomorfometricheskikhindeksovnizhneychelyustipodannymortopantomografiiidental'noyob"yemnoytomografii*Nauchno-prakticheskiyzhurnalInstitutStomatologii 46 pp 99-101

IOP Conf. Series: Materials Science and Engineering **412** (2018) 012096 doi:10.1088/1757-899X/412/1/012096

[5] Gahleitner A, Watzek G, Imhof H 2003*Dental CT: Imaging technique, anatomy, and pathologic conditions of the jaws*EurRadiol13 pp 366–76
[6] Kashapov R N, Kashapov L N, Kashapov N F 2017 *Analysis and development of methods for obtaining metallic powders for selective laser melting* IOP Conference Series: Materials Science and Engineering Vol 240 Is 1 **012071**
[7] Kashapov R N, Kashapov L N, Kashapov N F 2015 *The research of anodic microdischarges in plasma-electrolyte processing* IOP Conference Series: Materials Science and Engineering Vol 86 Is 1 **012019**
[8] Kashapov L N, Kashapov N F, Kashapov R N 2013 *Research of the impact acidity of electrolytic cathode on the course of the plasma-electrolytic process* Journal of Physics: Conference Series. Vol 479 Is 1 **012011**
[9] Kashapov L N, Kashapov N F, Kashapov R N 2014 *Influence of plasma-electrolyte discharge to the glass surface* Journal of Physics: Conference Series. Vol 567 Is 1 **012024**

Casting individual cervical telescopic implants for burn-out SLA-models

R M Kashapova[1], R N Kashapov[1,2], L N Kashapov[1], N F Kashapov[1], D V Chernyshov [1]
[1] Engineering Institute, Kazan Federal University, Kazan, 420008, Russia
[2] Laboratory of Radiation Physics, Kazan Physical-Technical Institute of the Kazan Scientific Center of the Russian Academy of Sciences, Kazan, 420029, Russia

kashramiln@gmail.com

Annotation. The production of individual implants, modeled specifically for a specific clinical case, is increasingly used every year in traumatology and oncology. Traditionally, individual implants are made on five coordinate milling CNC machines. However, the need to create individual implants with a mesh structure requires the search for new ways of production. In this paper, the feasibility of casting telescopic implants, obtained by laser stereolithography, has been studied. The process of burning photopolymer from a flask was studied. Inducast installation cast prototypes of cervical implants. The ineffectiveness of this technology in the manufacture of mesh collapsible telescopic implants has been established, but it can be used for structures with a thickness of 1 mm.

Introduction

Individual approach to the treatment of patients is becoming increasingly widespread. Starting from individual selection of medicines based on genetic analysis and ending with individual implants and orthoses [1,2,3]. The creation of individual products has always been a more expensive procedure in comparison with mass production. However, with the advent of additive production technologies, the creation of individual parts has become more accessible. Therefore, additive technologies began to be widely used in various fields of medicine. The task was set to study the combined method possibility of manufacturing individual cervical telescopic implants. Namely, how many complex designs can be obtained with the help of SLA technology of three-dimensional printing, and whether it will be possible to cast them out of metal [4,5]. This is very important, as there is a tendency to manufacture mesh implants that have a low weight and are not inferior in strength to the standard one. Another advantage of reticular implants is the possibility of filling them with various biomaterials for effective osseointegration. One of the common injuries is spinal injury. The greatest danger is a fracture or dislocation of the cervical spine. It is important to replace the damaged vertebra correctly. This is achieved only with the installation of individual telescopic implants, with the location of the upper platform changing possibility in the vertical and horizontal directions[6,7]. Therefore, in work on the example of this implant type, the possibility of using a combined casting method and SLA technology was investigated.

Content from this work may be used under the terms of the Creative Commons Attribution 3.0 licence. Any further distribution of this work must maintain attribution to the author(s) and the title of the work, journal citation and DOI.
Published under licence by IOP Publishing Ltd

IOP Conf. Series: Materials Science and Engineering **412** (2018) 012097 doi:10.1088/1757-899X/412/1/012097

Main part

To create an individual implant, you need to obtain digital data of the fracture site or dislocation. The surgeon decides to install a telescopic cervical implant. Obtaining a three-dimensional model of the fracture site is possible only in installations receiving a object layered structure, such type of equipment include tomographs. Digital data obtained with CT, MRI are needed to create a three-dimensional virtual model. Determine the location of installation and cervical implant attachment. Only after receiving this information will it be possible to begin modeling the geometry and calculating the component dimensions of the telescopic implant. With 3D Slicer software, the CT data was processed and 3D model of the patient cervical spine in STL format was created. In the MeshMagic program, we load the STL file of the fracture site and standard size models-blanks of telescopic cervical implants. We adapt the workpiece model to the fracture site. As a result we have individual three-dimensional virtual implant model in the STL format.

The technology of laser stereolithography is based on photoinitiated laser radiation or the emission of mercury lamp polymerization of the photopolymerizable composition (PPC). In a container with a liquid photopolymer, a mesh platform (elevator) is placed on which the prototype is "grown". Using this technology, a three-dimensional object projected on a computer is synthesized from liquid PPC by successive thin (0.03-0.2 mm) layers formed by laser radiation on a moving platform. The 3D model of the object is preliminarily converted from the STL-file format into a set of stratified sections with the required height step, the array of which is recorded in the executable file with the SLI extension. This file is a set of two-dimensional vector data that provides sequential laser beam orientation control by means of mirrors in the process of object synthesis, commands to turn on the laser and move the platform. A laser is activated that acts on those parts of the polymer that correspond to the target object walls, causing them to solidify. After that, the whole platform is dipped a little lower, by an amount equal to the thickness of the layer. After the printing is completed, the resulting object is immersed in a bath with special compositions to remove excess and cleaning. And, finally, the final irradiation with powerful ultraviolet light for final hardening. Like many other 3D prototyping techniques, SLA requires the construction of support structures that are manually deleted when printing is complete.

On the Uniontech RSPro 600, two types of samples were grown in pre-modeled models: with and without the mesh filling of the implant body (Figure 1). The maximum construction size is 600 x 600 x 400 mm (full bath), the accuracy is ± 0.1 mm (size <100 mm), ± 0.1% (size> 100 mm.) The accuracy may vary depending on the model parameters, geometry and the dimensions of the part, the orientation of the part, and the subsequent processing. For casting on burned models, a gating system was modeled, shown in Figure 2.

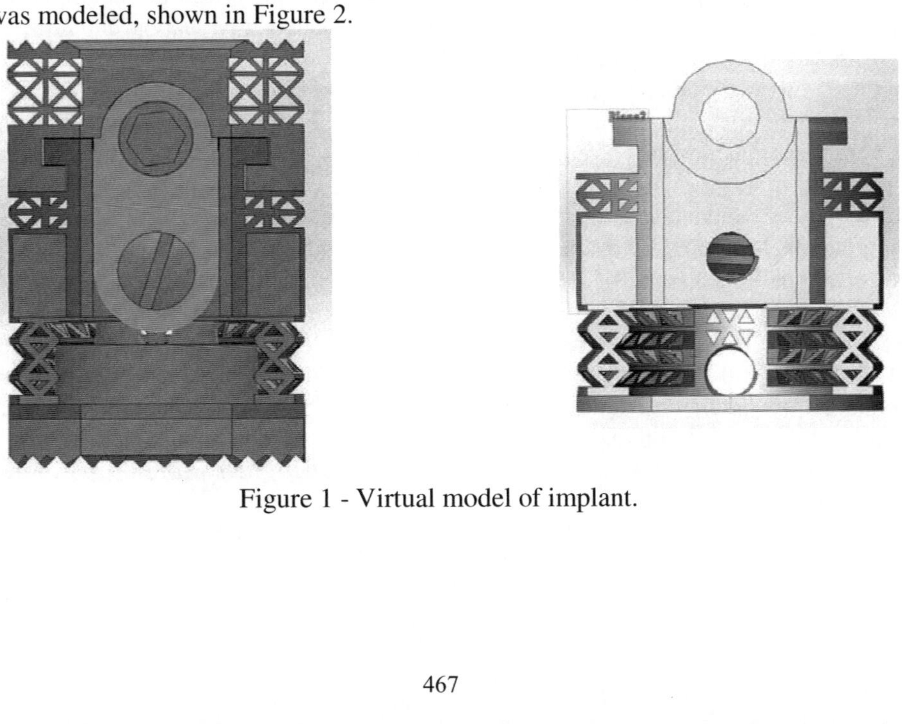

Figure 1 - Virtual model of implant.

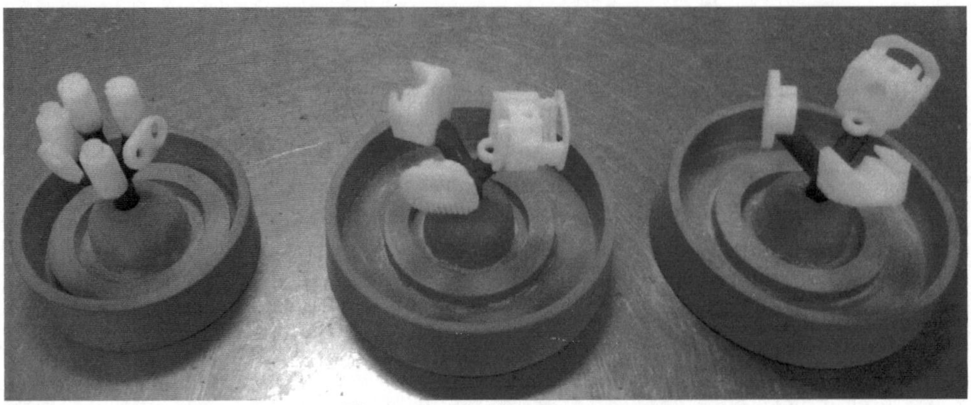

Figure 2 - Gating systems

The construction of the gating system begins with the establishment of each unit according to the feeder. Elements of the implant are supplied with sprue channels with a diameter of 5 mm. The channels number depends on the number of units cast. After building a gating system, it must be weighed. Based on the known mass of wax, the required amount of metal is determined. The calculation is made by the following formula: $M_{Me} = m_{wax} * 10 + 5$ g. Starbond CoS dental metal was used in the experiments, the following composition of Co - 59%, Cr - 25%, W - 9%, Mo - 3.5%, Si - 1%, other components (C, Fe, Mn, N) - maximum 1.5%. Ultimate tensile strength (Rp0.2) 650 MPa. The tensile strength is 910 MPa. The maximum elongation is 8%. The modulus of elasticity is 200 GPa. Vickers hardness 280 HV 10. Density 8,8 g / cm^3. Solidus-liquidus interval is 1305 - 1400 ° C. The casting temperature is 1500-1550 ° C. The thermal expansion coefficient is 14.0 μm / m. ° C.

In the second stage, the Yeti Expansion refractory molding (90 g) is prepared and kneaded with Yeti Liquid (22/20 ml) + distilled water (0/2 ml). To avoid the formation of air bubbles, a vacuum mixer Fox 88 from OMEC (Italy) was used. The kneading time was 30 seconds, the blade rotation speed was 400 rpm, and the pressure in the bowl was 0.8 Bar. The casting of the gate was made on a vibration table VB 1.1 with 2 vibration frequencies of 3000/6000 min^{-1}. After the molding mass has been filled in, wait 30 minutes before the molding material solidifies completely. The resultant hardening of the molding mass should be heated in the muffle furnace. In the experiments, the method of shock heating of the flask was used. In particular, the Programix 50 Ugin oven was preheated to a temperature of 900 ° C. A mold was used with an X6 size, the minimum warm-up time was 60 minutes. To effectively cast the metal, the crucible with the metal is also heated in a muffle furnace. In the next step, the heated crucible is placed in a casting chamber. The metal is heated by means of high frequency currents. Further, from the muffle furnace, a heated flask is extracted and vertically fixed to the crucible cover and pressed down from above with a clamping mechanism. After that, the vacuum chamber cover is closed, and the evacuation process of air begins with simultaneous high-frequency metal heating. During the heating process, the metal melts, resulting in the formation of a molten metal in the crucible, which is covered with an oxide film on top. At the rupture moment of the oxide film, the caster performs a casting process, namely, a vacuum-casting chamber is rotated, and compressed air is pumped at a rate of 3 bar/s to a pressure of 3.6 bar. As a result of the overturn, the metal from the crucible is poured into the flask and pressurized by excess pressure. The pressure provides the possibility of spilling metal into channels with a thickness more than 0.1 mm. In the experiments, a fore-casting installation was used. The vacuum-casting chamber is in an inverted state for 30 seconds, during which time the metal cools and crystallizes. Molding box must be cooled at room temperature, and mechanically clean the metal frame from the molding mass. A sandblasting unit was used, in which sand particles stream with a grain size of 250-300 μm, bombarded the surface of the sprue system and cleared the molding mass.

Figure 3 shows cast specimens of individual telescopic cervical implants.

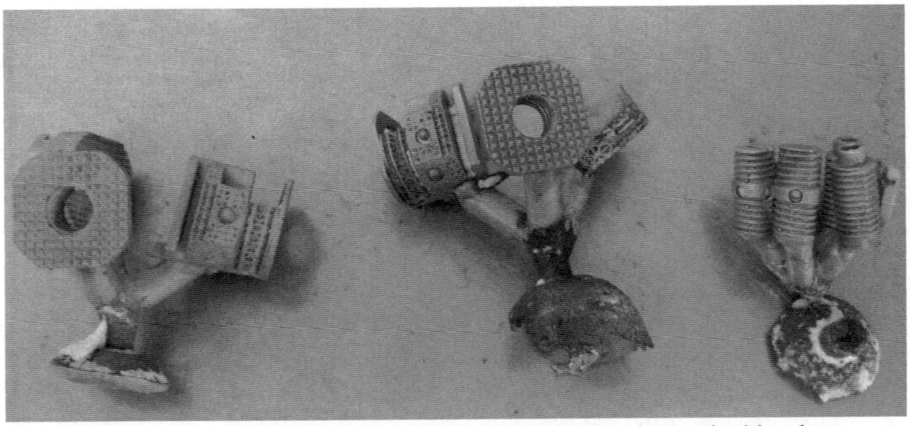

Figure 3 - Samples of castings of individual telescopic cervical implants.

As a result of the research, it has been established that implant molding without a mesh structure in the volume is successful. After additional machining, a ready-made disassembly implant can be obtained. When casting wire mesh there are two problems: the difficulty of molding associated with the air extraction from mesh volume and the difficulty of cleaning the molding mass is a mesh space in the implant volume.

This shows the relevance of creating new, more efficient methods for pouring the molding mass, which would allow it to effectively penetrate the interplanetary space of the implant. Otherwise, the formation of an air cavity in the intersection space and its subsequent filling with liquid metal is observed. The second demand is to develop an effective method of extracting the molding material after casting the metal. Without solving these two problems, casting implant mesh is difficult to implement.

Conclusion

The possibility of using the Uniontech RSPro 600 SLA-system for producing a burned model of a telescopic neck implant with a mesh structure was studied. The results of the model material influence on the molding mass are obtained. There is a cracking of the mold boz, which indicates the need to replace the "shock" heating with a "stepped" one.Inducast installation cast prototypes of cervical implants. The ineffectiveness of this technology in the manufacture of mesh collapsible telescopic implants has been established, but it can be used for structures with a thickness of 1 mm.

This work was funded by the subsidy allocated to Kazan Federal University for the state assignment in the sphere of scientific activities (11.1300.2017/4.6).

References

[1] Kashapov R N, Kashapov L N, Kashapov N F 2017 *Analysis and development of methods for obtaining metallic powders for selective laser melting* IOP Conference Series: Materials Science and Engineering Vol 240 Is 1 **012071**

[2] Kashapov R N, Kashapov L N, Kashapov N F 2015 *The research of anodic microdischarges in plasma-electrolyte processing* IOP Conference Series: Materials Science and Engineering Vol 86 Is 1 **012019**

[3] Kashapov L N, Kashapov N F, Kashapov R N 2013 *Research of the impact acidity of electrolytic cathode on the course of the plasma-electrolytic process* Journal of Physics: Conference Series. Vol 479 Is 1 **012011**

[4] Kashapov L N, Kashapov N F, Kashapov R N 2014 *Influence of plasma-electrolyte discharge to the glass surface* Journal of Physics: Conference Series. Vol 567 Is 1 **012024**

[5] Kashapov L N, Rudyk A N, Kashapov R N 2014 *Applying 3D-printing technology in planning operations of cancer patients* IOP Conference Series: Materials Science and Engineering 69 **012016**

[6] Kashapov L N, Kashapov N F, Kashapov R N, Denisov D G 2016 *Plasma electrolytic treatment of products after selective laser melting* Journal of Physics: Conference Series Vol 669 Is 1 **012029**
[7] Kashapov R N, Korobkina A I, Platonov E V, Saleeva G T 2014 *The method of manufacture of nylon dental partially removable prosthesis using additive technologies* IOP Conference Series: Materials Science and Engineering Vol 69 Is 1 **012026**

Institute of Physics
Dirac House, Temple Back
Bristol BS1 6BE UK

ISSN: 1757-8981
ISBN 978-1-5108-7368-1

9 781510 873681

CURRAN ASSOCIATES INC.
proceedings.com

International Conference on Frontier of Science and Society (ICFSS-2017)

IOP Conference Series: Earth and Environmental Science
Volume 156

Makassar, Indonesia
3 – 5 November 2017

IOP
Institute of Physics